NONLINEAR TIME SERIES ANALYSIS

WILEY SERIES IN PROBABILITY AND STATISTICS
Established by *Walter A. Shewhart and Samuel S. Wilks*

The ***Wiley Series in Probability and Statistics*** is well established and authoritative. It covers many topics of current research interest in both pure and applied statistics and probability theory. Written by leading statisticians and institutions, the titles span both state-of-the-art developments in the field and classical methods.

Reflecting the wide range of current research in statistics, the series encompasses applied, methodological and theoretical statistics, ranging from applications and new techniques made possible by advances in computerized practice to rigorous treatment of theoretical approaches. This series provides essential and invaluable reading for all statisticians, whether in academia, industry, government, or research.

NONLINEAR TIME SERIES ANALYSIS

Ruey S. Tsay
University of Chicago, Chicago, Illinois, United States

Rong Chen
Rutgers, The State University of New Jersey,
New Jersey, United States

WILEY

This edition first published 2019

Registered Office
John Wiley & Sons, Inc., 111 River Street, Hoboken, NJ 07030, USA

Editorial Office
John Wiley & Sons, Inc., 111 River Street, Hoboken, NJ 07030, USA

For details of our global editorial offices, customer services, and more information about Wiley products visit us at www.wiley.com.

Wiley also publishes its books in a variety of electronic formats and by print-on-demand. Some content that appears in standard print versions of this book may not be available in other formats.

Library of Congress Cataloging-in-Publication Data
Names: Tsay, Ruey S., 1951- author. | Chen, Rong, 1963- author.
Title: Nonlinear time series analysis / by Ruey S. Tsay and Rong Chen.
Description: Hoboken, NJ : John Wiley & Sons, 2019. | Series: Wiley series in probability and statistics | Includes index. |
Identifiers: LCCN 2018009385 (print) | LCCN 2018031564 (ebook) | ISBN 9781119264064 (pdf) | ISBN 9781119264071 (epub) | ISBN 9781119264057 (cloth)
Subjects: LCSH: Time-series analysis. | Nonlinear theories.
Classification: LCC QA280 (ebook) | LCC QA280 .T733 2019 (print) | DDC 519.5/5–dc23
LC record available at https://lccn.loc.gov/2018009385

Cover Design: Wiley
Cover Image: Background: © gremlin/iStockphoto;
Graphs: Courtesy of the author Ruey S. Tsay and Rong Chen

Set in 10/12.5pt TimesLTStd by Aptara Inc., New Delhi, India

MIX
Paper from responsible sources
FSC www.fsc.org FSC® C013604

10 9 8 7 6 5 4 3 2 1

To Teresa, Julie, Richard, and Victoria (RST)
To Danping, Anthony, and Angelina (RC)

CONTENTS

PREFACE

Time series analysis is concerned with understanding the dynamic dependence of real-world phenomena and has a long history. Much of the work in time series analysis focuses on linear models, even though the real world is not linear. One may argue that linear models can provide good approximations in many applications, but there are cases in which a nonlinear model can shed light far beyond where linear models can. The goal of this book is to introduce some simple yet useful nonlinear models, to consider situations in which nonlinear models can make significant contributions, to study basic properties of nonlinear models, and to demonstrate the use of nonlinear models in practice. Real examples from various scientific fields are used throughout the book for demonstration.

The literature on nonlinear time series analysis is enormous. It is too much to expect that a single book can cover all the topics and all recent developments. The topics and models discussed in this book reflect our preferences and personal experience. For the topics discussed, we try to provide a comprehensive treatment. Our emphasis is on application, but important theoretical justifications are also provided. All the demonstrations are carried out using R packages and a companion `NTS` package for the book has also been developed to facilitate data analysis. In some cases, a command in the `NTS` package simply provides an interface between the users and a function in another R package. In other cases, we developed commands that make analysis discussed in the book more user friendly. All data sets used in this book are either in the public domain or available from the book's web page.

The book starts with some examples demonstrating the use of nonlinear time series models and the contributions a nonlinear model can provide. Chapter 1 also discusses various statistics for detecting nonlinearity in an observed time series. We hope that the chapter can convince readers that it is worthwhile pursuing nonlinear modeling in analyzing time series data when nonlinearity is detected. In Chapter 2 we introduce some well-known nonlinear time series models available in the literature. The models discussed include the threshold autoregressive models, the Markov switching models, the smooth transition autoregressive models, and time-varying coefficient models. The process of building those nonlinear models is also addressed. Real examples are used to show the features and applicability of the models introduced. In Chapter 3 we introduce some nonparametric methods and discuss their applications in modeling nonlinear time series. The methods discussed include kernel smoothing, local polynomials, splines, and wavelets. We then consider nonlinear additive models, index models, and sliced inverse regression. Chapter 4 describes neural networks, deep learning, tree-based methods, and random forests. These topics are highly relevant in the current big-data environment, and we illustrate applications of these methods with real examples. In Chapter 5 we discuss methods and models for modeling non-Gaussian time series such as time series of count data, volatility models, and functional time series analysis. Poisson, negative binomial, and double Poisson distributions are used for count data. The chapter extends the idea of generalized linear models to generalized linear autoregressive and moving-average models. For functional time series, we focus on the class of convolution functional autoregressive models and employ sieve estimation with B-splines basis functions to approximate the true underlying convolution functions.

The book then turns to general (nonlinear) state space models (SSMs) in Chapter 6. Several models discussed in the previous chapters become special cases of this general SSM. In addition, some new nonlinear models are introduced under the SSM framework, including targeting tracking, among others. We then discuss methods for filtering, smoothing, prediction, and maximum likelihood estimation of the linear and Gaussian SSM via the Kalman filter. Special attention is paid to the linear Gaussian SSM as it is the foundation for further developments and the model can provide good approximations in many applications. Again, real examples are used to demonstrate various applications of SSMs. Chapter 7 is a continuation of Chapter 6. It introduces various extensions of the Kalman filter, including extended, unscented, and ensemble Kalman filters. The chapter then focuses on hidden Markov models (HMMs) to which the Markov switching model belongs. Filtering and estimation of HMMs are discussed in detail and real examples are used to demonstrate the applications. In Chapter 8 we introduce a general framework of sequential Monte Carlo methods that is designed to analyze nonlinear and non-Gaussian SSM. Some of the methods discussed are also referred to as particle filters in the literature. Implementation issues are discussed in detail and several applications are used for demonstration. We do not discuss multivariate nonlinear time series, even though many of the models and methods discussed can be generalized.

Some exercises are given in each chapter so that readers can practice empirical analysis and learn applications of the models and methods discussed in the book. Most of the exercises use real data so that there exist no true models, but good approximate models can always be found by using the methods discussed in the chapter.

Finally, we would like to express our sincere thanks to our friends, colleagues, and students who helped us in various ways during our research in nonlinear models and in preparing this book. In particular, Xialu Liu provided R code and valuable help in the analysis of convolution functional time series and Chencheng Cai provided R code of optimized parallel implementation of likelihood function evaluation. Daniel Peña provided valuable comments on the original draft. William Gonzalo Rojas and Yimeng Shi read over multiple draft chapters and pointed out various typos. Howell Tong encouraged us in pursuing research in nonlinear time series and K.S. Chan engaged in various discussions over the years. Last but not least, we would like to thank our families for their unconditional support throughout our careers. Their love and encouragement are the main source of our energy and motivation. The book would not have been written without all the support we have received.

The web page of the book is http://faculty.chicagobooth.edu/ruey.tsay/teaching/nts (for data sets) and www.wiley.com/go/tsay/nonlineartimeseries (for instructors).

R.S.T. *Chicago, IL*
R.C. *Princeton, NJ*

November 2017

CHAPTER 1

WHY SHOULD WE CARE ABOUT NONLINEARITY?

Linear processes and linear models dominate research and applications of time series analysis. They are often adequate in making statistical inference in practice. Why should we care about nonlinearity then? This is the first question that came to our minds when we thought about writing this book. After all, linear models are easier to use and can provide good approximations in many applications. Empirical time series, on the other hand, are likely to be nonlinear. As such, nonlinear models can certainly make significant contributions, at least in some applications. The goal of this book is to introduce some nonlinear time series models, to discuss situations under which nonlinear models can make contributions, to demonstrate the value and power of nonlinear time series analysis, and to explore the nonlinear world. In many applications, the observed time series are indirect (possibly multidimensional) observations of an unobservable underlying dynamic process that is nonlinear. In this book we also discuss approaches of using nonlinear and non-Gaussian state space models for analyzing such data.

Nonlinear Time Series Analysis, First Edition. Ruey S. Tsay and Rong Chen.

Companion website: www.wiley.com/go/tsay/nonlineartimeseries

To achieve our objectives, we focus on certain classes of nonlinear time series models that, in our view, are widely applicable and easy to understand. It is not our intention to cover all nonlinear models available in the literature. Readers are referred to Tong (1990), Fan and Yao (2003), Douc et al. (2014), and De Gooijer (2017) for other nonlinear time series models. The book, thus, shows our preference in exploring the nonlinear world. Efforts are made throughout the book to keep applications in mind so that real examples are used whenever possible. We also provide the theory and justifications for the methods and models considered in the book so that readers can have a comprehensive treatment of nonlinear time series analysis. As always, we start with simple models and gradually move toward more complicated ones.

1.1 SOME BASIC CONCEPTS

A scalar process x_t is a discrete-time time series if x_t is a random variable and the time index t is countable. Typically, we assume the time index t is equally spaced and denote the series by $\{x_t\}$. In applications, we consider mainly the case of x_t with $t \geq 1$. An observed series (also denoted by x_t for simplicity) is a realization of the underlying stochastic process.

A time series x_t is *strictly stationary* if its distribution is time invariant. Mathematically speaking, x_t is strictly stationary if for any arbitrary time indices $\{t_1, \ldots, t_m\}$, where $m > 0$, and any fixed integer k such that the joint distribution function of $(x_{t_1}, \ldots, x_{t_m})$ is the same as that of $(x_{t_1+k}, \ldots, x_{t_m+k})$. In other words, the shift of k time units does not affect the joint distribution of the series. A time series x_t is *weakly stationary* if the first two moments of x_t exist and are time invariant. In statistical terms, this means $E(x_t) = \mu$ and $\text{Cov}(x_t, x_{t+\ell}) = \gamma_\ell$, where E is the expectation, Cov denotes covariance, μ is a constant, and γ_ℓ is a function of ℓ. Here both μ and γ_ℓ are independent of the time index t, and γ_ℓ is called the lag-ℓ autocovariance function of x_t. A sequence of independent and identically distributed (iid) random variates is strictly stationary. A martingale difference sequence x_t satisfying $E(x_t \mid x_{t-1}, x_{t-2}, \ldots) = 0$ and $\text{Var}(x_t \mid x_{t-1}, x_{t-2}, \ldots) = \sigma^2 > 0$ is weakly stationary. A weakly stationary sequence is also referred to as a covariance-stationary time series. An iid sequence of Cauchy random variables is strictly stationary, but not weakly stationary, because there exist no moments. Let $x_t = \sigma_t \epsilon_t$, where $\epsilon_t \sim_{iid} N(0, 1)$ and $\sigma_t^2 = 0.1 + 0.2x_{t-1}^2$. Then x_t is weakly stationary, but not strictly stationary.

Time series analysis is used to explore the dynamic dependence of the series. For a weakly stationary series x_t, a widely used measure of serial dependence between x_t and $x_{t-\ell}$ is the lag-ℓ *autocorrelation function* (ACF) defined by

$$\rho_\ell = \frac{\text{Cov}(x_t, x_{t-\ell})}{\text{Var}(x_t)} \equiv \frac{\gamma_\ell}{\gamma_0}, \tag{1.1}$$

where ℓ is an integer. It is easily seen that $\rho_0 = 1$ and $\rho_\ell = \rho_{-\ell}$ so that we focus on ρ_ℓ for $\ell > 0$. The ACF defined in Equation (1.1) is based on the Pearson's correlation coefficient. In some applications we may employ the autocorrelation function using the concept of Spearman's rank correlation coefficient.

1.2 LINEAR TIME SERIES

A scalar process x_t is a linear time series if it can be written as

$$x_t = \mu + \sum_{i=-\infty}^{\infty} \psi_i a_{t-i}, \tag{1.2}$$

where μ and ψ_i are real numbers with $\psi_0 = 1$, $\sum_{i=-\infty}^{\infty} |\psi_i| < \infty$, and $\{a_t\}$ is a sequence of iid random variables with mean zero and a well-defined density function. In practice, we focus on the one-sided linear time series

$$x_t = \mu + \sum_{i=0}^{\infty} \psi_i a_{t-i}, \tag{1.3}$$

where $\psi_0 = 1$ and $\sum_{i=0}^{\infty} |\psi_i| < \infty$. The linear time series in Equation (1.3) is called a *causal* time series. In Equation (1.2), if $\psi_j \neq 0$ for some $j < 0$, then x_t becomes a non-causal time series. The linear time series in Equation (1.3) is weakly stationary if we further assume that $\text{Var}(a_t) = \sigma_a^2 < \infty$. In this case, we have $E(x_t) = \mu$, $\text{Var}(x_t) = \sigma_a^2 \sum_{i=0}^{\infty} \psi_i^2$, and $\gamma_\ell = \sigma_a^2 \sum_{i=0}^{\infty} \psi_i \psi_{i+\ell}$.

The well-known autoregressive moving-average (ARMA) models of Box and Jenkins (see Box et al., 2015) are (causal) linear time series. Any deviation from the linear process in Equation (1.3) results in a nonlinear time series. Therefore, the nonlinear world is huge and certain restrictions are needed in our exploration. Imposing different restrictions leads to different approaches in tackling the nonlinear world which, in turn, results in emphasizing different classes of nonlinear models. This book is no exception. We start with some real examples that exhibit clearly some nonlinear characteristics and employ simple nonlinear models to illustrate the advantages of studying nonlinearity.

1.3 EXAMPLES OF NONLINEAR TIME SERIES

To motivate, we analyze some real-world time series for which nonlinear models can make a contribution.

Example 1.1 Consider the US quarterly civilian unemployment rates from 1948.I to 2015.II for 270 observations. The quarterly rate is obtained by averaging the monthly rates, which were obtained from the Federal Reserve Economic Data (FRED) of the Federal Reserve Bank of St. Louis and were seasonally adjusted. Figure 1.1 shows the time plot of the quarterly unemployment rates. From the plot, it is seen that (a) the unemployment rate seems to be increasing over time, (b) the unemployment rate exhibits a cyclical pattern reflecting the business cycles of the US economy, and (c) more importantly, the rate rises quickly and decays slowly over a business cycle. As usual in time series analysis, the increasing trend can be handled by differencing. Let r_t be the quarterly unemployment rate and $x_t = r_t - r_{t-1}$ be the change series of r_t. Figure 1.2 shows the time plot of x_t. As expected, the mean of x_t appears to be stable over time. However, the asymmetric pattern in rise and decay of the unemployment rates in a business cycle shows that the rate is not time-reversible, which in turn suggests that the unemployment rates are nonlinear. Indeed, several nonlinear tests discussed later confirm that x_t is indeed nonlinear.

If a linear autoregressive (AR) model is used, the Akaike information criterion (AIC) of Akaike (1974) selects an AR(12) model for x_t. Several coefficients of the

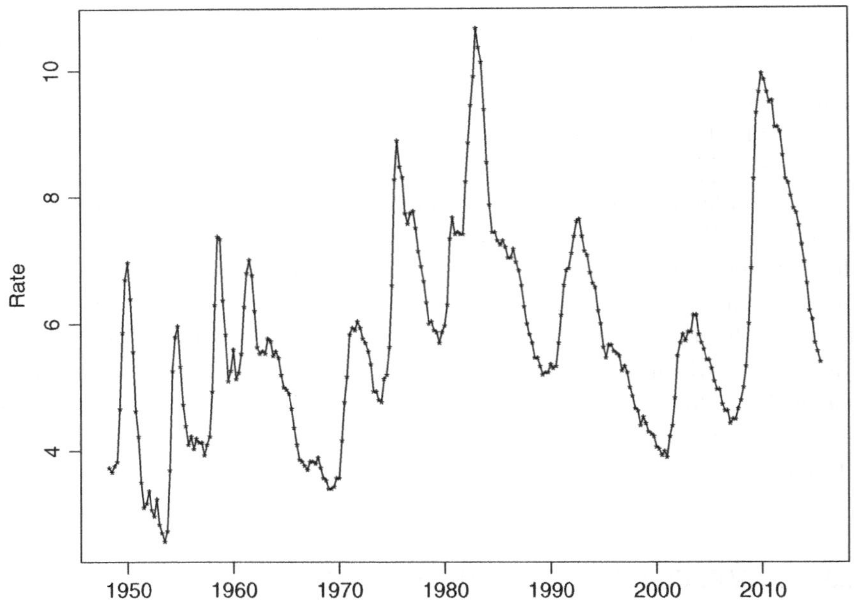

Figure 1.1 Time plot of US quarterly civilian unemployment rates, seasonally adjusted, from 1948.I to 2015.II.

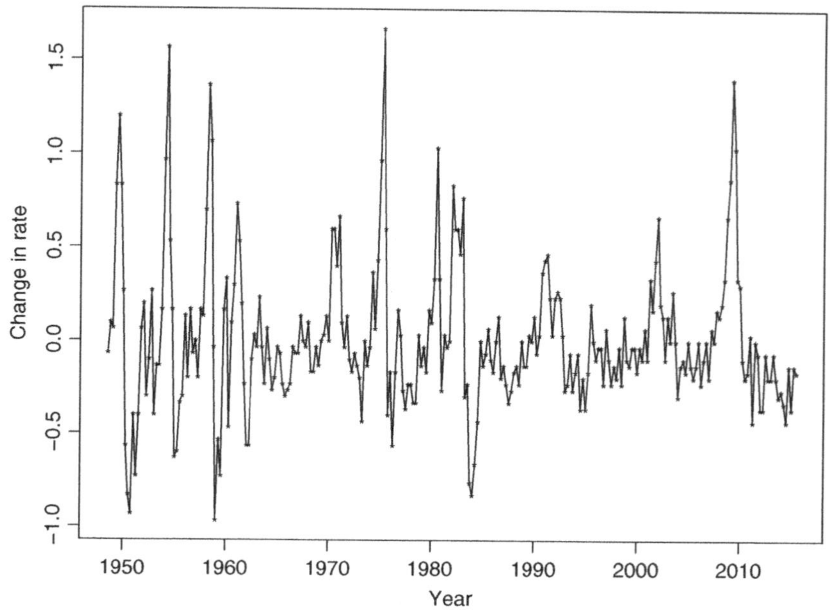

Figure 1.2 Time plot of the changes in US quarterly civilian unemployment rates, seasonally adjusted, from 1948.I to 2015.II.

fitted AR(12) model are not statistically significant so that the model is refined. This leads to a simplified AR(12) model as

$$x_t = 0.68x_{t-1} - 0.26x_{t-4} + 0.18x_{t-6} - 0.33x_{t-8} + 0.19x_{t-9} - 0.17x_{t-12} + a_t, \quad (1.4)$$

where the variance of a_t is $\sigma_a^2 = 0.073$ and all coefficient estimates are statistically significant at the usual 5% level. Figure 1.3 shows the results of model checking, which consists of the time plot of standardized residuals, sample autocorrelation function (ACF), and the p values of the Ljung–Box statistics $Q(m)$ of the residuals. These p values do not adjust the degrees of freedom for the fitted parameters, but they are sufficient in indicating that the fitted AR(12) model in Equation (1.4) is adequate.

On the other hand, one can use the self-exciting threshold autoregressive (TAR) models of Tong (1978, 1990) and Tong and Lim (1980) to describe the nonlinear characteristics of the data. The TAR model is one of the commonly used nonlinear time series models to be discussed in Chapter 2. Using the `TSA` package of R (R Development Core Team, 2015), we obtain the following two-regime TAR model

$$\begin{aligned} x_t &= 0.47x_{t-1} + 0.15x_{t-2} - 0.02x_{t-3} - 0.17x_{t-4} + a_{1t}, \quad \text{if } x_{t-1} \le \delta \qquad (1.5) \\ &= 0.85x_{t-1} - 0.08x_{t-2} - 0.22x_{t-3} - 0.29x_{t-4} + 0.23x_{t-5} + 0.36x_{t-6} - 0.14x_{t-7} \\ &\quad - 0.51x_{t-8} + 0.37x_{t-9} + 0.17x_{t-10} - 0.23x_{t-11} - 0.21x_{t-12} + a_{2t}, \ \text{if } x_{t-1} > \delta, \end{aligned}$$

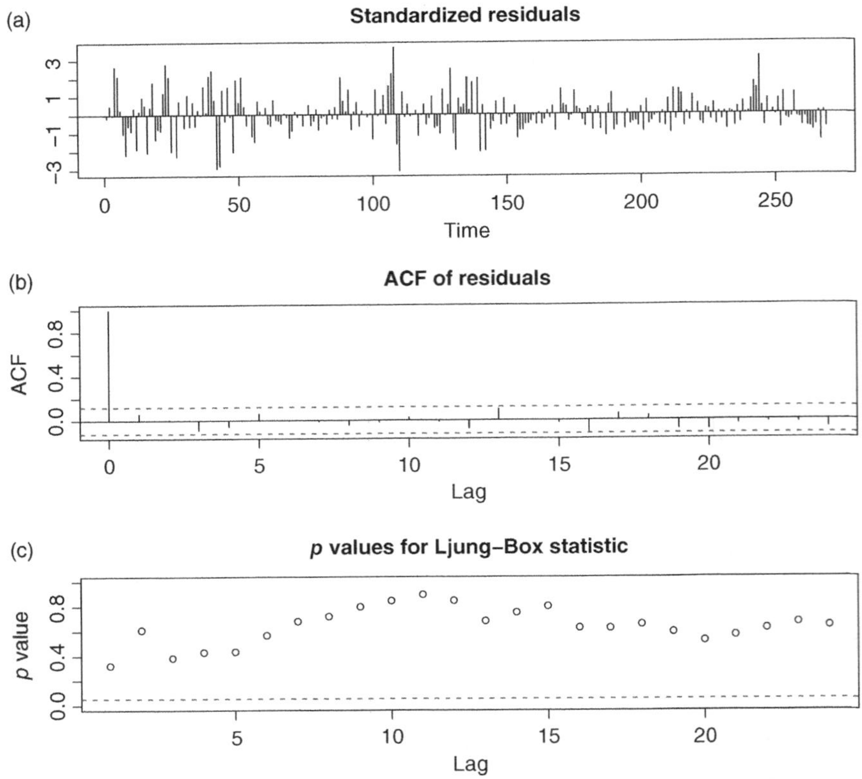

Figure 1.3 Model checking of the fitted AR(12) model of Equation (1.4) for the change series of US quarterly unemployment rates from 1948.II to 2015.II: (a) standardized residuals, (b) ACF of residuals, and (c) p values for Ljung–Box statistics.

where the threshold $\delta = -0.066666$ and the standard errors of a_{1t} and a_{2t} are 0.181 and 0.309, respectively. Some of the coefficient estimates are not statistically significant at the 5% level, but, for simplicity, we do not seek any further refinement of the model. Model checking shows that the fitted TAR model is adequate. Figure 1.4 shows the time plot of standardized residuals of model (1.5) and the sample autocorrelations of the standardized residuals. The Ljung–Box statistics of the standardized residuals show Q(12) = 6.91 (0.86) and Q(24) = 18.28(0.79), where the number in parentheses denotes the asymptotic p value. In this particular instance, x_{t-1} is the threshold variable and the fitted model shows that when the quarterly unemployment rate decreased by −0.07% or more, the dynamic dependence of the unemployment rates appears to be simpler than when the rate was increasing or changed mildly. In other words, the model implies that the dynamic dependence

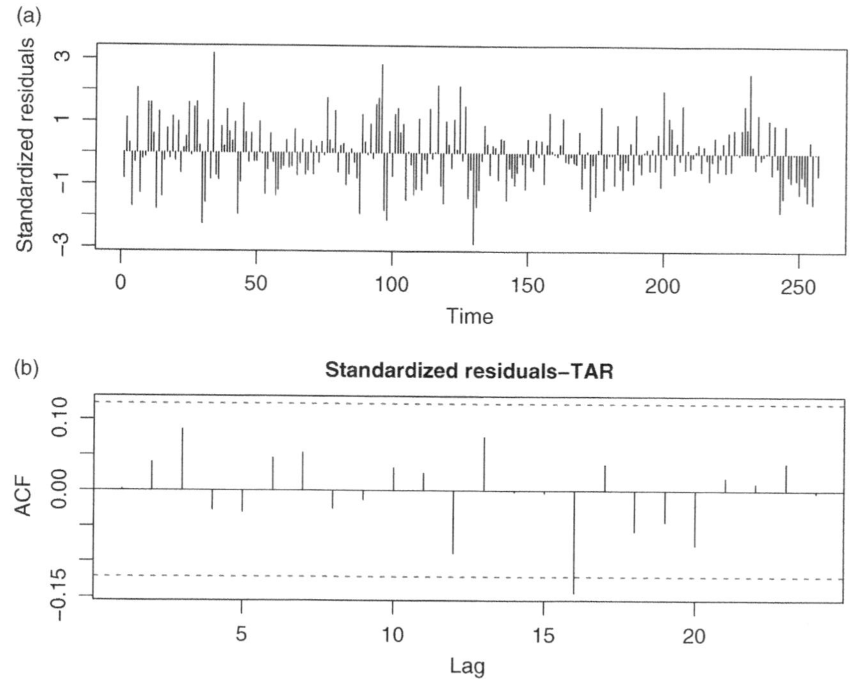

Figure 1.4 Model checking of the fitted TAR(4,12) model of Equation (1.5) for the change series of US quarterly unemployment rates from 1948.II to 2015.II. (a) The time plot of standardized residuals of model (1.5) and (b) the sample autocorrelations of the standardized residuals.

of the US quarterly unemployment rates depends on the status of the US economy. When the economy is improving, i.e. the unemployment rate decreased substantially, the unemployment rate dynamic dependence became relatively simple.

To compare the AR and TAR models in Equations (1.4) and (1.5) for the unemployment rate series, we consider model goodness-of-fit and out-of-sample predictions. For goodness of fit, Figure 1.5 shows the density functions of the standardized residuals of both fitted models. The solid line is for the TAR model whereas the dashed line is for the linear AR(12) model. From the density functions, it is seen that the standardized residuals of the TAR model are closer to the normality assumption. Specifically, the residuals of the TAR model are less skewed and have lower excess kurtosis. In this particular instance, the skewness and excess kurtosis of the standardized residuals of the linear AR(12) model are 0.318 and 1.323, respectively, whereas those of the TAR model are 0.271 and 0.256, respectively. Under the assumption that the standardized residuals are

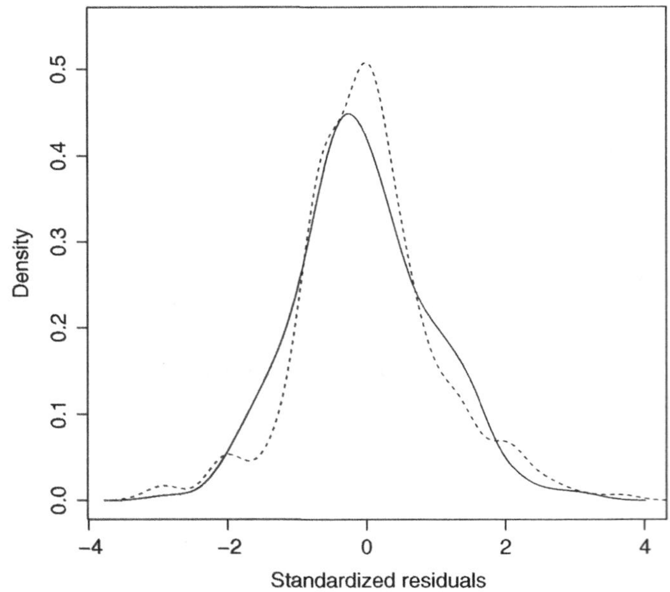

Figure 1.5 Density functions of standardized residuals of the linear AR(12) model in Equation (1.4) and the TAR(4,12) model of Equation (1.5) for the change series of US quarterly unemployment rates from 1948.II to 2015.II. The solid line is for the TAR model.

independent and identically distributed, the t ratios for excess kurtosis are 4.33 and 0.84, respectively, for the fitted AR(12) and the TAR model. Similarly, the t ratios for the skewness are 2.08 and 1.77, respectively, for AR(12) and the TAR model. If the 5% critical value of 1.96 is used, then one cannot reject the hypotheses that the standardized residuals of the fitted TAR model in Equation (1.5) are symmetric and do not have heavy tails. On the other hand, the standardized residuals of the linear AR(12) model are skewed and have heavy tails.

We also use rolling one-step ahead out-of-sample forecasts, i.e. back-testing, to compare the two fitted models. The starting forecast origin is $t = 200$ so that we have 69 one-step ahead predictions for both models. The forecasting procedure is as follows. Let n be the starting forecast origin. For a given model, we fit the model using the data from $t = 1$ to $t = n$; use the fitted model to produce a prediction for $t = n + 1$ and compute the forecasting error. We then advance the forecast origin by 1 and repeat the estimation–prediction process. We use root mean squared error (RMSE), mean absolute error (MAE) and (average) bias of predictions to quantify the performance of back-testing. In addition, we also classify the predictions based on the regime of the forecast origin. The results are given in Table 1.1. From the table, it is seen that while the nonlinear TAR model shows some improvement in

Criterion	Overall	Origins in Regime 1	Origins in Regime 2
(a) Linear AR(12) model			
RMSE	0.2166	0.1761	0.2536
MAE	0.1633	0.1373	0.1916
Bias	0.0158	0.0088	0.0235
(b) Threshold AR model			
RMSE	0.2147	0.1660	0.2576
MAE	0.1641	0.1397	0.1906
Bias	−0.0060	−0.1003	0.0968

Table 1.1 Performance of out-of-sample forecasts of the linear AR(12) model in Equation (1.4) and the threshold AR model in Equation (1.5) for the US quarterly unemployment rates from 1948.II to 2015.II. The starting forecast origin is 200, and there are 69 one-step ahead predictions.

out-of-sample predictions, the improvement is rather minor and seems to come from those associated with forecast origins in Regime 1.

In this example, a simple nonlinear model can help in prediction and, more importantly, the nonlinear model improves the model goodness of fit as shown by the properties of the residuals.

Example 1.2 As a second example, we consider the weekly crude oil prices from May 12, 2000 to August 28, 2015. The data used are part of the commodity prices available from Federal Reserve Bank of St. Louis and they are the crude oil prices, West Texas Intermediate, Cushing, Oklahoma. Figure 1.6 shows the time plot of the original crude oil prices and the first differenced series of the prices. The differenced series is used in our analysis as the oil prices exhibit strong serial dependence, i.e. unit-root nonstationarity. From the plots, it is seen that the variability of the time series varies over time. Let $x_t = p_t - p_{t-1}$, where p_t denotes the observed weekly crude oil price at week t. If scalar AR models are employed, an AR(8) model is selected by AIC and the fitted model, after removing insignificant parameters, is

$$x_t = 0.197x_{t-1} + 0.153x_{t-8} + a_t, \quad \sigma_a^2 = 6.43, \tag{1.6}$$

where the standard errors of both AR coefficients are around 0.034. This simple AR model adequately describes the dynamic correlations of x_t, but it fails to address the time-varying variability. Figure 1.7 shows the sample ACF of the

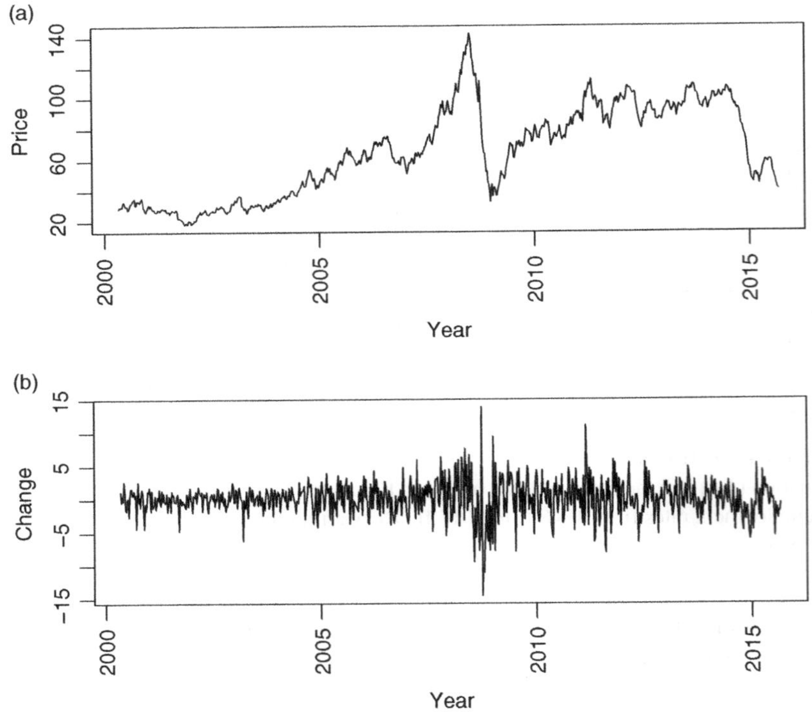

Figure 1.6 Time plot of weekly crude oil prices from May 12, 2000 to August 28, 2015: (a) oil prices and (b) the first differenced series. The prices are West Texas Intermediate, Cushing, Oklahoma, and obtained from FRED of Federal Reserve Bank, St. Louis

residuals and those of the squared residuals. From the plots, the residuals have no significant serial correlations, but the squared residuals have strong serial dependence. In finance, such a phenomenon is referred to as time-varying volatility or conditional heteroscedasticity.

One approach to improve the linear AR(8) model in Equation (1.6) is to use the generalized autoregressive conditional heteroscedastic (GARCH) model of Bollerslev (1986). GARCH models are nonlinear based on the linearity definition of Equation (1.2). In this particular instance, the fitted AR-GARCH model with Student-t innovations is

$$x_t \approx 0.227x_{t-1} + 0.117x_{t-8} + a_t, \tag{1.7}$$

$$a_t = \sigma_t \epsilon_t, \quad \epsilon_t \sim_{iid} t^*_{7.91},$$

$$\sigma_t^2 = 0.031 + 0.072a_{t-1}^2 + 0.927\sigma_{t-1}^2, \tag{1.8}$$

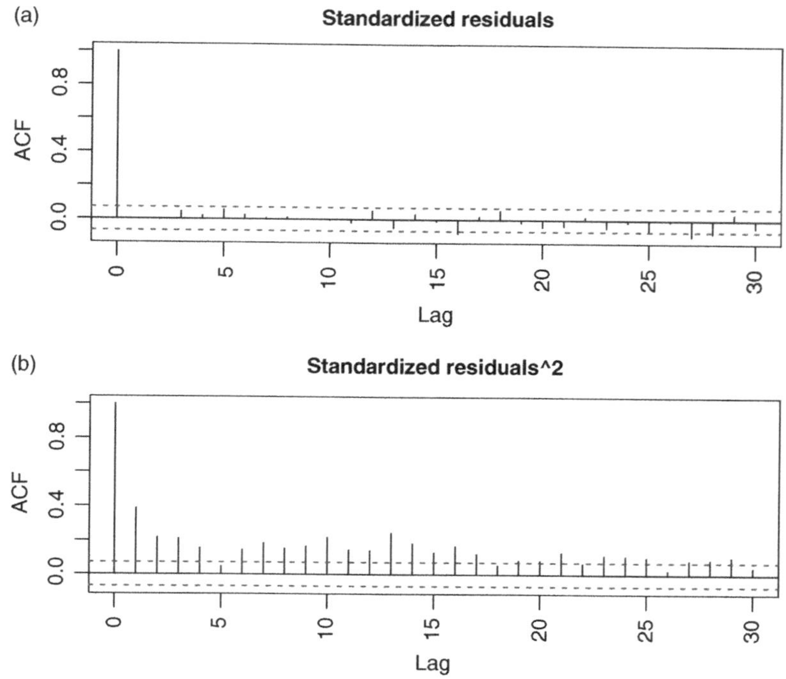

Figure 1.7 Sample autocorrelation functions of (a) the residuals of the AR(8) model in Equation (1.6) for the first differenced series of weekly crude oil prices and (b) the squared residuals.

where t_v^* denotes standardized Student-t distribution with v degrees of freedom. In Equation (1.7) we use the approximation $\approx$, since we omitted the insignificant parameters at the 5% level for simplicity. In Equation (1.8), the standard errors of the parameters are 0.024, 0.016, and 0.016, respectively. This volatility model indicates the high persistence in the volatility. For further details about volatility models, see Chapter 5 and Tsay (2010, Chapter 3). Model checking indicates that the fitted AR-GARCH model is adequate. For instance, we have Q(20) = 14.38 (0.81) and 9.89(0.97) for the standardized residuals and the squared standardized residuals, respectively, of the model, where the number in parentheses denotes the p value. Figure 1.8(a) shows the time plot of x_t along with point-wise two-standard errors limits, which further confirms that the model fits the data well. Figure 1.8(b) shows the quantile-to-quantile (QQ) plot of the standardized residuals versus the Student-t distribution with 7.91 degrees of freedom. The plot shows that it is reasonable to employ the Student-t innovations.

Compared with the linear AR(8) model in Equation (1.6), the fitted AR-GARCH model does not alter the serial dependence in the x_t series

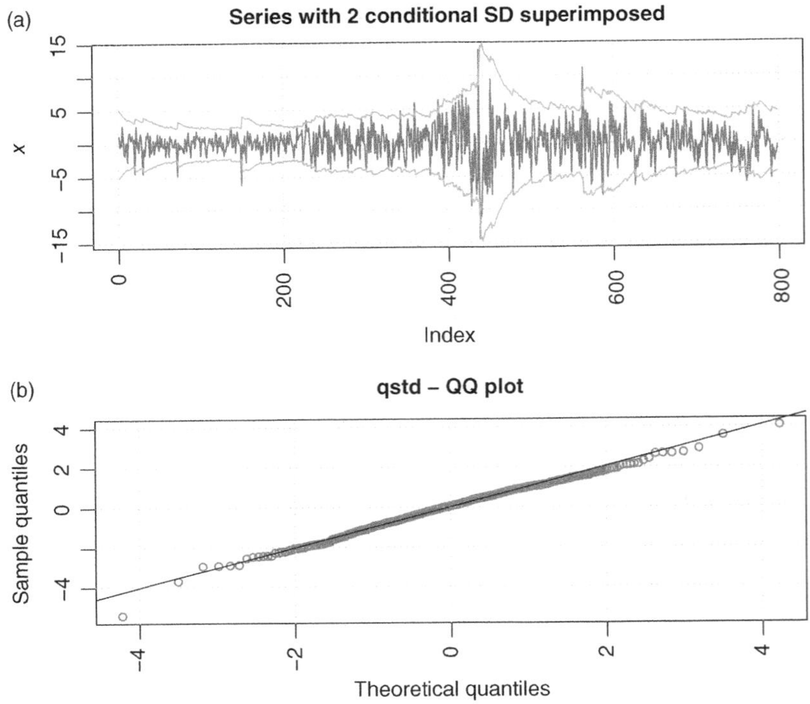

Figure 1.8 (a) The time plot of the first-differenced crude oil prices and point-wise two standard error limits of the model in Equations (1.7) and (1.8). (b) The quantile-to-quantile plot for the Student-t innovations.

because Equation (1.7) is relatively close to Equation (1.6). What the GARCH model does is to handle the time-varying volatility so that proper inference, such as interval predictions, can be made concerning the crude oil prices. In some finance applications volatility plays a key role and the fitted AR-GARCH nonlinear model can be used.

Alternatively, one can use the Markov switching model (MSM) to improve the linear AR(8) model. Details of the MSM are given in Chapters 2 and 7. In this particular application, the model used is

$$x_t = \begin{cases} \phi_{11}x_{t-1} + \phi_{12}x_{t-2} + \cdots + \phi_{18}x_{t-8} + \sigma_1\epsilon_t, & \text{if } s_t = 1, \\ \phi_{21}x_{t-1} + \phi_{22}x_{t-2} + \cdots + \phi_{28}x_{t-8} + \sigma_2\epsilon_t, & \text{if } s_t = 2, \end{cases} \tag{1.9}$$

where $\phi_{i,j}$ denotes the coefficient of state i at lag-j, ϵ_t are independent and identically distributed random variables with mean zero and variance 1, σ_i is the innovation standard error of state i, and s_t denotes the status of the state at time t. This is a

Par.	ϕ_{i1}	ϕ_{i2}	ϕ_{i3}	ϕ_{i4}	ϕ_{i5}	ϕ_{i6}	ϕ_{i7}	ϕ_{i8}	σ_i
					State 1				
Est.	−0.60	0.53	−0.43	−0.04	−0.08	−0.14	−0.08	0.12	3.04
s.e.	0.27	0.20	0.24	0.11	0.16	0.13	0.16	0.13	
					State 2				
Est.	0.37	−0.12	0.12	0.03	0.08	0.09	0.05	0.15	1.89
s.e.	0.05	0.04	0.05	0.04	0.04	0.05	0.04	0.04	

Table 1.2 Parameter estimates of the Markov switching model for the first-differenced series of weekly crude oil prices from May 19, 2000 to August 28, 2015. Par., parameter; Est., estimate; s.e., standard error.

simple two-state MSM and the states change over time according to the transition probability matrix

$$\boldsymbol{P} = \begin{bmatrix} p_{11} & p_{21} \\ p_{12} & p_{22} \end{bmatrix},$$

where p_{ij} denotes the probability of switching from state i at time $t-1$ to state j at time t. This notation is used in the `MSwM` package of R and in Hamilton (1994, Chapter 22). From the definition, the rows of $\boldsymbol{P}$ sum to 1 and p_{ii} denotes the probability of staying in state i from time $t-1$ to time t. Consequently, $1/(1-p_{ii})$ denotes the expected duration for state i.

For the first differenced series x_t of the crude oil prices, parameter estimates of the MSM are given in Table 1.2. The fitted transition matrix is

$$\widehat{\boldsymbol{P}} = \begin{bmatrix} 0.066 & 0.934 \\ 0.268 & 0.732 \end{bmatrix}.$$

Figure 1.9 shows the sample ACF and PACF of the residuals (a and b) and the squared residuals (c and d) of each state. From the plots, the serial correlations of x_t are well described by the fitted MSM in Equation (1.9). More interestingly, the ACF of squared residuals indicates that the MSM is capable of modeling the time-varying volatility. Figure 1.10 shows the time plots of smoothed and filtered probabilities for each state.

From Table 1.2, it is seen that state $s_t = 1$ corresponds to the volatile state and its coefficient estimates are more uncertain with larger standard errors. This is understandable since if x_{t-1} is in state 1, x_t stays in state 1 with a small probability

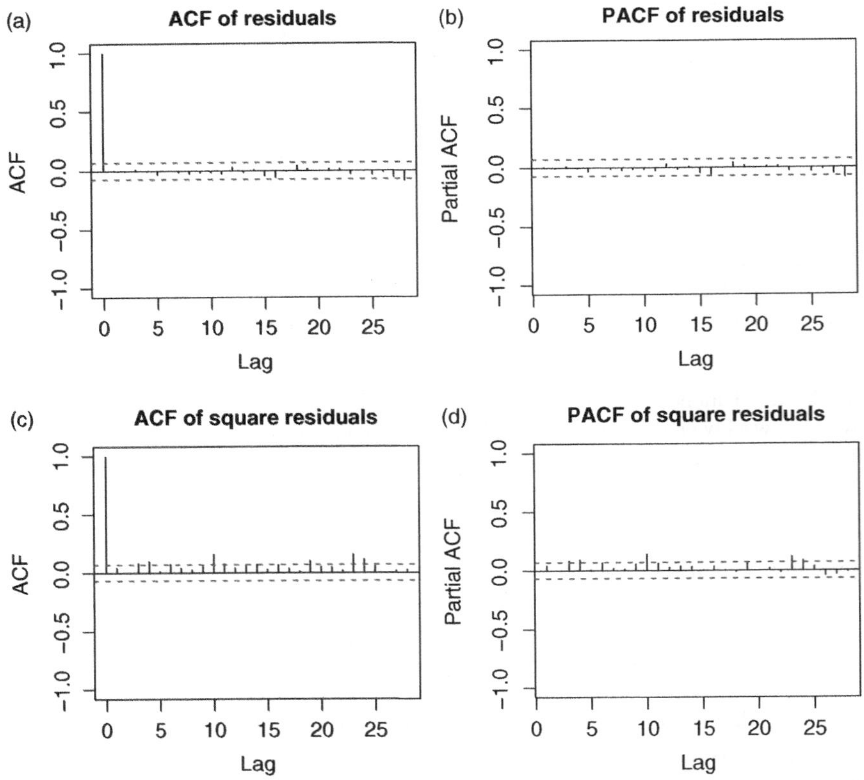

Figure 1.9 The sample autocorrelation functions of the residuals (a and b) and squared residuals (c and d), by state, for the Markov switching model of Equation (1.9) fitted to the first-differenced series of weekly crude oil prices from May 19, 2000 to August 28, 2015.

0.066 and it switches to state 2 about 93% of the time. The marginal probability that x_t is in state 1 is 0.223. The table also shows that the serial dependence of x_t depends on the state; the serial dependence can be modeled by an AR(3) model in state 1, but it requires an AR(8) model in state 2. Figure 1.10 confirms that x_t stays in state 2 often, implying that during the data span the crude oil prices encountered some volatile periods, but the high volatility periods are short-lived.

In this example, we show that nonlinear models can provide a deeper understanding of the dynamic dependence of a time series. In addition, various nonlinear models can be used to improve the fitting of a linear model. The Markov switching model allows the dynamic dependence of a time series to change according to the state it belongs to. It is also capable of handling time-varying volatility.

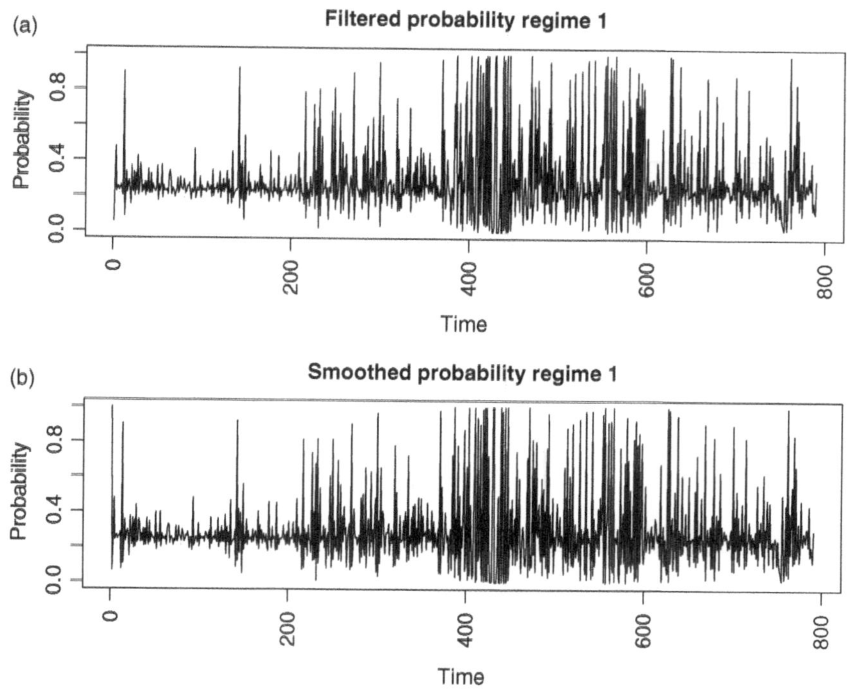

Figure 1.10 (a) Filtered and (b) smoothed probabilities for state 1 of the Markov switching model of Equation (1.9) fitted to the first-differenced series of weekly crude oil prices from May 19, 2000 to August 28, 2015.

Example 1.3 In this example, we revisit the well-known gas furnace series of Box and Jenkins (1976). The data consist of the input gas feed rate, x_t, and corresponding output CO_2 concentration, y_t, from a gas furnace. There are 296 observations taken at time intervals of 9 seconds. The top two plots of Figure 1.11 show the gas rate x_t and the output CO_2 concentration. These two series have been widely used in the literature to demonstrate the analysis of transfer function models. Following Box and Jenkins (1976), one can use linear transfer function models. The model for the gas feed rate is

$$x_t = 1.97x_{t-1} - 1.37x_{t-2} + 0.34x_{t-3} + \epsilon_t, \quad \sigma_\epsilon^2 = 0.035,$$

where the standard errors of the AR coefficients are 0.05, 0.10, and 0.05, respectively. The transfer function model used is

$$y_t = 53.37 - \frac{0.53B^3 + 0.38B^4 + 0.52B^5}{1 - 0.55B}x_t + \frac{1}{1 - 1.53B + 0.63B^2}a_t, \tag{1.10}$$

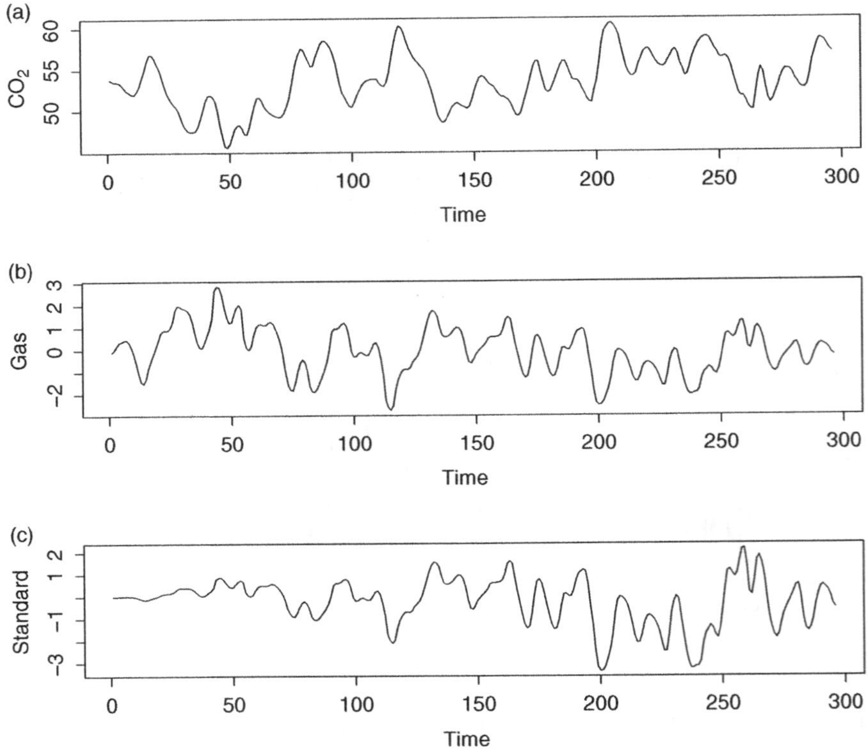

Figure 1.11 Time plots of the gas furnace data: (a) input gas feed rates, (b) output CO_2 concentrations, and (c) gas rates weighted by time trend $s_t = 2t/296$.

where B denotes the back-shift (or lag) operator such that $Bx_t = x_{t-1}$, the variance of a_t is 0.0576, and all parameter estimates are statistically significant at the usual 5% level. Residual analysis shows that the model in Equation (1.10) is adequate. For instance, the Ljung–Box statistics of the residuals gives Q(12) = 15.51 with p value 0.21.

On the other hand, nonlinear transfer function models have also been proposed in the literature, e.g. Chen and Tsay (1996). In this example, we follow the approach of Tsay and Wu (2003) by allowing the coefficients of the transfer function model to depend on a state variable s_t. In this particular instance, the state variable used is $s_t = 2t/296$, which simply represents the time sequence of the observations. The corresponding transfer function model would become

$$y_t = c_0 + \frac{\omega_3(s_t)B^3 + \omega_4(s_t)B^4 + \omega_5(s_t)B^5}{1 - \delta(s_t)B} x_t + \frac{1}{1 - \phi_1 B - \phi_2 B^2} a_t,$$

where $\omega_i(s_t)$ and $\delta(s_t)$ are smooth functions of the state variable s_t, c_0 and ϕ_i are constant. This is a special case of the functional-coefficient models of Chen and Tsay (1993). To simplify the model, Tsay and Wu (2003) used the first-order Taylor series approximations of the functional coefficients

$$\omega_i(s_t) \approx \omega_{i0} + \omega_{i1} s_t, \quad \delta(s_t) \approx \delta_0 + \delta_1 s_t,$$

where ω_{ij} and δ_j are constant, and simplified the model further to obtain the transfer function model

$$y_t = 52.65 - \frac{1.22B^3}{1 - 0.61B} x_t + 0.73 s_t + \frac{0.99B^3 - 0.99B^4}{1 - 0.65B}(s_t x_t) + \frac{1}{1 - 1.42B + 0.49B^2} a_t, \quad (1.11)$$

where the residual variance is $\sigma_a^2 = 0.0461$ and all coefficient estimates are statistically significant at the 5% level. The Ljung–Box statistics of the results gives Q(12) = 9.71 with p value 0.64. This model improves the in-sample fit as the residual variance drops from 0.0576 to 0.0431. Figure 1.11(c) shows the time plot of $s_t x_t$, from which the new variable $s_t x_t$ seems to emphasize the latter part of the x_t series.

Table 1.3 shows the results of out-of-sample forecasts of the two transfer function models in Equations (1.10) and (1.11). These summary statistics are based on 96 one-step ahead forecasts starting with initial forecast origin $t = 200$. As before, the models were re-estimated before prediction once a new observation was available. From the table it is easily seen that the model in Equation (1.11) outperforms the traditional transfer function model in all three measurements of out-of-sample prediction. The improvement in the root mean squared error (RMSE) is $(0.3674 - 0.3311)/0.3311 = 10.96\%$. This is a substantial improvement given that the model in Equation (1.10) has been regarded as a gold standard for the data set.

In this example we show that the functional-coefficient models can be useful in improving the accuracy of forecast. It is true that the model in Equation (1.11)

Model	Bias	RMSE	MAE
Equation (1.10)	0.0772	0.3674	0.2655
Equation (1.11)	0.0300	0.3311	0.2477

Table 1.3 Summary statistics of out-of-sample forecasts for the models in Equations (1.10) and (1.11). The initial forecast origin is 200 and the results are based on 96 one-step ahead predictions.

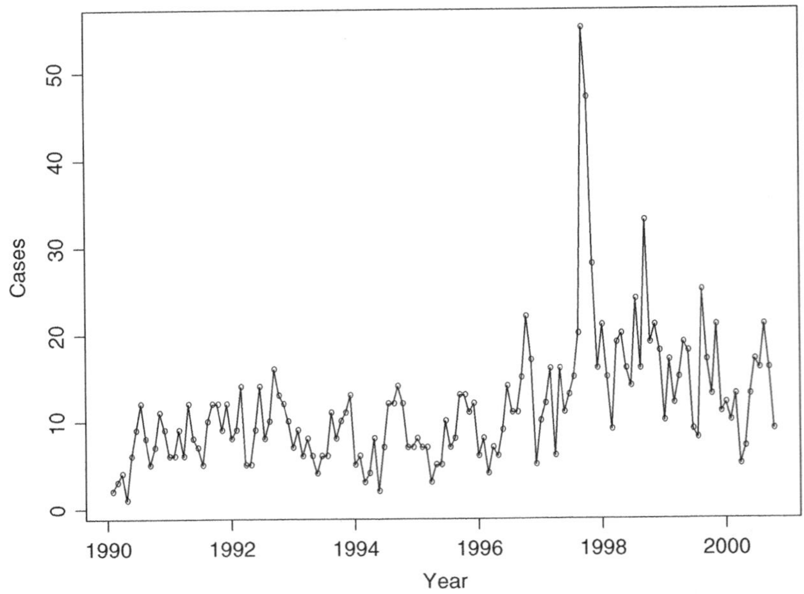

Figure 1.12 Time plot of the number of cases of Campylobacterosis infections in northern Quebec, Canada, in 4-week intervals from January 1990 to October 2000 for 140 observations.

remains linear because the added variable $s_t x_t$ can be treated as a new input variable. However, the model was derived by using the idea of functional-coefficient models, a class of nonlinear models discussed in Chapter 2.

Example 1.4 Another important class of nonlinear models is the generalized linear model. For time series analysis, generalized linear models can be used to analyze count data. Figure 1.12 shows the number of cases of Campylobacterosis infections in the north of the province Quebec, Canada, in 4-week intervals from January 1990 to the end of October 2000. The series has 13 observations per year and 140 observations in total. See Ferland et al. (2006) for more information. The data are available from the R package `tscount` by Liboschik et al. (2015). Campylobacterosis in an acute bacterial infectious disease attacking the digestive system. As expected, the plot of Figure 1.12 shows some seasonal pattern.

This is an example of time series of count data, which occur in many scientific fields, but have received relatively less attention in the time series literature. Similar to the case of independent data, Poisson or negative binomial distribution is often used to model time series of count data. Here one postulates that the time

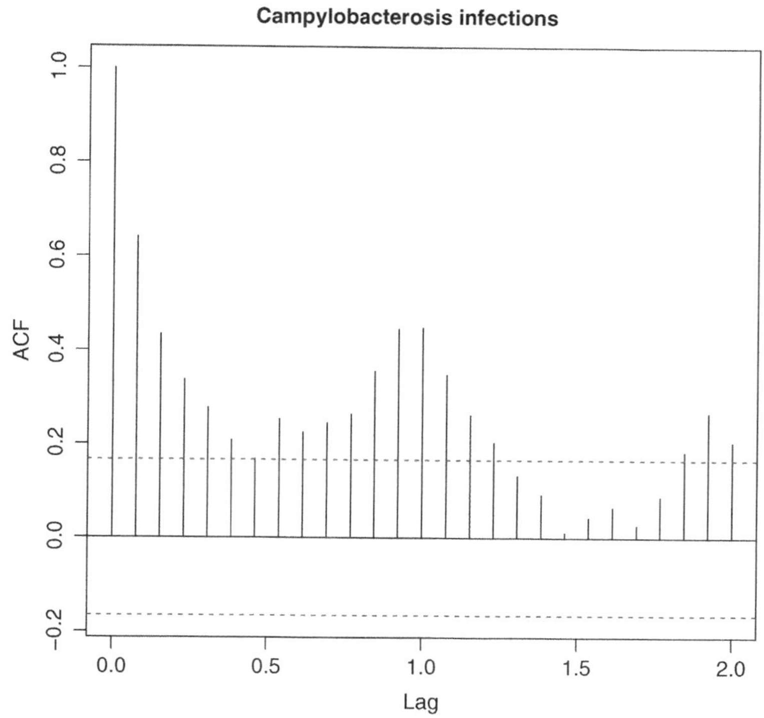

Figure 1.13 Sample autocorrelation function of the cases of Campylobacterosis infections in northern Quebec, Canada, in 4-week intervals from January 1990 to October 2000 for 140 observations.

series of count data is driven by an underlying intensity process λ_t. This latent process governs the time evolution of the conditional expectation of the data, has dynamic dependence, and is related to the data via a link function. In most applications, the link function is either the identify function or the log transformation. Details of modeling time series of count data are given in Chapter 5. Here we use the example to demonstrate yet another application of nonlinear time series analysis. Figure 1.13 shows the sample ACF of the data. Clearly, there is dynamic dependence with seasonality in the data.

Let x_t denote the number of cases of Campylobacterosis infections in week t and $\lambda_t = E(x_t|F_{t-1})$, where F_{t-1} denotes the information available at time $t-1$. A commonly used distribution for count data is

$$x_t|F_{t-1} \sim \text{Poi}(\lambda_t) \quad \text{or} \quad x_t|F_{t-1} \sim \text{NB}(\lambda_t, \alpha), \tag{1.12}$$

where Poi(λ_t) and NB(λ_t, α) denote, respectively, a Poisson distribution with mean λ_t and a negative binomial distribution with mean λ_t and dispersion parameter $\alpha > 0$. Specifically, for the Poisson distribution we have

$$P(x_t = k|\lambda_t) = \frac{1}{k!}e^{-\lambda_t}\lambda_t^k, \quad k = 0, 1, \ldots.$$

For the negative binomial distribution, there are several parameterizations available in the literature. We use the probability mass function

$$P(x_t = k|\lambda_t, \alpha) = \frac{\Gamma(\alpha + k)}{\Gamma(\alpha)\Gamma(k+1)}\left(\frac{\alpha}{\alpha + k}\right)^{\alpha}\left(\frac{\lambda_t}{\alpha + \lambda_t}\right)^{k}, \quad k = 0, 1, \ldots.$$

Under this parameterization, we have $E(x_t|F_{t-1}) = \lambda_t$ and $\text{Var}(x_t|F_{t-1}) = \lambda_t + \lambda_t^2/\alpha$. Here $\alpha \in (0, \infty)$ is the dispersion parameter and the negative binomial distribution approaches the Poisson distribution as $\alpha \to \infty$.

For the time series in Figure 1.12, one can introduce the dynamic dependence for the data by using the model

$$\lambda_t = \gamma_0 + \gamma_1 x_{t-1} + \delta_{13}\lambda_{t-13}, \tag{1.13}$$

where λ_{t-13} is used to describe the seasonality of the data. This model is similar to the GARCH model of Bollerslev (1986) for the volatility model and belongs to the class of observation-driven models in generalized linear models. See Ferland et al. (2006), among others. Using Equation (1.13) with negative binomial distribution, the quasi maximum likelihood estimation gives

$$\lambda_t = 2.43 + 0.594x_{t-1} + 0.188\lambda_{t-13},$$

where the dispersion parameter is 0.109 and the standard deviations of the coefficient estimates are 1.085, 0.092, and 0.123, respectively. Figure 1.14 shows various plots of model checking for the fitted model in Equation (1.13). From the time plot of Pearson residuals, it is seen that certain outlying observations exist so that the model can be refined. Comparing the autocorrelations of the data in Figure 1.13 and the residual autocorrelations in Figure 1.14, the simple model in Equation (1.13) does a decent job in describing the dynamic dependence of the count data. Details of model checking and refinement of the generalized linear models for time series of count data will be given in Chapter 5. The analysis can also be handled by the state space model of Chapter 6.

In this example, we demonstrate that the generalized linear models, which are nonlinear, can be used to analyze time series of count data.

1.4 NONLINEARITY TESTS

The flexibility of nonlinear models in data fitting may encounter the problem of finding spurious structure in a given time series. It is, therefore, of importance

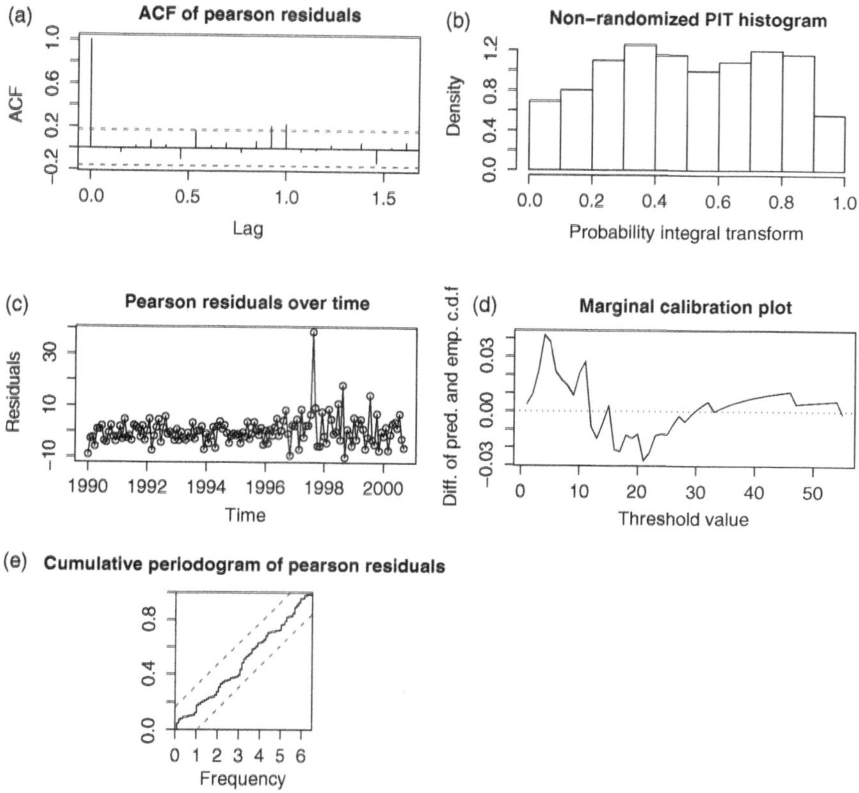

Figure 1.14 Various model diagnostic statistics for the model in Equation (1.13) with negative binomial distribution for the Campylobacterosis infection data. (a) ACF of Pearson residuals, (b) non-randomized PIT histogram, (c) Pearson residuals over time, (d) marginal calibration plot, and (e) cumulative periodogram of Pearson residuals.

to check the need for using nonlinear models. To this end, we introduce some nonlinearity tests for time series data. Both parametric and nonparametric tests are considered.

1.4.1 Nonparametric Tests

Several nonparametric statistics have been proposed in the literature for testing the nonlinearity of a stationary time series. We discuss some of those statistics that are useful and easily available for practical use.

The BDS test: Proposed by Brock et al. (1987) and later published by Brock et al. (1996), this test is widely used to verify the null hypothesis that a given time series consists of independent and identically distributed (iid) random variables. The test uses the idea of *correlation integral* popular in chaotic time series analysis. Roughly speaking, a correlation integral is a measure of the frequency with which temporal patterns repeat themselves in a data set. Consider a time series $\{x_t|t = 1,\dots,T\}$, where T denotes the sample size. Let m be a given positive integer. Define an m-history of the series as $\boldsymbol{x}_t^m = (x_t, x_{t-1}, \dots, x_{t-m+1})$ for $t = m, \dots, T$. Define the correlation integral at the embedding dimension m as

$$C(m,\epsilon) = \lim_{T_m \to \infty} \frac{2}{T_m(T_m - 1)} \sum \sum_{m \le s < t \le T} I(\boldsymbol{x}_t^m, \boldsymbol{x}_s^m | \epsilon) \tag{1.14}$$

where $T_m = T - m + 1$ is the number of constructed m-histories, ϵ is a given positive real number, and $I(\boldsymbol{u}, \boldsymbol{v}|\epsilon)$ is an indicator variable that equals to one if $\|\boldsymbol{u} - \boldsymbol{v}\| < \epsilon$ and zero otherwise, where $\|\cdot\|$ denotes the sup-norm of two vectors. For the m-histories, we have $I(\boldsymbol{u}, \boldsymbol{v}|\epsilon) = 1$ if $|u_i - v_i| < \epsilon$ for $i = 1, \dots, m$ and $=$ 0 otherwise. For a given ϵ, Equation (1.14) simply measures the probability of m-histories being within a distance ϵ of each other.

For testing purpose, the magnitude of the correlation integral in Equation (1.14) needs to be judged. To this end, the BDS test compares $C(m,\epsilon)$ with $C(1,\epsilon)$ under the null hypothesis. Intuitively, if $\{x_t\}$ are iid, then there exist no patterns in the data so that a probability of the m-history is simply the mth power of the corresponding probability of the 1-history. This is so because under independence $Pr(A \cap B) = Pr(A) \times Pr(B)$. In other words, under the iid assumption, we expect that $C(m,\epsilon) = C(1,\epsilon)^m$. The BDS test is then defined as

$$D(m,\epsilon) = \frac{\sqrt{T}[\hat{C}(m,\epsilon) - \{\hat{C}(1,\epsilon)\}^m]}{s(m,\epsilon)} \tag{1.15}$$

where $\hat{C}(k,\epsilon)$ is given by

$$\hat{C}(k,\epsilon) = \frac{2}{T_k(T_k - 1)} \sum \sum_{k \le s < t \le T} I(\boldsymbol{x}_t^k, \boldsymbol{x}_s^k | \epsilon), \quad k = 1, m,$$

and $s(m,\epsilon)$ denotes the standard error of $\hat{C}(m,\epsilon) - \{\hat{C}(1,\epsilon)\}^m$, which can be consistently estimated from the data under the null hypothesis. For details, readers may consult Brock et al. (1996) or Tsay (2010, Chapter 4). In practice, one needs to select the embedding dimension m and the distance ϵ.

The BDS test is available in the `fNonlinear` package of R under the command `bdsTest`. The user has the option to select the maximum embedding dimension m and the distance ϵ. The default maximum dimension is $m = 3$, which means the embedding dimensions used in the test are 2 and 3. The default choices of ϵ are $(0.5, 1, 1.5, 2)\hat{\sigma}_x$, where $\hat{\sigma}_x$ denotes the sample standard error of x_t. To demonstrate, we consider a simulation of iid random variables from $N(0, 1)$ and a daily log return series of the stock of International Business Machines (IBM) Corporation. Details are given below with output edited for simplicity.

R demonstration: BDS test using the package `fNonlinear`.

```
> require(fNonlinear)
> set.seed(1)
> x <- rnorm(300)
> bdsTest(x)
Title: BDS Test

Test Results:
  PARAMETER:
    Max Embedding Dimension: 3
    eps[1]: 0.482; eps[2]: 0.964
    eps[3]: 1.446; eps[4]: 1.927
  STATISTIC:
    eps[1] m=2: -1.1256;     eps[1] m=3: -1.4948
    eps[2] m=2: -0.7145;     eps[2] m=3: -1.1214
    eps[3] m=2: -0.6313;     eps[3] m=3: -0.8081
    eps[4] m=2: -0.7923;     eps[4] m=3: -1.2099
  P VALUE:
    eps[1] m=2: 0.2604;     eps[1] m=3: 0.135
    eps[2] m=2: 0.4749;     eps[2] m=3: 0.2621
    eps[3] m=2: 0.5278;     eps[3] m=3: 0.419
    eps[4] m=2: 0.4282;     eps[4] m=3: 0.2263

> require("quantmod")
> getSymbols("IBM",from="2010-01-02",to="2015-09-30")
> head(IBM)
           IBM.Open IBM.High IBM.Low IBM.Close IBM.Volume IBM.Adjusted
2010-01-04   131.18   132.97  130.85    132.45    6155300     117.6580
2010-01-05   131.68   131.85  130.10    130.85    6841400     116.2367
> rtn <- diff(log(as.numeric(IBM[,6])))
> ts.plot(rtn) # not shown
> Box.test(rtn,type="Ljung",lag=10)
        Box-Ljung test
data:  rtn
X-squared = 15.685, df = 10, p-value = 0.109 # No serial correlations
> bdsTest(rtn)
Title: BDS Test

Test Results:
  PARAMETER:
    Max Embedding Dimension: 3
    eps[1]: 0.006;     eps[2]: 0.012
    eps[3]: 0.018;     eps[4]: 0.024
  STATISTIC:
    eps[1] m=2: 4.4058;     eps[1] m=3: 5.3905
```

```
  eps[2] m=2: 4.2309;      eps[2] m=3: 5.429
  eps[3] m=2: 3.9341;      eps[3] m=3: 5.4255
  eps[4] m=2: 3.7848;      eps[4] m=3: 5.3304
P VALUE:
  eps[1] m=2: 1.054e-05;      eps[1] m=3: 7.026e-08
  eps[2] m=2: 2.327e-05;      eps[2] m=3: 5.666e-08
  eps[3] m=2: 8.349e-05;      eps[3] m=3: 5.78e-08
  eps[4] m=2: 0.0001538;      eps[4] m=3: 9.802e-08
```

From the output, it is clear that in this particular instance the BDS test performs well. It fails to reject the null hypothesis of iid for the random sample from $N(0, 1)$, but successfully rejects the null hypothesis that the daily log returns of IBM are iid.

Discussion: The BDS test is designed to test the null hypothesis of iid, as such one needs to remove any linear dynamic dependence before applying the test to detect nonlinearity in a time series. In other words, care must be exercised in using the BDS test to detect nonlinearity. In practice, the test is typically applied to the residuals of a fitted linear time series model. In addition, a rejection by the BDS test does not provide any specific information to improve the fitted model. Further analysis of the residuals is often needed to seek directions for model refinement after the null hypothesis is rejected by the test. As suggested by the default option, the distance ϵ used in the BDS test should be related to the standard error of x_t.

The McLeod–Li test: McLeod and Li (1983) proposed a general Portmanteau test for nonlinearity under the assumption that the time series x_t is fourth-order stationary, i.e. the x_t^2 process is weakly stationary. Similar to the BDS test, the proposed Portmanteau test is typically applied to the residual $\hat{a}_t$ of a fitted linear time series model. Define the lag-ℓ autocorrelation of the squared residuals as

$$\hat{\rho}_{aa}(\ell) = \frac{\sum_{t=\ell+1}^{T} \left(\hat{a}_t^2 - \hat{\sigma}^2\right)\left(\hat{a}_{t-\ell}^2 - \hat{\sigma}^2\right)}{\sum_{t=1}^{T} \left(\hat{a}_t^2 - \hat{\sigma}^2\right)},$$

where $\hat{\sigma}^2 = \sum_{t=1}^{T} \hat{a}_t^2/T$ and T is the sample size. McLeod and Li show that, for a fixed positive integer m, the joint distribution of

$$\sqrt{T}[\hat{\rho}_{aa}(1), \hat{\rho}_{aa}(2), \ldots, \hat{\rho}_{aa}(m)]'$$

is asymptotically multivariate normal with mean zero and identity covariance matrix provided that the fitted linear model is adequate for the x_t series. Using this result, McLeod and Li (1983) proposed the Portmanteau statistics

$$Q^*(m) = T(T+2)\sum_{\ell=1}^{m} \frac{\hat{\rho}_{aa}^2(\ell)}{T-\ell} \tag{1.16}$$

to detect nonlinearity in x_t. Under the assumption that x_t is fourth-order stationary and the fitted linear model is adequate, $Q^*(m)$ is asymptotically distributed as χ^2_m. This is essentially a Ljung–Box test on the x_t^2 series.

In the literature, the Portmanteau statistics of Equation (1.16) is often used to check for conditional heteroscedasticity. As a matter of fact, the test is asymptotically equivalent to the Lagrange multiplier test of Engle (1982) for the autoregressive conditional heteroscedastic (ARCH) model. To test for ARCH effects, Engle (1982) uses the AR(m) model

$$\hat{a}_t^2 = \beta_0 + \beta_1 \hat{a}_{t-1}^2 + \cdots + \beta_m \hat{a}_{t-m}^2 + \epsilon_t, \tag{1.17}$$

where ϵ_t denotes the error term, and considers the null hypothesis $H_0 : \beta_1 = \beta_2 = \cdots = \beta_m = 0$ versus the alternative hypothesis $H_a : \beta_i \neq 0$ for some $i \in \{1, \ldots, m\}$. The F-statistic of the linear regression in Equation (1.17) can be used to perform the test. Alternatively, one can use mF as a test statistic. Under the same conditions as those of McLeod and Li (1983), mF is asymptotically distributed as χ^2_m. The $Q^*(m)$ statistics of Equation (1.16) can be easily computed. We demonstrate this below.

R demonstration: McLeod–Li and Engle tests for nonlinearity.

```
> set.seed(15)
> xt <- rnorm(300)
> Box.test(xt,lag=10,type='Ljung')
        Box-Ljung test
data:  xt
X-squared = 11.095, df = 10, p-value = 0.3502 # No serial correlations
> Box.test(xt^2,lag=10,type='Ljung')
        Box-Ljung test
data:  xt^2
X-squared = 8.4539, df = 10, p-value = 0.5846 # Q-star test

> require(quantmod)
> getSymbols("MSFT",from="2009-01-02",to="2015-10-15",src="google")
> msft <- diff(log(as.numeric(MSFT$MSFT.Close)))  #log returns
> Box.test(msft,lag=10,type='Ljung')
        Box-Ljung test
data:  msft
X-squared = 17.122, df = 10, p-value = 0.0717 # No serial correlations
> Box.test(msft^2,lag=10,type='Ljung')
        Box-Ljung test
data:  msft^2
X-squared = 70.54, df = 10, p-value = 3.487e-11 ## Nonlinearity
### Engle's test with m = 10.
> nT <- length(msft)
> y <- msft[11:nT]^2
```

```
> X <- NULL
> for (i in 1:10)
+ X <- cbind(X,msft[(11-i):(nT-i)]^2)
+ }
> m2 <- lm(y ~X)  ## Linear regression
> anova(m2)
Analysis of Variance Table
Response: y
             Df     Sum Sq    Mean Sq F value    Pr(>F)
X            10 0.00003036 3.0358e-06    4.65 1.447e-06 ***
Residuals  1687 0.00110138 6.5286e-07
> qstar = 4.64*10
> pv = 1 -pchisq(qstar,10)
> print(c(qstar,pv))
[1] 4.65e+01 1.162495e-06  ### ARCH effect exists
```

From the output, the traditional Ljung–Box statistics confirm that both the series of iid $N(0, 1)$ random variables and the daily log returns of Microsoft (MSFT) stock from January 2010 to October 15, 2015 have no serial correlations. On the other hand, the McLeod–Li test statistic of (1.16) cannot reject the null hypothesis of linear time series for the $N(0, 1)$ iid random sample, but it clearly rejects linearity for the daily log returns of MSFT stock. The ARCH effect is also confirmed by the Lagrange multiplier test of Engle (1982).

Rank-based Portmanteau test: The McLeod–Li test requires the existence of the fourth moment of the underlying time series x_t. In some applications, empirical data may exhibit high excess kurtosis. In this situation, the performance of the McLeod–Li test may deteriorate. To overcome the impact of heavy-tails on the McLeod–Li test, one can apply the rank-based Ljung–Box statistics. See, for instance, Dufour and Roy (1986) and Tsay (2014, Chapter 7) and the references therein. Let R_i be the rank of $\hat{a}_i^2$ in the squared residuals $\{\hat{a}_t^2\}$. The lag-ℓ rank-based serial correlation is defined as

$$\tilde{\rho}_\ell = \frac{\sum_{t=\ell+1}^{T}(R_t - \bar{R})(R_{t-\ell} - \bar{R})}{\sum_{t=1}^{T}(R_t - \bar{R})^2}, \quad \ell = 1, 2, \ldots$$

where it can be shown that

$$\bar{R} = \sum_{t=1}^{T} R_t/T = (T + 1)/2$$

$$\sum_{t=1}^{T}(R_t - \bar{R})^2 = T(T^2 - 1)/12.$$

Furthermore, Dufour and Roy (1986) show that

$$E(\tilde{\rho}_\ell) = -(T-\ell)/[T(T-1)],$$

$$\text{Var}(\tilde{\rho}_\ell) = \frac{5T^4 - (5\ell+9)T^3 + 9(\ell-2)T^2 + 2\ell(5\ell+8)T + 16\ell^2}{5(T-1)^2T^2(T+1)}.$$

The rank-based Portmanteau statistic then becomes

$$Q_R(m) = \sum_{\ell=1}^{m} \frac{[\tilde{\rho}_\ell - E(\tilde{\rho}_\ell)]^2}{\text{Var}(\tilde{\rho}_\ell)}, \tag{1.18}$$

which is asymptotically distributed as χ^2_m if the $\{\hat{a}_t^2\}$ series has no serial correlations.

To compare the finite-sample performance of the McLeod–Li Portmanteau test in Equation (1.16) and its rank-based counterpart in Equation (1.18), we conduct a simple simulation. We generate iid sequences from $N(0,1)$ and t_5 distributions with 300 observations and compute the two test statistics for $m = 1, 5$, and 10. This process is repeated for 10,000 iterations. Some selected quantiles of the empirical distributions of $Q^*(1), Q^*(5), Q^*(10)$ and $Q_R(1), Q_R(5), Q_R(10)$ are given in Table 1.4 along with the quantiles of the corresponding χ^2_m distributions.

Figure 1.15 shows the empirical density functions $Q^*(10)$ and $Q_R(10)$ over the 10,000 iterations. The iid sequences were generated from $N(0,1)$ for 300 observations. The solid and dashed lines are for $Q^*(10)$ and $Q_R(10)$, respectively. The dotted line of the plot shows the density function of χ^2_{10}. Figure 1.16 shows similar density functions when the iid sequences were generated from a Student-t distribution with 5 degrees of freedom. From Table 1.4 and Figures 1.15 and 1.16, it is seen that both $Q^*(10)$ of McLeod–Li test in Equation (1.16) and the rank-based $Q_R(10)$ in Equation (1.18) follow reasonably well the asymptotic χ^2_{10} distribution under the normality. On the other hand, the $Q^*(10)$ statistic encounters some size distortion and power loss when the iid sequences are from the t_5 distribution whereas the rank-based Portmanteau statistic $Q_R(10)$ continues to perform well. See the shift of the empirical density of $Q_m(10)$ in Figure 1.16. Therefore, care must be exercised when one applies the McLeod–Li test to detect nonlinearity when the underlying process has heavy tails.

The Peña–Rodriguez test: Peña and Rodriguez (2002) proposed a modified Portmanteau test statistic that can be used for model checking of a fitted linear time series model, including nonlinearity in the residuals. Using simulation studies, the authors show that their modified test statistics perform well in finite samples. For simplicity, let z_t be a function of the residual series $\hat{a}_t$ of a fitted linear

Distribution	Statistics	0.025	0.05	0.5	0.95	0.975
	χ_1^2	0.00098	0.00393	0.455	3.842	5.024
$N(0,1)$	$Q^*(1)$	0.00096	0.00372	0.451	3.744	4.901
	$Q_R(1)$	0.00088	0.00392	0.463	3.762	5.056
t_5	$Q^*(1)$	0.00074	0.00259	0.288	2.936	4.677
	$Q_R(1)$	0.00098	0.00367	0.441	3.716	4.856
	χ_5^2	0.8312	1.1455	4.351	11.070	12.833
$N(0,1)$	$Q^*(5)$	0.7903	1.0987	4.129	11.045	13.131
	$Q_R(5)$	0.7884	1.1193	4.263	11.042	12.792
t_5	$Q^*(5)$	0.3634	0.5575	2.909	10.992	15.047
	$Q_R(5)$	0.8393	1.1287	4.254	11.123	13.002
	χ_{10}^2	3.2470	3.9403	9.342	18.307	20.483
$N(0,1)$	$Q^*(10)$	3.1077	3.7532	8.870	12.286	21.122
	$Q_R(10)$	3.1737	3.8524	9.184	18.163	20.516
t_5	$Q^*(10)$	1.1347	1.6916	6.685	19.045	23.543
	$Q_R(10)$	3.2115	3.8265	9.216	18.495	20.786

Table 1.4 Empirical quantiles of McLeod–Li Portmanteau statistics and the rank-based Portmanteau statistics for random samples of 300 observations. The realizations are generated from $N(0, 1)$ and Student-t distribution with 5 degrees of freedom. The results are based on 10,000 iterations.

model. For instance, $z_t = \hat{a}_t^2$ or $z_t = |\hat{a}_t|$. The lag-ℓ sample autocorrelation of z_t is defined as

$$\hat{\rho}_\ell = \frac{\sum_{t=\ell+1}^{T}(z_{t-\ell} - \bar{z})(z_t - \bar{z})}{\sum_{t=1}^{T}(z_t - \bar{z})^2}, \tag{1.19}$$

where, as before, T is the sample size and $\bar{z} = \sum_t z_t/T$ is the sample mean of z_t. As usual, $\hat{\rho}_\ell$ is a consistent estimate of the lag-ℓ autocorrelation ρ_ℓ of z_t under some regularity conditions. For a given positive integer m, the Peña and Rodriguez test statistic for testing the null hypothesis of $H_0 : \rho_1 = \cdots = \rho_m = 0$ is

$$\hat{D}_m = T[1 - |\hat{\boldsymbol{R}}_m|^{1/m}], \tag{1.20}$$

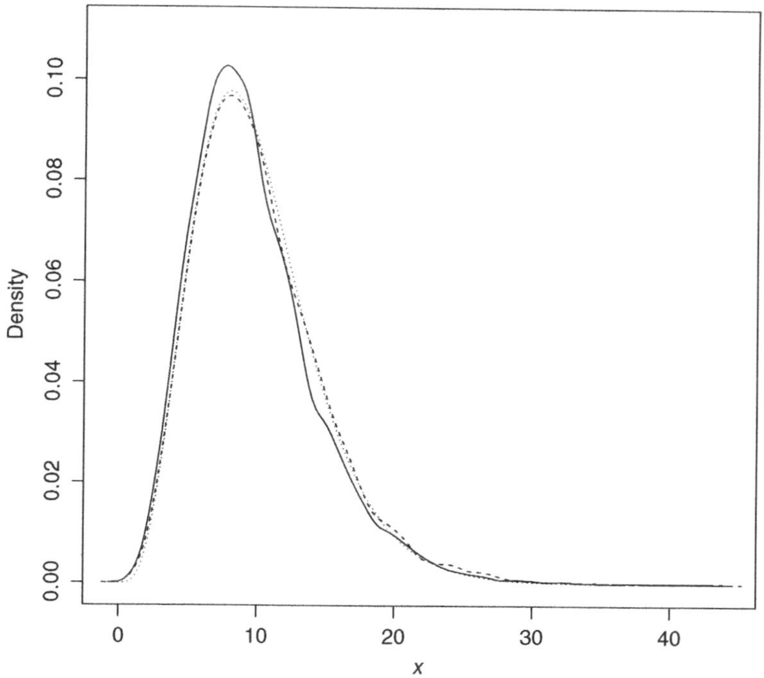

Figure 1.15 Empirical density functions of $Q^*(10)$ in Equation (1.16) (solid line) and $Q_R(10)$ in Equation (1.18) (dashed line) when the iid sequences are generated from $N(0, 1)$. The results are for sample size 300 and 10,000 iterations. The dotted line denotes the density function of χ^2_{10}.

where the $(m+1)$ by $(m+1)$ matrix $\widehat{\mathbf{R}}_m$ is defined below:

$$\widehat{\boldsymbol{R}}_m = \begin{bmatrix} 1 & \hat{\rho}_1 & \cdots & \hat{\rho}_m \\ \hat{\rho}_1 & 1 & \cdots & \hat{\rho}_{m-1} \\ \vdots & \vdots & \ddots & \vdots \\ \hat{\rho}_m & \hat{\rho}_{m-1} & \cdots & 1 \end{bmatrix}. \tag{1.21}$$

Using the idea of pseudo-likelihood, Peña and Rodriguez (2006) further modified the test statistic in Equation (1.20) to

$$D^*_m = -\frac{T}{m+1}\log(|\widehat{\boldsymbol{R}}_m|), \tag{1.22}$$

where the denominator $m+1$ is the dimension of $\widehat{\mathbf{R}}_m$. Under the assumption that the fitted ARMA(p, q) model is correctly specified, Peña and Rodriguez (2006)

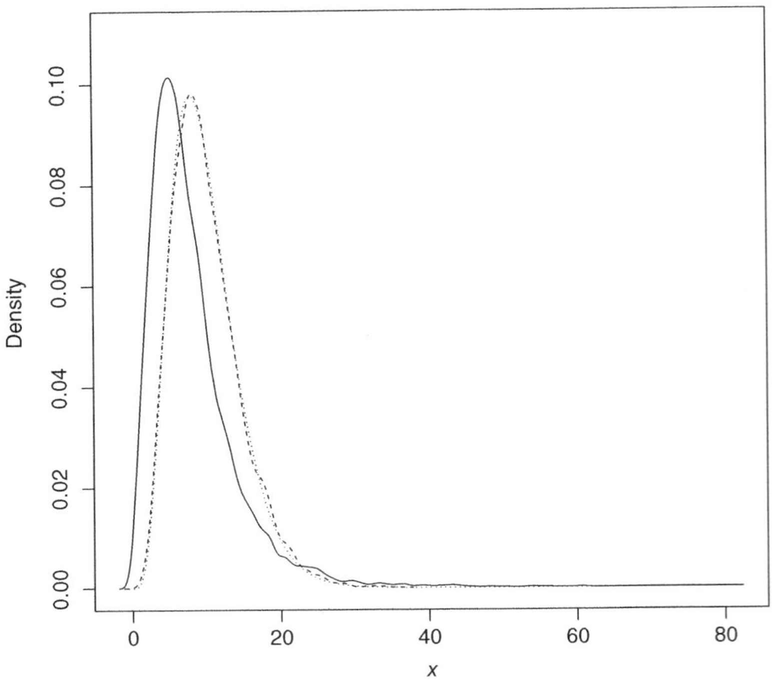

Figure 1.16 Empirical density functions of $Q^*(10)$ in Equation (1.16) (solid line) and $Q_R(10)$ in Equation (1.18) (dashed line) when the iid sequences are generated from a Student-t distribution with 5 degrees of freedom. The results are for sample size 300 and 10,000 iterations. The dotted line denotes the density function of χ^2_{10}.

show that the test statistic D^*_m of Equation (1.22) is asymptotically distributed as a mixture of m independent χ^2_1 random variates. The weights of the mixture are rather complicated. However, the authors derived two approximations to simplify the calculation. The first approximation of D^*_m is denoted by GD^*_m, which is approximately distributed as a gamma random variate, $\Gamma(\alpha, \beta)$, where

$$\alpha = \frac{3(m+1)[m-2(p+q)]^2}{2[2m(2m+1)-12(m+1)(p+q)]},$$

$$\beta = \frac{3(m+1)[m-2(p+q)]}{2m(2m+1)-12(m+1)(p+q)}.$$

The second approximation is

$$ND^*_m = (\alpha/\beta)^{-1/\lambda}(\lambda/\sqrt{\alpha})\left[(D^*_m)^{1/\lambda} - (\alpha/\beta)^{1/\lambda}\left(1 - \frac{\lambda-1}{2\alpha\lambda^2}\right)\right], \quad (1.23)$$

where α and β are defined as before, and

$$\lambda = \left[1 - \frac{2(m/2-(p+q))(m^2/(4(m+1))-(p+q))}{3(m(2m+1)/(6(m+1))-(p+q))^2}\right]^{-1}.$$

Asymptotically, ND_n^* is distributed as $N(0, 1)$. For large m, $\lambda \approx 4$. In practice, the authors recommend replacing $\hat{\rho}_\ell$ by $\tilde{\rho}_\ell = (T+2)\hat{\rho}_\ell/(T-\ell)$ in the $\hat{\boldsymbol{R}}_m$ matrix of Equation (1.21) for better performance in finite samples, especially when T is small.

To demonstrate the ND_m^* statistic of Equation (1.23), we consider the daily log returns of Microsoft stock from January 2, 2009 to October 15, 2015. The same series was used before in demonstrating the McLeod–Li test. The sample size is $T = 1708$ so that we select $m = \lfloor \log(T) \rfloor + 1 = 8$. The test, denoted by `PRnd`, is available in the package `NTS`.

R demonstration: Peña–Rodriguez test.

```
> require(quantmod); require(NTS)
> getSymbols("MSFT",from="2009-01-02",to="2015-10-15",src="google")
> msft <- diff(log(as.numeric(MSFT$MSFT.Close)))  # log returns
> log(length(msft))
[1] 7.443078
> PRnd(msft,m=8)
ND-stat & p-value  0.1828238 0.8549363
> PRnd(abs(msft),m=8)
ND-stat & p-value  2.302991 0.02127936
```

From the output, the daily log returns have no significant serial correlations, but their absolute values have serial correlations. Again, the results confirm the nonlinearity in the daily MSFT log returns.

1.4.2 Parametric Tests

Most parametric tests for nonlinearity are developed under the assumption of certain nonlinear models as the alternative. Roughly speaking, a well-behaved zero-mean stationary time series x_t can be written as a Volterra series

$$x_t = \sum_{i=1}^{t-1} \psi_i x_{t-i} + \sum_{i=1}^{t-1}\sum_{j=1}^{t-1} \psi_{ij} x_{t-i} x_{t-j} + \sum_i \sum_j \sum_k \phi_{ijk} x_{t-i} x_{t-j} x_{t-k} + \cdots + \epsilon_t, \quad (1.24)$$

where ϵ_t denotes the noise term and the ψs are real numbers. For a linear time series x_t, we have $\psi_{ij} = \psi_{ijk} = \cdots = 0$. Therefore, some of the higher-order coefficients are non-zero if x_t is nonlinear. If the third order and higher coefficients are zero, but $\psi_{ij} \neq 0$ for some i and j, then x_t becomes a bilinear process. Parametric tests for nonlinearity of x_t are statistics that exploit certain features of the Volterra series in Equation (1.24). We discuss some of the parametric tests in this section.

The RESET test: Ramsey (1969) proposed a specification test for linear least squares regression analysis. The test is referred to as a RESET test and is readily applicable to linear AR models. Consider the linear AR(p) model

$$x_t = \boldsymbol{X}'_{t-1}\boldsymbol{\phi} + a_t, \tag{1.25}$$

where $\boldsymbol{X}_{t-1} = (1, x_{t-1}, \ldots, x_{t-p})'$ and $\boldsymbol{\phi} = (\phi_0, \phi_1, \ldots, \phi_p)'$. The first step of the RESET test is to obtain the least squares estimate $\widehat{\boldsymbol{\phi}}$ of Equation (1.25) and compute the fitted value $\hat{x}_t = \boldsymbol{X}'_{t-1}\widehat{\boldsymbol{\phi}}$, the residual $\hat{a}_t = x_t - \hat{x}_t$, and the sum of squared residuals $SSR_0 = \sum_{t=p+1}^{T} \hat{a}_t^2$, where T is the sample size. In the second step, consider the linear regression

$$\hat{a}_t = \boldsymbol{X}'_{t-1}\boldsymbol{\alpha}_1 + \boldsymbol{M}'_{t-1}\boldsymbol{\alpha}_2 + v_t, \tag{1.26}$$

where $\boldsymbol{M}_{t-1} = (\hat{x}_t^2, \ldots, \hat{x}_t^{s+1})'$ for some $s \geq 1$, and compute the least squares residuals

$$\hat{v}_t = \hat{a}_t - \boldsymbol{X}'_{t-1}\widehat{\boldsymbol{\alpha}}_1 - \boldsymbol{M}'_{t-1}\widehat{\boldsymbol{\alpha}}_2$$

and the sum of squared residuals $SSR_1 = \sum_{t=p+1}^{T} \hat{v}_t^2$ of the regression. The basic idea of the RESET test is that if the linear AR(p) model in Equation (1.25) is adequate, then $\boldsymbol{\alpha}_1$ and $\boldsymbol{\alpha}_2$ of Equation (1.26) should be zero. This can be tested by the usual F statistic of Equation (1.26) given by

$$F = \frac{(SSR_0 - SSR_1)/g}{SSR_1/(T-p-g)} \quad \text{with} \quad g = s + p + 1, \tag{1.27}$$

which, under the linearity and normality assumption, has an F distribution with degrees of freedom g and $T - p - g$.

Remark: Because the variables $\hat{x}_t^k$ for $k = 2, \ldots, s+1$ tend to be highly correlated with $\boldsymbol{X}_{t-1}$ and among themselves, principal components of $\boldsymbol{M}_{t-1}$ that are not co-linear with $\boldsymbol{X}_{t-1}$ are often used in fitting Equation (1.26).

Keenan (1985) proposed a nonlinearity test for time series that uses $\hat{x}_t^2$ only and modifies the second step of the RESET test to avoid multicollinearity between $\hat{x}_t^2$ and $\boldsymbol{X}_{t-1}$. Specifically, the linear regression (1.26) is divided into two steps. In step 2(a), one removes linear dependence of $\hat{x}_t^2$ on $\boldsymbol{X}_{t-1}$ by fitting the regression

$$\hat{x}_t^2 = \boldsymbol{X}'_{t-1}\boldsymbol{\beta} + u_t$$

and obtaining the residual $\hat{u}_t = \hat{x}_t^2 - \boldsymbol{X}_{t-1}\hat{\boldsymbol{\beta}}$. In step 2(b), consider the linear regression

$$\hat{a}_t = \hat{u}_t\alpha + v_t,$$

and obtain the sum of squared residuals $SSR_1 = \sum_{t=p+1}^{T}(\hat{a}_t - \hat{u}_t\hat{\alpha})^2 = \sum_{t=p+1}^{T}\hat{v}_t^2$ to test the null hypothesis $\alpha = 0$.

The F-test: To improve the power of Keenan's and RESET tests, Tsay (1986) used a different choice of the regressor $\boldsymbol{M}_{t-1}$. Specifically, he suggested using $\boldsymbol{M}_{t-1} = vech(\boldsymbol{X}_{t-1}\boldsymbol{X}_{t-1}')$, where $vech(\boldsymbol{A})$ denotes the half-stacking vector of the matrix $\boldsymbol{A}$ using elements on and below the diagonal only. For example, if $p = 2$, then $\boldsymbol{M}_{t-1} = (x_{t-1}^2, x_{t-1}x_{t-2}, x_{t-2}^2)'$. The dimension of $\boldsymbol{M}_{t-1}$ is $p(p+1)/2$ for an AR(p) model. In practice, the test is simply the usual partial F statistic for testing $\boldsymbol{\alpha} = 0$ in the linear least squares regression

$$x_t = \boldsymbol{X}_{t-1}'\boldsymbol{\phi} + \boldsymbol{M}_{t-1}'\boldsymbol{\alpha} + e_t,$$

where e_t denotes the error term. Under the assumption that x_t is a linear AR(p) process, the partial F statistic follows an F distribution with degrees of freedom g and $T - p - g - 1$, where $g = p(p+1)/2$. We refer to this F test as the *F-test*. Luukkonen et al. (1988) further extended the test by augmenting $\boldsymbol{M}_{t-1}$ with cubic terms x_{t-i}^3 for $i = 1, \ldots, p$.

To demonstrate nonlinearity tests, consider the US quarterly unemployment rates of Example 1.1. In this particular instance, both Keenan test and F-test confirm that the series is nonlinear. The F-test is referred to as the `Tsay.test` in the `TSA` package.

R demonstration: Parametric tests.

```
> require(TSA)
> da=read.table("q-unrate.txt",header=T)
> rate <- da$rate
> Keenan.test(unrate)
$test.stat
[1] 7.428776
$p.value
[1] 0.006858382
$order
[1] 5
> Tsay.test(unrate)
$test.stat
[1] 2.626
$p.value
```

```
[1] 0.00108
$order
[1] 5
```

Remark: The dimension of $\boldsymbol{M}_{t-1}$ used in the F-test is $p(p+1)/2$, which can be high when the order p is not small compared with the sample size T. In this situation, the F-test might have low power in detecting nonlinearity. To improve the F-test, one can use the idea of thresholding. Specifically, one constructs $\boldsymbol{M}_{t-1}$ using only those lagged variables in $\boldsymbol{X}_{t-1}$ with t-ratio greater than some threshold in modulus. To illustrate, consider the well-known Canadian Lynx data, which is available in the `TSA` package. The series has 114 observations. With log-transformation, an AR(11) is selected by the `Keenan.test` and `Tsay.test` of the `TSA` package. In this particular case, the Keenan test shows that the log series is nonlinear, but the F-test fails. This is not surprising because the F-test fits $11(12)/2 = 66$ additional parameters whereas the effective sample size is only $114 - 11 = 103$. On the other hand, with threshold 1.645, the F-test also confirms that the log series is nonlinear. Here the threshold 1.645 selects five lagged variables to construct $\boldsymbol{M}_{t-1}$ in the F-test.

R demonstration:

```
> require(TSA)
> data(lynx)
> y <- log10(lynx)
> Keenan.test(y)
$test.stat
[1] 11.66997
$p.value
[1] 0.0009550016
$order
[1] 11
> Tsay.test(y)
$test.stat
[1] 1.316
$p.value
[1] 0.2256
$order
[1] 11
> require(NTS)
> F.test(y,thres=1.645)
$test.stat
[1] 1.971
$p.value
```

```
[1] 0.02858
$order
[1] 11
```

Threshold test: If the alternative model under study is the self-exciting threshold autoregressive (TAR) model shown in Example 1.1, one can derive specific test statistics to increase the power of detecting nonlinearity. One of the specific tests is the likelihood ratio statistic studied by Chan and Tong (1990) and Chan (1990, 1991). The TAR(p) model considered can be written as

$$x_t = \phi_0 + \sum_{i=1}^{p} \phi_i x_{t-i} + I(x_{t-d} > r)\left(\beta_0 + \sum_{i=1}^{p} \beta_i x_{t-i}\right) + a_t, \tag{1.28}$$

where a_t is a sequence of independent and identically distributed Gaussian random variables with mean zero and variance σ^2, ϕ_i and β_j are real valued parameters, $d > 0$ with x_{t-d} being the threshold variable, r is the threshold, and $I(x_{t-d} > r)$ is an indicator variable such that $I(x_{t-d} > r) = 1$ if $x_{t-d} > r$ and $= 0$ otherwise. The null hypothesis is $H_0 : \beta_i = 0$ for $i \in \{0, \ldots, p\}$ and the alternative hypothesis is $H_a : \beta_i \neq 0$ for some i. In Equation (1.28), we assume that the threshold r lies inside a known bounded closed interval $\tilde{R} \subseteq R$. For simplicity, we further assume that $d \leq p$. The likelihood ratio test statistic is then

$$\lambda = \frac{n\left(\hat{\sigma}_0^2 - \hat{\sigma}^2\right)}{\hat{\sigma}^2}, \tag{1.29}$$

where $n = T - p + 1$ with T being the sample size,

$$\hat{\sigma}^2 = \min_{r \in \tilde{R}, \theta, \phi} \left[\sum_{t=p+1}^{T} \left\{ x_t - \phi_0 - \sum_{i=1}^{p} \phi_i x_{t-i} - I(x_{t-d} > r)\left(\beta_0 + \sum_{i=1}^{p} \beta_i x_{t-i}\right)\right\}^2 \right],$$

$$\hat{\sigma}_0^2 = \min_{\phi} \left\{ \sum_{t=p+1}^{T} \left(x_t - \phi_0 - \sum_{i=1}^{p} \phi_i x_{t-i}\right)^2 \right\}.$$

The null hypothesis is rejected if and only if λ of Equation (1.29) is large. The limiting properties of the test statistic λ, however, are non-standard and more involved because the threshold r is undefined under the null hypothesis H_0. This is an issue of *nuisance parameter* under the null hypothesis. Interested readers are referred to Chan and Tong (1990). The percentage points of λ under the null hypothesis have been investigated by Chan (1990). In the literature, it is common to assume that $\tilde{R} = [v, u]$, where v and u are pre-specified real numbers with $v < u$. See, for instance, Davis (1987) and Andrews and Ploberger (1994).

Another specific threshold nonlinearity test is to transform the testing problem into a change point detection problem. The transformation is carried out via the *arranged autoregression*. See Tsay (1989, 1998) and the references therein. Specifically, the arranged autoregression transfers the TAR model under the alternative hypothesis H_a into a model-change problem with the threshold r being the change point. To see this, the TAR model in Equation (1.28) says that x_t follows essentially two linear models depending on whether $x_{t-d} < r$ or $x_{t-d} \geq r$. For a realization of T data points, say $\{x_t\}_{t=1}^{T}$, the threshold variable x_{t-d} can assume values $\{x_1, \ldots, x_{T-d}\}$. Let $x_{(1)} \leq x_{(2)} \leq \cdots \leq x_{(T-d)}$ be the ordered statistics of $\{x_t\}_{t=1}^{T-d}$. The TAR model can then be written as

$$x_{(j)+d} = \theta_0 + \sum_{i=1}^{p} \theta_i x_{(j)+d-i} + a_{(j)+d}, \quad j = 1, \ldots, T-d, \tag{1.30}$$

where $\theta_i = \phi_i$ if $x_{(j)} \leq r$, and $\theta_i = \phi_i + \beta_i$ if $x_{(j)} > r$, for $i = 0, \ldots, p$. Consequently, the threshold r is a change point for the linear regression in Equation (1.30), and we refer to Equation (1.30) as an arranged autoregression (in increasing order of the threshold variable x_{t-d}). Note that the arranged autoregression in Equation (1.30) does not alter the dynamic dependence of x_t on x_{t-i} for $i = 1, \ldots, p$ because $x_{(j)+d}$ still depends on $x_{(j)+d-i}$ for $i = 1, \ldots, p$. What is done is simply to present the TAR model in the threshold space instead of in the time space. That is, the equation with a smaller x_{t-d} appears before that with a larger x_{t-d}. The threshold test of Tsay (1989) is obtained as follows.

- Step 1: Fit Equation (1.30) using $j = 1, \ldots, m$, where m is a pre-specified positive integer (e.g., 30). Denote the least squares estimates of θ_i by $\hat{\theta}_{i,m}$, where m denotes the number of data points used in estimation.

- Step 2: Compute the predictive residual

$$\hat{a}_{(m+1)+d} = x_{(m+1)+d} - \hat{\theta}_{0,m} - \sum_{i=1}^{p} \hat{\theta}_{i,m} x_{(m+1)+d-i}$$

  and its standard error. Let $\hat{e}_{(m+1)+d}$ be the standardized predictive residual.

- Step 3: Use the recursive least squares method to update the least squares estimates to $\hat{\theta}_{i,m+1}$ by incorporating the new data point $x_{(m+1)+d}$.

- Step 4: Repeat Steps 2 and 3 until all data points are processed.

- Step 5: Consider the linear regression of the standardized predictive residual

$$\hat{e}_{(m+j)+d} = \alpha_0 + \sum_{i=1}^{p} \alpha_i x_{(m+j)+d-i} + v_t, \quad j = 1, \ldots, T-d-m \tag{1.31}$$

and compute the usual F statistic for testing $\alpha_i = 0$ in Equation (1.31) for $i = 0, \ldots, p$. Under the null hypothesis that x_t follows a linear AR(p) model, the F ratio has a limiting F distribution with degrees of freedom $p + 1$ and $T - d - m - p$.

We refer to the resulting F test as a *Tar-F test*. The idea behind the test is that under the null hypothesis there is no model change in the arranged autoregression in Equation (1.30) so that the standardized predictive residuals should be close to iid with mean zero and variance 1. In this case, they should have no correlations with the regressors $x_{(m+j)+d-i}$. For further details, including formulas for a recursive least squares method and some simulation study on performance of the Tar-F test, see Tsay (1989). The Tar-F test avoids the problem of nuisance parameters encountered by the likelihood ratio test. It does not require knowing the threshold r. It simply tests that the predictive residuals have no correlations with regressors if the null hypothesis holds. Therefore, the test does not depend on knowing the number of regimes in the alternative model. Yet the Tar-F test is not as powerful as the likelihood ratio test if the true model is indeed a two-regime TAR model with Gaussian innovations.

The likelihood ratio test for a two-regime Gaussian TAR model is available via the command `tlrt` of the `TSA` package, whereas the Tar-F test is available via the command `thr.test` of the `NTS` package. For the US quarterly unemployment rates of Example 1.1, both tests detect nonlinearity with $p = 5$ and $d = 5$ and default options. Details are given below.

R demonstration: Threshold tests.

```
> da <- read.table("q-unrate.txt",header=T)
> unrate <- da$rate
> require(NTS)
> require(TSA)
> thr.test(unrate,p=5,d=5)
SETAR model is entertained
Threshold nonlinearity test for (p,d):  5 5
F-ratio and p-value:  2.849921 0.01081263
> tlrt(unrate,p=5,d=5)
$percentiles
[1] 24.9 75.1
$test.statistic
    33.028
$p.value
    0.0003066953
```

1.5 EXERCISES

1.1 Consider the global land and ocean temperature anomalies from January 1880 to May 2017. The monthly data are available from the National Oceanic and Atmospheric Administration (NOAA) at www.ncdc.noaa.gov and are in the file `globaltemp1880-2017.csv`. The data are in degrees Celsius and the base period is 1901–2000. The temperature anomalies show an upward trend so consider the change series, i.e. the first difference. Is the differenced series nonlinear? Perform some tests and draw a conclusion. Does the series show threshold nonlinearity? Why?

1.2 Consider the daily log returns of the exchange-traded fund for the S&P 500 index with tick symbol `SPY` from January 3, 2007 to June 30, 2017. Daily log returns denote the first difference of log daily closing prices. The first column of the file `SPY0717.txt` contains the log return. Is the return series nonlinear? Perform some tests, including the BDS test to draw a conclusion.

1.3 Asset return series tend to exhibit conditional heteroscedasticity. The GARCH models are often used to handle conditional heteroscedasticity. Consider the SPY daily log returns of previous problem. The second column of the file `SPY0717.txt` shows the standardized residuals of a fitted AR(1)-GARCH(1,1) model with standardized Student-t innovations. Does the standardized residual series consist of independent and identically distributed random variables? Why?

1.4 Consider the grow rate series (first difference of the log index) of the US monthly producer price index from February 1913 to August 2017. The data can be obtained from FRED (PPIACO, 1982 index=100, not seasonally adjusted). The data are also in the file `PPIACO.csv`. Is the growth rate series linear? Perform some nonlinearity tests to draw a conclusion.

1.5 Consider again the growth rate series of the US monthly producer price index of the prior problem. Let x_t be the growth rate series. Use the `fGarch` package to fit the model:

```
require(fGarch)
m1 <- garchFit( arma(3,2)+garch(1,1),data=x_t,trace=F,
            cond.dist="std")
rt <- residuals(m1,standardize=T)
```

The `rt` denotes the standardized residuals. Apply the McLeod–Li test to detect any nonlinearity in the `rt` series and apply the BDS test to check whether the `rt` series consists of iid random variables.

REFERENCES

Akaike, H. (1974). A new look at the statistical model identification. *IEEE Transactions on Automatic Control* **AC-19**: 716–723.

Andrews, D. W. K. and Ploberger, W. (1994). Optimal tests when a nuisance parameter is present only under the alternative. *Econometrica* **62**: 1383–1414.

Bollerslev, T. (1986). Generalized autoregressive conditional heteroskedasticity. *Journal of Econometrics* **31**: 307–327.

Box, G. E. P. and Jenkins, G. M. (1976). *Time Series Analysis: Forecasting and Control*, revised ed. Holden-Day, San Francisco.

Box, G. E. P., Jenkins, G. M., Reinsel, G. C., and Ljung, G. M. (2015). *Time Series Analysis: Forecasting and Control*, 5th edition. Wiley, Hoboken, NJ.

Brock, W., Dechert, W.D., and Scheinkman, J. (1987). A test for independence based on the correlation dimension. Working paper, Department of Economics, University of Wisconsin-Madison.

Brock, W., Dechert, W. D., Scheinkman, J., and LeBaron, B. (1996). A test for independence based on the correlation dimension. *Econometric Reviews* **15**: 197–235.

Chan, K. S. (1990). Percentage points of likelihood ratio tests for threshold autoregression. *Journal of the Royal Statistical Society, Series B* **53**: 691–696.

Chan, K. S. (1991). Percentage points of likelihood ratio tests for threshold autoregression. *Journal of the Royal Statistical Society Series B* **53**: 691–696.

Chan, K. S. and Tong, H. (1990). On likelihood ratio tests for threshold autoregression. *Journal of the Royal Statistical Society, Series B* **52**: 469–476.

Chen, R. and Tsay, R. S. (1993). Functional-coefficient autoregressive models. *Journal of the American Statistical Association* **88**: 298-308.

Chen, R. and Tsay, R. S. (1996). Nonlinear transfer function. *Journal of Nonparametric Statistics* **6**: 193–204.

Davis, R. B. (1987). Hypothesis testing when a nuisance parameter is present only under the alternative. *Biometrika* **74**: 33–43.

De Gooijer, J. G. (2017). *Elements of Nonlinear Time Series Analysis and Forecasting*. Springer, Cham, Switzerland.

Douc, R., Moulines, E., and Stoffer, D. S. (2014). *Nonlinear Time Series: Theory, Methods, and Applications with R Examples*. CRC Press, Boca Raton, FL.

Dufour, J. M. and Roy, R. (1986). Generalized portmanteau statistics and tests of randomness. *Communications in Statistics – Theory and Methods* **15**: 2953–2972,

Engle, R. F. (1982). Autoregressive conditional heteroscedasticity with estimates of the variance of United Kingdom inflations. *Econometrica* **50**: 987–1007.

Fan, J. and Yao, Q. (2003). *Nonlinear Time Series: Nonparametric and Parametric Methods*. Springer, New York.

Ferland, R., Latour, A. and Oraichi, D. (2006). Integer-valued GARCH process. *Journal of Time Series Analysis* **27**: 923–942.

Hamilton, J. D. (1994). *Time Series Analysis*. Princeton University Press, Princeton, NJ.

Keenan, D. M. (1985). A Tukey non-additivity-type test for time series nonlinearity. *Biometrika* **72**: 39–44.

Liboschik, T., Fried, R., Fokianos, K., Probst, P., and Rathjens, J. (2015). *Analysis of Count Time Series*, R Package, http://tscount.r-forge.r-project.org.

Luukkonen, R., Saikkonen, P., and Teräsvirta (1988). Testing linearity against smooth transition autoregressive models. *Biometrika* **75**: 491–499.

McLeod, A. I. and Li, W. K. (1983). Diagnostic checking ARMA time series models using squared-residual autocorrelations. *Journal of Time Series Analysis* **4**: 269–273.

Peña, D. and Rodriguez, J. (2002). A powerful Portmanteau test of lack of fit for time series. *Journal of the American Statistical Association* **97**: 601–610.

Peña, D. and Rodriguez, J. (2006). The log of the determinant of the autocorrelation matrix for testing goodness of fit in time series. *Journal of Statistical Planning and Inference* **136**: 2706–2718.

R Development Core Team (2015). *R: A Language and Environment for Statistical Computing*. R Foundation for Statistical Computing, Vienna, Austria. ISBN 3-900051-07-0. http://www.R-project.org.

Ramsey, J. B. (1969). Tests for specification errors in classical linear least squares regression analysis. *Journal of the Royal Statistical Society Series B* **31**: 350–371.

Tong, H. (1978). On a threshold model. In C.H. Chen (ed.), *Pattern Recognition and Signal Processing*. Sijhoff & Noordhoff, Amsterdam.

Tong, H. (1990). *Non-Linear Time Series: A Dynamical System Approach*. Oxford University Press, Oxford.

Tong, H. and Lim, K. S. (1980). Threshold autoregression,limit cycles and cyclical data (with discussion). *Journal of the Royal Statistical Society, Series B* **42**: 245–292.

Tsay, R. S. (1986). Nonlinearity tests for time series. *Biometrika* **73**: 461–466.

Tsay, R. S. (1989). Testing and modeling threshold autoregressive processes. *Journal of the American Statistical Association* **84**: 231–240.

Tsay, R. S. (1998). Testing and modeling multivariate threshold models. *Journal of the American Statistical Association* **93**: 1188–1202.

Tsay, R. S. (2010). *Analysis of Financial Time Series*, 3rd edition. John Wiley, Hoboken, NJ.

Tsay, R. S. (2014). *Multivariate Time Series Analysis with R and Financial Applications*, John Wiley, Hoboken, NJ.

Tsay, R. S. and Wu, C. S. (2003). Forecasting with leading indicators revisited. *Journal of Forecasting* **22**: 603–617.

CHAPTER 2

UNIVARIATE PARAMETRIC NONLINEAR MODELS

In this chapter we introduce some univariate nonlinear time series models, discuss their properties, and demonstrate their applications. We also address the pros and cons of the models using real examples. The models introduced include threshold autoregressive (TAR) models, Markov switching models (MSM), smooth threshold autoregressive (STAR) models, and time-varying parameter models. We begin with some general setup for a dynamic system and the basic probability structure underling the system, hoping to provide readers with a deeper understanding of the models introduced.

2.1 A GENERAL FORMULATION

The goal of analyzing a time series x_t is to explore its dynamic dependence, and a general model for x_t is

$$x_t = f(x_{t-1}, x_{t-2}, \dots, a_t), \tag{2.1}$$

Nonlinear Time Series Analysis, First Edition. Ruey S. Tsay and Rong Chen.

Companion website: www.wiley.com/go/tsay/nonlineartimeseries

where $f(\cdot)$ is a well-defined real-valued function and a_t is an iid sequence of random variables with mean zero and variance σ_a^2. We often refer to a_t as the innovation of the system at time t and assume $0 < \sigma_a^2 < \infty$. For example, if $f(\cdot) = \phi_0 + \phi_1 x_{t-1} + a_t$, then x_t is an AR(1) time series. If $f(\cdot)$ is a nonlinear function, then x_t follows a nonlinear time series model. For many applications, the dynamic system $f(\cdot)$ only depends on a finite number of past values, and it can be written as $f(\cdot) = f(x_{t-1}, \ldots, x_{t-p}, a_t)$, where p is a positive integer.

To understand the properties of the dynamic system in Equation (2.1), it is often helpful to study the function $f(\cdot)$ recursively by dropping the innovation a_t. For instance, consider the AR(1) time series x_t with $\phi_0 = 0$. The $f(\cdot)$ function becomes $f(x_{t-1}) = \phi_1 x_{t-1}$. This is referred to as the *skeleton* of the system in Tong (1990). By recursive substitutions, we have $f^t(x_0) = f(f^{t-1}(x_0)) = \phi_1^t x_0$, where x_0 denotes the initial value of the system. Since x_0 can assume any real value, for $f^t(x_0) = \phi_1^t x_0$ to be meaningful as t increases we need $|\phi_1| < 1$. This leads to $\lim_{t\to\infty} f^t(x_0) = 0$, which is the mean of the underlying AR(1) series. It is well-known that the condition $|\phi_1| < 1$ is the necessary and sufficient condition for the weak stationarity of the AR(1) time series provided that $\sigma_a^2 < \infty$. The case for nonlinear models is much more involved, but the recursive substitutions of $f(\cdot)$ remain informative in studying the properties of time series x_t.

2.1.1 Probability Structure

For a given time index t, x_t is a random variable. We can study the properties of x_t by drawing random variates from its probability distribution function. Let $\{x_t^{(1)}, x_t^{(2)}, \ldots, x_t^{(m)}\}$ be a random sample of x_t, where m denotes the sample size. Then the mean value of x_t can be estimated by $\hat{\mu}_t = \frac{1}{m}\sum_{i=1}^m x_t^{(i)}$, provided that the mean $E(x_t) = \mu_t$ exists. Next, fixing the superscript and drawing a temporal series, we have a sample time series $\{x_1^{(1)}, x_2^{(1)}, x_3^{(1)}, \ldots\}$, denoted by $\{x_t^{(1)}\}$. This is called a *realization* of the time series $\{x_t\}$. By considering the m realizations jointly, we have an *ensemble* $\{x_t^{(i)} | t = 1, \ldots ; i = 1, \ldots, m\}$ of the time series x_t. Such an ensemble can be used to study properties of the series.

However, we do not have the luxury of owning an ensemble of a time series in real applications. What we have is just an observed realization $\{x_t | t = 1, \ldots, T\}$, where T denotes the length of the realization. How can we study the properties of x_t based on such a single realization? To this end, some additional concepts are needed. Suppose that the time series x_t is weakly stationary. Then, we have $E(x_t) = \mu$, which does not depend on t. Therefore, each data point x_t can be thought of as an estimate of the mean μ, and we can combine the estimates together to form a pooled estimate of μ, say $\hat{\mu} = \frac{1}{T}\sum_{t=1}^T x_t$, which is a time average. The statistical theory that provides justifications for pooling estimates over time is called the *ergodic theorem*. Ergodicity basically ensures that the law of large numbers continues to hold for the dependent time series data under certain conditions.

A weakly stationary time series x_t is said to be *ergodic in the mean* if the time average $\hat{\mu} = \frac{1}{T}\sum_{t=1}^{T} x_t$ converges in probability to $E(x_t)$ as $T \to \infty$. It can be shown that if the following condition holds

$$\sum_{i=0}^{\infty} |\gamma_i| < \infty, \tag{2.2}$$

where γ_i is the lag-i autocovariance of x_t, then x_t is ergodic in the mean. See, for instance, Hamilton (1994, Chapter 7). A weakly stationary time series x_t is said to be *ergodic in the second moment* if

$$\frac{1}{T-j}\sum_{t=j+1}^{T} (x_t - \mu)(x_{t-j} - \mu) \to_p \text{Cov}(x_t, x_{t-j}), \quad T \to \infty,$$

where $\to_p$ denotes convergence in probability. For a stationary (causal) linear time series x_t defined in Chapter 1, the condition in Equation (2.2) and the assumption that $E|a_t|^r < \infty$ for some $r > 2$ are sufficient for it to be ergodic in the second moment. If one further assumes that x_t is Gaussian, then the condition in Equation (2.2) implies that x_t is ergodic in all moments.

For nonlinear time series x_t, the condition for ergodicity is much harder to establish. The conditions tend to be model specific, but they can often be derived using the Markov chain theory as most nonlinear models considered have Markov properties. The appendix to this chapter gives some basic concepts of Markov chains. A good reference for studying the ergodicity and stability of a Markov chain is Meyn and Tweedie (2009). We make use of their results in this book, and shall revisit ergodicity briefly in Chapter 8 when we discuss the Markov chain Monte Carlo method.

2.2 THRESHOLD AUTOREGRESSIVE MODELS

The self-exciting TAR model is arguably the most widely used scalar nonlinear time series model. It is an extension of the piecewise linear regression model (or segmented linear regression model) with structural changes occurring in the threshold space. In the time series literature, the TAR model was proposed by Tong (1978) and has been widely used since the publication of Tong and Lim (1980). We start with the simple two-regime TAR model and discuss multiple-regime TAR models later.

2.2.1 A Two-regime TAR Model

A time series x_t follows a two-regime TAR model of order p with threshold variable x_{t-d} if it satisfies

$$x_t = \begin{cases} \phi_0 + \sum_{i=1}^{p} \phi_i x_{t-i} + \sigma_1 \epsilon_t, & \text{if } x_{t-d} \leq r, \\ \theta_0 + \sum_{i=1}^{p} \theta_i x_{t-i} + \sigma_2 \epsilon_t, & \text{if } x_{t-d} > r, \end{cases} \tag{2.3}$$

where ϵ_t is a sequence of iid random variables with mean zero and unit variance, θ_i and ϕ_j are real-valued parameters such that $\theta_i \neq \phi_i$ for some i, d is a positive integer denoting the delay, and r is the threshold. Often, we further assume that ϵ_t follows $N(0, 1)$. We use the same order p for both regimes. This is purely for simplicity as different orders can easily be used.

The TAR model in Equation (2.3) can also be written as

$$x_t = \phi_0 + \sum_{i=1}^{p} \phi_i x_{t-i} + \sigma_1 \epsilon_t + I(x_{t-d} > r)(\beta_0 + \sum_{i=1}^{p} \beta_i x_{t-i} + \gamma \epsilon_t), \tag{2.4}$$

where $I(x_{t-d} > r) = 1$ if $x_{t-d} > r$ and $= 0$ otherwise, $\beta_i = \theta_i - \phi_i$ for $i = 0, \ldots, p$ and $\gamma = \sigma_2 - \sigma_1$. In this way, the structural change of the model at $x_{t-d} = r$ becomes apparent with β_i denoting the change in the ith coefficient. This type of model occurs often in many scientific fields. For instance, let x_t denote the weight of a health-conscientious individual in week t. The person is likely to take certain action to reduce his weight when x_t exceeds some threshold. However, it takes time for his action to show its effect, resulting in a delay of d weeks to see changes in his weight. Consequently, a threshold model appears to be sensible for describing the time evolution of his weight.

To contrast the TAR models with the linear time series model, Figure 2.1 shows the time plot of a realization of the following two-regime TAR model

$$x_t = \begin{cases} -1.5 x_{t-1} + \epsilon_t, & \text{if } x_{t-1} \leq 0, \\ 0.6 x_{t-1} + \epsilon_t, & \text{if } x_{t-1} > 0, \end{cases} \tag{2.5}$$

with 400 observations, where ϵ_t are iid standard Gaussian random variates. The horizontal line of the plot denotes zero. This simulated example shows clearly several features of nonlinear models. First, the plot shows that the model in Equation (2.5) is weakly stationary, yet one of the lag-1 coefficients is greater than 1 in modulus. Second, the plot shows that the mean of the series is positive, yet the model has no constant term in either regime. Third, there are more positive realizations than negative ones. In other words, the time series x_t cannot stay long in the negative region. Why? To understand these features, we need to investigate the properties of the TAR models.

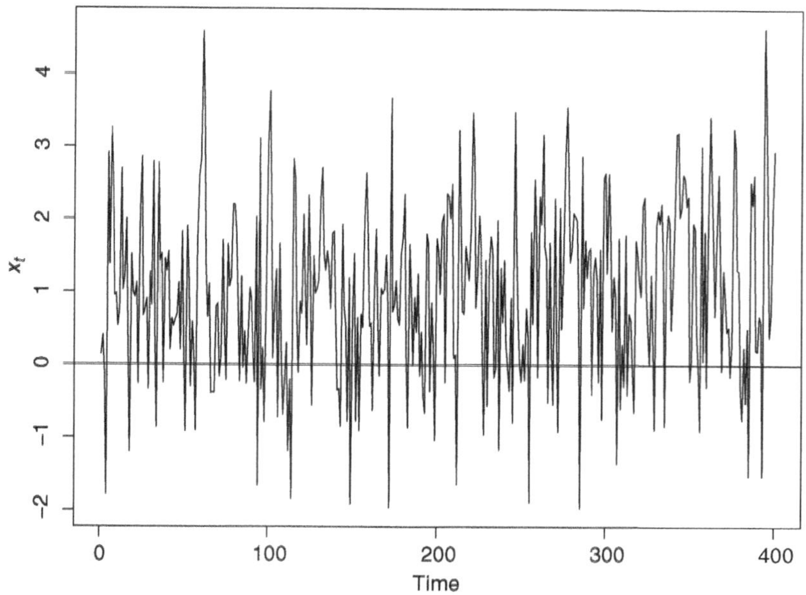

Figure 2.1 A simulated realization with 400 observations from the model in Equation (2.5). The horizontal line denotes zero.

2.2.2 Properties of Two-regime TAR(1) Models

Consider the simplest two-regime TAR(1) model

$$x_t = \begin{cases} \phi_1 x_{t-1} + \sigma_1 \epsilon_t, & \text{if } x_{t-1} \le 0, \\ \theta_1 x_{t-1} + \sigma_2 \epsilon_t, & \text{if } x_{t-1} > 0, \end{cases} \tag{2.6}$$

where the delay is $d = 1$, threshold is zero, and ϵ_t is an iid sequence of continuous random variables with mean zero, variance 1, and a positive probability density function. The skeleton of this simple model is

$$f(x_{t-1}) = \begin{cases} \phi_1 x_{t-1}, & \text{if } x_{t-1} \le 0, \\ \theta_1 x_{t-1}, & \text{if } x_{t-1} > 0. \end{cases} \tag{2.7}$$

Since x_0 can assume any real number, it is easily seen that for the series x_t to be well-defined, we need $\phi_1 < 1$ and $\theta_1 < 1$. Otherwise, we can find some x_0 for which recursive substitutions of $f(\cdot)$ lead to explosive values. This is not surprising as the stationary region for a linear AR(1) model is $-1 < \phi_1 < 1$. On the other hand, the switching between the two regimes creates a new opportunity for TAR models. It turns out that the (geometrically) ergodic region of the two-regime

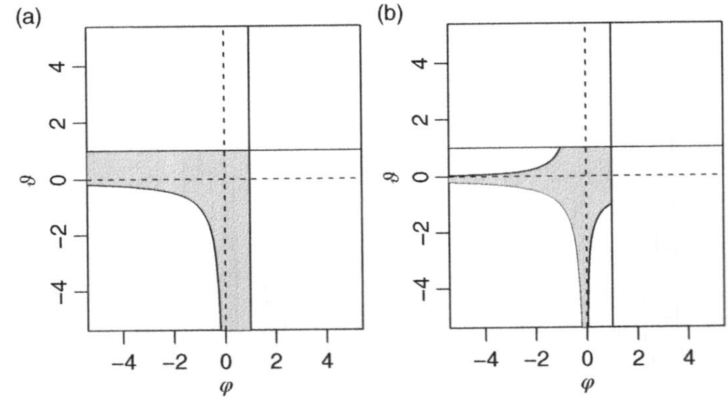

Figure 2.2 Ergodic region of a two-regime TAR(1) model: (a) model with delay $d = 1$ and (b) model with delay $d = 2$.

TAR(1) model in Equation (2.6) is given by the conditions that (a) $\phi_1 < 1$, (b) $\theta_1 < 1$, and (c) $\phi_1\theta_1 < 1$. The condition (c) ensures that the corresponding skeleton of the model in Equation (2.7) remains stable for any arbitrary starting value. For illustration, consider the simulated TAR model in Equation (2.5) for which the three ergodicity conditions hold. In this particular instance it is true that the model in Regime 1 is explosive with coefficient satisfying $|\phi_1| = |-1.5| > 1$. However, in this regime, we have $x_{t-1} \leq 0$ so that $-1.5x_{t-1} > 0$ provided that $x_{t-1} \neq 0$. Consequently, we have $P(x_t > 0) > P(x_t \leq 0)$, implying that with high probability the system would switch to Regime 2. Once x_t is in Regime 2, it follows a stationary AR(1) model with coefficient 0.6. In this way, the explosive nature of the system in Regime 1 would not lead to any issue of non-stability for x_t. In other words, for the model in Equation (2.5), the dynamic dependence in Regime 2 would pull back any explosive value of x_t created by Regime 1. This expanding and pulling phenomenon makes the nonlinear TAR model interesting. Figure 2.2(a) shows the geometrically ergodic region of the TAR model in Equation (2.6). This result was shown in Petruccelli and Woolford (1984).

The skeleton of a two-regime TAR model also depends on the delay parameter d. Consider the general two-regime TAR(1) model

$$x_t = \begin{cases} \phi_1 x_{t-1} + \sigma_1\epsilon_t, & \text{if } x_{t-d} \leq 0, \\ \theta_1 x_{t-1} + \sigma_2\epsilon_t, & \text{if } x_{t-d} > 0, \end{cases} \tag{2.8}$$

where $d > 1$. For this particular model, Chen and Tsay (1991) show that the geometrically ergodic conditions of the model are (a) $\phi_1 < 1$, (b) $\theta_1 < 1$, and

Size	Mean	Med.	Var.	Range	Prob.	$\hat{\rho}_1$	$\hat{\rho}_2$	$\hat{\phi}_{22}$
400	0.84	0.88	1.20	(−2.01,3.80)	0.23	0.04	−0.02	−0.02
4000	0.81	0.78	1.43	(−3.94,6.80)	0.25	0.14	−0.01	−0.03
40000	0.83	0.80	1.38	(−3.53,6.44)	0.24	0.14	0.01	−0.01
400000	0.83	0.80	1.38	(−4.35,7.83)	0.24	0.14	0.01	−0.01

Table 2.1 Summary statistics of the two-regime TAR(1,1) model in Equation (2.5) with different sample sizes, where "Prob." denotes $P(x_t \leq 0)$, $\hat{\rho}_i$ denotes Lag-i sample autocorrelation and $\hat{\phi}_{22}$ Lag-2 sample partial autocorrelation.

(c) $\phi_1^{s(d)}\theta_1^{t(d)} < 1$ and $\phi_1^{t(d)}\theta_1^{s(d)} < 1$, where $s(d)$ and $t(d)$ are functions of the delay parameter d. For instance, we have $s(d) = t(d) = 1$ if $d = 1$ and $s(d) = 1$, $t(d) = 2$ for $d = 2$. The ergodicity regime of the model in Equation (2.8) with $d = 2$ is shown in Figure 2.2(b). In general, the ergodicity conditions for a TAR model are hard to obtain. On the other hand, the skeleton of a given TAR model can be obtained easily by recursive substitutions.

Statistical properties of stationary TAR models are hard to obtain, even for the simple model in Equation (2.6). However, one can use simulations to gain some insights. To demonstrate, Table 2.1 provides some empirical summary statistics for the TAR model in Equation (2.5). From the table, it is seen that the summary statistics, except the range, become stable when the sample size is sufficiently large. Figure 2.3 shows the empirical density functions of x_t for the realizations given in Table 2.1, where the black line is for sample size 400, the dashed line is for sample size 4000, and the dotted line is for sample size 40,000. From the plots, the marginal density functions of x_t are almost identical for the sample sizes 4000 and 40,000, confirming the weak stationarity of the model.

The nonlinear characteristics of the realizations shown in Table 2.1 are easy to see even for the case of sample size 400. Figure 2.4 shows the scatter plot of x_t versus x_{t-1}. The solid line of the plot corresponds to the fitted line by a local smoothing technique, via the command `loess` in R. See Chapter 3 for local smoothing. From the scatter plot and the fitted line, it is clear that the dependence of x_t on x_{t-1} is nonlinear. As a matter of fact, the fitted line shows a V-shape function, which is indicative of the two different slopes of the two-regime TAR(1) model of Equation (2.5).

Remark: The simulations of the TAR models used are obtained by the command `uTAR.sim` of the `NTS` package.

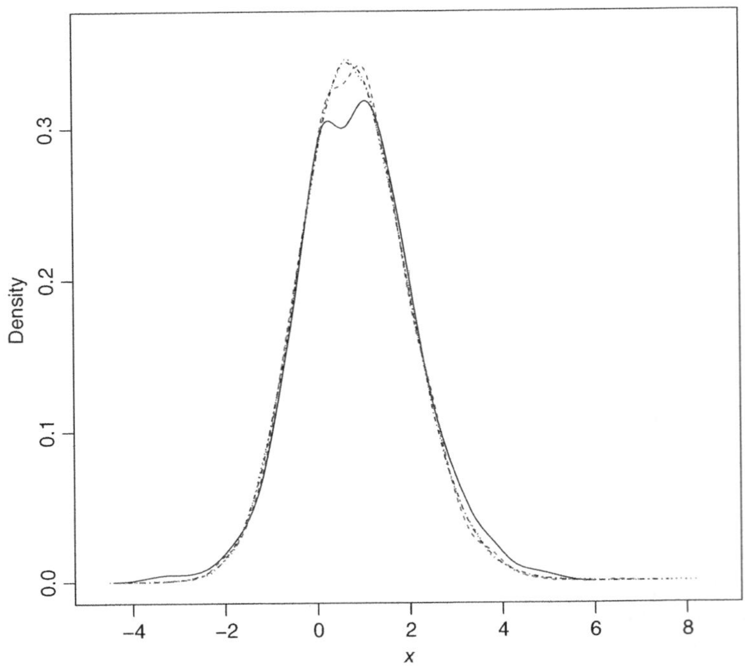

Figure 2.3 Empirical (marginal) density functions for the TAR model of Equation (2.5) with sample sizes given in Table 2.1. The solid line is for sample size 400, the dashed line is for sample size 4000, and the dotted line is for sample size 40,000.

2.2.3 Multiple-regime TAR Models

The generalization of two-regime TAR models to multiple regimes is straightforward. For a positive finite integer $m > 1$, let $\{r_i | i = 0, \dots, m\}$ be a sequence of real numbers satisfying

$$r_0 = -\infty < r_1 < r_2 < \cdots < r_{m-1} < r_m = \infty,$$

where, for simplicity, we treat $-\infty$ and ∞ as real numbers. A time series x_t follows an m-regime self-exciting TAR(p) model with threshold variable x_{t-d}, where $d > 0$, if it satisfies

$$x_t = \begin{cases} \phi_{0,1} + \sum_{i=1}^{p} \phi_{i,1} x_{t-i} + \sigma_1 \epsilon_t, & \text{if } x_{t-d} \le r_1, \\ \phi_{0,2} + \sum_{i=1}^{p} \phi_{i,2} x_{t-i} + \sigma_2 \epsilon_t, & \text{if } r_1 < x_{t-d} \le r_2, \\ \cdots & \\ \phi_{0,m} + \sum_{i=1}^{p} \phi_{i,m} x_{t-i} + \sigma_m \epsilon_t, & \text{if } r_{m-1} < x_{t-d}, \end{cases} \tag{2.9}$$

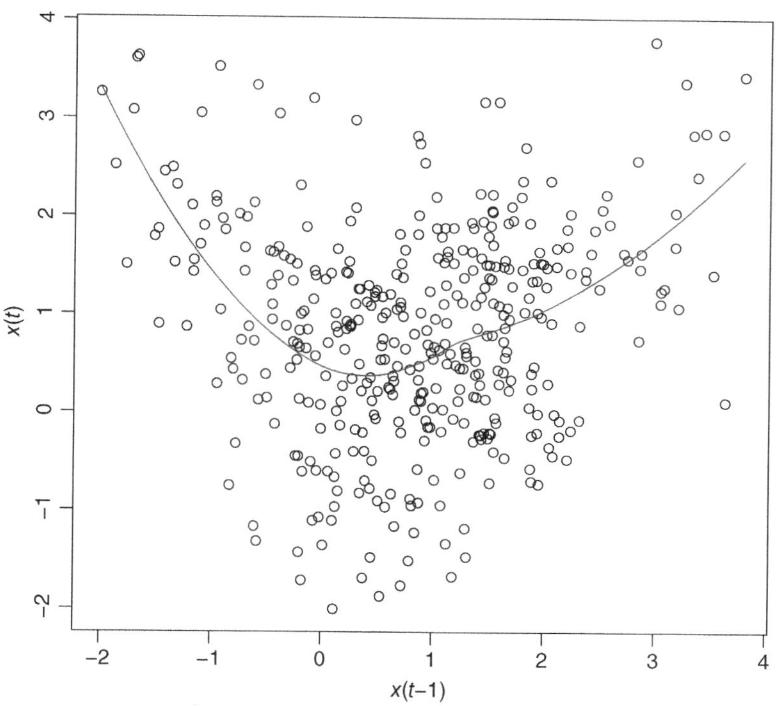

Figure 2.4 Scatter plot of x_t versus x_{t-1} for a realization of 400 observations from the two-regime TAR(1) model in Equation (2.5). The solid line denotes the fitted value via the `loess` command in R.

where σ_i are positive real numbers, ϵ_t are iid $N(0, 1)$, and $\phi_{i,j}$ are real parameters. The polynomial $\phi_j(B) = 1 - \phi_{1,j}B - \cdots - \phi_{p,j}B^p$ is referred to as the AR polynomial of the jth regime. The model in Equation (2.9) can be written in a compact form as

$$x_t = \phi_{0,j} + \phi_{1,j}x_{t-1} + \cdots + \phi_{p,j}x_{t-p} + \sigma_j\epsilon_t, \quad \text{if } r_{j-1} < x_{t-d} \leq r_j \tag{2.10}$$

where $j = 1, \ldots, m$. Let $\boldsymbol{\phi}_j = (\phi_{0,j}, \phi_{1,j}, \ldots, \phi_{p,j}, \sigma_j)'$ denote the parameter vector associated with the jth regime. For the model in Equation (2.10) to be meaningful, we require that $\boldsymbol{\phi}_i \neq \boldsymbol{\phi}_j$ if $i \neq j$. The conditions for geometric ergodicity of the m-regime TAR(p) model are relatively complicated, but the assumption that ϵ_t is a continuous random variable with support $(-\infty, \infty)$ simplifies the matter. The basic idea is that the time series x_t should visit each regime infinitely often so that the stationary distribution of x_t is non-degenerated. Chen and Tsay (1993) provided the following sufficient condition. Let $c_i = \max\{|\phi_{i,j}|, j = 1, \ldots, m\}$ for $i = 1, \ldots, p$, and the probability density function of ϵ_t is positive everywhere on the real line R^1, then the TAR process x_t in Equation (2.10) is geometrically ergodic provided that all the roots of the characteristic function

$$\lambda^p - c_1\lambda^{p-1} - \cdots - c_{p-1}\lambda - c_p = 0$$

are inside the unit circle (see Section 2.5). Most of the applications of TAR models focus on $m = 2$ or $m = 3$, however.

2.2.4 Estimation of TAR Models

Estimation of TAR models is often carried out by the nonlinear least squares method. Properties of the resulting estimators are hard to establish, however. As a matter of fact, limiting properties of the least squares estimates are only available for the two-regime TAR(p) model. It is believed that similar results can be shown for multi-regime TAR models.

Chan (1993) considered the estimation of the two-regime TAR(p) model of Equation (2.3). Let $\boldsymbol{\phi} = (\phi_0, \phi_1, \ldots, \phi_p)'$ and $\boldsymbol{\theta} = (\theta_0, \theta_1, \ldots, \theta_p)'$ be the parameter vectors of the two regimes, respectively. Let $\boldsymbol{\Theta} = (\boldsymbol{\phi}', \boldsymbol{\theta}, r, d)'$ be the coefficient vector of the model in Equation (2.3) and $\boldsymbol{\Theta}_0$ be the true coefficient vector. Suppose that the realizations $\{x_1, \ldots, x_T\}$ are available, where T denotes the sample size. Let $E_{\Theta}(x_t \mid \boldsymbol{F}_{t-1})$ be the conditional expectation of the model given coefficient vector $\boldsymbol{\Theta}$ and the past observations $\boldsymbol{F}_{t-1} = (x_{t-1}, x_{t-2}, \ldots, x_1)$. The objective function of the conditional least squares estimation is

$$L_T(\boldsymbol{\Theta}) = \sum_{t=p+1}^{T} [x_t - E_{\Theta}(x_t \mid \boldsymbol{F}_{t-1})]^2.$$

In practice, the objective function $L_T(\boldsymbol{\Theta})$ is minimized by minimizing $\boldsymbol{S}(r, d)$, where $\boldsymbol{S}(r, d)$ is the minimum residual sum of squares of the model given fixed threshold r and delay d. $\boldsymbol{S}(r, d)$ can be obtained by minimizing the sum of squares of residuals using data in each regime. Specifically, given r and d, we let $S_1 = \{t \mid x_{t-d} \leq r\}$ and $S_2 = \{t \mid x_{t-d} > r\}$ be the collections of time indexes for Regimes 1 and 2, respectively. Then, $\boldsymbol{\phi}$ is estimated by the ordinary least-squares linear regression

$$x_t = \phi_0 + \sum_{i=1}^{p} \phi_i x_{t-i} + e_t, \quad t \in S_1,$$

and $\boldsymbol{\theta}$ by the ordinary least-squares linear regression

$$x_t = \theta_0 + \sum_{i=1}^{p} \theta_i x_{t-i} + e_t, \quad t \in S_2,$$

where e_t denotes the error term. The variances σ_1^2 and σ_2^2 are then estimated by the residual mean squared errors of the prior two linear regressions, respectively. Then the objective function $L_T(\boldsymbol{\Theta})$ can be minimized by minimizing $S(r, d)$.

Since any r between two consecutive ordered observations $(x_{(i)}, x_{(i+1)})$ yields the same value of the conditional least squares criterion $L_T(\boldsymbol{\Theta})$, where $x_{(i)}$ is the ith order statistic of the sample, it is often assumed that r only takes finite possible values. Chan (1993) assumed that $r = x_{(n)}$, where n is within a pre-specified range, e.g. $0.05T \leq n \leq 0.95T$. In addition, d is also assumed to be in a finite set, e.g. $d \in \{1, \ldots, p\}$. Therefore, minimizing $S(r, d)$ is carried out in finite steps.

Let $\widehat{\boldsymbol{\Theta}}$ be the resulting least squares estimate of $\boldsymbol{\Theta}$. Suppose that the two-regime TAR(p) model of Equation (2.3) is stationary and ergodic, having finite second moments and $1 \leq d \leq p$, and that the density function of the stationary distribution of $(x_1, \ldots, x_p)'$ is positive everywhere. Chan (1993) showed that $\widehat{\boldsymbol{\Theta}}$ is strongly consistent as $T \to \infty$. That is, $\widehat{\boldsymbol{\Theta}} \to \boldsymbol{\Theta}_0$ almost surely as $T \to \infty$. In addition, $\hat{\sigma}_i^2$ $(i = 1, 2)$ are also strongly consistent.

To obtain the limiting distribution of $\widehat{\boldsymbol{\Theta}}$, Chan (1993) used properties of the Markov chain. Let $\boldsymbol{x}_t = (x_t, x_{t-1}, \ldots, x_{t-p+1})'$. It is easy to see that $\{\boldsymbol{x}_t\}$ is a Markov chain. Denote the ℓ-step transition probability of $\boldsymbol{x}_t$ by $\boldsymbol{P}^{\ell}(\boldsymbol{x}, A) = \boldsymbol{P}(\boldsymbol{x}_{t+\ell} \in A \mid \boldsymbol{x}_t = \boldsymbol{x})$, where $\boldsymbol{x} \in R^p$ (the p-dimensional Euclidean space) and A is a Borel set. The following four conditions are needed:

- C1: $\boldsymbol{x}_t$ has a unique invariant measure $\pi(\cdot)$ such that there exist positive real values K and $\rho < 1$ such that for every $\boldsymbol{x} \in R^p$ and every positive integer n, $\|\boldsymbol{P}^n(\boldsymbol{x}, \cdot) - \pi(\cdot)\| \leq K(1 + |\boldsymbol{x}|)\rho^n$, where $\| \cdot \|$ and $| \cdot |$ denote the total variation norm and the Euclidean norm, respectively.
- C2: ϵ_t is absolutely continuous with a uniformly continuous and positive probability density function (pdf). In addition, $E(\epsilon_t^4) < \infty$.
- C3: x_t is stationary with marginal pdf $\eta(\cdot)$ and $E(x_t^4) < \infty$.
- C4: The autoregressive function is discontinuous, that is, there exists $\boldsymbol{X}^* = (1, x_{p-1}, x_{p-2}, \ldots, x_0)'$ such that $(\boldsymbol{\phi} - \boldsymbol{\theta}) \cdot \boldsymbol{X}^* \neq 0$ and $x_{p-d} = r$.

Assume that Conditions C1 to C4 hold. Chan (1993) showed that $T(\hat{r} - r)$ converges in distribution to a random quantity M, which is determined by a compound Poisson process, where $\hat{r}$ is the least squares estimate of the threshold r. Furthermore, $T(\hat{r} - r)$ is asymptotically independent of $\sqrt{T}(\hat{\boldsymbol{\phi}} - \boldsymbol{\phi}, \hat{\boldsymbol{\theta}} - \boldsymbol{\theta})$ and the latter is asymptotically normal with a distribution the same as that for the case when r is known.

The convergence rate of the threshold estimate $\hat{r}$ is T, implying that the threshold estimate is super consistent. This is not surprising under Condition C4, which says that the threshold r corresponds to a jump in the autoregressive function. In practice, time series data of continuous variables tend to be informative about the locations of jumps. Chan and Tsay (1998) studied the case under which Condition C4 fails, i.e. the autoregressive function of the TAR model is smooth. The estimate $\hat{r}$ of the threshold then is only $\sqrt{T}$ consistent.

2.2.5 TAR Modeling

TAR modeling has been extensively studied in the literature. See, for instance, Tsay (1989) and the reference therein. For two-regime TAR models, one often starts the model selection procedure by postulating an order for the autoregression, say p, and assumes that the delay d satisfies $1 \le d \le p$. Furthermore, one also assumes that the threshold r assumes certain value of the ordered statistics of the data. Denote the order statistics of the data by

$$x_{(1)} \le x_{(2)} \le \cdots \le x_{(n)} \le x_{(n+1)} \le \cdots \le x_{(T)}.$$

One assumes that $r \in \{x_{(l)}, x_{(l+1)}, \ldots, x_{(u)}\}$, where $l = \lfloor \alpha T \rfloor$ and $u = \lfloor (1-\alpha)T \rfloor$ with $0 < \alpha < 0.3$ and $\lfloor z \rfloor$ denoting the integer part of the real number z. The default choice of α for the TAR modeling in the `TSA` package of R is $\alpha = 0.05$. The choice of α may also depend on balancing the number of observations in each regime.

With pre-specified maximum AR order p and α, one can select the delay d, the threshold r, the AR order p_i in each regime, and the autoregressive coefficients by using information criterion, e.g. the Akaike information criterion (AIC), of fitted candidate models. See Akaike (1974). The computation involved might be intensive, but can be done relatively easily with certain recursive procedures. Let T_i be the number of data points in Regime S_i, where $i = 1$ and 2. Under normality, the maximum likelihood estimate of the innovation variance is

$$\tilde{\sigma}_i^2 = \frac{1}{T_i} \sum_{t \in S_i} (x_t - \hat{\phi}_{0,i} - \sum_{j=1}^{p_i} \hat{\phi}_{j,i} x_{t-j})^2,$$

where p_i is the AR order of Regime i, $\hat{\phi}_{j,i}$ is the least squares estimate of $\phi_{j,i}$, and it is understood that $\phi_{0,i}$ can be dropped if no constant term is needed in Regime i. Then, the overall AIC for the fitted two-regime TAR model is

$$AIC = T_1 \ln\left(\tilde{\sigma}_1^2\right) + T_2 \ln\left(\tilde{\sigma}_2^2\right) + T[1 + \ln(2\pi)] + 2n_p,$$

where n_p denotes the number of parameters used in the model. For instance, $n_p = (p_1 + 1) + (p_2 + 1) + 1$ if both regimes contain the constant term.

On the other hand, building multi-regime TAR models could become difficult. For a given data set, care must be exercised because there exists a trade off between the number of regimes and the autoregressive model of each regime. Intuitively, the more the number of regimes, the lower the autoregressive order of individual regime. In practice, the number of regimes is often assumed to be small such as $m \leq 3$. If a three-regime model becomes plausible, one can use various tools of explanatory data analysis to seek possible locations of the thresholds. For instance, Tsay (1989) uses the recursive least squares method to estimate arranged autoregression and plots the resulting autoregressive coefficient estimates to locate possible threshold values. The basic idea of such a method is that thresholds are locations of model changes under the arranged autoregression. An alternative approach is to apply some nonparametric method, such as the local polynomial fitting, to x_t versus the threshold variable x_{t-d}. Ideally, the thresholds correspond to change points of the fitted model.

To illustrate, we consider the annual copper prices from 1800 to 1996. The data are from Hyndman (2016) and the prices are adjusted prices after removing a long-term trend. Figure 2.5 shows the time plot of the price series and its sample autocorrelation function. As expected, the price evolves over time and has strong serial dependence. Let x_t denote the copper price of the tth year in the data. Figure 2.6 shows the scatter plot of x_t versus x_{t-1}. The solid line denotes the fitted value by the local smoothing method `loess` of R. From the plot and local fit, the dependence of x_t on x_{t-1} appears to be nonlinear. In particular, the slope of the fitted line seems to change when x_{t-1} is slightly less than 1.0. From the plot,

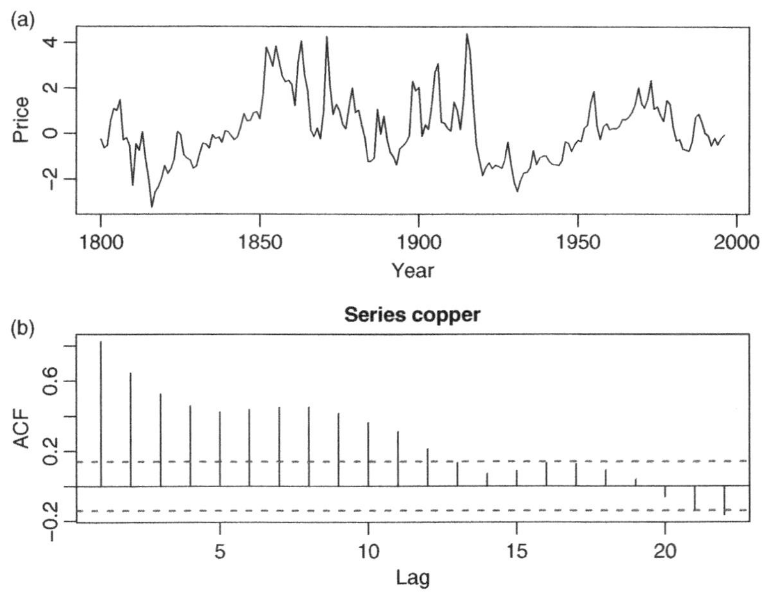

Figure 2.5 (a) The time plot of the annual copper prices from 1800 to 1996. The price is the adjusted price after removing a long-term trend. (b) The sample autocorrelation function of the copper price series.

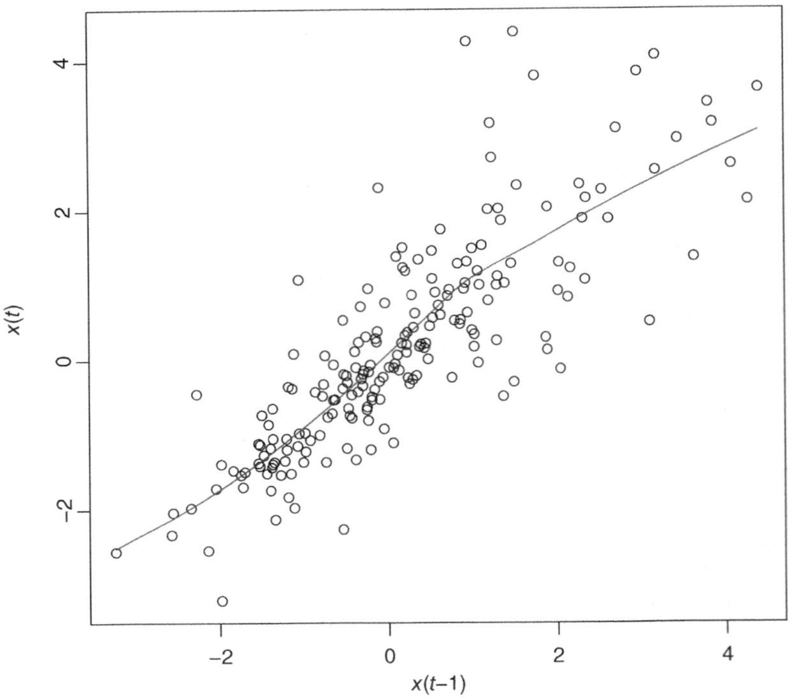

Figure 2.6 The scatter plot of copper price versus its lagged value. The solid line is the fitted value via the local smoothing method `loess` of R.

it seems that a two-regime TAR model might be suitable for the series of annual copper prices.

R commands used: Annual copper prices.

```
> copper <- scan(file="copper1800t1996.txt")  # load data into R
> tdx <- c(1800:1996)  # years
> par(mfcol=c(2,1))
> plot(tdx,copper,xlab='year',ylab='price',type='l')
> acf(copper)
> y <- copper[2:197]; x <- copper[1:196]
> m1 <- loess(y~x)  ## local smoothing
> sx <- sort(x,index=T)  ## sorting the threshold variable
> par(mfcol=c(1,1))
> ix <- sx$ix ## index for order-statistics
> plot(x,y,xlab='x(t-1)',ylab='x(t)')
> lines(x[ix],m1$fitted[ix],col="red")
```

2.2.6 Examples

We demonstrate TAR modeling by considering two examples. The data used are relatively easy to analyze, but they show the procedure we use to build TAR models in practice.

Example 2.1 Consider again the annual adjusted copper prices of Figure 2.5. If a linear time series model is used, an AR(12) model is selected by the AIC. After refinements, the fitted model is

$$x_t = 0.90x_{t-1} - 0.16x_{t-2} + 0.16x_{t-6} + 0.17x_{t-11} - 0.22x_{t-12} + a_t, \quad (2.11)$$

where $\hat{\sigma}_a^2 = 0.57$. The AIC of the model is 461.13 and the model fits the data reasonably well with Ljung–Box statistics of the residuals showing $Q(10) = 5.33$, which corresponds to a p value of 0.38 based on the χ_5^2 reference distribution. All coefficient estimates of Equation (2.11) are statistically significant at the 5% level. An alternative linear model for the series is an ARMA(3,2) model

$$x_t = -0.59x_{t-1} + 0.31x_{t-2} + 0.66x_{t-3} + a_t + 1.57a_{t-1} + 0.96a_{t-2}, \quad (2.12)$$

where $\hat{\sigma}_a^2 = 0.58$ and the AIC of the model is 465.48. This model also fits the data well with $Q(10) = 9.53$, which gives a p value of 0.09 compared with χ_5^2.

Following the indication of Figure 2.6, we use a two-regime TAR model. The `tar` command of the `TSA` package can automatically remove insignificant autoregressive coefficients and selects the model

$$x_t = \begin{cases} 0.08 + 0.93x_{t-1} + \sigma_1\epsilon_t & \text{if } x_{t-1} \le 0.946, \\ 0.73 + 0.73x_{t-1} - 0.29x_{t-2} + \sigma_2\epsilon_t & \text{if } x_{t-1} > 0.946, \end{cases} \quad (2.13)$$

where $\hat{\sigma}_1^2 = 0.36$ and $\hat{\sigma}_2^2 = 1.30$. This model fits the data well with $Q(10) = 9.47$, which gives a p value of 0.09 based on the χ_5^2 distribution. The sample sizes for the two regimes are 144 and 51, respectively. The overall AIC of model (2.13) is 427.8. For this nonlinear model, the constant term 0.08 of the first regime is not significant at the 5% level, but we keep it for simplicity.

Discussion: It is interesting to compare the fitted models for the series of annual copper prices. First, between the two linear models, the AR(12) model in Equation (2.11) is slightly preferred because its AIC and $\hat{\sigma}_a^2$ are lower. Second, the AIC of the TAR model appears to be lower than those of the linear models. However, care must be exercised because different sample sizes are used in calculating the AIC values. In this particular instance, the linear models use $T = 197$ observations, but the TAR model uses $T = 195$ observations because the maximum autoregressive order is 2 and the delay is 1. It is important to adjust for the effective sample size used in calculating information criteria when different classes of models are used. For the copper price series, the AIC for the TAR model would become

$427.8 \times 197/195 = 432.2$, which remains lower than those of the linear models. Third, part of the contribution of the TAR model in this particular instance arises from the fact that the residual variances of the two regimes are rather different. This phenomenon can also be seen from the scatter plot of Figure 2.6.

R demonstration: The copper price series. Output shortened.

```
> copper <- scan("copper1800t1996.txt")
> m0 <- ar(copper,method="mle")
> m0$order
[1] 12
> m1 <- arima(copper,order=c(12,0,0))
> m1
Call: arima(x = copper, order = c(12, 0, 0))

Coefficients:
         ar1      ar2     ar3     ar4      ar5     ar6      ar7     ar8
      0.8999  -0.1981  0.0399  0.0459  -0.0855  0.1931  -0.0691  0.1418
s.e.  0.0689   0.0930  0.0938  0.0936   0.0936  0.0945   0.0948  0.0949
          ar9     ar10    ar11     ar12  intercept
      -0.0416  -0.0518  0.2037  -0.2324     0.1695
s.e.   0.0950   0.0954  0.0963   0.0706     0.3360

sigma^2 estimated as 0.5542:  log likelihood = -222.47,  aic = 472.94
> c1 <- c(NA,NA,0,0,0,NA,0,0,0,0,NA,NA,0)
> m1a <- arima(copper,order=c(12,0,0),fixed=c1)
> m1a
Call: arima(x = copper, order = c(12, 0, 0), fixed = c1)

Coefficients:
         ar1      ar2  ar3  ar4  ar5     ar6  ar7  ar8  ar9  ar10    ar11
      0.8984  -0.1596    0    0    0  0.1559    0    0    0     0  0.1671
s.e.  0.0674   0.0696    0    0    0  0.0456    0    0    0     0  0.0704
         ar12  intercept
      -0.2166          0
s.e.   0.0696          0

sigma^2 estimated as 0.5666:  log likelihood = -224.57,  aic = 461.13
> require(forecast)   ### For automatic selection of ARIMA models.
> auto.arima(copper)
Series: copper
ARIMA(3,0,2) with zero mean
Coefficients:
          ar1     ar2     ar3     ma1     ma2
      -0.5852  0.3070  0.6622  1.5709  0.9630
s.e.   0.0710  0.0962  0.0850  0.0564  0.0265
```

```
sigma^2 estimated as 0.5769:  log likelihood=-226.74
AIC=465.48   AICc=465.92   BIC=485.18
> m2 <- arima(copper,order=c(3,0,2),include.mean=F)
> m2
Call: arima(x = copper, order = c(3, 0, 2), include.mean = F)

Coefficients:
          ar1     ar2     ar3     ma1     ma2
      -0.5852  0.3070  0.6622  1.5709  0.9630
s.e.   0.0710  0.0962  0.0850  0.0564  0.0265

sigma^2 estimated as 0.5769:  log likelihood = -226.74,  aic = 465.48
> Box.test(m2$residuals,lag=10,type='Ljung')
        Box-Ljung test
data:  m2$residuals
X-squared = 9.5266, df = 10, p-value = 0.483
> Box.test(m1a$residuals,lag=10,type='Ljung')
        Box-Ljung test
data:  m1a$residuals
X-squared = 5.334, df = 10, p-value = 0.8678
> pv <- 1-pchisq(5.334,5)  ## Compute p-value, adjusting df.
> pv
[1] 0.3764918
> pv <- 1-pchisq(9.53,5)
> pv
[1] 0.08970192
> require(TSA)  # Load TSA package
> m3 <- tar(copper,12,12,1)
> m3$thd
          0.9459653
> m3$qr1$coefficients
intercept-copper      lag1-copper
       0.0809485        0.9519411
> m3$qr2$coefficients
intercept-copper      lag1-copper      lag2-copper
       0.8258597        0.7184221       -0.3023816
> m4 <- tar(copper,2,2,1)  ### Lower the AR orders to 2.
> m4$thd
          0.9459653
> m4$qr1$coefficients
intercept-copper      lag1-copper
      0.07536253       0.92867483
> m4$qr2$coefficients
intercept-copper      lag1-copper      lag2-copper
       0.7303377        0.7336741       -0.2881161
```

```
> m4$AIC
427.8
> m4$rms1
0.3628672
> m4$rms2
1.303754
> Box.test(m4$residuals,lag=10,type='Ljung')
        Box-Ljung test
data:  m4$residuals
X-squared = 9.4695, df = 10, p-value = 0.4882
> pv <- 1 - pchisq(9.47,5)
> pv
[1] 0.09172323
```

Example 2.2 In this example, we consider the monthly precipitation (in mm) of London, United Kingdom, from January 1983 to April 1994 for 136 observations. Again, the data are downloaded from Hyndman (2016). Figure 2.7(a) shows the time plot of the data whereas Figure 2.7(b) gives the sample partial autocorrelation function (PACF). From the plots, the series appears to be weakly stationary with

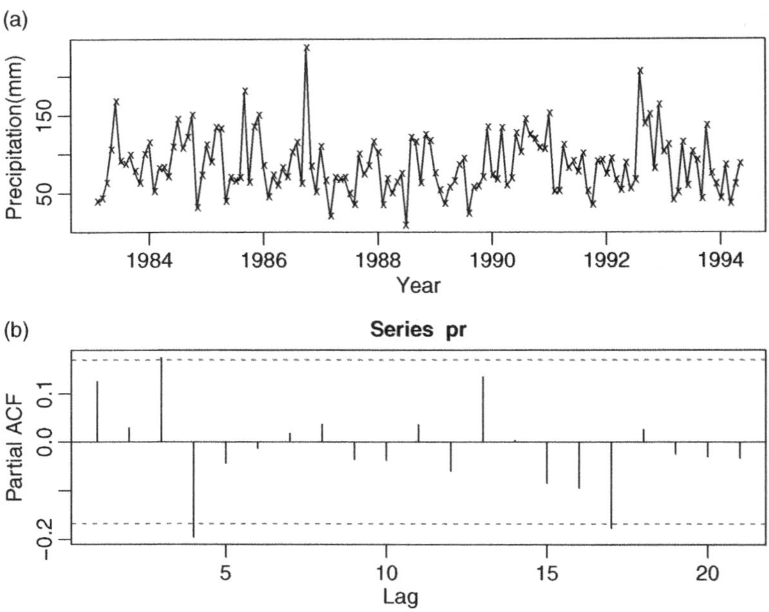

Figure 2.7 (a) The time plot of the monthly precipitation (in mm) of London, United Kingdom, from January 1983 to April 1993. (b) The sample partial autocorrelation function of the precipitation series.

some minor serial dependence. The PACF indicates that an AR(4) model might be suitable for the series. The fitted model, after removing some insignificant parameters, is

$$x_t = 71.87 + 0.16x_{t-1} + 0.20x_{t-3} - 0.20x_{t-4} + a_t, \tag{2.14}$$

where x_t is the monthly precipitation and $\hat{\sigma}_a^2 = 1280$. This simple model fits the data well, as shown by the model checking statistics in Figure 2.8.

On the other hand, with $p = 4$, the Keenan test of Chapter 1 strongly rejects the null hypothesis of linearity. The test statistic is 29.23 with p value close to zero. Next, we use TAR models. With $p = 4$ and the delay $d \in \{1, 2, 3, 4\}$, we applied the threshold nonlinearity test of Tsay (1989). The test statistics and their p values (in parentheses) are 0.36(0.87), 1.37(0.24), 0.37(0.87), and 2.09(0.07), respectively, for $d = 1, \ldots, 4$. The threshold nonlinearity is only marginally supported when $d = 4$. Consequently, we use $d = 4$ for the delay. Figure 2.9 shows the scatter plot of x_t versus x_{t-4} for the precipitation series. The solid line is the fitted value using

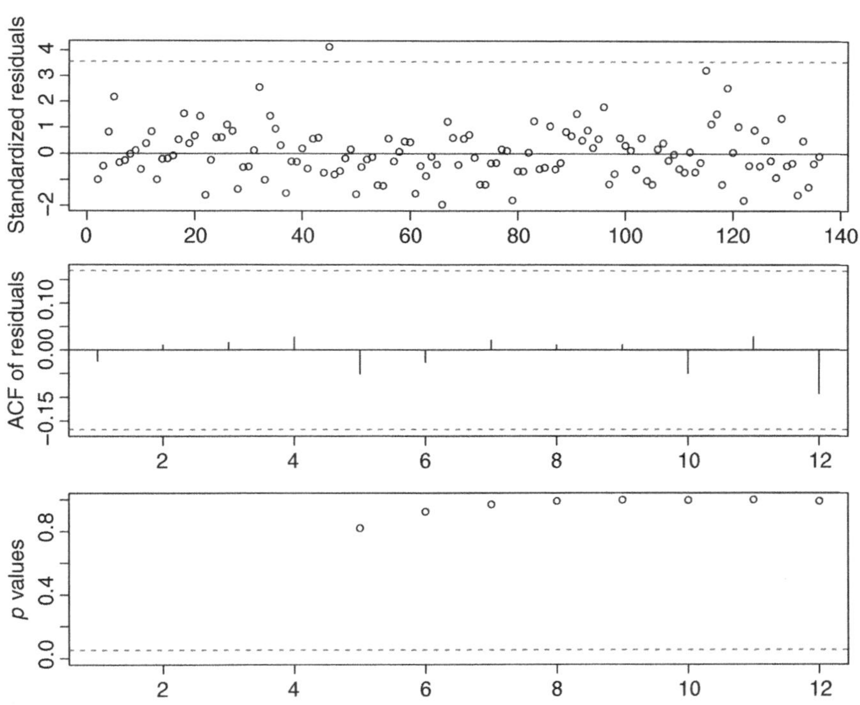

Figure 2.8 Model checking of the fitted AR(4) model for the monthly precipitation series of London, United Kingdom, from January 1983 to April 1993.

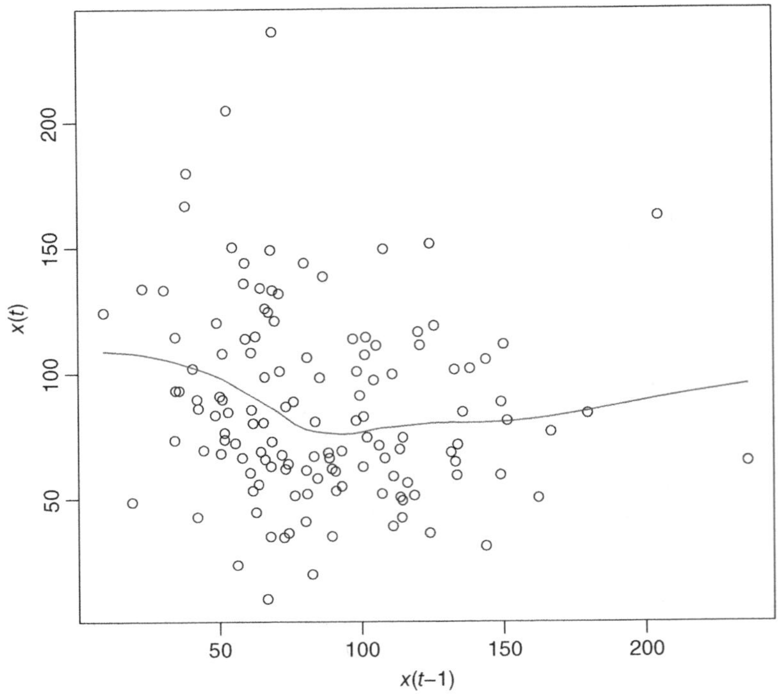

Figure 2.9 The scatter plot of monthly precipitation versus its lag-4 value. The solid line is the fitted value via the local smoothing method `loess`.

the local smoothing method `loess`. The fitted line provides some justification for using a two-regime TAR model with a threshold around 75.0 mm.

The fitted TAR model is

$$x_t = \begin{cases} 98.39 + \sigma_1 \epsilon_t, & \text{if } x_{t-4} \leq 71.5, \\ 41.27 + 0.19x_{t-1} - 0.05x_{t-2} + 0.26x_{t-3} + \sigma_2 \epsilon_t, & \text{if } x_{t-4} > 71.5, \end{cases} \tag{2.15}$$

where $\hat{\sigma}_1^2 = 1826.6$ and $\hat{\sigma}_2^2 = 823.4$. The overall AIC of the model is 1313. The sample sizes for the two regimes are 57 and 75, respectively. The Ljung–Box statistics show that the residuals of the fitted TAR model have no serial correlations. For instance, $Q(12) = 3.18$, which corresponds to a p value of 0.67 compared with the χ_7^2 distribution. Finally, for this particular instance, AIC for the linear AR(4) model in Equation (2.14) is 1369.25 whereas that of the TAR model in Equation (2.15) is 1313.

Discussion: The limiting properties of the parameter estimates for the TAR model in Equation (2.15) are, strictly speaking, unknown, because the delay $d = 4$ is greater than the maximum AR order of the model. The fitted TAR model confirms that the serial dependence in the precipitation is weak. In particular, the precipitation was not serially correlated when $x_{t-4} \leq 71.5$ mm.

R demonstration: Precipitation series of London. Output abridged.

```
> pr <- scan(file="m-precipitationLondon.txt")
> Keenan.test(pr,4)
$test.stat
[1] 29.23289
$p.value
[1] 3.099153e-07
> require(NTS)
> thr.test(pr,4,d=1)
SETAR model is entertained
Threshold nonlinearity test for (p,d):  4 1
F-ratio and p-value:  0.3664116 0.8702775
....
> thr.test(pr,4,d=4)
SETAR model is entertained
Threshold nonlinearity test for (p,d):  4 4
F-ratio and p-value:  2.091034 0.07409199
> ### TAR modeling
> m5 <- tar(pr,4,4,d=4)
> names(m5)
 [1] "dxy1"      "dxy2"    "p1"          "q1"           "d"
 [6] "qr1"       "qr2"     "x.regime1"   "y.regime1"    "x.regime2"
[11] "y.regime2" "thd"     "thdindex"    "qr1"          "qr2"
[16] "i1"        "i2"      "x"           "m"            "rss1"
[21] "rss2"      "n1"      "n2"          "std.res"      "p1"
[26] "p2"        "rms1"    "rms2"        "is.constant1" "is.constant2"
[31] "residuals" "AIC"     "aic.no.thd"  "y"            "like"
[36] "method"
> m5$thd
    71.5
> m5$qr1$coefficients
intercept-pr
    98.39298
> m5$qr2$coefficients
intercept-pr      lag1-pr       lag2-pr       lag3-pr
 41.27491106   0.19071082  -0.04809674   0.26076648
> m5$rms1; m5$rms2
```

```
1826.553;    823.3881
> m5$AIC
1313
> m5$n1
[1] 57
> m5$n2
[1] 75
> Box.test(m5$residuals,lag=12,type='Ljung')
        Box-Ljung test
data:  m5$residuals
X-squared = 3.1796, df = 12, p-value = 0.9941
```

2.2.7 Predictions of TAR Models

Predictions of a TAR model can be obtained by simulation. We use the two-regime TAR model in Equation (2.3) in our discussion. Let n be the forecast origin. Typically, $n = T$ is the sample size. Let ℓ be the forecast horizon, i.e. we are interested in predicting $x_{n+\ell}$ given the model and the data $\{x_n, x_{n-1}, \dots\}$. Let $x_n(\ell)$ denote the point forecast of $x_{n+\ell}$ and $e_n(\ell)$ the associated forecast error. For simplicity, we ignore the model and parameter uncertainty in our discussion and use the minimum mean squared errors criterion to obtain the forecasts. Other loss functions can be used if needed.

One-step ahead prediction: Given that $d \geq 1$, the one-step ahead prediction of a TAR model is straightforward to compute. This is so because the value of the threshold variable x_{t+1-d} is observed. If $x_{t+1-d} \leq r$, then we have

$$x_n(1) = \phi_0 + \sum_{i=1}^{p} \phi_i x_{n+1-i}, \quad e_n(1) = \sigma_1 \epsilon_{n+1}.$$

Under normality, a 95% interval prediction of x_{n+1} is $x_n(1) \pm 1.96\sigma_1$. On the other hand, if $x_{n+1-d} > r$, then we have

$$x_n(1) = \theta_0 + \sum_{i=1}^{p} \theta_i x_{n+1-i}, \quad e_n(1) = \sigma_2 \epsilon_{n+1},$$

and a 95% interval prediction of x_{n+1} is $x_n(1) \pm 1.96\sigma_2$.

Multi-step ahead prediction: Consider a two-step ahead prediction. If the threshold value x_{n+2-d} is observed and $x_{n+2-d} \leq r$, then

$$x_n(2) = \phi_0 + \phi_1 x_n(1) + \sum_{i=2}^{p} \phi_i x_{n+2-i}, \quad e_n(2) = \phi_1 e_n(1) + \sigma_1 \epsilon_{n+2}.$$

If $x_{n+2-d} > r$, then we have

$$x_n(2) = \theta_0 + \theta_1 x_n(1) + \sum_{i=2}^{p} \theta_i x_{n+2-i}, \quad e_n(2) = \theta_1 e_n(1) + \sigma_2 \epsilon_{n+2}.$$

On the other hand, if x_{n+2-d} is not observed, then simulation is needed to compute the forecasts.

In general, it is more convenient to use simulation to compute forecasts of a TAR model. Multiple iterations are often needed to obtain reliable forecasts. The procedure is as follows.

Forecasting procedure: Let m be the number of iterations. For $j = 1, \ldots, m$

1. Draw ℓ random variates from the distribution of ϵ_t. Typically, we assume $\epsilon_t \sim N(0, 1)$. Denote the random variates by $\epsilon_{n+1}^{(j)}, \ldots, \epsilon_{n+\ell}^{(j)}$.
2. For $t = n + 1, \ldots, n + \ell$, use the TAR model and $\{\epsilon_{n+i}^{(j)}\}$ to generate realizations $x_{n+1}^{(j)}, \ldots, x_{n+\ell}^{(j)}$ of the jth iteration.

Finally, compute the point forecasts and associated variances for $i = 1, \ldots, \ell$ by

$$\tilde{x}_n(i) = \frac{1}{m} \sum_{j=1}^{m} x_{n+i}^{(j)}, \quad \text{Var}[e_n(i)] = \frac{1}{m-1} \sum_{j=1}^{m} [x_{n+i}^{(j)} - \tilde{x}_n(i)]^2.$$

Self-exciting TAR models can be extended to the multivariate case. See, for instance, Tsay (1998). Threshold moving-average models have been studied by Ling and Tong (2005) and the references therein.

2.3 MARKOV SWITCHING MODELS

Another class of nonlinear models widely used in econometrics is the Markov switching (MS) model. The idea of Markov switching has a long history in the literature. See, for instance, Tong (1978) and Neftci (1984). The model attracted much attention in econometrics after the publication of Hamilton (1989).

Consider a time series x_t. Let S_t denote the state of the process at time t. For a two-state Markov switching model (MSM), S_t assumes two possible values, say $S_t = 1$ or $S_t = 2$. For example, if x_t is the growth rate of US quarterly domestic gross product (GDP), then S_t may represent the status of the US economy at time t with $S_t = 1$ representing expansion and $S_t = 2$ contraction. The series x_t follows a two-state autoregressive MSM if it satisfies

$$x_t = \begin{cases} \phi_{0,1} + \phi_{1,1} x_{t-1} + \cdots + \phi_{p,1} x_{t-p} + \sigma_1 \epsilon_t & \text{if } S_t = 1, \\ \phi_{0,2} + \phi_{1,2} x_{t-1} + \cdots + \phi_{p,2} x_{t-p} + \sigma_2 \epsilon_t & \text{if } S_t = 2, \end{cases} \tag{2.16}$$

where $\phi_{i,j}$ are real numbers, $\sigma_i > 0$, and $\{\epsilon_t\}$ is a sequence of iid random variables with mean zero and variance 1.0. Let $\phi_j(B) = 1 - \phi_{1,j}B - \cdots - \phi_{p,j}B^j$ be the autoregressive polynomial of the j state. We often require that the roots of $\phi_j(B) = 0$ are outside the unit circle for $j = 1$ and 2. The state transition of the model is governed by the transition probabilities

$$P(S_t = 2 \mid S_{t-1} = 1) = \eta_1, \quad P(S_t = 1 \mid S_{t-2} = 2) = \eta_2,$$

where $0 < \eta_j < 1$. In matrix form, we have the transition probability matrix

$$\boldsymbol{P} = \begin{bmatrix} 1-\eta_1 & \eta_1 \\ \eta_2 & 1-\eta_2 \end{bmatrix}.$$

Let $\boldsymbol{\phi}_j = (\phi_{0,j}, \phi_{1,j}, \ldots, \phi_{p,j})'$ be the coefficient vector of state j. For the autoregressive MSM in Equation (2.16) to be meaningful, we require either $\boldsymbol{\phi}_1 \neq \boldsymbol{\phi}_2$ or $\sigma_1 \neq \sigma_2$ or both. This model was used in McCulloch and Tsay (1994), who also considered a logistic function for η_i.

From the definition, a two-state MSM is similar to a two-regime TAR model. The main difference lies in the switching between the two autoregressive models. For the TAR model, the switching is governed by the threshold variable whereas the switching is based on transition probabilities in the MSM. In some sense, one can say that the switching between regimes of a TAR model is "deterministic" as the regime is determined once the value of the threshold variable x_{t-d} is known. On the other hand, the switching is "stochastic" for an MSM because it is governed by a probability law. This difference becomes apparent in applications because the actual value of the state S_t is rarely certain for a given MSM. What one can infer is the probability of being in a given state, such as $P(S_t = 1 \mid \text{data, model})$. In fact, an attractive feature of using MSM is that it enables users to estimate both filtered and smoothed state probabilities. However, predictions using MSM depend solely on the transition probability matrix without the aid of an observed threshold variable, hence they tend to have higher uncertainty. Wu and Chen (2007) considered a hybrid model between TAR and MSM by assuming the transition is stochastic but based on a "strength" of the threshold variable. Specifically, it is assumed that $P(S_t = 1) = exp\{a + b(x_{t-d} - r)\}/(1 + exp\{a + b(x_{t-d} - r)\})$ for $b > 0$, so that when $|x_{t-d} - r|$ is large, the regime can be identified with high certainty, but when x_{t-d} is close to the threshold value r, the regime identity becomes highly uncertain. The model is closely related to the *smooth transition autoregressive model* of Section 2.4. McCulloch and Tsay (1994) used a logistic regression to model the transition probabilities.

Some discussions of MSM are in order. First, it is straightforward to generalize the model to have more than two states. Let $S = \{1, \dots, k\}$ denote the possible values of state variable S_t. A k-state Markov switching autoregressive model can be written as

$$x_t = \phi_{0,S_t} + \phi_{1,S_t} x_{t-1} + \cdots + \phi_{p,S_t} x_{t-p} + \sigma_{S_t} \epsilon_t, \quad S_t \in \{1, \dots, k\}, \quad (2.17)$$

where ϵ_t is a sequence of iid random variables with mean zero and variance 1, $\phi_{i,j}$ are real numbers, $\sigma_i > 0$, and the state transition is governed by a $k \times k$ transition matrix $\boldsymbol{P} = [p_{ij}]$ such that $P(S_t = j \mid S_{t-1} = i) = p_{ij}$. The Markovian property is used here so that $P(S_t = j \mid S_{t-1} = i, S_{t-2}, \dots, S_0) = P(S_t = j \mid S_{t-1} = i)$. For further information on the Markov chain, see the appendix to this chapter. Similar to the multi-regime TAR models, multi-state Markov switching models are hard to handle in applications. Our discussions focus mainly on the two-state models. Second, MSM has an inherent labeling issue. Consider the model in Equation (2.16). The labeling of a state is rather arbitrary. As a matter of fact, one can easily switch the labels between state 1 and state 2 and obtain the same model. For instance, consider the US quarterly GDP growth rates. State 1 can be used to represent either "expansion" or "contraction". This type of model identifiability is harmless provided that proper care is exercised in estimation to uniquely identify the model. For the GDP example, it is natural to require that the conditional mean of the growth rate is higher in the expansion state. The labeling issue becomes more complex as the number of states increases. For a k-state MSM, there are $k!$ equivalent representations induced by labeling. Third, the autoregressive model of a given state can be expanded to include some explanatory variables. The resulting model is flexible and can be useful in many applications. Fourth, since state S_t is not directly observable, it is a latent variable. Thus, MSM belongs to the statistical family of latent variables as such it requires heavier computation in estimation. Finally, the Markov switching model is a special case of the hidden Markov model of Chapter 7 and can be estimated accordingly.

Remark: The Markov switching model in Equation (2.16) is rather general. The model can also be written as

$$x_t - \mu_{S_t} = \phi_{1,S_t}(x_{t-1} - \mu_{S_{t-1}}) + \cdots + \phi_{p,S_t}(x_{t-p} - \mu_{S_{t-p}}) + \sigma_{S_t} \epsilon_t, \quad (2.18)$$

where $\{S_v\}$ is the sequence of state variables, which assume the value 1 or 2. This model representation highlights the flexibility of MSM. It shows that the model can have different expectations as well as different dynamic structures for different states. Some special cases of the model (2.18) have been widely used in the literature. For instance, Hamilton (1989) used the special case of $\phi_{i,1} = \phi_{i,2}$ for $i = 1, \ldots, p$ but $\phi_{0,1} \neq \phi_{0,2}$ to study the business cycle dynamic of US GDP growth rates. For further references, see Frühwirth-Schnatter (2006, Chapter 12).

2.3.1 Properties of Markov Switching Models

The conditions for strict stationarity and existence of high-order moments of an MSM are rather complicated. Interested readers are referred to Holst et al. (1994), Krolzig (1997), Yao and Attali (2000), and Francq and Zakoian (2001). For stationary Markov switching autoregressive processes, Timmermann (2000) derived certain moments and proposed ways to compute those moments. In general, an MSM can introduce dynamic dependence from either the autoregressive structure or the hidden Markov process, or both.

2.3.2 Statistical Inference of the State Variable

An important application of MSM is to make inference about state S_t given the data and the model. Let $\boldsymbol{x}^t = (x_1, \ldots, x_t)$ denote the observations available at time t and model M. Two types of statistical inference of S_t are of interest in practice. The first type is concerned with the probability $P(S_t = i \mid \boldsymbol{x}^t, M)$ and is called *filtered* state probability, where $i = 1, \ldots, k$ for a k-state MSM. It assesses the probability of S_t being in a given state based on the model and currently available data. For instance, an economist might be interested in the probability that an economy is in recession given the currently available data and the proposed model. The second type of inference is concerned with the probability $P(S_t = i \mid \boldsymbol{x}^T, M)$ and is referred to as the *smoothed* state probability, where T is the sample size and $T > t$. The issue here is to make inference about the state at time t when more data become available. For instance, the exact date of an economy dropping into recession is often made after more quarterly economic data become available.

For MSM, various algorithms have been developed in the literature for filtering and smoothing the state probability. See, for instance, Frühwirth-Schnatter (2006, Chapter 11). The algorithms are discrete versions of the general filtering and smoothing methods in state space models with the Kalman filter, as discussed in Chapters 6 and 7 and in Tsay (2010, Chapter 11). We introduce some algorithms below that can be derived by using the Bayes theorem or conditional distributions of normal random variables in statistics.

2.3.2.1 *Filtering State Probabilities* Filtered state probabilities can be obtained sequentially by the following steps for $t = 1, \ldots, T$.

1. One-step ahead prediction probability: for $j = 1, \dots, k$,

$$\begin{aligned} P(S_t = j \mid \boldsymbol{x}^{t-1}, M) \\ &= \sum_{i=1}^{k} P(S_t = j \mid S_{t-1} = i, \boldsymbol{x}^{t-1}, M) P(S_{t-1} = i \mid \boldsymbol{x}^{t-1}, M) \\ &= \sum_{j=1}^{k} p_{ij} P(S_{t-1} = i \mid \boldsymbol{x}^{t-1}, M), \end{aligned} \tag{2.19}$$

where the second equality holds because the Markov chain of the MSM used is homogeneous with time-invariant transition probability matrix $\boldsymbol{P} = [p_{ij}]$.

2. Filtering state probability:

$$P(S_t = j \mid \boldsymbol{x}^t, M) = \frac{p(x_t \mid S_t = j, \boldsymbol{x}^{t-1}, M) P(S_t = j \mid \boldsymbol{x}^{t-1}, M)}{p(x_t \mid \boldsymbol{x}^{t-1}, M)}, \tag{2.20}$$

where $p(x_t \mid S_t = j, \boldsymbol{x}^{t-1}, M)$ denotes the probability density function of x_t given state j, data $\boldsymbol{x}^{t-1}$, and the model M, and

$$P(x_t \mid \boldsymbol{x}^{t-1}, M) = \sum_{i=1}^{k} p(x_t \mid S_t = i, \boldsymbol{x}^{t-1}, M) P(S_t = i \vert \boldsymbol{x}^{t-1}, M). \tag{2.21}$$

At time $t = 1$, the filter starts with some initial distribution $P(S_0 = i \mid \boldsymbol{P})$, e.g. the invariant state distribution if exists, so that

$$P(S_1 = j \mid \boldsymbol{x}^0, M) = \sum_{i=1}^{k} p_{ij} P(S_0 = i \mid \boldsymbol{P}).$$

With the initial values $P(S_0 = i \mid \boldsymbol{P})$, where $i = 1, \dots, k$, the prior algorithm is simply an adaptive procedure for updating information when new data x_t becomes available. Given the data $\boldsymbol{x}^{t-1}$ and the model M, Equation (2.19) acts as a prior that summarizes information about state S_t. Once the new data x_t is available, Equation (2.20) updates the information about state S_t via the posterior distribution.

Equation (2.19) follows the concept of total probability whereas Equation (2.20) can be obtained by Bayes' theorem. More specifically, writing $\boldsymbol{x}^t = (x_t, \boldsymbol{x}^{t-1})$, one can apply Bayes' theorem, $P(A \mid B) = P(B \mid A)P(A)/P(B)$ with $A = \{S_t = j \mid \boldsymbol{x}^{t-1}, M\}$ and $B = \{x_t \mid \boldsymbol{x}^{t-1}, M\}$, to obtain Equation (2.20).

Turn to the smoothed state probabilities. Several algorithms have also been proposed in the literature. See, for instance, Chib (1996).

2.3.2.2 Smoothing State Probabilities

To begin, one applies the algorithm of **filtering state probabilities** so that the filtered probabilities $P(S_t = j \mid \boldsymbol{x}^t, M)$ are available for $j = 1, \dots, k$ and $t = 1, \dots, T$. Then, the following backward algorithm is carried out sequentially for $t = T, T-1, \dots, 0$.

1. Starting values: $P(S_T = j \mid \boldsymbol{x}^T, M)$, where $j = 1, \dots, k$.
2. Smoothed probabilities: for $t = T-1, T-2, \cdots$,

$$P(S_t = j \mid \boldsymbol{x}^T, M) = \sum_{i=1}^{k} \frac{p_{ji}(t)P(S_t = j \mid \boldsymbol{x}^t, M)P(S_{t+1} = i \mid \boldsymbol{x}^T, M)}{\sum_{v=1}^{k} p_{vi}(t)P(S_t = v \mid \boldsymbol{x}^t, M)}$$

$$= \sum_{i=1}^{k} \frac{p_{ji}P(S_t = j \mid \boldsymbol{x}^t, M)P(S_{t+1} = i \mid \boldsymbol{x}^T, M)}{\sum_{v=1}^{k} p_{vi}P(S_t = v \mid \boldsymbol{x}^t, M)} \tag{2.22}$$

where $p_{ji}(t) = P(S_{t+1} = i \mid S_t = j, \boldsymbol{x}^t, M)$, which reduces to p_{ji} for the MSM with time-invariant transition probability matrix $\boldsymbol{P} = [p_{ij}]$.

From Equation (2.22), it is seen that to obtain the smoothed probabilities $P(S_t = j \mid \boldsymbol{x}^T, M)$ one needs to know the filtered probabilities $P(S_t = i \mid \boldsymbol{x}^t, M)$ at time t and the smoothed probabilities $P(S_{t+1} = v \mid \boldsymbol{x}^T, M)$ at time $t+1$. When $t = T-1$, the required probabilities are available from the filtering algorithm discussed in Section 2.3.2.1. Consequently, the smoothed state probabilities are obtained by algorithms of forward filtering and backward smoothing.

It remains to verify Equation (2.22). To this end, by total probability, we have

$$P(S_t = j \mid \boldsymbol{x}^T, M) = \sum_{i=1}^{k} P(S_t = j, S_{t+1} = i \mid \boldsymbol{x}^T, M) \tag{2.23}$$

$$= \sum_{i=1}^{k} P(S_t = j \mid S_{t+1} = i, \boldsymbol{x}^T, M)P(S_{t+1} = i \mid \boldsymbol{x}^T, M)$$

Next, by Bayes' theorem $P(A \mid B) \propto P(B \mid A)P(A)$, we have

$$P(S_t = j \mid S_{t+1} = i, \boldsymbol{x}^T, M)$$
$$\propto p(x_{t+1}, \dots, x_T | S_t = j, S_{t+1} = i, \boldsymbol{x}^t, M)P(S_t = j | S_{t+1} = i, \boldsymbol{x}^t, M). \tag{2.24}$$

Note that, under the MSM, the joint distribution of $(x_{t+1}, \dots, x_T)$ is independent of S_t when S_{t+1} is given. Consequently, Equation (2.24) reduces to

$$P(S_t = j \mid S_{t+1} = i, \boldsymbol{x}^T, M) \propto P(S_t = j \mid S_{t+1} = i, \boldsymbol{x}^t, M).$$

Next, applying Bayes' theorem once more we have

$$P(S_t = j \mid S_{t+1} = i, \boldsymbol{x}^t, M) \propto P(S_{t+1} = i \mid S_t = j, \boldsymbol{x}^t, M)P(S_t = j \mid \boldsymbol{x}^t, M)$$
$$= p_{ji}P(S_t = j \mid \boldsymbol{x}^t, M). \tag{2.25}$$

Since Equation (2.25) must be normalized so that the total probability is 1, we obtain

$$P(S_t = j \mid S_{t+1} = i, \boldsymbol{x}^T, M) = \frac{p_{ji} P(S_t = j \mid \boldsymbol{x}^t, M)}{\sum_{v=1}^{k} p_{vi} P(S_t = v \mid \boldsymbol{x}^t, M)}. \tag{2.26}$$

Combining Equations (2.23) and (2.26), we obtain the smoothing Equation (2.22).

2.3.3 Estimation of Markov Switching Models

We divide the estimation of MSMs into two main steps. In the first step, we assume that the states are observable, i.e. $\boldsymbol{S} = (S_0, S_1, \dots, S_T)'$ is known, where S_0 is the initial state. In practice, $\boldsymbol{S}$ is, of course, unobservable, but the division simplifies the discussion and the results obtained will be used later in iterative estimation procedures.

Suppose that x_t follows a k-state MSM with transition probability matrix $\boldsymbol{P} = [p_{ij}]$ and a realization $\boldsymbol{x}^T = (x_1, \dots, x_T)'$ is available. For ease in notation, we write $\boldsymbol{x} = \boldsymbol{x}^T$ in the rest of this section. The autoregressive order p and the number of states k of the model are given. Let $\boldsymbol{\phi}_j = (\phi_{0,j}, \phi_{1,j}, \dots, \phi_{p,j})'$ be the AR parameter vector of state j, where $j = 1, \dots, k$. Let $\boldsymbol{\vartheta}$ be the collection of all parameters of the model, i.e. $\boldsymbol{\vartheta} = (\boldsymbol{\phi}_1, \dots, \boldsymbol{\phi}_k, \sigma_1, \dots, \sigma_k, \boldsymbol{P})$.

2.3.3.1 *The States are Known*

Given the data $\boldsymbol{x}$ and the states $\boldsymbol{S}$, the likelihood function of the model can be written as

$$L(\boldsymbol{x}, \boldsymbol{S} \mid \boldsymbol{\vartheta}) = p(\boldsymbol{x} \mid \boldsymbol{S}, \boldsymbol{\vartheta}) P(\boldsymbol{S} \mid \boldsymbol{\vartheta}), \tag{2.27}$$

where $p(\boldsymbol{x} \mid \boldsymbol{S}, \boldsymbol{\vartheta})$ is the joint probability density function of the data given the states $\boldsymbol{S}$ and $P(\boldsymbol{S} \mid \boldsymbol{\vartheta})$ is the joint probability density of the states. Equation (2.27) is referred to as the complete-data likelihood in the literature. For the MSM used, $P(\boldsymbol{S} \mid \boldsymbol{\vartheta})$ only depends on the transition matrix $\boldsymbol{P}$. Specifically, we have

$$\begin{aligned} P(\boldsymbol{S} \mid \boldsymbol{\vartheta}) = P(\boldsymbol{S} \mid \boldsymbol{P}) &= \left[\prod_{t=1}^{T} P(S_t \mid S_{t-1}, \boldsymbol{P})\right] P(S_0 \mid \boldsymbol{P}) \\ &= \left[\prod_{t=1}^{T} p_{S_{t-1}, S_t}\right] P(S_0 \mid \boldsymbol{P}) \\ &= P(S_0 \mid \boldsymbol{P}) \prod_{i=1}^{k} \prod_{j=1}^{k} p_{ij}^{n_{ij}}, \end{aligned} \tag{2.28}$$

where n_{ij} denotes the number of transitions from state i to state j in $\boldsymbol{S}$, i.e.

$$n_{ij} = \#\{t \mid S_{t-1} = i, S_t = j\}, \quad 1 \leq i, j \leq k.$$

For the MSM used, we also have

$$p(\boldsymbol{x} \mid \boldsymbol{S}, \boldsymbol{\vartheta}) = \prod_{t=1}^{T} p(x_t \mid S_t, \boldsymbol{x}^{t-1}, \boldsymbol{\vartheta}). \tag{2.29}$$

For the MSM, the probability density function of x_t is known once S_t is given. Equation (2.29), thus, can be written, under normality, as

$$\begin{aligned} p(\boldsymbol{x} \mid \boldsymbol{S}, \boldsymbol{\vartheta}) &= \prod_{j=1}^{k} \prod_{t:S_t=j} p(x_t \mid \boldsymbol{\phi}_j, \sigma_j, \boldsymbol{x}^{t-1}) \\ &= \prod_{j=1}^{k} \prod_{t:S_t=j} \frac{1}{\sqrt{2\pi}\sigma_j} \exp\left[\frac{-1}{2\sigma_j^2}(x_t - \phi_{0,j} - \sum_{i=1}^{p} \phi_{i,j} x_{t-i})^2\right] \end{aligned} \tag{2.30}$$

where $\prod_{t:S_t=j}$ denotes the product over t satisfying $p + 1 \leq t \leq T$ and $S_t = j$. Using Equations (2.27), (2.28), and (2.30), one can easily obtain the maximum likelihood estimates of the MSM under the complete data setting. For instance, we have $\hat{p}_{ij} = n_{ij}/T$ and $\hat{\boldsymbol{\phi}}_j$ and $\hat{\sigma}_j^2$ are the same as those of the usual AR estimation using data with $S_t = j$ only.

Bayesian estimates under conjugate priors: For Bayesian estimation, we have

$$p(\boldsymbol{\vartheta} \mid \boldsymbol{x}, \boldsymbol{S}) \propto L(\boldsymbol{x}, \boldsymbol{S} \mid \boldsymbol{\vartheta}) p(\boldsymbol{\vartheta}),$$

where $p(\boldsymbol{\vartheta})$ denotes the prior distribution. Often one uses product priors such that

$$p(\boldsymbol{\vartheta}) = \prod_{j=1}^{k} p(\boldsymbol{\phi}_j, \sigma_j) p(\boldsymbol{P}). \tag{2.31}$$

For instance, $p(\boldsymbol{\phi}_j, \sigma_j) = p(\boldsymbol{\phi}_j) p(\sigma_j)$, where the usual conjugate priors of autoregressions can be used. Keep in mind that the parameters in $\boldsymbol{P}$ satisfy the conditions that $\sum_{j=1}^{k} p_{ij} = 1$ for $i = 1, \ldots, k$.

In what follows, we provide Bayesian estimates of MSM parameters under the complete-data setting with conjugate priors. The results are widely available in the literature. See, for instance, Tsay (2010, Chapter 12).

Assume that the priors for $\boldsymbol{\phi}_j$ and σ_j^2 are

$$\boldsymbol{\phi}_j \sim N(\boldsymbol{\phi}_{j,o}, \boldsymbol{\Sigma}_{j,o}), \qquad \frac{v_j \lambda_j}{\sigma_j^2} \sim \chi^2_{v_j},$$

where $\boldsymbol{\phi}_{j,o}$, $\boldsymbol{\Sigma}_{j,o}$, v_j, and λ_j are hyper-parameters for state j. From Equation (2.30), the sample estimates of $\boldsymbol{\phi}_j$ and σ_j^2 are

$$\hat{\boldsymbol{\phi}}_j = \left(\sum_{t:S_t=j} \boldsymbol{X}_t\boldsymbol{X}_t'\right)^{-1}\left(\sum_{t:S_t=j} \boldsymbol{X}_t x_t\right), \quad \hat{\sigma}_j^2 = \frac{1}{T_j}\sum_{t:S_t=j}(x_t - \boldsymbol{X}_t'\hat{\boldsymbol{\phi}}_j)^2,$$

where $\boldsymbol{X}_t = (1, x_{t-1}, \ldots, x_{t-p})'$ and $T_j = \#\{t \mid S_t = j, p+1 \le t \le T\}$. It is understood that $T_j > p$ for the estimates to exist. The posterior estimates are

$$\tilde{\boldsymbol{\phi}}_j \sim N(\boldsymbol{\phi}_{j,*}, \boldsymbol{\Sigma}_{j,*}), \quad \frac{v_j\lambda_j + \sum_{t:S_t=j}(x_t - \boldsymbol{X}_t'\tilde{\boldsymbol{\phi}}_j)^2}{\tilde{\sigma}_j^2} \sim \chi^2_{v_j+T_j-p-1}, \tag{2.32}$$

where

$$\boldsymbol{\Sigma}_{j,*}^{-1} = \frac{\sum_{t:S_t=j}\boldsymbol{X}_t\boldsymbol{X}_t'}{\hat{\sigma}_j^2} + \boldsymbol{\Sigma}_{j,o}^{-1}, \quad \tilde{\boldsymbol{\phi}}_{j,*} = \boldsymbol{\Sigma}_{j,*}\left(\frac{\sum_{t:S_t=j}\boldsymbol{X}_t\boldsymbol{X}_t'}{\hat{\sigma}_j^2}\hat{\boldsymbol{\phi}}_j + \boldsymbol{\Sigma}_{j,o}^{-1}\boldsymbol{\phi}_{j,o}\right).$$

Let $\boldsymbol{P}_{i.}$ denote the ith row of the transition matrix $\boldsymbol{P}$. We assume that the priors for $\boldsymbol{P}_{i.}$ are independent for different i (i.e., row independent). Since each row sums to 1, we assume that

$$\boldsymbol{P}_{i.} \sim D(e_{i1}, \ldots, e_{ik}), \quad i = 1, \ldots, k,$$

where $D(\cdot)$ denotes a Dirichlet distribution with hyper parameters e_{ij}. Using Equation (2.28), the posterior distribution of $\boldsymbol{P}_{i.}$ is also independent for different i, and is given by

$$\boldsymbol{P}_{i.} \mid \boldsymbol{S} \sim D(e_{i1} + n_{i1}, \ldots, e_{ik} + n_{ik}), \tag{2.33}$$

where n_{ij} denotes the number of transition from state i to state j. For $k = 2$, the prior posterior distribution reduces to the beta distribution.

2.3.3.2 *The States are Unknown* For simplicity, we assume for now that the number of states k is known. The selection of k will be discussed later. With k fixed, there are k^{T+1} possible configurations for the states $\boldsymbol{S} = (S_0, S_1, \ldots, S_T)'$. Let $\mho$ be the collection of all k^{T+1} possible state configurations. The likelihood function of the data then becomes a finite mixture of the complete-data likelihood function of Equation (2.27), that is,

$$\begin{aligned} L(\boldsymbol{x} \mid \boldsymbol{\vartheta}) &= \sum_{\boldsymbol{S}\in\mho} p(\boldsymbol{x} \mid \boldsymbol{S}, \boldsymbol{\vartheta})P(\boldsymbol{S} \mid \boldsymbol{P}) \\ &= \sum_{\boldsymbol{S}\in\mho}\prod_{t=p+1}^{T} p(x_t \mid \boldsymbol{x}^{t-1}, \boldsymbol{\phi}_{S_t}, \sigma_{S_t})\prod_{i=1}^{k}\prod_{j=1}^{k} p_{ij}^{n_{ij}(\boldsymbol{S})}P(S_0 \mid \boldsymbol{P}), \end{aligned} \tag{2.34}$$

where $n_{ij}(\boldsymbol{S})$ denotes the number of transitions from state i to state j for the given state configuration $\boldsymbol{S}$, and p is the AR order of the model. Strictly speaking, this is a conditional likelihood function as we treat $x_1, \dots, x_p$ as fixed.

Maximizing the likelihood function of Equation (2.34) is non-trivial because k^{T+1} could be very large, even for $k = 2$. In the literature, two general approaches have been proposed to obtain the maximum likelihood estimates, especially under the normality assumption. Both approaches use the idea of data augmentation and require many iterations to achieve convergence in estimation. The first method uses the EM algorithm. See, for instance, Hamilton (1990). We briefly describe the estimation via EM algorithm later. Asymptotic properties of the maximum likelihood estimators of MSM can be found in Cappé et al. (2005). The second approach is to use Markov chain Monte Carlo (MCMC) methods such as those discussed in McCulloch and Tsay (1994) and Frühwirth-Schnatter (2006, Chapter 11).

Remark: Note that the likelihood function in Equation (2.34) is invariant under permutations of the label of states. There are ways to deal with this issue. See, for instance, the discussion in Frühwirth-Schnatter (2006). For the autoregressive MSM discussed in this book, one can apply certain constraints to uniquely identify the states. For instance, McCulloch and Tsay (1994) required that the mean of the AR model in state 1 is higher than that in state 2 when a two-state MSM is applied to the growth rates of US quarterly GDP. In this case, state 1 represents the case where the US economy is growing.

MCMC estimation: This estimation method also involves two main steps to obtain Bayesian estimates of the MSM using Equation (2.34). It starts with some prior specification for the parameters in $\boldsymbol{\vartheta}$, such as the independent priors of Equation (2.31). Furthermore, conjugate priors are used to simplify the random draws from the conditional posterior distributions. The first main step is to draw random variates from the conditional posterior distributions of $\boldsymbol{\vartheta}$ under the assumption that state $\boldsymbol{S}$ is known. For instance, one can use the posterior distributions discussed in Section 2.3.3.1. The second main step is to draw state $\boldsymbol{S}$ conditioned on the parameters drawn in the first step.

There are two general approaches available to draw state $\boldsymbol{S}$. The first approach is to draw a single state S_t conditioned on $\boldsymbol{\vartheta}, \boldsymbol{x}$, and other states, i.e. $\boldsymbol{S}_{-t}$, where $\boldsymbol{S}_{-t} = (S_0, \dots, S_{t-1}, S_{t+1}, \dots, S_T)'$. This approach is referred to as a single-move sampling approach and is used in McCulloch and Tsay (1994). The second approach is to draw $\boldsymbol{S}$ jointly conditioned on $\boldsymbol{\vartheta}$ and $\boldsymbol{x}$. This approach is referred to as a multi-move sampling approach. See, for instance, Chib (1996) and Krolzig (1997). In general, the multi-move sampling approach is preferred, especially when the state transition is sticky (or persistent). On the other hand, the single-move sampling approach is easier to understand and implement.

Single-move Sampling

The conditional posterior probability distribution $P(S_t = j \mid \boldsymbol{S}_{-t}, \boldsymbol{x}, \boldsymbol{\vartheta})$ of state S_t can be obtained by Bayes' theorem,

$$P(S_t \mid \boldsymbol{S}_{-t}, \boldsymbol{x}, \boldsymbol{\vartheta}) \propto P(\boldsymbol{x} \mid \boldsymbol{S}, \boldsymbol{\vartheta}) P(\boldsymbol{S} \mid \boldsymbol{P}) \tag{2.35}$$

$$\propto \prod_{t=1}^{T} p(x_t \mid \boldsymbol{x}^{t-1}, \boldsymbol{S}^t, \boldsymbol{\vartheta}) \prod_{t=1}^{T} P(S_t \mid S_{t-1}, \boldsymbol{x}^{t-1}, \boldsymbol{\vartheta}) P(S_0 \mid \boldsymbol{P}),$$

where $\boldsymbol{S}^t = (S_0, S_1, \ldots, S_t)'$. By Markov properties, S_t only depends on S_{t-1} and S_{t+1}, and we can drop terms that are independent of S_t to obtain that, for $1 \leq t \leq T-1$,

$$P(S_t = j \mid \boldsymbol{S}_{-t}, \boldsymbol{x}, \boldsymbol{\vartheta}) \propto p(x_t \mid \boldsymbol{x}^{t-1}, \boldsymbol{S}^{t-1}, S_t = j, \boldsymbol{\vartheta}) p_{S_{t-1}, j} p_{j, S_{t+1}},$$

where it is understood that the state transition is homogeneous. For the two extreme time points, we have

$$P(S_0 = j \mid \boldsymbol{S}_{-0}, \boldsymbol{x}, \boldsymbol{\vartheta}) \propto p_{j, S_1} P(S_0 = j \mid \boldsymbol{P}),$$

$$P(S_T = j \mid \boldsymbol{S}^{T-1}, \boldsymbol{x}, \boldsymbol{\vartheta}) \propto p(x_T \mid \boldsymbol{x}^{T-1}, \boldsymbol{S}^{T-1}, S_T = j, \boldsymbol{\vartheta}) p_{S_{T-1}, j}.$$

Since $S_t \in \{1, \ldots, k\}$, the above formulas of the conditional posterior probability sum to 1 so that one can draw random variates of S_t easily using the corresponding multinomial distribution.

Multi-move sampling

The joint posterior distribution $P(\boldsymbol{S} \mid \boldsymbol{x}, \boldsymbol{\vartheta})$ can be written as

$$P(\boldsymbol{S} \mid \boldsymbol{x}, \boldsymbol{\vartheta}) = \left[\prod_{t=0}^{T-1} P(S_t \mid S_{t+1}, \ldots, S_T, \boldsymbol{x}, \boldsymbol{\vartheta})\right] P(S_T \mid \boldsymbol{x}, \boldsymbol{\vartheta}), \tag{2.36}$$

where $P(S_T \mid \boldsymbol{x}, \boldsymbol{\vartheta})$ is the filtered probability distribution at $t = T$. Furthermore, the conditional probability distribution $P(S_t \mid S_{t+1}, \ldots, S_T, \boldsymbol{x}, \boldsymbol{\vartheta}) = P(S_t \mid S_{t+1}, \boldsymbol{x}, \boldsymbol{\vartheta})$, which is given in Equation (2.26). To summarize, we outline the procedure of multi-move sampling of the states below.

Multi-move sampling procedure of the states: To sample the state path $\boldsymbol{S}^{(v)}$ for the vth iteration of MCMC estimation, we perform the following:

(a) Apply the algorithm of filtered state probabilities to obtain $P(S_t = j \mid \boldsymbol{x}^t, \boldsymbol{\vartheta})$ for $j = 1, \ldots, k$ and $t = 1, \ldots, T$.

(b) Sample $S_T^{(v)}$ from the filtered state probability distribution $P(S_T = j \mid \boldsymbol{x}, \boldsymbol{\vartheta})$.

(c) For $t = T-1, T-2, \ldots, 0$, sample $S_t^{(v)}$ from the conditional probability distribution

$$P(S_t = j \mid S_{t+1}^{(v)} = m, \boldsymbol{x}^t, \boldsymbol{\vartheta}) = \frac{p_{jm} P(S_t = j \mid \boldsymbol{x}^t, \boldsymbol{\vartheta})}{\sum_{i=1}^{k} p_{im} P(S_t = i \mid \boldsymbol{x}^t, \boldsymbol{\vartheta})}.$$

From the steps, it is clear that for each time t, $P(S_t = j \mid S_{t+1}^{(v)} = m, \boldsymbol{x}^t, \boldsymbol{\vartheta})$ must be computed for $j = 1, \ldots, k$. This, in turn, requires the filtered state probabilities $P(S_t = i \mid \boldsymbol{x}^t, \boldsymbol{\vartheta})$ for $i = 1, \ldots, k$, but those probabilities are available in step (a).

EM algorithm estimation: Hamilton (1990) provides details of estimating MSMs via the EM algorithm. To simplify the estimation, he considers the conditional likelihood function assuming that $x_1, \ldots, x_p$ are fixed. In addition, he assumes that the conditional probability distribution of $(S_1, \ldots, S_p)$ satisfies

$$p(S_p = i_p, S_{p-1} = i_{p-1}, \ldots, S_1 = i_1 \mid \boldsymbol{x}^p) = \xi_{i_p, i_{p-1}, \ldots, i_1},$$

which is independent of $\boldsymbol{\vartheta}$. These new probabilities sum to 1 and are parameters to be estimated jointly with $\boldsymbol{\vartheta}$. If the probability $\xi_{i_p, i_{p-1}, \ldots, i_1}$ is dependent of the transition matrix $\boldsymbol{P}$, the estimation becomes more complicated. See the next section for details.

Under Hamilton's setting, the parameters are $(\boldsymbol{\vartheta}, \boldsymbol{\xi})$, where $\boldsymbol{\xi} = \{\xi_{i_p, i_{p-1}, \ldots, i_1}\}$ and $\boldsymbol{\xi}$ has k^p possible values that sum to 1. Also, with the assumption on the first p states, the initial state S_0 is not needed. The likelihood function then becomes

$$L(\boldsymbol{x} \mid \boldsymbol{\vartheta}, \boldsymbol{\xi}) = \sum_{\boldsymbol{S}} p(\boldsymbol{x} \mid \boldsymbol{S}, \boldsymbol{\vartheta}, \boldsymbol{\xi}) P(\boldsymbol{S} \mid \boldsymbol{P}, \boldsymbol{\xi}), \tag{2.37}$$

where $\boldsymbol{S} = (S_1, \ldots, S_T)'$. The idea of the EM algorithm is to evaluate the log likelihood function of Equation (2.37) by replacing $P(\boldsymbol{S} \mid \boldsymbol{P}, \boldsymbol{\xi})$ by its value given the data and the parameters of the previous iteration. In other words, the algorithm replaces $P(\boldsymbol{S} \mid \boldsymbol{P}, \boldsymbol{\xi})$ by its expected value given the data and previous parameter estimates. Let $\boldsymbol{\Theta} = (\boldsymbol{\vartheta}, \boldsymbol{\xi})$ be the collection of parameters and $\boldsymbol{\Theta}^{(v)}$ be the parameter estimates of the vth iteration. Then, the EM algorithm updates the estimates $\boldsymbol{\Theta}^{(v+1)}$ via

$$p_{ij}^{(v+1)} = \frac{\sum_{t=p+1} p(S_t = j, S_{t-1} = j \mid \boldsymbol{x}, \boldsymbol{\Theta}^{(v)})}{\sum_{t=p+1} p(S_{t-1} = i \mid \boldsymbol{x}, \boldsymbol{\Theta}^{(v)})}, \quad i,j = 1, \ldots, k. \tag{2.38}$$

Obtain $\vartheta^{(v+1)}$ by solving

$$\sum_{t=p+1}^{T}\sum_{S_t=1}^{k}\cdots\sum_{t-p=1}^{k}\frac{\partial \log p(x_t \mid S_t, \Theta)}{\partial \vartheta}\Big|_{\vartheta=\vartheta^{(v+1)}} \times P(S_t, \dots, S_{t-p} \mid \boldsymbol{x}, \Theta^{(v)}) = 0. \tag{2.39}$$

Finally, the initial probabilities are

$$\xi_{i_p,i_{p-1},\dots,i_1} = P(S_p = i_p, \dots, S_1 = i_1 \mid \boldsymbol{x}, \Theta^{(v)}), \quad i_1, \dots, i_m = 1, \dots, k. \tag{2.40}$$

In Equations (2.38) to (2.40), the state probabilities needed are smoothed state probabilities that can be computed recursively, as discussed in Section 2.3.2.2, and the algorithm starts with some arbitrary initial parameter values.

2.3.3.3 *Sampling the Unknown Transition Matrix* It remains to consider the role played by $P(S_0 \mid \boldsymbol{P})$ in the estimation. It turns out that the assumption made concerning the distribution of S_0 affects the way random draws of $\boldsymbol{P}$ are taken. If the distribution of S_0 is independent of the transition matrix $\boldsymbol{P}$, then the traditional Gibbs sampling can be used to draw rows of $\boldsymbol{P}$, as discussed in Equation (2.33). On the other hand, if S_0 follows the invariant distribution associated with $\boldsymbol{P}$, then posterior distributions of the rows of $\boldsymbol{P}$ are no longer independent. In this case, one needs to use the Metropolis–Hastings algorithm to draw random variates of rows of $\boldsymbol{P}$. Often one can use certain Dirichlet distribution as a proposal density. See also Chapter 7.

2.3.4 Selecting the Number of States

When the number of states k is unknown, one can use multiple possible values for k and use marginal likelihoods to select k. For instance, assume $k \in \{1, \dots, K\}$ with, K being a pre-determined positive integer. The marginal likelihoods of interest are $\log p(\boldsymbol{x} \mid k = i)$ for $i = 0, \dots, K$, where $k = 0$ denotes model without Markov switching. One selects k that corresponds to the maximum marginal likelihood. Several methods have been proposed in the literature to compute the marginal likelihood, especially when MCMC methods are used in estimation. For details, readers are referred to Frühwirth–Schnatter (2006, Chapter 5). In applications, meaningful interpretations of the states can also play a role in selecting the number of states k.

2.3.5 Prediction of Markov Switching Models

A simple way to obtain predictions of a given MSM is via simulation. Assume that the forecast origin is T and the forecast horizon is h. That is, we are interested in forecasting $x_{T+1}, \dots, x_{T+h}$ given the model and the data $\boldsymbol{x}^T$. If explanatory variable z_t is also used, then we assume that $z_{T+1}, \dots, z_{T+h}$ are also available. In some cases, a time series model for z_t can be built to obtain predictions of $z_{T+1}, \dots, z_{T+h}$.

From the given model M, which includes the transition matrix $\boldsymbol{P}$, and the data $\boldsymbol{x}^T$, one can apply the filtering procedure of Section 2.3.2.1 to obtain the filtered state probability $P(S_T = j \mid \boldsymbol{x}^T, M)$ for $j = 1, \dots, k$. Let N be the number of iterations used in the simulation. For the ℓth iteration ($\ell = 1, \dots, N$), perform the following procedure:

Simulation procedure

1. Draw h random variates from the distribution of ϵ_t. Denote these random draws by $\{\epsilon_{T+1}^{(\ell)}, \dots, \epsilon_{T+h}^{(\ell)}\}$.
2. Draw the state S_T using the filtered state probability $P(S_T = j \mid \boldsymbol{x}^T, M)$. Denote the random state as $S_T = v$.
3. Given $S_T = v$, do the following for $i = 1, \dots, h$:
   (a) Conditioned on S_{T+i-1}, draw the state S_{T+i} using the transition probability matrix $\boldsymbol{P}$.
   (b) Compute x_{T+i} based on the model M, the data $\boldsymbol{x}^T$, and available predictions, say $\{x_{T+1}, \dots, x_{T+i-1}\}$.
4. Denote the resulting predictions by $x_{T+1}^{(\ell)}, \dots, x_{T+h}^{(\ell)}$.

The above procedure gives N sets of predictions, namely $\{x_{T+1}^{(\ell)}, \dots, x_{T+h}^{(\ell)} \mid \ell = 1, \dots, N\}$. One can use these predictions to obtain point and interval forecasts. Specifically, the point forecasts are the averages,

$$x_T(1) = \frac{1}{N}\sum_{\ell=1}^{N} x_{T+1}^{(\ell)}, \dots, \quad x_T(h) = \frac{1}{N}\sum_{\ell=1}^{N} x_{T+h}^{(\ell)},$$

where $x_T(i)$ denotes the i-step ahead prediction of x_{T+i} at the forecast origin $t = T$. The sample variance of $\{x_{T+i}^{(\ell)} | \ell = 1, \dots, N\}$ can be used as the variance of the i-step ahead forecast errors. In practice, N should be sufficiently large.

2.3.6 Examples

We demonstrate MSMs by considering some examples. In this section, we use the `MSwM` package of R, which uses the EM algorithm of Hamilton (1990) to perform estimation. In Chapter 7, we use the package `HiddenMarkov` to estimate the Markov switching models.

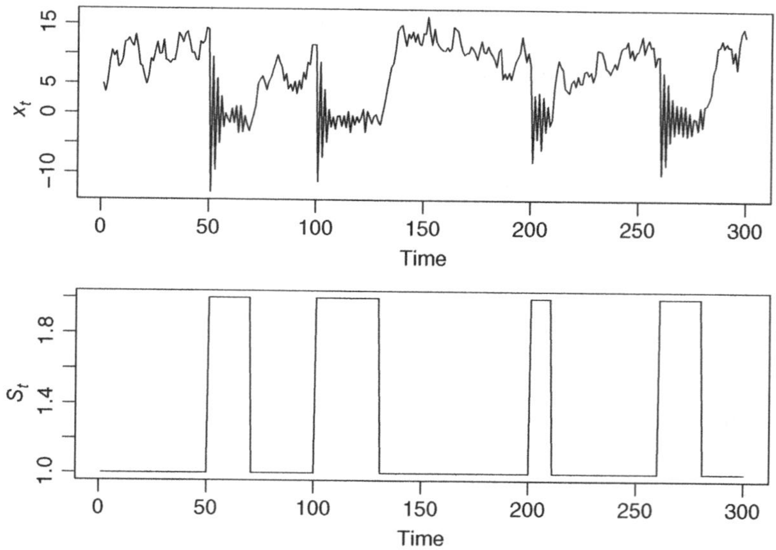

Figure 2.10 Time plots of (a) the simulated time series x_t and (b) the associated state variable S_t. The data are generated from the model in Equation (2.41).

Example 2.3 To introduce the `MSwM` package and MSMs, we start with a simple simulated example. We generated 300 observations from the following model

$$x_t = \begin{cases} 2.0 + 0.8x_{t-1} + 1.5\epsilon_t & \text{if } S_t = 1, \\ -2.0 + 0.6z_t - 0.8x_{t-1} + \epsilon_t & \text{if } S_t = 2, \end{cases} \tag{2.41}$$

where ϵ_t are iid $N(0, 1)$ and z_t are iid $U[0, 1]$. State S_t is given below

$$S_t = \begin{cases} 1 & \text{if } t \in [1:50;\ 71:100;\ 131:200;\ 211:250;\ 271:300] \\ 2 & \text{if } t \in [51:70;\ 101:130;\ 201:210;\ 251:270]. \end{cases}$$

Figure 2.10 shows the time plot of x_t and the state variable S_t. In this example, the exogenous variable z_t affects x_t in state 2. The two-state feature of the model is clearly seen from the time plot of x_t.

For this particular instance, the sample correlation between x_t and z_t is −0.03, which is not significant at the 5% level. This is not surprising as z_t only appears in state 2, which has a smaller sample size. The fitted linear regression model is

$$x_t = 6.76 - 0.51z_t + e_t, \quad \hat{\sigma}_e = 5.79.$$

Assuming that $p = 1$ and $k = 2$ are known, we can write the model as

$$x_t = \beta_{0,S_t} + \beta_{1,S_t} z_t + \phi_{S_t} x_{t-1} + \sigma_{S_t} \epsilon_t,$$

where $S_v = 1$ or 2. This formulation is used in the `MSwM` package and the four-state dependent parameters are $(\beta_{0,S_t}, \beta_{1,S_t}, \phi_{S_t}, \sigma_{S_t})'$, which will be used to demonstrate the flexibility of the package. More specifically, the package allows users to select which parameters may switch from one state to another and which parameters remain constant across states. In this particular case, all four parameters are state-dependent so that we use the subcommand sw = (T,T,T,T) with "T" denoting state-dependent and "F" state-invariant. See the subcommand `sw` in the R output below where we allow all four parameters to switch.

The fitted MSM is

$$x_t = \begin{cases} 2.0 - 0.40z_t + 0.83x_{t-1} + 1.54\epsilon_t, & \text{if } S_t = 1, \\ -1.75 + 0.34z_t - 0.80x_{t-1} + 0.91\epsilon_t, & \text{if } S_t = 2, \end{cases} \tag{2.42}$$

where the standard errors of the coefficient estimates are 0.34, 0.36, and 0.03 for state 1 and 0.20, 0.35, and 0.02 for state 2. The fitted transition probability matrix is

$$\hat{\boldsymbol{P}} = \begin{bmatrix} 0.982 & 0.018 \\ 0.052 & 0.948 \end{bmatrix}.$$

From the results, it is seen that the parameter estimates are reasonable except the coefficient of z_t. The estimated switching probabilities are $P(S_t = 2 \mid S_{t-1} = 1) = 0.018$ and $P(S_t = 1 \mid S_{t-1} = 2) = 0.052$. These low probabilities are also reasonable as there is only a small number of switches in the data-generating model.

We now turn to model checking. Figure 2.11 shows the ACF and PACF of the residuals and their squares of the fitted model in Equation (2.42). There exist no serial correlations in either the residuals or the squared residuals. Figure 2.12 shows the Q–Q plot and time plot of the residuals. Both plots fail to indicate any major model discrepancy. Thus, the model fits the data well. Finally, Figure 2.13 shows the filtered (a) and smoothed (b) probabilities for state 1 of the fitted model. It is interesting to compare the smoothed state probabilities with the actual states shown in Figure 2.10. The classification of states works nicely in this simple example.

R demonstration: Markov switching model and the `MSwM` package.

```
#### Generating MS process x(t) and explanatory variable z(t)
set.seed(51)
n <- 300; np1 <- n+1
St <- c(rep(1,51),rep(2,20),rep(1,30),rep(2,30),rep(1,70),rep(2,10),
        rep(1,50),rep(2,20),rep(1,20))
zt <- runif(np1,min=0,max=1) ## regressor
at <- rnorm(np1)  ## epsilon(t)
```

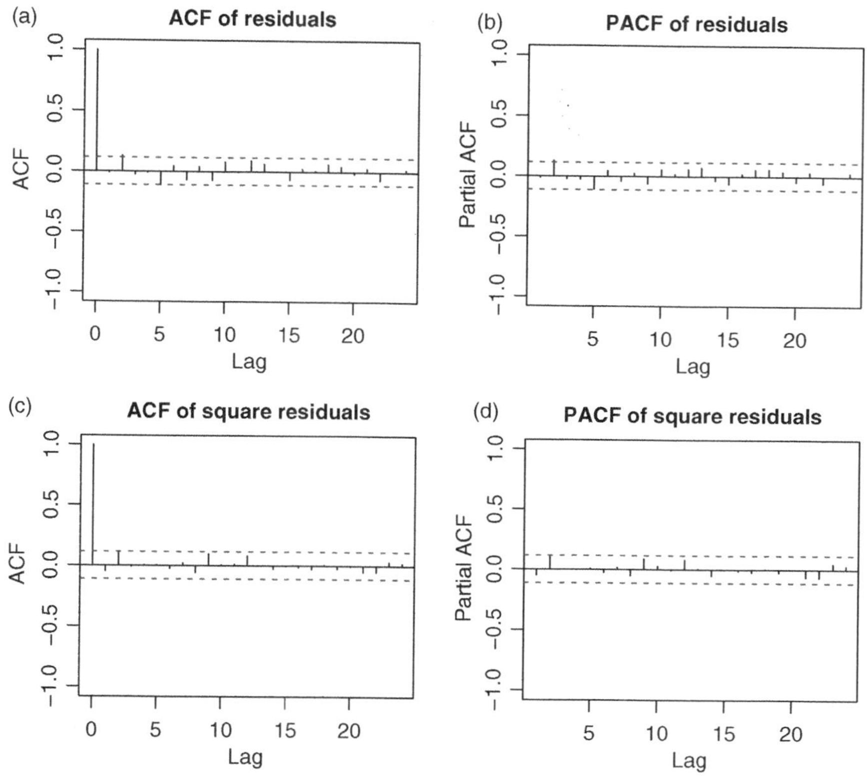

Figure 2.11 Autocorrelation function and partial autocorrelation function of the residuals (a and b) and squared residuals (c and d) of the fitted Markov switching model in Equation (2.42).

```
x <- at[1]   ### set initial value
for (i in 2:np1){
 tmp = 0
 if(St[i]==1){
  tmp = 2.0 + 0.8*x[i-1] + 1.5*at[i]
 }
 else{
  tmp = -2.0+0.6*zt[i]-0.8*x[i-1]+at[i]
 }
 x <- c(x,tmp)
}
xt <- x[-1]; zt <- zt[-1]; St <- St[-1]   ## Remove the initial value
> x <- cbind(xt,rep(1,300),zt) # Start model fitting
> colnames(x) <- c("xt","cnst","zt")
```

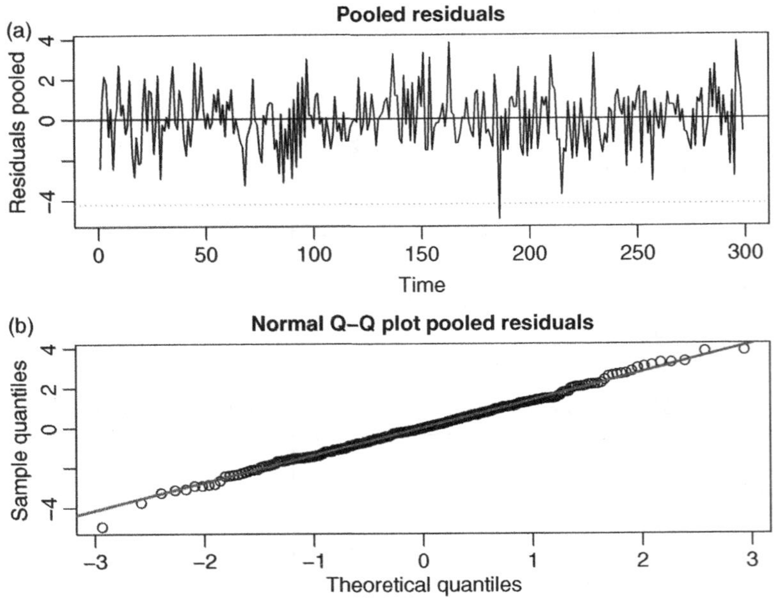

Figure 2.12 (a) Time plot and (b) normal probability plot of the residuals of the fitted Markov switching model in Equation (2.42).

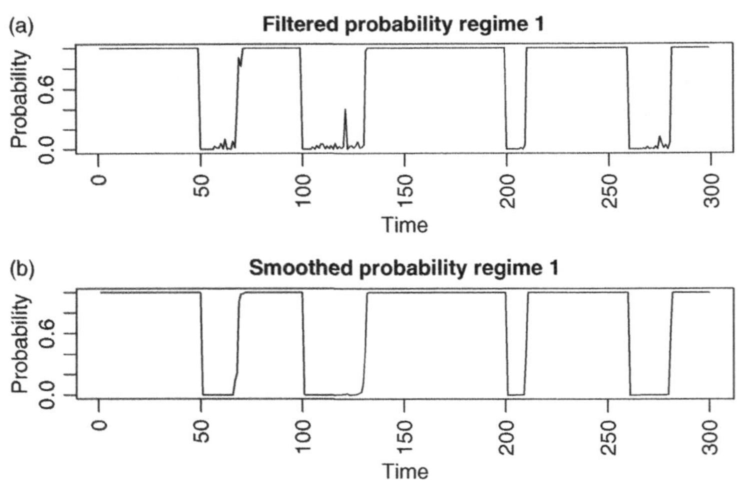

Figure 2.13 (a) Filtered and (b) smoothed state probabilities for Regime 1 of the fitted Markov switching model in Equation (2.42).

```
> X <- data.frame(x)
> m1 <- lm(xt~-1+cnst+zt,data=X)
> summary(m1)
Call: lm(formula = xt ~ -1 + cnst + zt, data = X)
Coefficients:
     Estimate Std. Error t value Pr(>|t|)
cnst   6.7610     0.6269  10.784   <2e-16 ***
zt    -0.5134     1.1343  -0.453    0.651
---
Signif. codes:  0 *** 0.001 ** 0.01 * 0.05 . 0.1   1
Residual standard error: 5.785 on 298 degrees of freedom

> require(MSwM) # Load the package
> m2 <- msmFit(m1,k=2,p=1,sw=c(T,T,T,T))
> summary(m2)
Markov Switching Model
Call: msmFit(object = m1, k = 2, sw = c(T, T, T, T), p = 1)
       AIC      BIC    logLik
  1101.089 1157.494 -544.5443

Coefficients:
Regime 1
---------
        Estimate Std. Error t value  Pr(>|t|)
cnst(S)   2.0030     0.3374  5.9366  2.91e-09 ***
zt(S)    -0.4008     0.3574 -1.1214    0.2621
xt_1(S)   0.8265     0.0300 27.5500 < 2.2e-16 ***
---
Residual standard error: 1.542166

Regime 2
---------
        Estimate Std. Error  t value Pr(>|t|)
cnst(S)  -1.7467     0.1953  -8.9437   <2e-16 ***
zt(S)     0.3449     0.3454   0.9986    0.318
xt_1(S)  -0.8043     0.0215 -37.4093   <2e-16 ***
---
Residual standard error: 0.9059516

Transition probabilities:
            Regime 1  Regime 2
Regime 1 0.98171408 0.0517276
Regime 2 0.01828592 0.9482724
> plotDiag(m2)  # Residual ACF and PACF
> par(mfcol=c(2,1))
```

```
> plotDiag(m2,which=1) # Residual plot
> plotDiag(m2,which=2) # Q-Q plot of residuals
> slotNames(m2)
> slotNames(m2@Fit)
> dim(m2@Fit@filtProb)
> par(mfrow=c(2,1))
> plot(m2@Fit@filtProb[,1],type='l',main='Filtered Probability Regime 1',
    ylab='prob',xlab='time')
> plot(m2@Fit@smoProb[,1],type='l',main='Smoothed Probability Regime 1',
    ylab='prob',xlab='time')
# Figure 2.13
```

Example 2.4 Consider the quarterly growth rates of US real gross domestic product (GDP) from the second quarter of 1947 to the second quarter of 2015. The growth rate is defined as $x_t = \ln(X_t) - \ln(X_{t-1})$, where X_t is the quarterly US GDP in billions of 2009 chained dollars. The data are obtained from FRED of the Federal Reserve Bank of St. Louis. This series (or its subsets) has been widely used to demonstrate Markov switching models. Figure 2.14 shows the time plot of the growth rates. The apparent change in volatility of the series after the 1980s is referred to as the *Great Moderation* in the literature.

The AIC selects an AR(3) model for the series x_t of growth rates. Therefore, we start the analysis with a two-state AR(3) MSM and obtain the fitted model

$$x_t = \phi_{0,S_t} + \phi_{1,S_t}x_{t-1} + \phi_{2,S_t}x_{t-2} + \phi_{3,S_t}x_{t-3} + \sigma_{S_t}\epsilon_t, \quad S_t = 1 \text{ or } 2, \quad (2.43)$$

where the parameter estimates and their standard errors (in parentheses) are

State	ϕ_0	ϕ_1	ϕ_2	ϕ_3	σ
1	0.0052(0.0013)	0.37(0.08)	0.09(0.09)	−0.09(0.08)	0.0110
2	0.0044(0.0009)	0.20(0.10)	0.29(0.09)	−0.10(0.08)	0.0046

and the transition probability matrix is

$$\hat{\boldsymbol{P}} = \begin{bmatrix} 0.9904 & 0.0096 \\ 0.0199 & 0.9801 \end{bmatrix}.$$

Figure 2.15 shows the autocorrelation and partial autocorrelation functions of the residuals and squared residuals of model (2.43). The model seems to fit the data reasonably well as there exists no significant serial dependence in the residuals. Figure 2.16 shows the filtered and smoothed state probabilities for Regime 1 of the fitted model. From the probability plot, it is seen that the fitted model essentially captures the dominating feature of the GDP growth rates, namely the change in

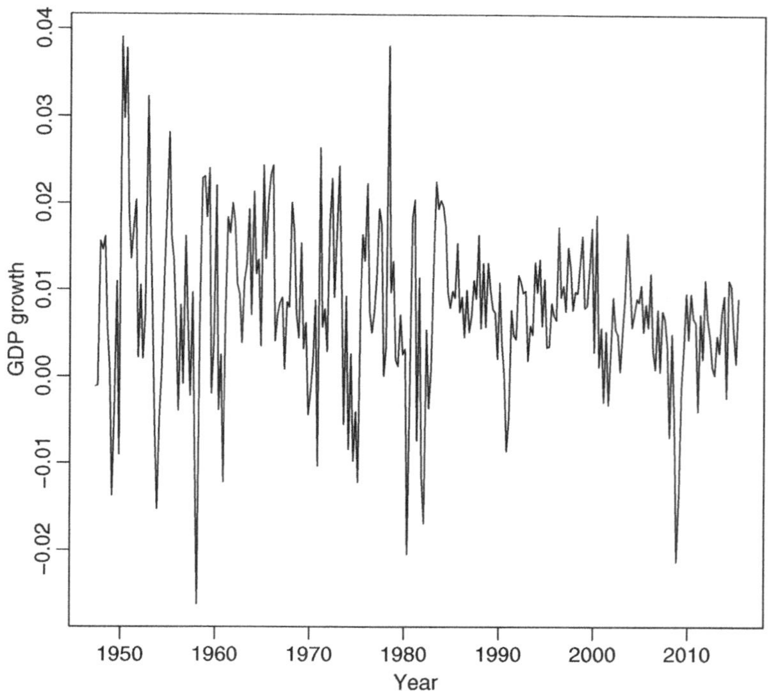

Figure 2.14 Time plot of the quarterly growth rates of US real gross domestic product from the second quarter of 1947 to the second quarter of 2015.

volatility after the 1980s. While it is reassuring to see that the Markov switching model can describe the main feature of the data, it remains unsatisfactory to see that the model fails to show other characteristics of the data.

There are ways to improve the fitted model in Equation (2.43). One can divide the series into two subseries based on the Great Moderation and analyze the subseries separately. One can also increase the flexibility of the fitted model. We adopt the latter approach by increasing the number of states to three, illustrating that the characteristics of the data may help to select the number of states.

The fitted three-state MSM is in the same form as that of Equation (2.43) except that S_t now assumes three values denoted by $\{1, 2, 3\}$. The parameter estimates and their standard errors (in parentheses) are given below:

State	ϕ_0	ϕ_1	ϕ_2	ϕ_3	σ
1	0.0052(0.0013)	0.32(0.09)	0.09(0.09)	−0.07(0.09)	0.0111
2	0.0081(0.0007)	0.57(0.09)	−0.01(0.06)	−0.08(0.07)	0.0025
3	0.0002(0.0008)	0.62(0.09)	0.30(0.07)	−0.26(0.06)	0.0033

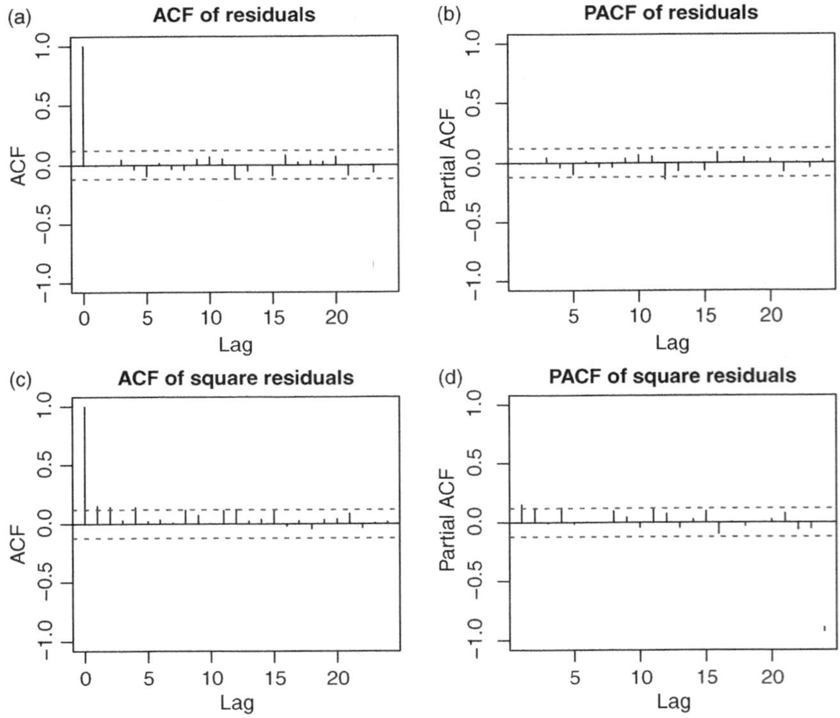

Figure 2.15 Autocorrelations and partial autocorrelations of residuals (a and b) and squared residuals (c and d) of the two-state AR(3) model in Equation (2.43) for the growth rates of US quarterly real GDP.

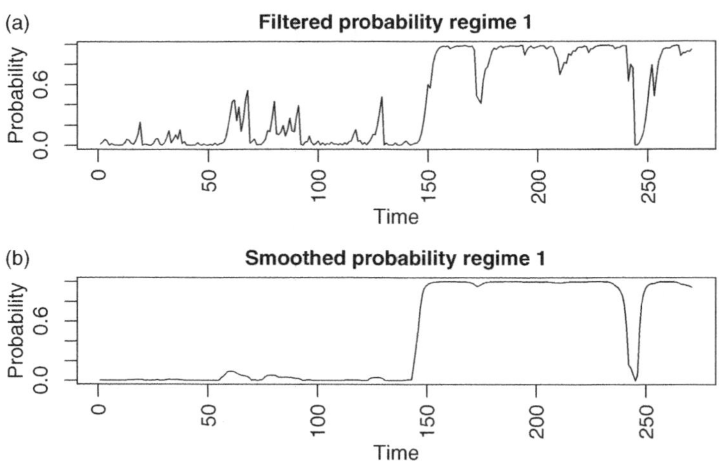

Figure 2.16 (a) Filtered and (b) smoothed state probabilities of Regime 1 of the fitted two-state AR(3) Markov switching model in Equation (2.43) for the growth rates of US real quarterly GDP from the second quarter of 1947 to the second quarter of 2015.

and the transition probability matrix is

$$\widehat{\boldsymbol{P}} = \begin{bmatrix} 0.951 & 0.030 & 0.019 \\ 0.086 & 0.008 & 0.906 \\ 0.033 & 0.563 & 0.404 \end{bmatrix}.$$

Figures 2.17 and 2.18 show the model checking for the fitted three-state AR(3) MSM. As expected, the residuals of the fitted model do not have strong serial dependence, and the residual plot shows no surprise in view of the Great Moderation. On the other hand, the Q–Q plot indicates that the normality assumption is questionable for the series.

Some discussions of the fitted three-state AR(3) MSM are in order. First, the model of state 3 differ markedly from those of the other two states. In particular, the expected GDP growth rate of state 3 is essentially zero, whereas the expected GDP growths are positive in states 1 and 2. Using $\mu = 4 \times \phi_0/(1 - \phi_1 - \phi_2 - \phi_3)$, we

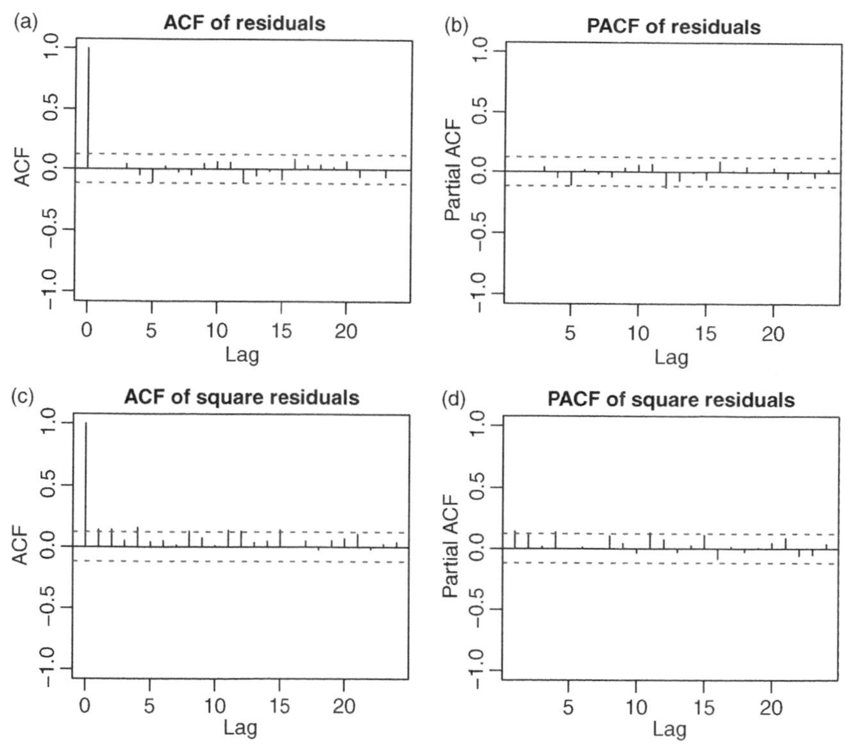

Figure 2.17 Autocorrelations and partial autocorrelations of residuals (a and b) and squared residuals (c and d) of the fitted three-state AR(3) model for the growth rates of US quarterly real GDP.

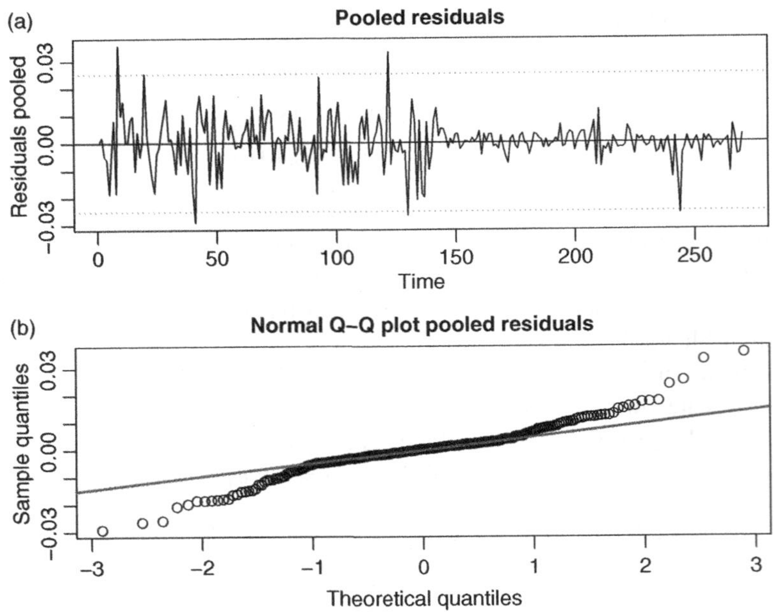

Figure 2.18 (a) Time plot and (b) normal probability plot of the residuals of the fitted three-state AR(3) model for the growth rates of US quarterly real GDP.

obtain that the annualized expected growth rates are approximately 3.2%, 6.2%, and 0.2%, respectively, for state 1 to state 3. Based on these expected values, the three states can be classified as representing the "moderate", "high", and "zero"" growth periods of the US economy. Second, the fitted AR models of states 1 and 2 are basically an AR(1) model with a small or moderate coefficient, indicating that the GDP growth rate reverts to its mean quickly in these two states. On the other hand, for state 3, all three AR coefficients are significantly different from zero at the usual 5% level. In addition, the fitted AR polynomial is approximately $\phi(B) = 1 - 0.6197B - 0.3003B^2 + 0.2649B^3$, which has one real and two complex roots. A simple calculation shows that the complex conjugate roots give rise to an average period of 15.99 quarters for business cycles. For discussion of AR polynomial roots, see, for instance, Tsay (2010, Chapter 2). Thus, in addition to state switching, the fitted MSM also supports the pattern of business cycles of the US economy. Third, the fitted transition matrix also reveals some interesting characteristics of the model:

1. The probability of staying in state 1 is high at 95.1%. In addition, when the US economy grew at a moderate rate, it had a slightly higher probability to jump into the high-growth than zero-growth state (3% versus 1.9%).

2. When the US economy grew rapidly, it was almost certain to contract into the zero-growth state. The probability is about 90.6%.
3. When the US economy was in trouble (in the zero-growth state), it either jumped out with a rapid growth or remained in the zero growth. The probabilities are approximately 56% and 40%, respectively.

Fourth, the probabilities of switching to other states are 0.049, 0.992, and 0.596, respectively, for state 1 to state 3. Consequently, the expected durations for the states are 20.4, 1.0, and 1.7 quarters, respectively, implying that the rapid-growth and zero-growth periods of the US economy were rather short on average. This is not surprising.

Fifth, Figure 2.19 shows the filtered and smoothed state probabilities of the fitted three-state model. From the plots, it is seen that, as expected, the US economy was most often in the moderate-growth state (state 1). However, the transitions among the states increased markedly during the Great Moderation era. This implies that, with reduced volatility in the GDP growth rates, the status of US economy seems to change quickly.

Finally, the R output for the three-state MSM model is given below.

R commands and output for three-state AR(3) MSM: The `MSwM` package.

```
> da=read.table("GDPC1.txt",header=T)
> gdp=diff(log(da[,4]))
> length(gdp)
[1] 273
> x <- cbind(gdp,rep(1,length(gdp)))
> colnames(x) <- c("gdp","cnst")
> X <- data.frame(x)
> m1 <- lm(gdp~-1+cnst,data=X)  # Basic model for the expected growth
> summary(m1)
Call:   lm(formula = gdp ~ -1 + cnst, data = X)
Coefficients:
      Estimate Std. Error t value Pr(>|t|)
cnst 0.0078125  0.0005784   13.51   <2e-16 ***
---
> require(MSwM)
> m2 <- msmFit(m1,k=3,p=3,sw=c(T,T,T,T,T))
> summary(m2)
Markov Switching Model
Call: msmFit(object = m1, k = 3, sw = c(T, T, T, T, T), p = 3)
        AIC       BIC   logLik
  -1839.701 -1729.339 931.8506

Coefficients:
```

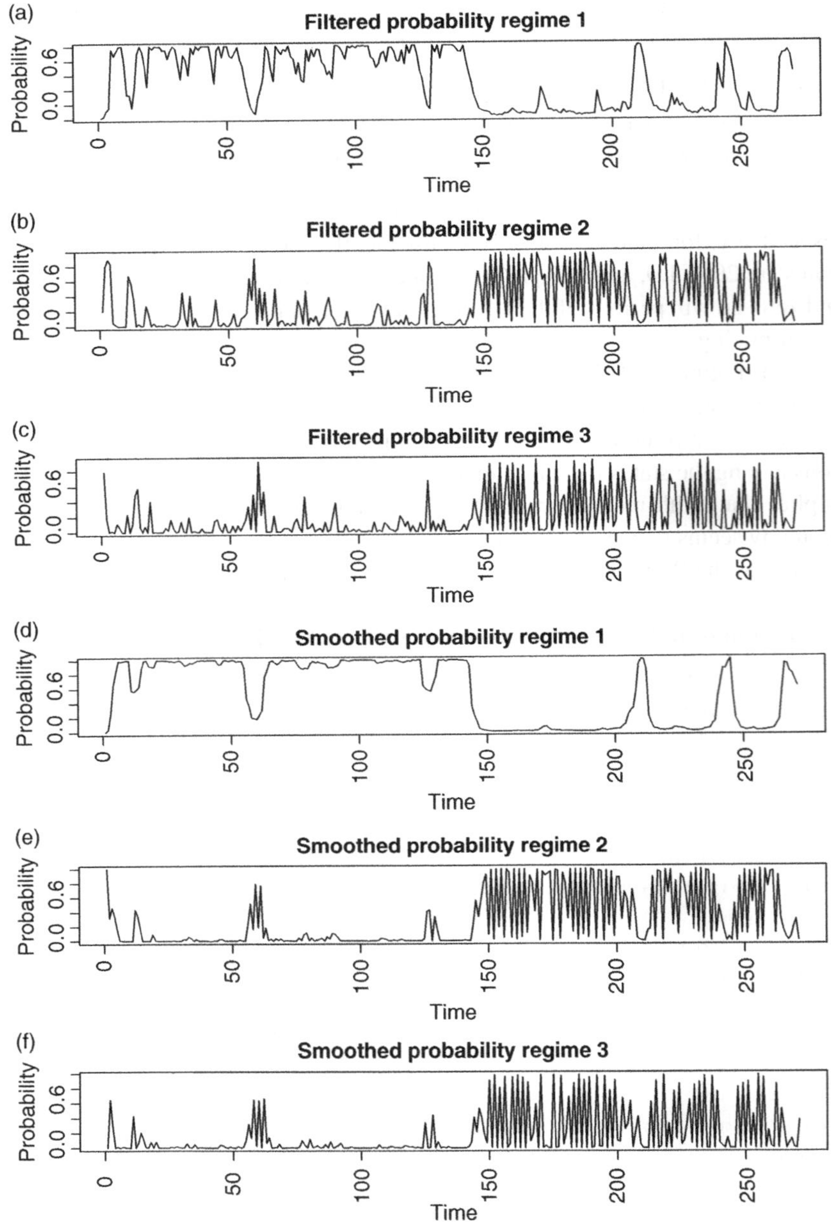

Figure 2.19 Filtered (a–c) and smoothed (d–f) state probabilities of the fitted three-state AR(3) MSM for the growth rates of US real quarterly GDP from the second quarter of 1947 to the second quarter of 2015.

```
Regime 1
---------
          Estimate Std. Error t value  Pr(>|t|)
cnst(S)     0.0052     0.0013  4.0000 6.334e-05 ***
gdp_1(S)    0.3176     0.0873  3.6380 0.0002748 ***
gdp_2(S)    0.0908     0.0940  0.9660 0.3340442
gdp_3(S)   -0.0691     0.0934 -0.7398 0.4594214
---
Signif. codes:  0  ***  0.001  **  0.01  *  0.05  .  0.1    1

Residual standard error: 0.011107

Regime 2
---------
          Estimate Std. Error t value  Pr(>|t|)
cnst(S)     0.0081     0.0008 10.1250 < 2.2e-16 ***
gdp_1(S)    0.5686     0.0899  6.3248 2.536e-10 ***
gdp_2(S)   -0.0096     0.0594 -0.1616    0.8716
gdp_3(S)   -0.0825     0.0652 -1.2653    0.2058
---
Residual standard error: 0.002533775

Regime 3
---------
          Estimate Std. Error t value  Pr(>|t|)
cnst(S)     0.0002     0.0008  0.2500    0.8026
gdp_1(S)    0.6197     0.0937  6.6137 3.748e-11 ***
gdp_2(S)    0.3003     0.0687  4.3712 1.236e-05 ***
gdp_3(S)   -0.2649     0.0607 -4.3641 1.276e-05 ***
---
Residual standard error: 0.003278711

Transition probabilities:
           Regime 1    Regime 2   Regime 3
Regime 1 0.95122395 0.086451577 0.03303734
Regime 2 0.02950221 0.007717656 0.56273477
Regime 3 0.01927384 0.905830767 0.40422789
> plotDiag(m2)  ## Residual ACF and PACF
> par(mfcol=c(2,1))
> plotDiag(m2,which=1) # Residual plot
> plotDiag(m2,which=2) # Residual Q-Q plot
> plotProb(m2,which=1)  # State probabilities (Not shown)
> plotProb(m2,which=2) # State-1 probabilities (Not shown)
> plotProb(m2,which=4) # State-3 probabilities (No shown)
> par(mfrow=c(3,1))
```

```
> plot(m2@Fit@filtProb[,1],type='l',main='Filtered Probability Regime 1',
    ylab='prob',xlab='time')
> plot(m2@Fit@filtProb[,2],type='l',main='Filtered Probability Regime 2',
    ylab='prob',xlab='time')
> plot(m2@Fit@filtProb[,3],type='l',main='Filtered Probability Regime 3',
    ylab='prob',xlab='time')
# Figure 2.19 upper three
> plot(m2@Fit@smoProb[,1],type='l',main='Smoothed Probability Regime 1',
   ylab='prob',xlab='time')
> plot(m2@Fit@smoProb[,2],type='l',main='Smoothed Probability Regime 2',
   ylab='prob',xlab='time')
> plot(m2@Fit@smoProb[,3],type='l',main='Smoothed Probability Regime 3',
   ylab='prob',xlab='time')
# Figure 2.19 lower three
> p1 = c(1, -0.6197, -.3003, .2649)
> s1 <- polyroot(p1)
> s1
[1]  1.394760+0.57827i -1.655885-0.00000i  1.394760-0.57827i
> Mod(s1)
[1] 1.509885 1.655885 1.509885
> kk=2*pi/acos(1.3948/1.5099)
> kk
[1] 15.98833
```

Remark: One can also use the switching regression format to fit the autoregressive MSM in the `MSwM` package. We illustrate the commands and output below. As expected, except for changes in label, the results are essentially the same.

R demonstration: An alternative representation of MSM.

```
> X <- cbind(gdp[4:273],rep(1,270),gdp[3:272],gdp[2:271],gdp[1:270])
> colnames(X) <- c("gdp","cnst","gdpl1","gdpl2","gdpl3")
> X <- data.frame(X)
> m1 <- lm(gdp~-1+.,data=X) ## Linear AR(3) fit
> summary(m1)
Call:  lm(formula = gdp ~ -1 + ., data = X)

Coefficients:
         Estimate Std. Error t value Pr(>|t|)
cnst    0.0048627  0.0008053   6.038 5.22e-09 ***
gdpl1   0.3465507  0.0608520   5.695 3.26e-08 ***
gdpl2   0.1284203  0.0639292   2.009   0.0456 *
gdpl3  -0.0958892  0.0607570  -1.578   0.1157
---
Residual standard error: 0.008842 on 266 degrees of freedom
```

```
> m2 <- msmFit(m1,k=3,sw=c(T,T,T,T,T))
> summary(m2)
Markov Switching Model
Call: msmFit(object = m1, k = 3, sw = c(T, T, T, T, T))
        AIC       BIC   logLik
  -1840.463 -1730.101 932.2315

Coefficients:
Regime 1
---------
         Estimate Std. Error t value  Pr(>|t|)
cnst(S)    0.0082     0.0008 10.2500 < 2.2e-16 ***
gdpl1(S)   0.5855     0.1283  4.5635 5.031e-06 ***
gdpl2(S)   0.0068     0.0862  0.0789    0.9371
gdpl3(S)  -0.1048     0.1129 -0.9283    0.3533
---
Residual standard error: 0.002392347

Regime 2
---------
         Estimate Std. Error t value  Pr(>|t|)
cnst(S)    0.0052     0.0013  4.0000 6.334e-05 ***
gdpl1(S)   0.3272     0.0866  3.7783 0.0001579 ***
gdpl2(S)   0.0951     0.0909  1.0462 0.2954687
gdpl3(S)  -0.0749     0.0886 -0.8454 0.3978875
---
Residual standard error: 0.01110063

Regime 3
---------
         Estimate Std. Error t value  Pr(>|t|)
cnst(S)    0.0002     0.0009  0.2222   0.82416
gdpl1(S)   0.6197     0.0860  7.2058 5.771e-13 ***
gdpl2(S)   0.2634     0.1303  2.0215   0.04323 *
gdpl3(S)  -0.2378     0.0934 -2.5460   0.01090 *
---
Residual standard error: 0.003247124

Transition probabilities:
            Regime 1   Regime 2   Regime 3
Regime 1 0.003303693 0.02108811 0.53566037
Regime 2 0.064955893 0.95537940 0.04115438
Regime 3 0.931740414 0.02353249 0.42318526
```

2.4 SMOOTH TRANSITION AUTOREGRESSIVE MODELS

Another class of nonlinear time series models is smooth transition autoregressive (STAR) models. See Tong (1990, p. 107), Teräsvirta (1994), and the survey article of Van Dijk, Teräsvirta, and Frances (2002). The STAR model is similar to the threshold autoregressive model. The main difference between these two models is the mechanism governing the transition between regimes. We use a two-regime model in our discussion. The concept can be extended to the case with more than two regimes. To make the connection between different classes of nonlinear model clear, the two-regime TAR(p) model discussed in Section 2.2 can be rewritten as

$$\begin{aligned} x_t = {} & (\phi_{0,1} + \phi_{1,1}x_{t-1} + \cdots + \phi_{p,1}x_{t-p} + \sigma_1\epsilon_t)[1 - I(x_{t-d} > r)] \\ & + (\phi_{0,2} + \phi_{1,2}x_{t-1} + \cdots + \phi_{p,2}x_{t-p} + \sigma_2\epsilon_t)I(x_{t-d} > r), \end{aligned}$$

where $I(y)$ denotes the indicator variable such that $I(y) = 1$ if y holds and $= 0$ otherwise. Here the transition of the model from one regime to another is governed by the step function $I(x_{t-d} > r)$, where d is the delay, x_{t-d} is the threshold variable, and r is the threshold. The STAR model, on the other hand, uses a smooth (continuous) transition function with range between 0 and 1.

Consider a time series x_t. Let s_t be the threshold variable. For instance, $s_t = x_{t-d}$ for some $d > 0$. The series x_t follows a two-regime STAR model with transition function $G(s_t \mid \gamma, c)$ if it satisfies

$$\begin{aligned} x_t = {} & (\phi_{0,1} + \phi_{1,1}x_{t-1} + \cdots + \phi_{p,1}x_{t-p})[1 - G(s_t \mid \gamma, c)] \\ & + (\phi_{0,2} + \phi_{1,2}x_{t-1} + \cdots + \phi_{p,2}x_{t-p})G(s_t \mid \gamma, c) + a_t \end{aligned} \tag{2.44}$$

where a_t is an iid sequence of random noises with mean zero and variance $\sigma_a^2 > 0$. In this model, we have $0 \leq G(s_t \mid \gamma, c) \leq 1$, where γ and c are referred to as the scale and location parameters of the transition function. The two AR models are assumed to satisfy the usual conditions for weak stationarity. From the definition, the STAR series x_t follows the AR(p) model with AR polynomial $\phi_1(B) = 1 - \phi_{1,1}B - \cdots - \phi_{p,1}B^p$ if $G(s_t \mid \gamma, c) = 0$. On the other extreme, it follows another AR(p) model with AR polynomial $\phi_2(B) = 1 - \phi_{1,2}B - \cdots - \phi_{p,2}B^p$ if $G(s_t \mid \gamma, c) = 1$. For $0 < G(s_t \mid \gamma, c) < 1$, x_t is a weighted average between two AR models with weights governed by $G(s_t \mid \gamma, c)$. Clearly, the parameters γ and c are only defined when the two AR polynomials are different. It is easy to extend the STAR model in Equation (2.44) to have different innovation variances. However, a common variance is often used in the literature.

Different choices of the transition function $G(s_t \mid \gamma, c)$ result in different types of STAR models. For instance, one can use $G(s_t \mid \gamma, c) = \Phi[(s_t - c)/\gamma]$, where

$\Phi(z)$ is the cumulative distribution function of the standard Gaussian random variable. In the literature, two transition functions are commonly used. They are the logistic and exponential functions. The logistic function is defined as

$$G(s_t \mid \gamma, c) = \frac{1}{1 + \exp[-\gamma(s_t - c)]}, \tag{2.45}$$

and the resulting model is referred to as a logistic STAR (or LSTAR) model. The exponential function is defined as

$$G(s_t \mid \gamma, c) = 1 - \exp[-\gamma(s_t - c)^2], \quad \gamma > 0, \tag{2.46}$$

and the resulting model is called an exponential STAR (ESTAR) model. The difference between LSTAR and ESTAR models is that the exponential function in Equation (2.46) approaches 1 when $s_t \to \pm\infty$. On the other hand, the logistic function in Equation (2.45) approaches 1 only when $\gamma(s_t - c) \to \infty$.

Figure 2.20 shows three logistic transition functions $G(x \mid \gamma, 0)$ for $\gamma = 0.5$, 1, and 2. The solid line is for $\gamma = 1$, the dashed line for $\gamma = 0.5$, and the dotted line

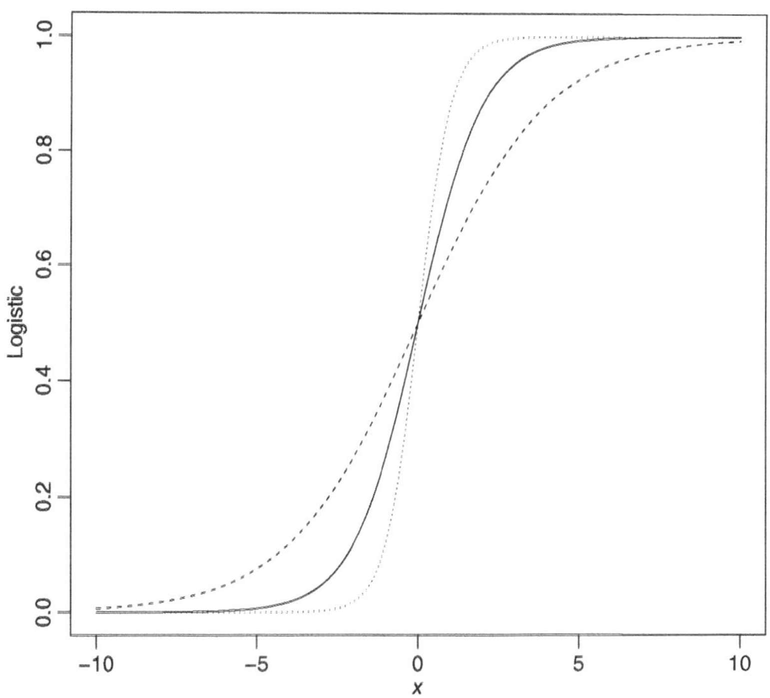

Figure 2.20 Logistic transition functions for $c = 0$ and $\gamma = 0.5$ (dashed), 1.0 (solid), and 2.0 (dotted), respectively.

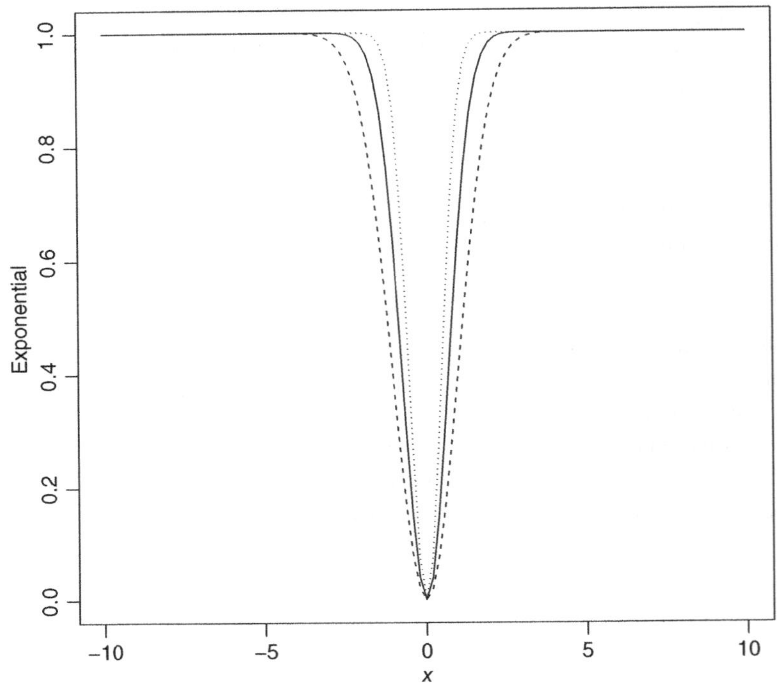

Figure 2.21 Exponential transition functions for $c = 0$ and $\gamma = 0.5$ (dashed), 1.0 (solid), and 2.0 (dotted), respectively.

for $\gamma = 2$. From the plots, it is seen that the transition approaches the step function used in the TAR model when $\gamma \to \infty$. Figure 2.21 shows the three corresponding exponential transition functions. From the plots, the transition approaches an (inverted) singular function at $x = 0$ when $\gamma \to \infty$.

The idea of smooth transition is appealing as abrupt changes in the dynamic dependence of a time series are relatively rare in application. Changes are likely to evolve gradually over time. However, the STAR model often encounters some difficulties in estimating the transition parameters γ and c, regardless of the choice of transition function. See, for instance, the examples used in this section and in Van Dijk et al. (2002). This is so because one seldom has a sufficient number of observations to provide accurate estimates of the transition parameters in real applications. To avoid this difficulty, one may fix a priori one of the transition parameters, e.g. γ.

Example 2.5 Consider the US monthly industrial production (IP) index from January 1919 to February 2016. The data are available from FRED and are

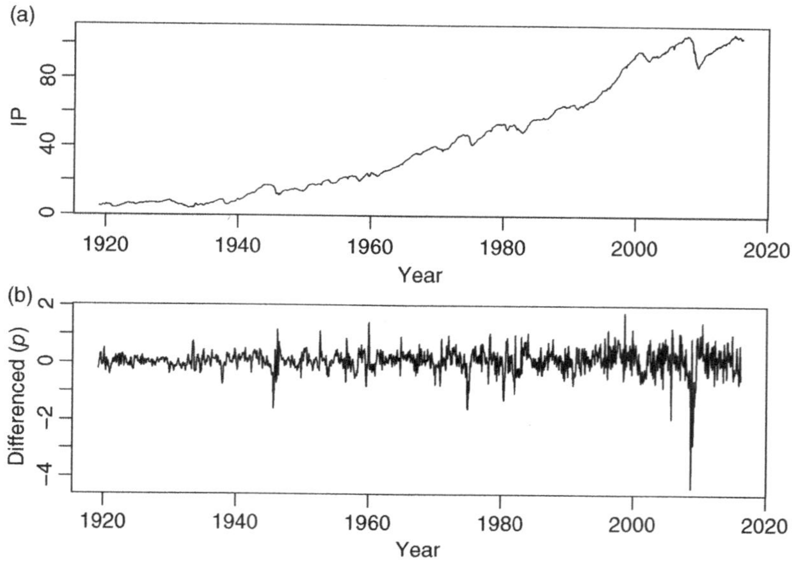

Figure 2.22 Time plots of US monthly industrial production index from January 1919 to February 2016: (a) industrial production index and (b) its first differenced series.

seasonally adjusted. Figure 2.22(a) shows the time plot of the monthly IP index. As expected, an upward trend exists in the series and, hence, we focus on the first differenced series, i.e. the monthly increments in the industrial production index. Let $x_t = p_t - p_{t-1}$, where p_t is the US monthly IP index. Figure 2.22(b) shows the time plot of the x_t series.

Linear time series analysis indicates that the residuals of an ARMA(4,2) model for the x_t series have significant serial correlations at multiple seasonal lags. Therefore, the following seasonal model is used:

$$(1 + 0.39B - 0.50B^2 - 0.19B^3 - 0.18B^4)(1 - 0.45B^{12})(x_t - 0.085)$$
$$= (1 + 0.60B - 0.25B^2)(1 - 0.60B^{12})a_t, \qquad (2.47)$$

where $\sigma_a^2 = 0.134$ and the standard errors of the seasonal AR and MA coefficients are 0.095 and 0.084, respectively. To simplify the analysis, we apply seasonal operator $(1 - 0.45B^{12})/(1 - 0.6B^{12})$ to x_t and denote the seasonally filtered data by y_t, i.e., $y_t = [(1 - 0.45B^{12})/(1 - 0.6B^{12})]x_t$. Figure 2.23 shows the time plot and sample autocorrelation function of the seasonally filtered series y_t. The serial correlations of seasonal lags are no longer statistically significant. Therefore, our analysis focuses on the filtered series y_t. The procedure of seasonal filtering is given in the R demonstration below.

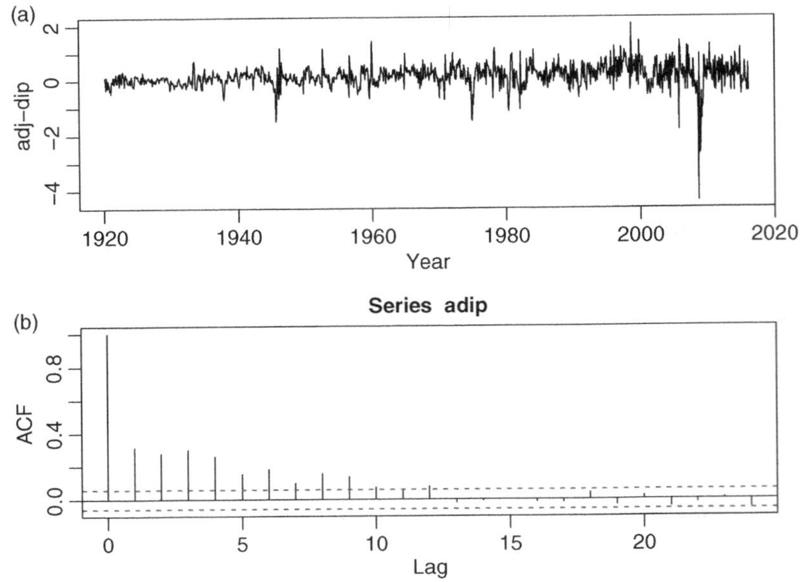

Figure 2.23 Time plot and sample autocorrelation function of the seasonally filtered series of the US monthly industrial production index.

In our analysis, we use the package `tsDyn` to fit LSTAR models. Among several models, the following model seems to fit the y_t series well:

$$\begin{aligned} y_t = {} & (0.31 + 0.69y_{t-1} - 0.45y_{t-2} + 0.23y_{t-3} + 0.27y_{t-4} + 0.25y_{t-5} - 0.25y_{t-6} \\ & - 0.27y_{t-7} + 0.03y_{t-8})[1 - G(y_{t-4}|2.72, -0.49)] \\ & + (-0.35 - 0.65y_{t-1} + 0.72y_{t-2} - 0.07y_{t-3} - 0.25y_{t-4} - 0.19y_{t-5} \\ & + 0.39y_{t-6} + 0.29y_{t-7} + 0.06y_{t-8})G(y_{t-4}|2.72, -0.49) + \epsilon_t \end{aligned} \tag{2.48}$$

where $\sigma_\epsilon^2 = 0.128$ and $G(y_{t-4}|2.72, -0.49) = (1 + \exp[-2.72(y_{t-4} + 0.49)])^{-1}$ is the logistic function in Equation (2.45). The residual ACF of the LSTAR model in Equation (2.48) is shown in Figure 2.24 from which the serial correlations of the residuals are all small, if any. Compared with the linear model in Equation (2.47), the LSTAR model has smaller residual variance. Finally, one can use the fitted LSTAR to produce forecasts as shown in the R demonstration.

R demonstration: STAR model. Output edited to shorten presentation.

```
> da <- read.table("INDPRO.txt",header=T)
> ip <- da[,2]
```

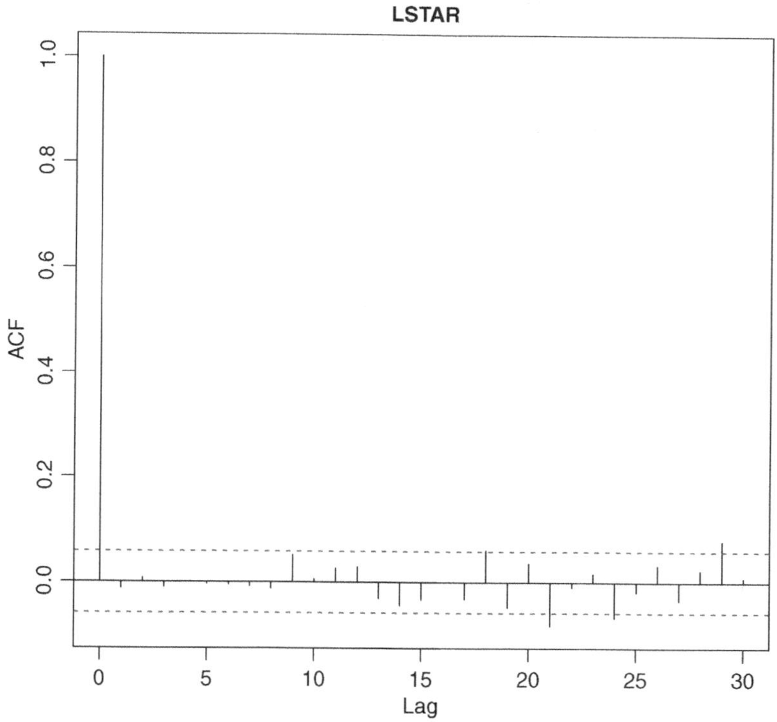

Figure 2.24 Residual autocorrelation function of the LSTAR model in Equation (2.48) for the seasonally filtered series of the US monthly industrial production index.

```
> dip <- diff(ip)
> tdx <- c(1:nrow(da))/12+1919
> par(mfcol=c(2,1))
> plot(tdx,ip,xlab="year",ylab='ip',type='l')
> plot(tdx[-1],dip,xlab="year",ylab='diff(ip)',type='l')
> m4<- arima(dip,order=c(4,0,2),seasonal=list(order=c(1,0,1),period=12))
> m4
Call:arima(x=dip,order=c(4,0,2),seasonal=list(order=c(1,0,1),period=12))

Coefficients:
          ar1     ar2     ar3     ar4     ma1      ma2    sar1     sma1
      -0.3982  0.4951  0.1925  0.1813  0.6016  -0.2501  0.4517  -0.6033
s.e.   0.1210  0.1040  0.0625  0.0420  0.1228   0.1251  0.0949   0.0836
      intercept
         0.0848
s.e.     0.0199
```

```
sigma^2 estimated as 0.1341:  log likelihood = -483.16,  aic = 986.32
> wt <- dip[13:1165]-0.4517*dip[1:1153]
> f1 <- rep(0,12)
> f1[12] <- .6033
> adip <- filter(adip,f1,method="r",ini=rep(0,12))
> plot(tdx[-c(1:13)],adip,xlab='year',ylab='adj-dip',type='l')
> acf(adip,lag=24)
> require(tsDyn)
> n5 <- lstar(adip,m=8,mL=8,mH=8,thDelay=4)
Optimization algorithm converged
Optimized values fixed for regime 2: gamma=2.7217, th=-0.48605; SSE=147.069
> summary(n5)
Non linear autoregressive model
LSTAR model
Coefficients:
Low regime:
   const.L     phiL.1     phiL.2     phiL.3     phiL.4     phiL.5     phiL.6
 0.3115624  0.6870931 -0.4545742  0.2301197  0.2654436  0.2460963 -0.2653893
    phiL.7     phiL.8
-0.2528939  0.0281774

High regime:
   const.H     phiH.1     phiH.2     phiH.3     phiH.4     phiH.5     phiH.6
-0.3474566 -0.6539442  0.7211033 -0.0691959 -0.2485699 -0.1878756  0.2913649
    phiH.7     phiH.8
 0.3911484  0.0579544

Smoothing parameter: gamma = 2.722

Threshold
Variable: Z(t) = + (0)X(t) + (0)X(t-1)+ (0)X(t-2)+ (0)X(t-3)+ (1)X(t-4)
+ (0)X(t-5)+ (0) X(t-6)+ (0) X(t-7)

Value: -0.486

Fit: residuals variance = 0.1276,  AIC = -2334, MAPE = 339.8%

Coefficient(s):
          Estimate  Std. Error  t value Pr(>|z|)
const.L   0.311562    0.249085   1.2508 0.210996
phiL.1    0.687093    0.284827   2.4123 0.015851 *
phiL.2   -0.454574    0.364479  -1.2472 0.212328
phiL.3    0.230120    0.132841   1.7323 0.083222 .
phiL.4    0.265444    0.087511   3.0333 0.002419 **
phiL.5    0.246096    0.181475   1.3561 0.175072
phiL.6   -0.252894    0.112054  -2.2569 0.024015 *
```

```
phiL.7  -0.265389   0.117760  -2.2536 0.024219 *
phiL.8   0.028177   0.082941   0.3397 0.734060
const.H -0.347457   0.312239  -1.1128 0.265799
phiH.1  -0.653944   0.330864  -1.9765 0.048101 *
phiH.2   0.721103   0.396752   1.8175 0.069138 .
phiH.3  -0.069196   0.163429  -0.4234 0.672003
phiH.4  -0.248570   0.136231  -1.8246 0.068058 .
phiH.5  -0.187876   0.185756  -1.0114 0.311820
phiH.6   0.391148   0.131696   2.9701 0.002977 **
phiH.7   0.291365   0.129243   2.2544 0.024171 *
phiH.8   0.057954   0.110205   0.5259 0.598974
gamma    2.721685   1.622500   1.6775 0.093452 .
th      -0.486046   0.306333  -1.5867 0.112590
---
Non-linearity test of full-order LSTAR model against full-order AR model
 F = 6.8874 ; p-value = 7.0957e-09

 > predict(n5,n.ahead=5)
Time Series:
Start = 1166
End = 1170
Frequency = 1
[1] 0.030077216 0.014749998 0.113773221 0.003575012 0.139365484
```

2.5 TIME-VARYING COEFFICIENT MODELS

Time-varying coefficient (TVC) models are perhaps the most commonly used models when weak stationarity conditions or linearity conditions become questionable. They are nonlinear and flexible. As a matter of fact, the way in which the coefficients evolve over time plays a critical role in determining properties of TVC models. In this section we introduce some simple TVC models, discuss their properties, and demonstrate their applications.

2.5.1 Functional Coefficient AR Models

Let $\boldsymbol{S}_t$ denote a random vector observable at time t. In practice, $\boldsymbol{S}_t$ is expected to be related to x_t and often includes lagged values of x_t. A general functional coefficient autoregressive (FAR) model can be written as

$$x_t = \phi_0(\boldsymbol{S}_t) + \sum_{i=1}^{p} \phi_i(\boldsymbol{S}_t)x_{t-i} + \epsilon_t, \tag{2.49}$$

where $\phi_i(\boldsymbol{S}_t)$ is a smooth function of $\boldsymbol{S}_t$ for $i = 0, 1, \ldots, p$ and $\{\epsilon_t\}$ is a sequence of iid random variables with mean zero and variance σ^2. See Chen and Tsay (1993).

This model contains several nonlinear models available in the literature as special cases. For instance, if $\phi_i(\mathbf{S}_t) = a_i + b_i \exp(-c_i x_{t-d}^2)$, then the model becomes the exponential autoregressive model of Haggan and Ozaki (1981), where a_i, b_i and c_i are constants. In addition, if $\phi_i(\mathbf{S}_t) = \phi_{i,1} I(x_{t-d} \leq c) + \phi_{i,2} I(x_{t-d} > c)$, then the model reduces to a TAR model, where $I(s)$ denotes the indicator variable for the statement s. The model itself is a special case of the state-dependent models of Priestley (1980), in which $\mathbf{S}_t$ is the state vector at time t.

Some probabilistic properties of the FAR models in Equation (2.49) have been studied by Chen and Tsay (1993). For instance, if $|\phi_i(\mathbf{S}_t)| \leq c_i$ for $i = 0, \ldots, p$, where c_i is a constant, and the probability density function of ϵ_t is positive everywhere on the real line R^1, then the FAR process x_t in Equation (2.49) is geometrically ergodic provided that all the roots of the characteristic function

$$\lambda^p - c_1 \lambda^{p-1} - \cdots - c_{p-1}\lambda - c_p = 0$$

are inside the unit circle. Furthermore, if $\phi_i(\mathbf{S}_t) = g_i(\mathbf{S}_t) + h_i(\mathbf{S}_t)$ such that both $g_i(\cdot)$ and $h_i(\cdot)$ are bounded functions with $|g_i(\cdot)| < c_i$ and $h_i(\mathbf{S}_t)x_i$ is uniformly bounded, then the aforementioned conditions also imply the geometric ergodicity of the FAR process in Equation (2.49). See Theorems 1.1 and 1.2 of Chen and Tsay (1993). Cai et al. (2000) studied functional-coefficient regression models for nonlinear time series.

FAR are flexible, and the final model used often depends on the domain knowledge of the series under study or prior experience of the analyst. Example 1.3 of Chapter 1 shows that a simple FAR model fares better in transfer function modeling. In this section, we use the filtered monthly industrial production index of Example 2.5 to demonstrate the applicability of FAR models.

Example 2.6 Consider again the monthly US industrial production index from January 1919 to February 2016. Similar to Example 2.5, we use the seasonally filtered series y_t to remove the seasonality remaining in the data. In this particular instance of FAR modeling, we use two possible choices of $\mathbf{S}_t$. In the first case, we use $\mathbf{S}_t = y_{t-1}$ and assume that

$$\phi_i(\mathbf{S}_t) = \phi_i(y_{t-1}) = \beta_i + \gamma_i y_{t-1},$$

where β_i and γ_i are constant parameters. The AIC criterion selects an AR(9) model for y_t, and, hence, our modeling begins with a FAR model of order 9. However, estimation results show that the parameter estimates associated with y_{t-9} are statistically insignificant and we arrive at the model

$$\begin{aligned} y_t = {} & 0.05 + 0.22y_{t-1} + (0.12 + 0.09y_{t-1})y_{t-2} + (0.18 - 0.12y_{t-1})y_{t-3} \\ & + (0.04 - 0.16y_{t-1})y_{t-4} + (0.07 + 0.06y_{t-1})y_{t-6} - (0.07 - 0.16y_{t-1})y_{t-7} \\ & + (0.08 - 0.20y_{t-1})y_{t-8} + \epsilon_t \end{aligned} \tag{2.50}$$

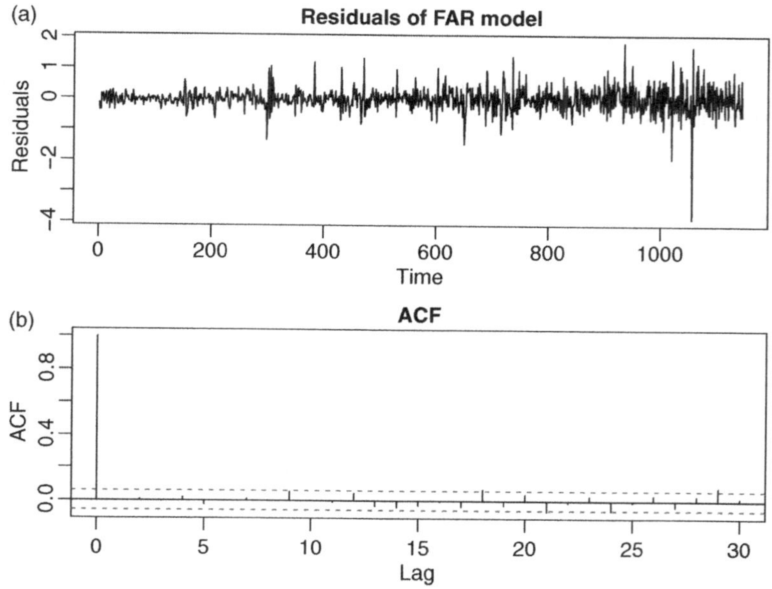

Figure 2.25 Time plot and sample autocorrelation function of the residuals of the functional-coefficient model in Equation (2.50).

where $\hat{\sigma} = 0.3613$ and all coefficient estimates, but 0.04 of y_{t-4} and 0.06 of $y_{t-1}y_{t-6}$, are statistically significant at the 5% level. Figure 2.25 shows the time plot and sample autocorrelation function of the residuals of the model in Equation (2.50). From the plot, the residuals have no significant serial correlations. In fact, the Ljung–Box statistics show Q(24) = 31.02 with p value approximately 0.15.

In the second case, we use $S_t = y_{t-4}$ as suggested by the LSTAR model of the prior section and assume that

$$\phi_i(S_t) = \phi_i(y_{t-4}) = \beta_i + \gamma_i y_{t-4}.$$

After removing insignificant parameter estimates, we obtain the model

$$\begin{aligned} y_t &= 0.05 + (0.18 - 0.24y_{t-4})y_{t-1} + (0.13 + 0.16y_{t-4})y_{t-2} \\ &\quad + (0.16 - 0.12y_{t-4})y_{t-3} + 0.07y_{t-4} + 0.06y_{t-6} - (0.05 - 0.12y_{t-4})y_{t-7} \\ &\quad + 0.06y_{t-8} + \epsilon_t \end{aligned} \tag{2.51}$$

where $\hat{\sigma} = 0.3618$ and all coefficients are statistically significant at the 5% level except for the coefficient −0.05 of y_{t-7}, which has a p value of 0.072. Residual series of the model in Equation (2.51) are similar to those of the model in Equation

(2.50) and have no serial correlations. For instance, the Ljung–Box statistics show $Q(24) = 29.08$ with p value 0.22.

The fitted FAR models in Equations (2.50) and (2.51) belong to bilinear autoregressive models or quadratic AR models. They can be fitted by the command `lm` of the linear model in R provided that the regressors are properly formed. Details are given in the R demonstration. The two fitted FAR models are close to each other. They demonstrate the flexibility of FAR models. To compare the two fitted models, we can use cross-validation because the models can be casted in the linear model framework. For a given model and iteration, the cross-validation procedure is as follows:

1. Select a random subsample from the design matrix, without replacement.
2. Refit the model using the selected subsample.
3. Use the fitted model to predict the remaining data and compute the forecast errors.
4. Compute the root mean squares of forecast errors (RMSE) and the mean absolute forecast errors (MAE).

Repeat the prior procedure for many iterations, and compute the averages of RMSE and MAE over the iterations for model comparison.

For the monthly industrial production data, the cross-validation slightly prefers the FAR model in Equation (2.50). Details are given in the R demonstration.

R demonstration: FAR models. Output edited.

```
da=read.table("INDPRO.txt",header=T)
xt <- diff(da$VALUE)
wt <- xt[13:length(xt)]-0.4517*xt[1:(length(xt)-12)]
f1 <- rep(0,12); f1[12]=0.6033
yt <- filter(wt,f1,method="r",ini=rep(0,12))
ist <- 9; nT <- length(yt)
y <- yt[ist:nT]; X <- NULL
for (i in 1:8){
X <- cbind(X,yt[(ist-i):(nT-i)])}
colnames(X) <- paste("lag'',1:8,sep="'')
Z <- model.matrix(y~lag1*(lag2+lag3+lag4+lag6+lag7+lag8),data=data.frame(X))
m1 <- lm(y~-1+Z)
summary(m1)
Call: lm(formula = y ~ -1 + Z)
Coefficients:
               Estimate Std. Error t value Pr(>|t|)
Z(Intercept)   0.05058    0.01258   4.021 6.18e-05 ***
Zlag1          0.22153    0.03157   7.016 3.92e-12 ***
```

```
Zlag2          0.11574     0.03137    3.689 0.000236 ***
Zlag3          0.18077     0.02997    6.032 2.19e-09 ***
Zlag4          0.04270     0.03127    1.366 0.172360
Zlag6          0.06628     0.03021    2.193 0.028476 *
Zlag7         -0.07221     0.02979   -2.424 0.015520 *
Zlag8          0.08205     0.02975    2.758 0.005917 **
Zlag1:lag2     0.09464     0.03879    2.440 0.014856 *
Zlag1:lag3    -0.12200     0.04488   -2.719 0.006658 **
Zlag1:lag4    -0.15602     0.03090   -5.050 5.15e-07 ***
Zlag1:lag6     0.05693     0.05841    0.975 0.329944
Zlag1:lag7     0.15932     0.04988    3.194 0.001441 **
Zlag1:lag8    -0.19915     0.05522   -3.607 0.000324 ***
---
Residual standard error: 0.3613 on 1131 degrees of freedom
Multiple R-squared:  0.2855,    Adjusted R-squared:  0.2767
F-statistic: 32.28 on 14 and 1131 DF,  p-value: < 2.2e-16
Box.test(m1$residuals,lag=24,type="Ljung")
        Box-Ljung test
data:  m1$residuals
X-squared = 31.201, df = 24, p-value = 0.1481
### Model comparison
> m1f <- cvlm(y,Z,subsize=1000,iter=300)
Average RMSE:  0.1405556 Average MAE:  0.2469571
> m1f <- cvlm(y,Z,subsize=1000,iter=500)
Average RMSE:  0.1389267 Average MAE:  0.2464725
> m1f <- cvlm(y,Z,subsize=1000,iter=1000)
Average RMSE:  0.1386875 Average MAE:  0.2470499
##
Z<-model.matrix(y~lag6+lag8+lag4*(lag1+lag2+lag3+lag7),data=data.frame(X))
m2 <- lm(y~-1+Z)
summary(m2)
Call: lm(formula = y ~ -1 + Z)
Coefficients:
             Estimate Std. Error t value Pr(>|t|)
Z(Intercept)  0.05250    0.01259   4.170 3.28e-05 ***
Zlag6         0.06115    0.02981   2.052 0.040421 *
Zlag8         0.05909    0.02902   2.036 0.041960 *
Zlag4         0.06998    0.03162   2.213 0.027104 *
Zlag1         0.17990    0.02899   6.206 7.62e-10 ***
Zlag2         0.12606    0.02940   4.287 1.96e-05 ***
Zlag3         0.16464    0.03155   5.218 2.15e-07 ***
Zlag7        -0.05460    0.03031  -1.801 0.071906 .
Zlag4:lag1   -0.24395    0.03566  -6.841 1.29e-11 ***
Zlag4:lag2    0.15998    0.04772   3.353 0.000827 ***
Zlag4:lag3   -0.11864    0.03743  -3.169 0.001569 **
Zlag4:lag7    0.12208    0.03289   3.712 0.000216 ***
```

```
---
Residual standard error: 0.3618 on 1133 degrees of freedom
Multiple R-squared:  0.2821,    Adjusted R-squared:  0.2745
Box.test(m2$residuals,lag=24,type="Ljung")
        Box-Ljung test
data:  m2$residuals
X-squared = 29.079, df = 24, p-value = 0.2172
> m2f <- cvlm(y,Z,subsize=1000,iter=300)
Average RMSE:  0.145679 Average MAE:  0.2506342
> m2f <- cvlm(y,Z,subsize=1000,iter=500)
Average RMSE:  0.1438586 Average MAE:  0.2492642
> m2f <- cvlm(y,Z,subsize=1000,iter=1000)
Average RMSE:  0.1423353 Average MAE:  0.2491681
```

2.5.2 Time-varying Coefficient AR Models

A time-varying coefficient AR (tvAR) model can be written as

$$x_t = \phi_{0,t} + \sum_{i=1}^{p} \phi_{i,t} x_{t-i} + \epsilon_t, \tag{2.52}$$

where $\phi_{j,t}$, for $j = 0, \ldots, p$, are random variables and $\{\epsilon_t\}$ is a sequence of iid random variables with mean zero and variance σ^2, and $\{\phi_{j,t}\}$ are independent of ϵ_t. There are many ways to specify the time evolution of the random coefficients $\phi_{j,t}$, resulting in various tvAR models. In this section we assume that each $\{\phi_{j,t}\}$ process follows a random-walk model and $\{\phi_{j,t}\}$ and $\{\phi_{k,t}\}$ are mutually independent if $j \neq k$. The independence assumption between ϕ_{it} is for simplicity. It can be relaxed at the expense of increased complexity in estimation.

A simple way to understand the tvAR model is to cast it in a state space model. Specifically, let $\boldsymbol{\phi}_t = (\phi_{0,t}, \phi_{1t}, \ldots, \phi_{pt})'$ be the $(p+1)$-dimensional vector of random coefficients at time t, and $\boldsymbol{F}_t = (1, x_{t-1}, \ldots, x_{t-p})$ be a $(p+1)$-dimensional row vector of regressors. Then, we have

$$x_t = \boldsymbol{F}_t \boldsymbol{\phi}_t + \epsilon_t, \tag{2.53}$$

$$\boldsymbol{\phi}_t = \boldsymbol{G} \boldsymbol{\phi}_{t-1} + \boldsymbol{w}_t, \tag{2.54}$$

where $\boldsymbol{G}$ is the $(p+1)$-dimensional identify matrix, $\{\epsilon_t\}$ are iid $N(0, \sigma^2)$, $\boldsymbol{w}_t = (w_{0t}, w_{1t}, \ldots, w_{pt})'$ is a $(p+1)$-dimensional random vector such that $\{\boldsymbol{w}_t\}$ forms a sequence of iid Gaussian random vectors with $E(\boldsymbol{w}_t) = \boldsymbol{0}$ and $\text{cov}(\boldsymbol{w}_t) = \boldsymbol{W} = \text{diag}\{\sigma_{ii}^2 | i = 0, \ldots, p\}$, a diagonal matrix. Clearly, the variance σ_{ii}^2 controls the variability of the coefficient ϕ_{it}. In particular, ϕ_{it} reduces to a constant if $\sigma_{ii} = 0$. Kitagawa (2010, Chapter 13) provides a good introduction of the tvAR models via the state space model representation.

Equation (2.53) is refereed to as the *observation equation* and Equation (2.54) is the *state transition* equation with $\boldsymbol{\phi}_t$ being the state vector. The noise ϵ_t becomes the observational noise and $\boldsymbol{w}_t$ is the state transition noise. The initial state $\boldsymbol{\phi}_0$ is a random vector with mean $\boldsymbol{\phi}_*$ and covariance matrix $\boldsymbol{\Sigma}_\phi$. Both $\boldsymbol{\phi}_*$ and $\boldsymbol{\Sigma}_\phi$ are assumed to be known. In this section, we let $\boldsymbol{\phi}_*$ be the least squares estimates of the coefficients of a linear AR(p) model and $\boldsymbol{\Sigma}_\phi$ an arbitrary diagonal matrix. The unknown parameters of the state space model in Equations (2.53) and (2.54) are σ^2 and the diagonal elements of $\boldsymbol{W}$. In application, these parameters can be estimated by the maximum likelihood method. Inference concerning the time-varying coefficients $\boldsymbol{\phi}_t$ can be obtained by *filtering* and *smoothing* of state space model. More specifically, inference of the model can be obtained via the well-known Kalman filter. For further information on the state space model, see Tsay (2010, Chapter 11) and Chapter 6.

A useful special case of the tvAR model is the local level model given by

$$x_t = \phi_t + \epsilon_t, \quad \epsilon_t \sim_{iid} N(0, \sigma_\epsilon^2) \tag{2.55}$$

$$\phi_t = \phi_{t-1} + w_t, \quad w_t \sim_{iid} N(0, \sigma_w^2). \tag{2.56}$$

Here $\boldsymbol{F}_t = 1$ and $\boldsymbol{G} = 1$, and we have $E(x_t \mid \mathcal{F}_{t-1}) = \phi_{t-1}$, where $\mathcal{F}_{t-1}$ denotes the information available at time $t-1$. Therefore, the conditional mean of x_t follows a random walk. The model is also known as an exponential smoothing model because x_t follows an autoregressive integrated moving-average of order (0,1,1) model, i.e. ARIMA(0,1,1). See Tsay (2010, Chapter 11).

Remark: The R package `dlm` can be used to estimate the tvAR model discussed in this section. The command `tvAR` of the `NTS` package makes use of functions of the `dlm` package to perform parameter estimation of the tvAR model. The command `tvARFiSm` also uses the `dlm` package to perform filtering and smoothing of the state variables. In the `tvAR` command, variance parameters are parameterized in log scale.

Example 2.7 Consider the daily realized volatility of Alcoa stock from January 2, 2003 to May 7, 2004 for 340 observations. The realized volatility is obtained by using intraday 10-minute log returns, and the data have been used in Example 11.1 of Tsay (2010). Let x_t denote the logarithm of the realized volatility. Figure 2.26 shows the time plot of x_t (most variable line). In this particular instance, the initial state ϕ_0 is assumed to have $E(\phi_0) = \mu$ and $\text{cov}(\phi_0) = \exp(1)$. The maximum likelihood estimates of σ_ϵ and σ_w are 0.48 and 0.074, respectively. As expected, these estimates are close to the values shown in Example 11.1 of Tsay (2010, Chapter11). Figure 2.26 also shows the filtered and smoothed state ϕ_t of the fitted

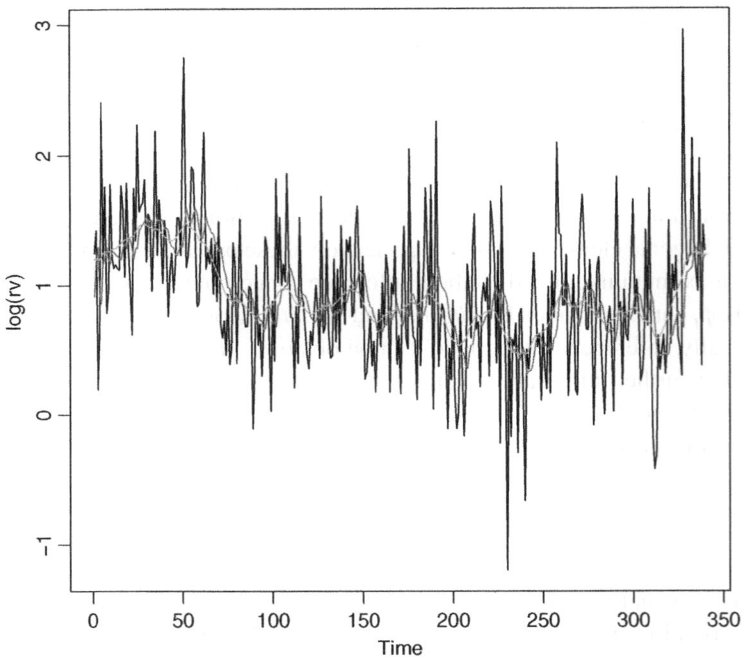

Figure 2.26 Time plot, filtered (medium variability), and smoothed (smooth line) states of the realized volatility of Alcoa stock.

model. From the plots, the tvAR model describes adequately the movement of the realized volatility.

Example 2.8 Consider again the filtered series of the US monthly industrial production index (see Example 2.6). For simplicity, we use a tvAR(4) model in Equation (2.52) for the series y_t. Here $\boldsymbol{\phi}_t$ is a five-dimensional state vector and $\boldsymbol{F}_t = (1, y_{t-1}, \ldots, y_{t-4})$; see the state space model in Equations (2.53) and (2.54). Here the mean of the initial state $\boldsymbol{\phi}_0$ consists of the coefficient estimates of the AR(4) model and the covariance matrix of $\boldsymbol{\phi}_0$ is a diagonal matrix with elements exp(1). The parameter estimates are $\hat{\sigma} = 0.298$ and $\hat{\sigma}_{ii} = 0.025$, 0.249, 0.005, 0.006, and 0.042, respectively, for $i = 0, \ldots, 4$. From the estimates, it is seen that the coefficient of x_{t-1} varies markedly over time. Figure 2.27 shows the time plots of the filtered states whereas Figure 2.28 shows the time plots of the smoothed state variables. In both plots, the upper panel shows ϕ_{0t} and ϕ_{1t} of the states whereas the lower panel shows ϕ_{2t} to ϕ_{4t} of the states. Figure 2.29 shows the time plot of the one-step ahead predicted residuals of the fitted tvAR(4) model. Except for a large outlying value, the predicted residuals look reasonable.

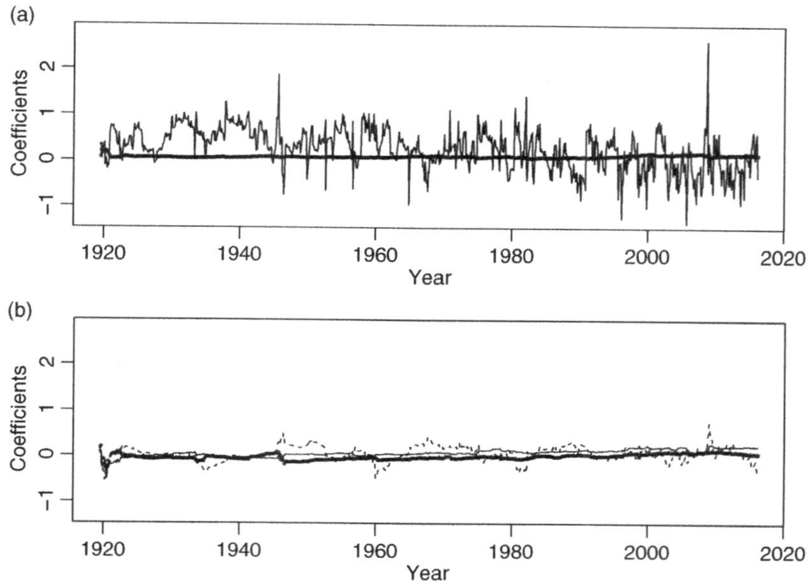

Figure 2.27 Filtered state variables of a tvAR(4) model for the filtered US monthly industrial production index. (a) ϕ_{0t} (bold) and ϕ_{1t}, and (b) ϕ_{2t} (bold), ϕ_{3t} (line), and ϕ_{4t}.

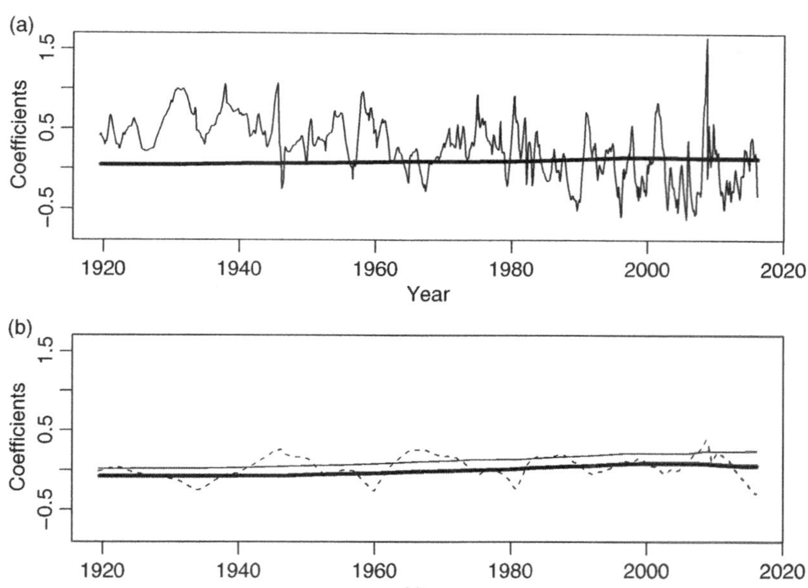

Figure 2.28 Smoothed state variables of a tvAR(4) model for the filtered US monthly industrial production index. (a) ϕ_{0t} (bold) and ϕ_{1t}, and (b) ϕ_{2t} (bold), ϕ_{3t} (line), and ϕ_{4t}.

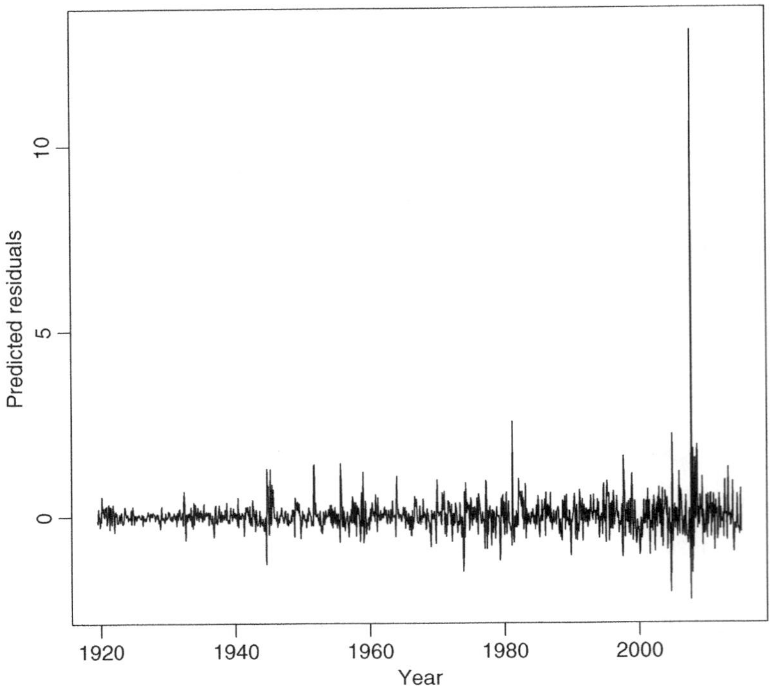

Figure 2.29 Time plot of filtered residuals of a tvAR(4) model for the US monthly industrial production index.

R demonstration: Time-varying AR models. Output edited.

```
> require(NTS); require(dlm)
> da <- read.table("aa-3rv.txt") ## Example 7
> xt <- log(da[,2])
> m1 <- tvAR(xt,lags=NULL)
par: -1.467084 -5.219068
> sqrt(exp(m1$par))
[1] 0.48020505 0.07356883
> m1k <- tvARFiSm(xt,lags=NULL,par=m1$par)
> fil <- m1k$filter; sm <- m1k$smooth
> plot(xt,xlab='time',ylab='log(rv)',type='l')
> lines(fil$m,col="blue"); lines(sm$s,col="red")
###
> adip <- scan("m-adip.txt")
> mk <- tvAR(adip,lags=c(1:4))
par: -2.419878 -7.369506 -2.779889 -10.71095 -10.26296 -6.325027
```

```
> sqrt(exp(mk$par))
[1] 0.298215 0.025103 0.24909 0.00472 0.00591 0.04232
> mkk <- tvARFiSm(adip,lags=c(1:4),par=mk$par)
> names(mkk)
[1] "filter" "smooth"
> filter <- mkk$filter; smooth <- mkk$smooth
> names(filter)
[1] "y"    "mod" "m"    "U.C" "D.C" "a"    "U.R" "D.R" "f"
> resi <- adip[5:1153]-filter$f
> m <- filter$m; s <- smooth$s
> range(m)
[1] -1.265166  2.750762
> range(s)
[1] -0.7726883  1.5985201
> tdx <- c(2:1154)/12+1919
> tdx <- tdx[-c(1:3)]
> par(mfcol=c(2,1))
> plot(tdx,m[,1],xlab='year',ylab='coefs',ylim=c(-1.3,2.8),pch="*",cex=0.5)
> lines(tdx,m[,2],lty=1)
> plot(tdx,m[,3],xlab='year',ylab='coefs',ylim=c(-1.3,2.8),pch="*",cex=0.5)
> lines(tdx,m[,4])
> lines(tdx,m[,5],lty=2)
> plot(tdx,s[,1],xlab='year',ylab='coefs',ylim=c(-0.8,1.61),pch="*",cex=0.5)
> lines(tdx,s[,2],lty=1)
> plot(tdx,s[,3],xlab='year',ylab='coefs',ylim=c(-0.8,1.61),pch="*",cex=0.5)
> lines(tdx,s[,4])
> lines(tdx,s[,5],lty=2)
```

Remark: Consider the tvAR model in Equation (2.52). For simplicity, we assume $\phi_{0t} = 0$ for all t, that is, x_t is mean-adjusted. If one assumes that each AR coefficient ϕ_{it} follows a random distribution, e.g. $\phi_{it} \sim N(\phi_i, \sigma_{ii}^2)$, and the distributions are mutually independent, then the tvAR model reduces to a random-coefficient AR (RCA) model. In this particular case, let $\eta_{it} = \phi_{it} - \phi_i$ for $i = 1, \ldots, p$, then the model becomes

$$x_t = \sum_{i=1}^{p} \phi_i x_{t-i} + (x_{t-1}, \ldots, x_{t-p})\boldsymbol{\eta}_t + \epsilon_t,$$

where $\boldsymbol{\eta}_t = (\eta_{1t}, \ldots, \eta_{pt})'$ is a random vector satisfying $E(\boldsymbol{\eta}_t) = \mathbf{0}$ and $\text{cov}(\boldsymbol{\eta}_t) = \text{diag}\{\sigma_{ii}^2 | i = 1, \ldots, p\}$. Let $\boldsymbol{F}_{t-1}$ denote the information available at time $t-1$, then we

$$E(x_t \mid \boldsymbol{F}_{t-1}) = \sum_{i=1}^{p} \phi_i x_{t-i}, \quad \text{cov}(x_t \mid \boldsymbol{F}_{t-1}) = \sigma^2 + \sum_{i=1}^{p} x_{t-i}^2 \sigma_{ii}^2.$$

Consequently, the model has conditional heteroscedasticity. For further relationship between RCA models and autoregressive conditional heteroscedastic (ARCH) models, readers are referred to Tsay (1987) and the references therein.

Example 2.9 To demonstrate the RCA modeling, we consider again the filtered US monthly industrial production index of Example 2.8. The sample mean of the y_t series is 0.117 and $\tilde{y}_t = y_t - 0.117$ is the mean-adjusted series. We fit an RCA(4) model to the data and obtain

$$\tilde{y}_t = 0.22\tilde{y}_{t-1} + 0.17\tilde{y}_{t-2} + 0.18\tilde{y}_{t-3} + 0.04\tilde{y}_{t-4} + (\tilde{y}_{t-1}, \tilde{y}_{t-2}, \tilde{y}_{t-3}, \tilde{y}_{t-4})\boldsymbol{\eta}_t + \epsilon_t, \tag{2.57}$$

where $\boldsymbol{\eta}_t$ is a sequence of iid Gaussian random vectors with mean zero and covariance matrix $\text{cov}(\boldsymbol{\eta}_t) = \text{diag}\{0.269, 0.203, 0.119, 0.063\}$, and σ^2 of ϵ_t is 0.041. Figure 2.30 shows the time plot of standardized residuals of the fitted RCA(4) model in Equation (2.57) and the sample autocorrelation function of the squared standardized residuals. From the plot and ACF, the fitted model seems reasonable.

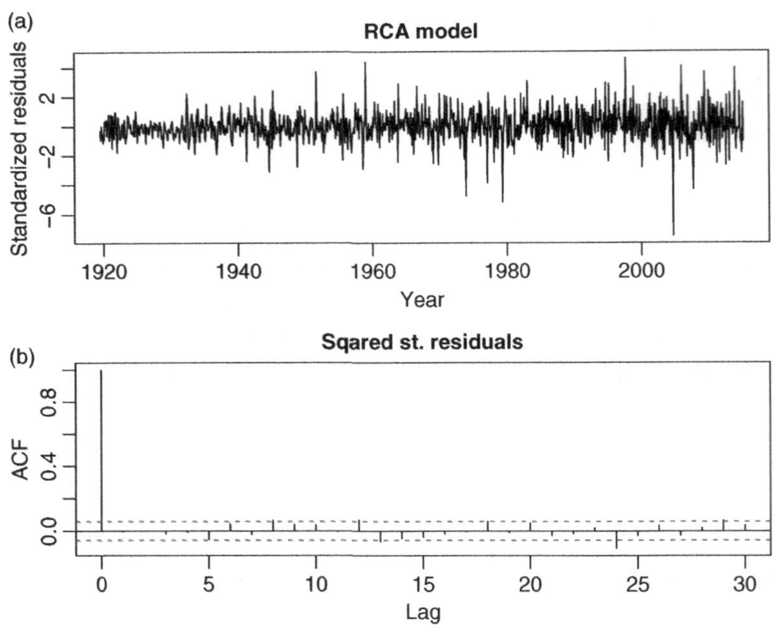

Figure 2.30 Time plot of standardized residuals of (a) the random-coefficient AR(4) model in Equation (2.57) and (b) the sample autocorrelation function of the squared standardized residuals.

In fact, the conditional heteroscedasticity is well modeled by the RCA(4) model. For instance, the Ljung–Box statistics of the squared standardized residuals show Q(12) = 6.61 with p value 0.88.

R demonstration: RCA model. Output edited.

```
> mm <- rcAR(adip,lags=c(1:4)) # NTS package
Sample mean:  0.1166405
Estimates:
        est se.est   tratio
X1  0.2216 0.0379   5.8515
X2  0.1653 0.0377   4.3801
X3  0.1811 0.0354   5.1228
X4  0.0441 0.0327   1.3476
   -1.3136 0.1563  -8.4028
   -1.5962 0.2426  -6.5794
   -2.1289 0.3376  -6.3050
   -2.7627 0.4419  -6.2512
   -3.1882 0.1098 -29.0300
> names(mm)
[1] "par"          "se.est"      "residuals"   "sresiduals"
> exp(mm$par[5:9])
0.26885490 0.20265937 0.11896990 0.06312199 0.04124666
> Box.test(mm$sresiduals^2,lag=12,type='Ljung')
        Box-Ljung test
data:  mm$sresiduals^2
X-squared = 6.616, df = 12, p-value = 0.8819
```

2.6 APPENDIX: MARKOV CHAINS

In this appendix we briefly introduce some important concepts of Markov chains. The discussion is restricted to finite-state Markov chains $\{S_t, t = 1, \dots\}$, where S_t assumes values in a finite set. The concepts can easily be extended to general spaces.

Let $S = \{s_1, s_2, \dots, s_k\}$ be a collection of k possible values. We refer to S as the state space, and each element s_i of S is called a state. Let $\boldsymbol{P}$ be a $k \times k$ real-valued matrix defined by

$$\boldsymbol{P} = \begin{bmatrix} p_{11} & p_{12} & p_{13} & \cdots & p_{1k} \\ p_{21} & p_{22} & p_{23} & \cdots & p_{2k} \\ \vdots & \vdots & \vdots & \ddots & \vdots \\ p_{k1} & p_{k2} & p_{k3} & \cdots & p_{kk} \end{bmatrix},$$

where $p_{ij} \geq 0$ and $\sum_{j=1}^{k} p_{ij} = 1$ for a fixed i. Consider a sequence of random variables $\{S_0, S_1, \dots\}$, where S_i assumes a value in the state space S with S_0 being the initial state. This sequence forms a Markov chain if the value of S_{t+1} is determined by the following transition equation

$$P(S_{t+1} = s_j \mid S_0 = i_0, \dots, S_{t-1} = i_{t-1}, S_t = s_i) = P(S_{t+1} = s_j \mid S_t = s_i) = p_{ij},$$

where $i_j \in S$ for $j = 0, \dots, t-1$. More precisely, such a sequence $\{S_t\}$ is called a (time-homogeneous) Markov chain with transition probability $\boldsymbol{P}$ on the state space S. If k is finite, we have a finite-state Markov chain. From the definition, S_{t+1} assumes a value in S according to the transition probability $\boldsymbol{P}$ and the value of S_t. The history $\{S_0, \dots, S_{t-1}\}$ plays no role in determining S_{t+1} once S_t is given. This feature is referred to as the *memorylessness* of Markov chains. Since $\boldsymbol{P}$ is fixed, we have

$$\begin{aligned} P(S_{t+2} = s_j \mid S_t = s_i) &= \sum_{v=1}^{k} P(S_{t+1} = s_v \mid S_t = s_i) P(S_{t+2} = s_j \mid S_{t+1} = s_v) \\ &= \sum_{v=1}^{k} p_{iv} p_{vj}, \end{aligned}$$

which is the (i,j)th element of the matrix $\boldsymbol{P}^2$. This feature applies to any power of the matrix $\boldsymbol{P}$ with $\boldsymbol{P}^0 = \boldsymbol{I}$, the $k \times k$ identity matrix. Following the literature, we use th notation $p_{ij}^{(n)}$ to denote the (i,j)th element of $\boldsymbol{P}^n$. Properties of the Markov chain are thus determined by the transition probability matrix $\boldsymbol{P}$. For instance, we are often interested in knowing the stationary distribution of S_t as $t \to \infty$ for a given $\boldsymbol{P}$.

Two important properties of Markov chains are *irreducibility* and *aperiodicity*. The key result is that an aperiodic and irreducible Markov chain has a stationary distribution. See below for further details. The state s_j is said to be *accessible* from state s_i if the chain started in state s_i can transition into state s_j in finite steps. Mathematically, state s_j is accessible from state s_i if there exists a nonnegative integer n, which may depend on s_i and s_j, such that

$$P(S_n = s_j \mid S_0 = s_i) = p_{ij}^{(n)} > 0.$$

We use the notation $s_i \to s_j$ to denote that s_j is accessible from s_i. If $s_i \to s_j$ and $s_j \to s_i$ hold simultaneously, then s_i and s_j communicate with each other (denoted by $s_i \leftrightarrow s_j$). A subset C of S is a communicating class if every pair of states in C communicates with each other. A communicating class C is said to be *close* if the probability of leaving the class is zero. In other words, if C is close and $s_i \in C$, but $s_j \notin C$, then s_j is not accessible from s_i. A Markov chain is said to be *irreducible* if its state space S forms a single communicating class. In other words, the Markov chain is irreducible if it is possible to get to any state s_j from any state s_i.

The period ℓ_i of state s_i of a Markov chain is defined as

$$\ell_i = \gcd\{n > 0 \ s.t. \ P(S_n = s_i \mid S_0 = s_i) > 0\}, \tag{2.58}$$

where "gcd" stands for the greatest common divisor, provided that the set is not empty. The period ℓ_i is undefined if the set in Equation (2.58) is empty. Since the period ℓ_i is defined in term of gcd, state s_i may not be reached in ℓ_i steps. For example, if s_i appears in steps $\{4, 6, 8, 10, 12, 14, \cdots\}$, then we have $\ell_i = 2$, but S_2 did not assume the value s_i. The state s_i is said to be *aperiodic* if $\ell_i = 1$. A Markov chain is *aperiodic* if every state is aperiodic. An irreducible Markov chain only needs one state to be aperiodic to imply that it is aperiodic.

A state s_i of a Markov chain is said to be an *absorbing state* if it is not possible to leave s_i. Mathematically, we have $p_{ii} = 1$ and $p_{ij} = 0$ for $i \neq j$. If every state can reach an absorbing state, then the Markov chain is an absorbing chain.

A state s_i is said to be *transient* if, given that the chain starts in state s_i, there is a non-zero probability that the chain will never return to s_i. To be more precise, define the random variable

$$T_i = \inf\{n \geq 1 : S_n = s_i | S_0 = s_i\}$$

which is called the *hitting time* of state s_i. The quantity

$$h_{ii}^{(n)} = P(T_i = n)$$

is the probability that the chain returns to state s_i for the first time after n steps. Then, state s_i is transient if

$$P(T_i < \infty) = \sum_{n=1}^{\infty} h_{ii}^{(n)} < 1.$$

A state s_i is *recurrent* if it is not transient. If s_i is recurrent, then $P(T_i < \infty) = 1$ so that it has a finite hitting time with probability 1. It can be shown that state s_i is recurrent if and only if

$$\sum_{n=0}^{\infty} p_{ii}^{(n)} = \infty.$$

Both transience and recurrence are class properties in the sense that every state in a communicating class either holds that property or does not hold that property.

The mean recurrence time of state s_i is defined as

$$M_i = E(T_i) = \sum_{n=1}^{\infty} n h_{ii}^{(n)}.$$

State s_i is *positive recurrent* if M_i is finite. Otherwise, s_i is *null recurrent*.

A state s_i of a Markov chain is said to be *ergodic* if it is aperiodic and positive recurrent. In other words, s_i is ergodic if it is recurrent, has a period of 1, and has a finite mean recurrence time. If all states in an irreducible Markov chain are ergodic, then the chain is ergodic. It can be shown that a finite-state irreducible Markov chain is ergodic if it has an aperiodic state.

Consider a probability distribution $\boldsymbol{\pi} = (\pi_1, \ldots, \pi_k)$ over the state space S. This distribution is called a stationary distribution of a Markov chain with time-invariant transition probability matrix $\boldsymbol{P} = [p_{ij}]$, if (a) $\pi_i \geq 0$, (b) $\sum_{i=1}^{k} \pi_i = 1$, and (c)

$$\pi_j = \sum_{i=1}^{k} \pi_i p_{ij}, \quad j = 1, \ldots, k.$$

Treating $\boldsymbol{\pi}$ as a row vector, the prior condition can be written as $\boldsymbol{\pi} = \boldsymbol{\pi P}$. This means that the stationary distribution of a Markov chain, if it exists, is a normalized left-eigenvector of the transition matrix $\boldsymbol{P}$ associated with eigenvalue 1. Of course, one can use

$$\boldsymbol{P}'\boldsymbol{\pi}' = \boldsymbol{\pi}'$$

so that $\boldsymbol{\pi}'$ is an eigenvector of the matrix $\boldsymbol{P}'$ associated with unit eigenvalue.

An irreducible Markov chain has a stationary distribution if and only if all of its states are positive recurrent. In this case, $\boldsymbol{\pi}$ is unique and is related to the mean expected return time M_i via

$$\pi_i = \frac{C}{M_i},$$

where C is a normalization constant. Furthermore, if the positive recurrent chain is both irreducible and aperiodic, then it has a limiting distribution

$$\lim_{n\to\infty} p_{vi}^{(n)} = \frac{C}{M_i},$$

for any states s_i and s_j. For further information on Markov chains, see Karlin and Taylor (1975).

2.7 EXERCISES

Again, all tests use the 5% type-I error.

2.1 Use the following commands to simulate a two-regime TAR(1) model with 400 observations.

```
require(NTS)
set.seed(1)
phi <- matrix(c(0.8,-0.7),2,1)
m1 <- uTAR.sim(400,c(1,1),phi)
xt <- m1$series
```

(a) Obtain a time plot for the data and its sample autocorrelation function with 12 lags.

(b) Obtain a scatter plot of x_t versus x_{t-1} and draw a smooth line on the plot using the `loess` local smoothing.

(c) Apply the threshold tests of Chapter 1 to confirm the threshold nonlinearity of the time series.

(d) Finally, build a two-regime TAR model for the series and write down the fitted model.

2.2 Consider the monthly mean duration of unemployment in the USA. The data were seasonally adjusted from January 1948 to August 2017. They are available from FRED and are also in the file `Unempduration.csv`. The mean duration shows an upward trend so let x_t be the first difference of the original data.

(a) Build a linear AR model for x_t. Write down the fitted model and perform model checking.

(b) Perform the threshold tests of Chapter 1 to confirm that x_t exhibits threshold nonlinearity.

(c) Use $p = 5$ and $d \in \{1, \dots, 5\}$, where d denotes the delay, to select the threshold variable x_{t-d}. You may use the p value of the threshold test to select d.

(d) Build a two-regime TAR(5) model for x_t. Perform model checking and write down the fitted model.

(e) Compare the linear AR model and the TAR model.

2.3 Consider the monthly value-weighted SMALL-LoBM portfolio returns of Fama-French data, denoted by x_t. See the web: mba.tuck.dartmonth.edu/pages/faculty/ken.french/data_library.html. Also, consider the monthly Fama-French three factors, also available from the same web. Denote the factors by Mrk.RF, SMB, and HML, respectively. The data span is from July 1926 to July 2017 for 1903 observations. See also the files `F-F_Factors.csv` and `F-F_6_Portfolios.csv`.

(a) Fit the linear regression model:

$$x_t = \beta_0 + \beta_1 \text{Mrk.RF}_t + \beta_2 \text{SMB}_t + \beta_3 \text{HML}_t + \epsilon_t.$$

Write down the model and perform model checking.

(b) Fit a two-state Markov switching model to the data and write down the fitted model.

(c) Compare the two fitted models in (a) and (b).

2.4 Again, consider the change series of the US monthly unemployment duration from February 1948 to August 2017. Fit a two-state Markov switching autoregressive model to the series. Perform model checking and write down the fitted model. Compare the Markov switching model with the threshold model built previously. Comment on the two models.

2.5 Again, consider the change series of the US monthly unemployment duration from February 1948 to August 2017. Fit a LSTAR model to the series. Perform model checking and write down the fitted model. Use the fitted model to produce one-step to five-step ahead predictions for x_t at the forecast origin August 2017.

2.6 Again, consider the change series of the US monthly unemployment duration from February 1948 to August 2017. Fit a time-varying coefficient AR model to the data. Perform model checking and write down the fitted model. Plot the filtered and smooth state variables.

REFERENCES

Akaike, H. (1974). A new look at the statistical model identification. *IEEE Transactions on Automatic Control* **AC-19**: 716–723.

Cai, Z., Fan, J., and Yao, Q. (2000). Functional-coefficient regression models for nonlinear time series. *Journal of the American Statistical Association* **95**: 941–956.

Cappé, Q., Moulines, E., and Rydén, T. (2005). *Inference in Hidden Markov Models*. Springer Series in Statistics. Springer, New York.

Chan, K. S. (1993). Consistency and limiting distribution of the least squares estimator of a threshold autoregressive model. *The Annals of Statistics* **21**: 520–533.

Chan, K. S. and Tsay, R. S. (1998). Limiting properties of the conditional least squares estimator of a continuous TAR model. *Biometrika* **85**: 413–426.

Chen, R. and Tsay, R. S. (1991). On the ergodicity of TAR(1) processes. *Annals of Applied Probability* **1**: 613–634.

Chen, R. and Tsay, R. S. (1993). Functional-coefficient autoregressive models. *Journal of the American Statistical Association* **88**: 298–308.

Chib, S. (1996). Calculating posterior distributions and modal estimates in Markov mixture models. *Journal of Econometrics* **75**: 79–97.

Francq, C. and Zakoian, J. (2001). Stationarity of multivariate Markov switching ARMA models. *Journal of Econometrics* **102**: 339–364.

Frühwirth-Schnatter, S. (2006). *Finite Mixture and Markov Switching Models*. Springer, New York.

Haggan, V. and Ozaki, T. (1981). Modeling nonlinear vibrations using an amplitude-dependent autoregressive time series model. *Biometrika* **68**: 189–196.

Hamilton, J. D. (1989). A new approach to the economic analysis of nonstationary time series and the business cycle. *Econometrica* **57**: 357–384.

Hamilton, J. D. (1990). Analysis of time series subject to changes in regime. *Journal of Econometrics* **45**: 39–70.

Hamilton, J. D. (1994). *Time Series Analysis*. Princeton University Press, Princeton, NJ.

Holst, U., Lindgren, G., Holst, J., and Thuvesholmen, M. (1994). Recursive estimation in switching autoregressions with a Markov regime. *Journal of Time Series Analysis* **15**: 489–506.

Hyndman, R. J. (2016). Time Series Data Library. http://data.is/TSDLdemo. Accessed March, 2016.

Karlin, S. and Taylor, H. M. (1975). *A First Course in Stochastic Processes*, 2nd edition. Academic Press, San Diego.

Kitagawa, G. (2010). *Introduction to Time Series Modeling*. CRC Press, Chapman and Hall, Boca Raton, FL.

Krolzig, H. M. (1997). *Markov Switching Vector Autoregressions: Modeling, Statistical Inference, and Applications to Business Cycle Analysis*. Lecture Notes in Economics and Mathematical Systems. Springer, New York.

Ling, S. and Tong, H. (2005). Testing for a linear MA model against threshold MA models. *Annals of Statistics* **33**: 2529–2552.

McCulloch, R. E. and Tsay, R. S. (1994). Statistical inference of macroeconomic time series via Markov switching models. *Journal of Time Series Analysis* **15**: 523–539.

Meyn, S. and Tweedie, R. (2009). *Markov Chains and Stochastic Stability*, 2nd edition. Cambridge University Press, Cambridge.

Neftci, S. N. (1984). Are economic time series asymmetric over the business cycles? *Journal of Political Economy* **92**: 307–328.

Petruccelli, J. and Woolford, S. W. (1984). A threshold AR(1) model. *Journal of Applied Probability* **21**: 270–286.

Priestley, M. B. (1980). State-dependent models: a general approach to nonlinear time series analysis. *Journal of Time Series Analysis* **1**: 47–71.

Teräsvirta, T. (1994). Specification, estimation, and evaluation of smooth transition autoregressive models. *Journal of the American Statistical Association* **89**: 208–218.

Timmermann, A. (2000). Moments of Markov switching models. *Journal of Econometrics* **96**: 75–111.

Tong, H. (1978). On a threshold model. In C.H. Chen (ed.), *Pattern Recognition and Signal Processing*. Sijhoff & Noordhoff, Amsterdam.

Tong, H. (1990). *Non-Linear Time Series: A Dynamical System Approach*. Oxford University Press, Oxford.

Tong, H. and Lim, K. S. (1980). Threshold autoregression,limit cycles and cyclical data (with discussion). *Journal of the Royal Statistical Society, Series B* **42**: 245–292.

Tsay, R. S. (1987). Conditional heteroscedastic time series models. Processes. *Journal of the American Statistical Association* **82**: 590–604.

Tsay, R. S. (1989). Testing and modeling threshold autoregressive processes. *Journal of the American Statistical Association* **84**: 231–240.

Tsay, R. S. (1998). Testing and modeling multivariate threshold models. *Journal of the American Statistical Association* **93**: 1188–1202.

Tsay, R. S. (2010). *Analysis of Financial Time Series*, 3rd edition. John Wiley, Hoboken, NJ.

Van Dijk, D., Teräsvirta, T., and Frances, F. P. (2002). Smooth transition autoregressive models: a survey of recent developments. *Econometric Reviews* **21**: 1–47.

Wu, S. and Chen, R. (2007) Threshold variable selection and threshold variable driven switching autoregressive models. *Statistica Sinica* **17**: 241–264.

Yao, J. and Attali, J. (2000). On stability of nonlinear AR processes with Markov switching. *Advances in Applied Probabilities* **32**: 394–407.

CHAPTER 3

UNIVARIATE NONPARAMETRIC MODELS

In this chapter we introduce nonparametric modeling of univariate time series. We start with kernel smoothing, local polynomial methods, B-splines, and smoothing splines, and use the methods discussed to explore nonlinearity in a time series and to introduce nonlinear additive autoregressive (NAAR) models. The nonparametric approach to time series analysis increases the flexibility in modeling nonlinearity embedded in the data. We also introduce some wavelet methods, including thresholding, and discuss their applications in time series analysis. Finally, we consider index models and sliced inverse regression for time series analysis. For a review of nonparametric time series analysis, see Härdle et al. (1997). Several R packages are used in this chapter in empirical analysis.

3.1 KERNEL SMOOTHING

The kernel method has a long history in statistics, dating back to Nadaraya (1964) and Watson (1964). Consider a dependent variable Y of interest and an independent

Nonlinear Time Series Analysis, First Edition. Ruey S. Tsay and Rong Chen.

Companion website: www.wiley.com/go/tsay/nonlineartimeseries

variable X. The model is $Y = m(X) + \epsilon$, where ϵ is the error term with zero mean and a variance that may depend on X. Denote the conditional mean and variance of Y given $X = x$ as

$$E(Y \mid X = x) = m(x), \quad \text{Var}(Y \mid X = x) = \sigma^2(x). \tag{3.1}$$

Many statistical applications are concerned with estimating the mean function $m(x)$ and the variance function $\sigma^2(x)$ under some regularity conditions. Assume that a random sample $\{(y_1, x_1), \ldots, (y_n, x_n)\}$ of n observations is available. The well-known Nadaraya–Watson estimator of the mean function $m(x)$ is given by

$$\hat{m}_h(x) = \frac{\sum_{i=1}^n K_h(x_i - x)y_i}{\sum_{i=1}^n K_h(x_i - x)} \equiv \sum_{i=1}^n w_i(x)y_i \tag{3.2}$$

where $K_h(\cdot) = K(./h)/h$, $K(\cdot)$ is a kernel function, which is usually a symmetric probability density, h is the *bandwidth* (or *smoothing parameter*), which is a nonnegative real number controlling the size of the local neighborhood of x. The denominator of Equation (3.2) is used for normalization purpose so that the weights $w_i(x)$ sum to one. Since the kernel $K(\cdot)$ is a density function, we have $K(x) \geq 0$ and $\int K(z)dz = 1$. In Equation (3.2), we have

$$w_i(x) = \frac{K_h(x_i - x)}{\sum_{i=1}^n K_h(x_i - x)}. \tag{3.3}$$

Thus, the Nadaraya–Watson estimator is a linear estimator and uses a weighted average of y_i to estimate the conditional mean function at $X = x$ with more weights given to y_i if x_i is closer to x.

The commonly used kernels include the Gaussian kernel $K(z) = \frac{\exp(-z^2/2)}{\sqrt{2\pi}}$ and the symmetric Beta family,

$$K(z) = \frac{1}{\text{Beta}(1/2, \gamma + 1)}(1 - z^2)_+^\gamma, \quad \gamma = 0, 1, 2, \ldots \tag{3.4}$$

where $(\cdot)_+$ denotes the positive part of the given function. The choices of $\gamma = 0$, 1, 2, and 3 result in using the *uniform*, the *Epanechnikov* (Epanechnikov, 1969), the *bi-weight* and the *tri-weight* kernel functions, respectively. Note that Beta(0.5,γ + 1) = 2, 4/3, 16/15, 32/35 for $\gamma = 0, \ldots, 3$, respectively. Figure 3.1 shows the plots of Gaussian kernel (dashed), uniform kernel (horizontal), Epanechnikov kernel (dotted), bi-weight kernel, and tri-weight kernel (solid), respectively. The Gaussian kernel is not re-scaled to the domain $[-1, 1]$. If one re-scales the domain to $[-1, 1]$, then the Gaussian kernel becomes the limiting curve of the symmetric Beta family in Equation (3.4) as $\gamma \to \infty$.

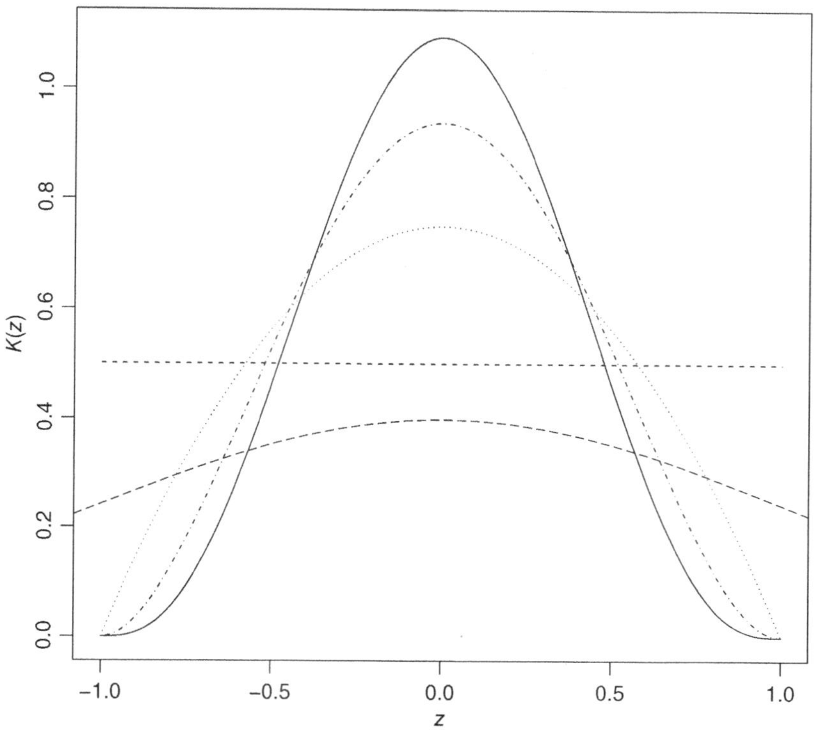

Figure 3.1 Plots of commonly used kernel functions, including Gaussian (dashed), uniform, Epanechnikov (dotted), bi-weight, and tri-weight (solid).

Properties of kernel estimators of the mean function $\hat{m}(x)$ have been investigated in the literature. See, for instance, Härdle (1990). The criteria used to measure the goodness of fit of a Kernel estimator are the mean squared error (MSE) and the mean integrated squared error (MISE). The MSE is defined as

$$\text{MSE}(x) = E[\{\hat{m}(x) - m(x)\}^2 \mid \boldsymbol{x}],$$

where $\boldsymbol{x} = (x_1, x_2, \ldots, x_n)$. This criterion is used if the objective of the analysis is to estimate the mean function at the point x. The MISE is defined as

$$\text{MISE} = \int \text{MSE}(x)w(x)dx,$$

where $w(\cdot) \geq 0$ is a weight function. From its definition, MISE is used if the objective of the estimation is to recover the whole function $m(x)$. The choice of bandwidth h also plays a role in applications. It controls the smoothness of the estimated $m(x)$ and the trade off between bias and variance of the estimator. There are various proposals available in the literature to choose the bandwidth for a given random sample, including the plug-in method and cross-validation. The most commonly used cross-validation method is the leave-one-out procedure. The criterion is

$$\mathrm{CV}(h) = \frac{1}{n}\sum_{j=1}^{n}[y_i - \hat{m}_{h(j)}(x_j)]^2,$$

where the subscript (j) indicates that the jth data point (x_j, y_j) is left out of the fitting. One selects the h that minimizes this CV(h) criterion. For simplicity, we often use the default options in R packages to select the bandwidth and the kernel function.

It is computationally expensive to calculate CV(h) on a dense grid of h. Often the generalized cross-valuation (GCV) criterion is used. Specifically,

$$\mathrm{GCV}(h) = \frac{\sum_{j=1}^{n}[y_j - \hat{m}_h(x_j)]^2}{n[1 - \mathrm{tr}(S(h))/n]^2},$$

where $S(h)$ is an $n \times n$ matrix with the (i,j)th element being $w_j(x_i, h)$ defined in Equation (3.3), $\hat{m}_h(x_j)$ denotes the fitted value of the function $m(x_j)$ using bandwidth h. This is computationally much faster since $\hat{m}_h(\cdot)$ and tr$(S(h))$ are evaluated using all observations.

Another kernel estimator available is the Gasser–Müller estimator, defined as

$$\hat{m}_h(x) = \sum_{i=1}^{n}\left[\int_{s_{i-1}}^{s_i} K_h(u-x)du\right] y_i, \tag{3.5}$$

where $s_i = (x_i + x_{i+1})/2$, $x_0 = -\infty$, and $x_{n+1} = \infty$. A special feature of this estimator is that the weights are summed to one and, hence, no denominator is needed. The estimator is a modification of the one proposed by Priestley and Chao (1972). Properties of the Gasser–Müller estimator are studied in Müller (1988). From the definitions, the weights of the kernel estimator in Equation (3.2) are proportional to the heights of re-scaled kernel whereas those of Gasser–Müller estimator in Equation (3.5) are determined by the areas under the kernel function and bounded by the mid-points of x_i close to x.

Remark: One can use the command `ksmooth` in R to obtain the Nadaraya–Watson estimator. However, the kernel functions available are uniform and Gaussian only. The command `hcv` of the package `sm` uses cross-validation to select bandwidth.

Example 3.1 A widely used example in the statistical literature for nonparametric analysis is the Old Faithful data from Yellowstone National Park. The data set is available from R and consists of two variables, namely the waiting time (Y) and

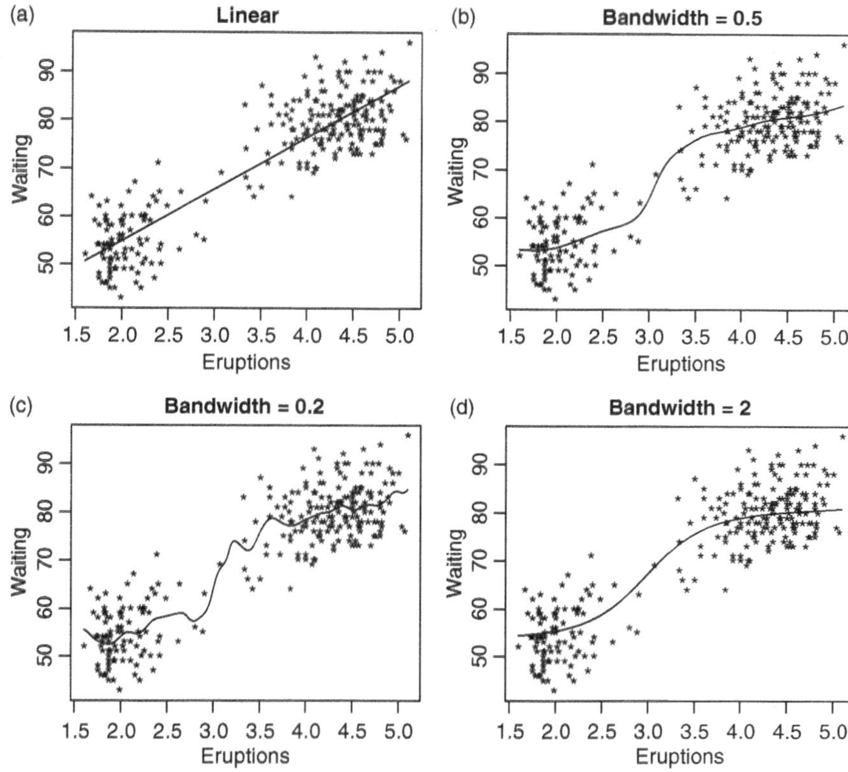

Figure 3.2 (a) Linear model and (b–d) Nadaraya–Watson estimators with different choices of bandwidth for the Old Faithful data.

duration of an eruption (X), with 272 observations. We use the data to demonstrate kernel smoothing of the model

$$y_i = m(x_i) + \epsilon_i,$$

where $\{\epsilon_i\}$ is an iid sequence of random noises with mean zero and homoscedastic variance σ^2. Figure 3.2(a) shows the scatter plot of the data with the fitted linear regression line. The other three plots in Figure 3.2 are the scatter plot with the Nadaraya–Watson estimator $\hat{m}(x)$ obtained using the Gaussian kernel and bandwidths 0.2, 0.5, and 2.0, respectively. From the plots, the best choice of bandwidth for the data seems to be 0.5. The plot also suggests that the mean function $m(x)$ is likely to be nonlinear. To confirm the bandwidth selection, we use cross-validation to select the optimal bandwidth. The selected bandwidth is 0.42, which is close to 0.5. Figure 3.3 shows the result of cross-validation (a) and the resulting Nadaraya–Watson estimator (b).

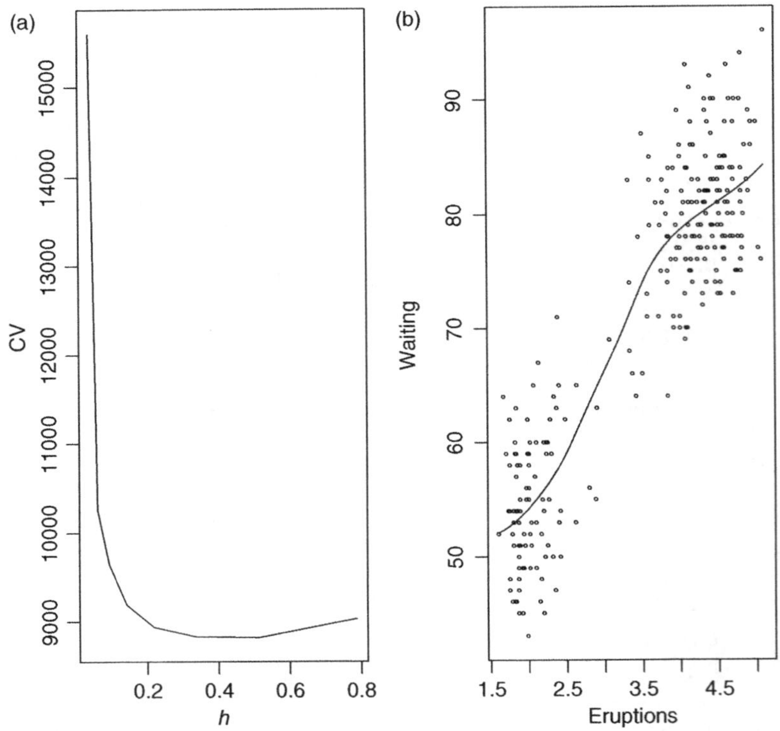

Figure 3.3 Result of (a) cross-validation and (b) the associated Nadaraya–Watson estimator for the Old Faithful data.

R demonstration: Kernel smoothing.

```
> data(faithful)
> dim(faithful)
[1] 272   2
> par(mfcol=c(2,2))
> plot(waiting~ eruptions,data=faithful,main="linear",pch="*")
> m1 <- lm(waiting~ eruptions,data=faithful)
> lines(faithful$eruptions,m1$fitted.values)
> plot(waiting~ eruptions,data=faithful,main="bandwidth=0.2",pch="*")
> lines(ksmooth(faithful$eruptions,faithful$waiting,"normal",0.2))
> plot(waiting~ eruptions,data=faithful,main="bandwidth=0.5",pch="*")
> lines(ksmooth(faithful$eruptions,faithful$waiting,"normal",0.5))
> plot(waiting~ eruptions,data=faithful,main="bandwidth=2",pch="*")
> lines(ksmooth(faithful$eruptions,faithful$waiting,"normal",2))
> par(mfcol=c(1,2))
> require(sm)
> h <- hcv(faithful$eruptions,faithful$waiting,display="lines")
> h
```

```
[1] 0.4243375
> sm.regression(faithful$eruptions,faithful$waiting,h=h,xlab="eruptions",
ylab="waiting")
```

3.2 LOCAL CONDITIONAL MEAN

A direct application of kernel methods in nonlinear time series analysis is to estimate the conditional mean function of the general nonlinear autoregressive model

$$x_t = f(x_{t-1}, \dots, x_{t-p}) + \epsilon_t, \tag{3.6}$$

where p is a positive integer and ϵ_t is an iid sequence of random variates with mean zero and variance $0 < \sigma^2 < \infty$. Let $\boldsymbol{X}_{t-1} = (x_{t-1}, \dots, x_{t-p})'$ be the vector of past p lagged values. Let $\boldsymbol{y} = (y_1, \dots, y_p)'$ be a p-dimensional vector on which the conditional mean of x_t, i.e. $E(x_t \mid \boldsymbol{X}_{t-1} = \boldsymbol{y}) = f(y_1, \dots, y_p)$, is of interest. Suppose the available sample is $\{x_1, \dots, x_T\}$. A simple approach to estimate the conditional mean via nonparametric kernel methods is as follows. Let $N_i(\boldsymbol{y}) = \{t \mid \|\boldsymbol{X}_{t-1} - \boldsymbol{y}\| < \delta_i,\ \ t = p, \dots, T\}$, where δ_i is a given positive real number and $\| \cdot \|$ denotes the Euclidean norm. In other words, $N_i(\boldsymbol{y})$ consists of the past p-dimensional vectors $\boldsymbol{X}_{t-1}$ that are in the δ_i-neighborhood of $\boldsymbol{y}$. The simple estimator is then

$$\hat{f}(y_1, \dots, y_p) = \frac{1}{\#N_i(\boldsymbol{y})} \sum_{t \in N_i(\boldsymbol{y})} x_t, \tag{3.7}$$

where $\#N_i(\boldsymbol{y})$ denotes the number of the p-dimensional vectors in $N_i(\boldsymbol{y})$. In other words, one simply uses the average of all x_t for which $\boldsymbol{X}_{t-1}$ is in the δ_i-neighborhood of $\boldsymbol{y}$. This estimator can be thought of a kernel estimate using the uniform kernel. Truong (1993) proved strong consistency and asymptotic normality of $\hat{f}(\boldsymbol{x})$ in Equation (3.7) and derived the optimal rate of convergence for suitable sequences $\delta_i \to 0$.

Many authors used the Nadaraya–Watson estimator with product kernels to estimate the conditional mean $f(x_1, \dots, x_p)$ of the general nonlinear AR model in Equation (3.6). See Härdle et al. (1997) and the references therein. The estimator is

$$\hat{f}(y_1, \dots, y_p) = \frac{\sum_{t=p+1}^{T} \prod_{i=1}^{p} K[(y_i - x_{t-i})/h_i] x_t}{\sum_{t=p+1}^{T} \prod_{i=1}^{p} K[(y_i - x_{t-i})/h_i]}, \tag{3.8}$$

where $K(\cdot)$ is a chosen kernel and h_i is the bandwidth for the ith lagged variable. Here the effective sample size is $n = T - p$. Limiting properties of the kernel estimator in Equation (3.8) have been derived by several authors under certain mixing conditions that govern the dynamic dependence of the x_t series. Robinson (1983) and Masry and Tjϕstheim (1995) proved strong consistency and asymptotic normality for α-mixing series x_t. Bierens (1983, 1987) and Collomb and Härdle (1986) showed the uniform consistency of the estimator for ϕ-mixing series x_t. The asymptotic results of kernel smoothing for time series data are almost the same as those for independent observations (e.g., nonparametric regression), due to the *whitening by window phenomenon*. It is noticed that under the mixing conditions, the observations within a small window in the space of $\boldsymbol{X}_{t-1}$ (for which we take averages) are almost independent, since the time gaps among these observations tend to be large though they are close in space.

Mixing conditions are often used in the time series literature to control the dependence between x_t and x_{t-j} as the time distance between the two variables j increases. Specifically, x_t is said to be α-mixing if

$$\sup_{A\in\mathcal{F}_1^n, B\in\mathcal{F}_{n+k}^\infty} |P(A\cap B)-P(A)P(B)| \leq \alpha_k, \tag{3.9}$$

where $\alpha_k \to 0$ as $k\to\infty$ and $\mathcal{F}_i^j$ is the σ-field generated by $\{x_i, x_{i+1}, \ldots, x_j\}$. See Robinson (1983). The series x_t is said to be ϕ-mixing if

$$|P(A\cap B)-P(A)P(B)| \leq \phi_k P(A), \tag{3.10}$$

for any $A\in\mathcal{F}_1^n$ and $B\in\mathcal{F}_{n+k}^\infty$ and $\phi_k\to 0$ as $k\to\infty$. The ϕ-mixing is stronger and is also known as uniformly mixing. The rate at which α_k and ϕ_k go to zero plays an important role in deriving the asymptotic properties of the kernel estimators. In practice, mixing conditions are hard to check, but some commonly used time series models are known to satisfy these conditions. See Tweedie (1975).

Example 3.2 Consider the quarterly unemployment durations of the USA from 1948.I to 2017.II. The original data are obtained from FRED and are monthly and seasonally adjusted. We compute the quarterly series by taking the mean of the three monthly durations. Figure 3.4(a) shows the time plot of the quarterly unemployment duration, which exhibits an upward trend. Figure 3.4(b) shows the first differenced series of the quarterly durations. The monthly unemployment duration series has been analyzed in the Exercises of Chapter 2.

Let x_t be the first differenced series of the quarterly unemployment durations. If linear models are used, an AR(1) model provides a decent fit. The model is $x_t = 0.52x_{t-1} + a_t$, with $\hat{\sigma}_a^2 = 0.566$. However, the threshold nonlinear test shows that x_t is nonlinear with test statistic 4.88 and p value 0.008. Therefore, we apply the local conditional mean estimator of Equation (3.8) with $p = 1$ to the series. The conditional means are estimated at observed x_{t-1}. Figure 3.5 shows the scatter plot of x_t versus x_{t-1} and the fitted conditional mean function (solid line). The estimation is carried by a Gaussian kernel with bandwidth 0.416. From the plot, it is seen that the dependence of x_t on x_{t-1} is indeed nonlinear. For comparison,

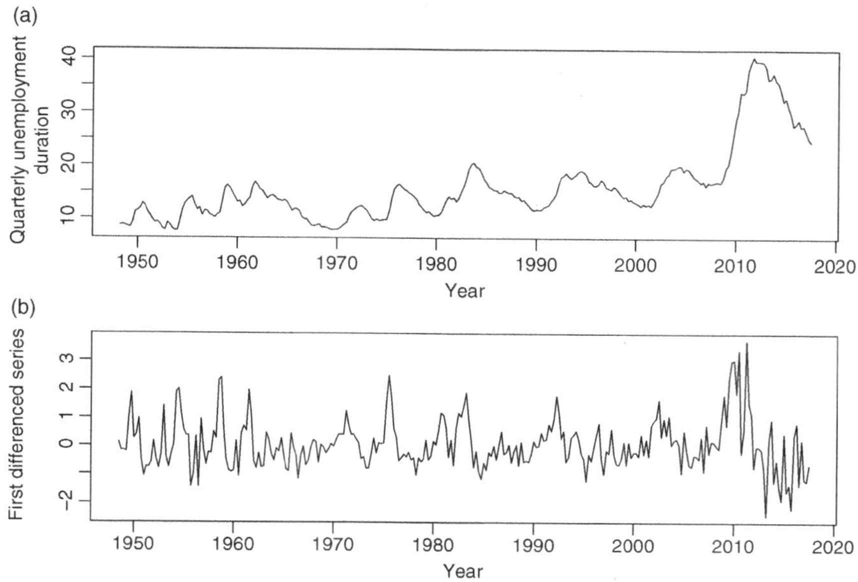

Figure 3.4 Time plots of US quarterly unemployment durations from 1948.I to 2017.II. (a) Quarterly unemployment durations and (b) the first differenced series.

the dashed line of Figure 3.5 shows the fitted line via the `lowess` procedure of R. See the discussion in the next section for the method. The dashed line also shows certain nonlinearity.

Remark: For the local conditional mean estimation, we used the package `np` of R. The bandwidth was selected automatically by the command `npreg`.

R demonstration: The `np` package of R.

```
> da <- read.table("q-unempdur.txt",header=T)
> y <- da[,2]
> dy <- diff(y)
> ts.plot(dy)
> m1 <- arima(dy,order=c(1,0,0),include.mean=F)
> m1
Call:arima(x = dy, order = c(1, 0, 0), include.mean = F)
Coefficients:
         ar1
      0.5238
s.e.  0.0510

sigma^2 estimated as 0.566:  log likelihood = -314.38,  aic = 632.77
> k1 <- lowess(dy[2:277]~ dy[1:276]) # lowess fitting
```

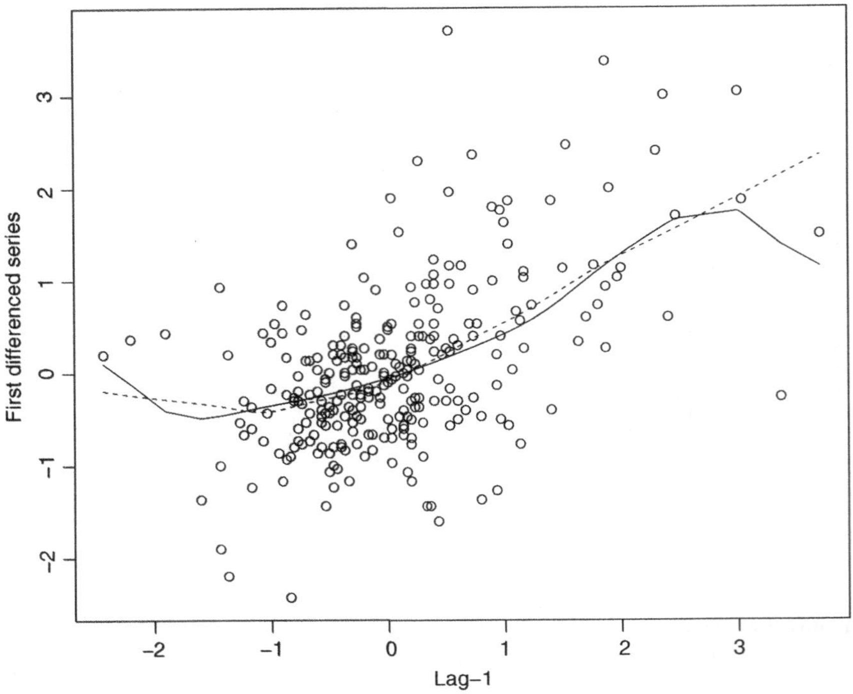

Figure 3.5 Scatter plot of the first differenced series of US quarterly unemployment duration versus its lagged values. The solid line is the local conditional mean fit via Gaussian kernel and the dashed line shows the `lowess` fit. The sample period is from 1948.II to 2017.II.

```
> require(np)
> Y <- dy[2:277]
> X <- matrix(dy[1:276],276,1)
> m2 <- npreg(txdat=X,tydat=Y,residuals=T)
> names(m2)
 [1] "bw"         "bws"         "pregtype"    "xnames"      "ynames"
 [6] "nobs"       "ndim"        "nord"        "nuno"        "ncon"
[11] "pscaling"   "ptype"       "pckertype"   "pukertype"   "pokertype"

[16] "eval"       "mean"        "merr"        "grad"        "gerr"
[21] "resid"      "ntrain"      "traininseval" "gradients"  "residuals"
[26] "R2"         "MSE"         "MAE"         "MAPE"        "CORR"
[31] "SIGN"       "rows.omit"   "nobs.omit"   "call"
> m2$bw
[1] 0.4163248
> m2$pckertype
```

```
[1] "Second-Order Gaussian"
> fit <- Y-m2$resid
> k2 <- sort(X[,1],index=T)
> tdx <- c(1:278)/4+1948
> par(mfcol=c(2,1))
> plot(tdx,y,xlab='year',ylab='q-dur',type='l')
> plot(tdx[-1],dy,xlab='year',ylab='diff(dur)',type='l')
> par(mfcol=c(1,1))
> plot(X[,1],Y,xlab='Lag-1',ylab='diff(dur)')
> lines(X[,1][k2$ix],fit[k2$ix])
> lines(k1$x,k1$y,lty=2)
```

3.3 LOCAL POLYNOMIAL FITTING

Local polynomial fitting also has a long history in statistics; see, for instance, Stone (1977) and Cleveland (1979). Consider the model in Equation (3.1) and let n be the effective sample size. If the conditional mean function $m(x)$ is differentiable, then, by Taylor series expansion, one can approximate $m(z)$ in the neighborhood of x by

$$m(z) \approx \sum_{j=0}^{p} \frac{m^{(j)}(x)}{j!}(z-x)^j \equiv \sum_{j=0}^{p} \beta_j (z-x)^j$$

where $m^{(j)}(x)$ denotes the jth derivative of $m(\cdot)$ evaluated at x. Therefore, one can estimate β_j by minimizing the objective function of a locally weighted polynomial regression

$$L(\boldsymbol{\beta}) = \sum_{i=1}^{n} \left[y_i - \sum_{j=0}^{p} \beta_j (x_i - x)^j \right]^2 K_h(x_i - x), \tag{3.11}$$

where $K(\cdot)$ is a kernel function, h is a bandwidth, and $\boldsymbol{\beta} = (\beta_0, \beta_1, \ldots, \beta_p)'$. Denote the estimate of β_j by $\hat{\beta}_j$ obtained by minimizing Equation (3.11). Then, an estimator of $m^{(v)}(x)$ is

$$\widehat{m}^{(v)}(x) = v!\hat{\beta}_v. \tag{3.12}$$

By varying x in a suitably chosen domain, one can obtain an estimated curve $\widehat{m}^{(v)}(\cdot)$ for the function $m^{(v)}(\cdot)$.

To demonstrate, consider the case of $p = 1$. In this case, Equation (3.11) becomes

$$L(\beta_0, \beta_1) = \sum_{i=1}^{n} [y_i - \beta_0 - \beta_1 (x_i - x)]^2 K_h(x_i - x). \tag{3.13}$$

Taking partial derivatives of $L(\beta_0, \beta_1)$ with respect to β_j, equating the derivatives to zero, and letting $W_i(x) = K_h(x_i - x)$, we obtain the first-order equations

$$\sum_{i=1}^{n} W_i(x) y_i = \beta_0 \sum_{i=1}^{n} W_i(x) + \beta_1 \sum_{i=1}^{n} (x_i - x) W_i(x)$$

$$\sum_{i=1}^{n} y_i (x_i - x) W_i(x) = \beta_0 \sum_{i=1}^{n} (x_i - x) W_i(x) + \beta_1 \sum_{i=1}^{n} (x_i - x)^2 W_i(x).$$

Next, define

$$S_{n,j} = \sum_{i=1}^{n} W_i(x)(x_i - x)^j, \quad j = 0, 1, 2.$$

The first-order equations become

$$\begin{bmatrix} S_{n,0} & S_{n,1} \\ S_{n,1} & S_{n,2} \end{bmatrix} \begin{bmatrix} \beta_0 \\ \beta_1 \end{bmatrix} = \begin{bmatrix} \sum_{i=1}^{n} W_i(x) y_i \\ \sum_{i=1}^{n} W_i(x)(x_i - x) y_i \end{bmatrix}.$$

Consequently, we have

$$\hat{\beta}_0 = \frac{S_{n,2} \sum_{i=1}^{n} W_i(x) y_i - S_{n,1} \sum_{i=1}^{n} (x_i - x) W_i(x) y_i}{S_{n,0} S_{n,2} - S_{n,1}^2},$$

which is an estimate of the conditional mean function $m(x)$. Some algebraic calculations show that

$$\hat{\beta}_0 = \frac{\sum_{i=1}^{n} \{W_i(x)[S_{n,2} - (x_i - x) S_{n,1}]\} y_i}{\sum_{i=1}^{n} W_i(x)[S_{n,2} - (x_i - x) S_{n,1}]} \equiv \frac{\sum_{i=1}^{n} w_i y_i}{\sum_{i=1}^{n} w_i} \qquad (3.14)$$

where the weight is now given by $w_i = W_i(x)[S_{n,2} - (x_i - x) S_{n,1}]$. Thus, the local polynomial estimate of $m(x)$ with $p = 1$ is also a weighted average of y_i. The weights, however, depend on the choice of the order p. A nice feature of the estimate in Equation (3.14) is that

$$\sum_{i=1}^{n} (x_i - x) w_i = S_{n,2} S_{n,1} - S_{n,1} S_{n,2} = 0.$$

Also, if one uses $p = 0$ in Equation (3.11), then the local polynomial estimate of $m(x)$ reduces to the Nadaraya–Watson estimator in Equation (3.2). To avoid the numerical difficulty of zero denominator in Equation (3.14), one often uses the modified version

$$\hat{m}(x) = \frac{\sum_{i=1}^{n} w_i y_i}{1/n + \sum_{i=1}^{n} w_i}.$$

For further information on local polynomial fitting, see Fan and Gijbels (1996).

Remark: Under the assumption that the underlying $m(x)$ function is continuously differentiable to the pth order, the local polynomial estimator is superior to the kernel estimator, with smaller bias, faster convergence rate, and smaller minimax risk. It also performs better on the boundary of the observed data. However, in practice one is never sure about the degrees of smoothness of the underlying function. Often a local linear estimator is used. See Fan and Yao (2003).

Cleveland (1979) proposes a robust version of locally weighted least squares method called LOcally Weighted Scatterplot Smoothing (LOWESS). Instead of fixing bandwidth, LOWESS uses a fixed number of nearest neighbors for smoothing. The kernel function used by LOWESS is

$$K(z) = \frac{70}{81}(1 - |z|^3)^3 I_{[-1,1]}(z), \tag{3.15}$$

where $I_A(z)$ is the indicator variable of the interval A, e.g, $I_{[-1,1]}(z) = 1$ if and only if $z \in [-1, 1]$. The basic idea of LOWESS is to start the fitting with a local polynomial estimate using a fixed number of nearest neighbors, and uses the fitted values to compute residuals. Then, assign weights to the residuals based on their magnitudes with large residuals receiving lower weights. One can then repeat the local polynomial fitting with new weights that are the products of the weights of the original local fitting and the residual weights. By so doing, large residuals are down-weighted to reduce their impacts on the model fitting. The procedure is iterated for several iterations to obtain robust local polynomial fitting.

In what follows, we briefly describe the LOWESS procedure as the estimator is often used in applications. Consider, again, the random sample with n observations $\{(x_1, y_1), \ldots, (x_n, y_n)\}$. Let v be a selected integer satisfying $v \in (0, n]$. Focus on the observation (x_k, y_k). Select a bandwidth h_k so that there are exactly v nearest neighbors in the window $[x_k - h_k, x_k + h_k]$. For a selected kernel function $K(\cdot)$ and the bandwidth h_k, we can obtain a local polynomial estimate for $m(x_k)$ by assigning weight

$$K_{h_k}(x_i - x_k) = h_k^{-1} K[h_k^{-1}(x_i - x_k)], \quad i = 1, \ldots, n.$$

For a given order p, one can obtain a local estimate $\hat{\beta}_0$ by using Equation (3.11). Denote the resulting estimate by

$$\hat{y}_k = \hat{\beta}_0 = \sum_{i=1}^{n} w_i(x_k) y_i,$$

where $w_i(x_k)$ is the weight assigned to the data point (x_i, y_i). This estimate is carried out for each x_k, $k = 1, \ldots, n$. Also, define the residual

$$r_k = y_k - \hat{y}_k, \quad k = 1, \ldots, n. \tag{3.16}$$

Next, let M be the median of the absolute residuals, i.e. $M = \text{median}\{|r_1|, \ldots, |r_n|\}$. One defines the robustness weight by

$$\eta_i = B(r_i/(6M)), \quad i = 1, \ldots, n,$$

where

$$B(z) = (1 - |z|^2)^2 I_{[-1,1]}(z)$$

is the bi-weight kernel of Equation (3.4) with $\gamma = 2$. Now, one can repeat the local pth order polynomial fitting using the product weights $\eta_i K_{h_k}(x_i - x_k)$. The additional weight η_i takes into account the heteroscedasticity in the noise variance, similar to the weighted least squares approach. For example, if $r_i > 6M$, then $\eta_i = 0$. That is, (x_i, y_i) is considered as an outlier, hence it is assigned a zero weight and is not included in the analysis. Larger r_i indicates larger variance of ε_i, hence receiving smaller weight.

The procedure is iterated multiple times to obtain stable estimates.

Example 3.3 Consider the annual mean sunspot numbers from 1700 to 2006. The data are downloaded from WDC-SILSO, Royal Observatory of Belgium, Brussels. Figure 3.6 shows the time plot of the series. The sunspot data have been widely used in the literature as an example of nonlinear time series. See,

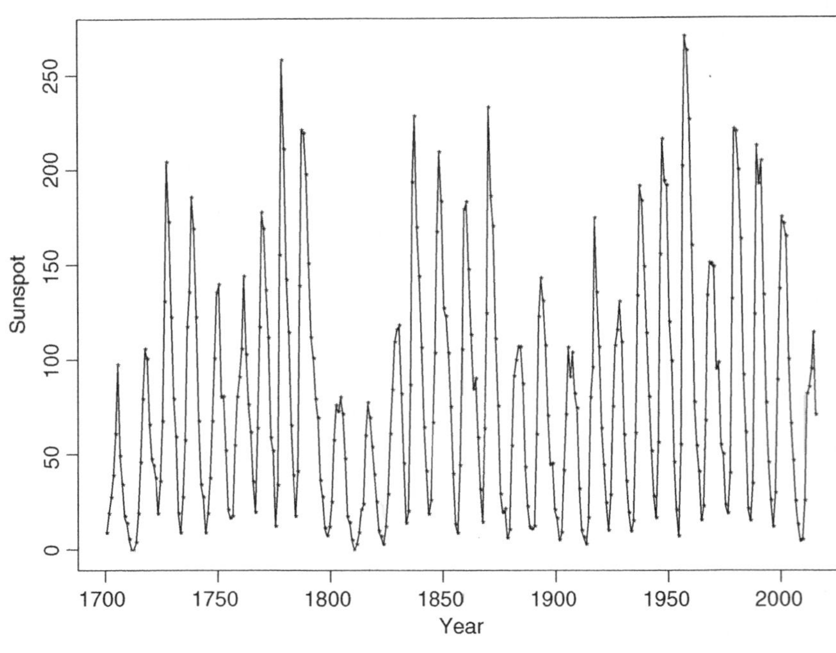

Figure 3.6 Time plot of annual mean sunspot numbers from 1700 to 2006.

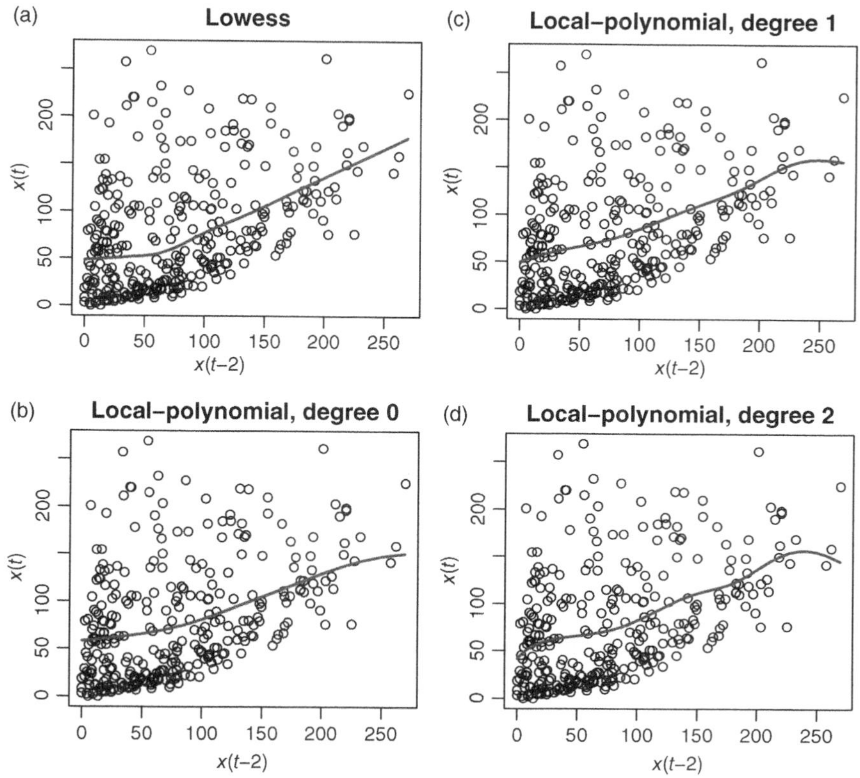

Figure 3.7 Scatter plot of x_t versus x_{t-2} of the annual sunspot numbers: (a) the LOWESS line and (b)–(d) local polynomial fits of degrees 0, 1, and 2, respectively. The bandwidth of local polynomial fitting is 50% of the sample standard deviation of the data.

for instance, Tong (1990) and the references therein. In this example, we focus on the relationship between x_t and x_{t-2}, where x_t denotes the annual sunspot number of year $t + 1699$. In other words, we consider $x_t = m(x_{t-2}) + \epsilon_t$, where ϵ_t is a sequence of random noises. Figure 3.7 shows the scatter plot of x_t versus x_{t-2} and several estimates of $m(x_{t-2})$. The line in part (a) is obtained via LOWESS with default options, namely $v = 2/3$ and three iterations. The lines in parts (b) to (d) are from the local polynomial fitting with degrees 0, 1, and 2, respectively. The bandwidth is $h = 31$, which is approximately $0.5\hat{\sigma}$, with $\hat{\sigma}$ being the sample standard deviation of x_t. Thus, part (b) shows the traditional Nadaraya–Watson estimate. From the plots, the degree used in local polynomial fitting may affect the boundaries of the fit. The robustness of the LOWESS procedure is clearly seen when x_{t-2} is approximately 50.

R demonstration: Local polynomial fitting via the `KernSmooth` package.

```
da <- read.table("SN_y_tot_V2.0.txt",header=T)
sn <- da[,2]; n <- length(sn)
s <- sqrt(var(sn))
y <- sn[3:n]; x <- sn[1:(n-2)]
require(KernSmooth)
par(mfcol=c(2,2))
plot(x,y,xlab="x(t-2)",ylab="x(t)",main="(a) Lowess")
lines(lowess(x,y),col="red")
m3 <- locpoly(x,y,degree=0,bandwidth=0.5*s)
plot(x,y,xlab="x(t-2)",ylab="x(t)",main="(b) Local polynomial, degree 0")
lines(m3$x,m3$y,col="red")
m4 <- locpoly(x,y,degree=1,bandwidth=0.5*s)
plot(x,y,xlab="x(t-2)",ylab="x(t)",main="(c) Local polynomial, degree 1")
lines(m4$x,m4$y,col="red")
m5 <- locpoly(x,y,degree=2,bandwidth=0.5*s)
plot(x,y,xlab="x(t-2)",ylab="x(t)",main="(d) Local polynomial, degree 2")
lines(m5$x,m5$y,col="red")
```

3.4 SPLINES

The local polynomial fitting of Section 3.3 is motivated by Taylor series expansion. By increasing the degree of the polynomial used, we increase the flexibility in model fitting. However, the method is still treating the domain of the X variable locally, as the function is estimated at each point individually using a local window centering around it. An alternative approach is to estimate the function globally using an increasingly larger set of basis functions to approximate the underlying function. The entire function is then estimated in one step under a global optimization criterion. Global basis functions such as polynomials are not stable and inflexible. Using local basis functions such as *piecewise polynomials* increases flexibility. Consider the simple case in which the dependent variable Y is a function of a scalar variable X,

$$Y = f(X) + \epsilon,$$

where ϵ denotes the noise term. One can divide the domain of X into contiguous intervals and approximate $f(X)$ by a separate polynomial in each interval. To demonstrate, we use the smooth function $f(X) = 2\sin(2\pi X)$, and let $\{x_i\}$ be the 51 equally spaced points in [0,1]. The observed dependent variable is $y_i = f(x_i) + \epsilon_i$, where ϵ_i are iid $N(0, 1)$. We divide the range of x_i into three non-overlapping contiguous intervals, namely

$$I_1 = [0, 0.32), \quad I_2 = [0.32, 0.66), \quad I_3 = [0.66, 1].$$

These intervals are defined by the points $\xi_0 = 0$, $\xi_1 = 0.32$, $\xi_2 = 0.66$, and $\xi_4 = 1$. In the literature, these $\{\xi_i\}$ are called *knots*. The data and the true function $f(x)$

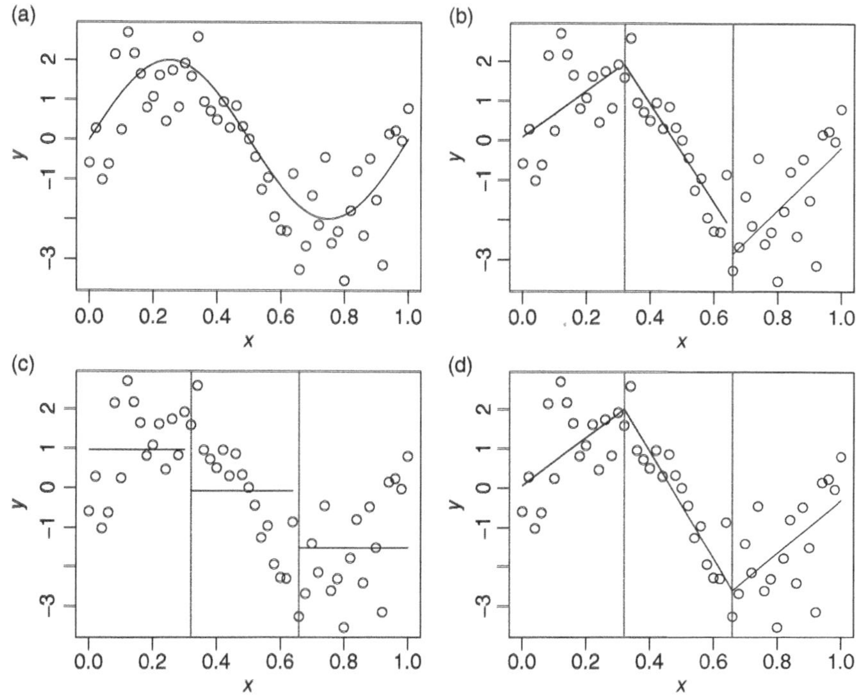

Figure 3.8 Illustration of piecewise polynomial fits: (a) the data and the underlying true function, (b) the fit of piecewise polynomials of degree 1, (c) the fit of piecewise polynomials of degree zero, and (d) the fit of piecewise polynomials of degree 1 with the continuity constraints at knots.

are shown in Figure 3.8(a). Figure 3.8(c) shows the piecewise polynomial fit with degree zero, i.e. the sample mean of each interval. Figure 3.8(b) shows the results of piecewise linear regression, i.e. polynomial of degree 1. Clearly, there is improvement when we increase the degree of the piecewise polynomial from zero to one, and the improvement is achieved by using three additional slope parameters. However, the discontinuity of the fitted curve at the knots ξ_1 and ξ_2 is in sharp contrast with the idea that the underlying function is smooth. Figure 3.8(d) shows the fit of piecewise polynomials with degree 1 and the two continuity constraints at knots ξ_1 and ξ_2. The constrained fitting appears to be sensible.

The three fitted models of Figure 3.8 can be described as follows:

1. Piecewise constant:

$$y_i = \mu_i + \epsilon_i, \quad \text{if} \quad x_i \in [\xi_{i-1}, \xi_i).$$

2. Piecewise linear regression:

$$y_i = \alpha_i + \beta_i x_i + \epsilon_i, \quad \text{if} \quad x_i \in [\xi_{i-1}, \xi_i).$$

3. Piecewise linear regression with continuity constraints: The same model as Case 2 with the constraints

$$\alpha_1 + \beta_1 \xi_1 = \alpha_2 + \beta_2 \xi_1 \quad \text{and} \quad \alpha_2 + \beta_2 \xi_2 = \alpha_3 + \beta_3 \xi_2.$$

The numbers of coefficients of the three models are 3, 6, and 4, respectively. A more direct way to describe the three models is to think that we are constructing certain functions of the X variable to approximate the function $f(X)$. The piecewise constant model is achieved by creating three indicator functions

$$b_1(X) = I[X < \xi_1], \quad b_2(X) = I[\xi_1 \leq X < \xi_2], \quad b_3(X) = I[\xi_2 \leq X],$$

and we simply fit the model

$$y_i = \mu_1 b_1(x_i) + \mu_2 b_2(x_i) + \mu_3 b_3(x_i) + \epsilon_i.$$

The piecewise linear regression is achieved by adding three more functions

$$b_4(X) = X b_1(X), \quad b_5(X) = X b_2(X), \quad b_6(X) = X b_3(X),$$

and approximating $f(X)$ by a linear combination of $\{b_j(X) | j = 1, \ldots, 6\}$. Finally, the constrained linear regression is achieved by using

$$b_1(X) = 1, \quad b_2(X) = X, \quad b_3(X) = (X - \xi_1)_+, \quad b_4(X) = (X - \xi_2)_+,$$

where $(z)_+ = \max(0, z)$ is the positive part of z. Here $b_3(X)$ and $b_4(X)$ are simply ramp functions.

While the piecewise linear regression (polynomial of order 1) with continuity constraints provides a continuous fitted function, the resulting curve is not smooth. The curve is not differentiable at the knots. On the other hand, we often prefer smooth functions in applications. This can be achieved by increasing the degree the piecewise polynomial used and requiring smoothness in certain derivatives of the fitted function. Figure 3.9 shows the fits of some smooth splines to the illustrative data set. The four plots differ in the degrees of freedom used. Figure 3.9(d) shows a reasonable fit in this particular instance.

R demonstration: Smoothing spline in Figure 3.9.

```
> x <- seq(0,1,0.02); f <- 2*sin(2*pi*x)
> set.seed(11)
> y <- f + rnorm(51)
```

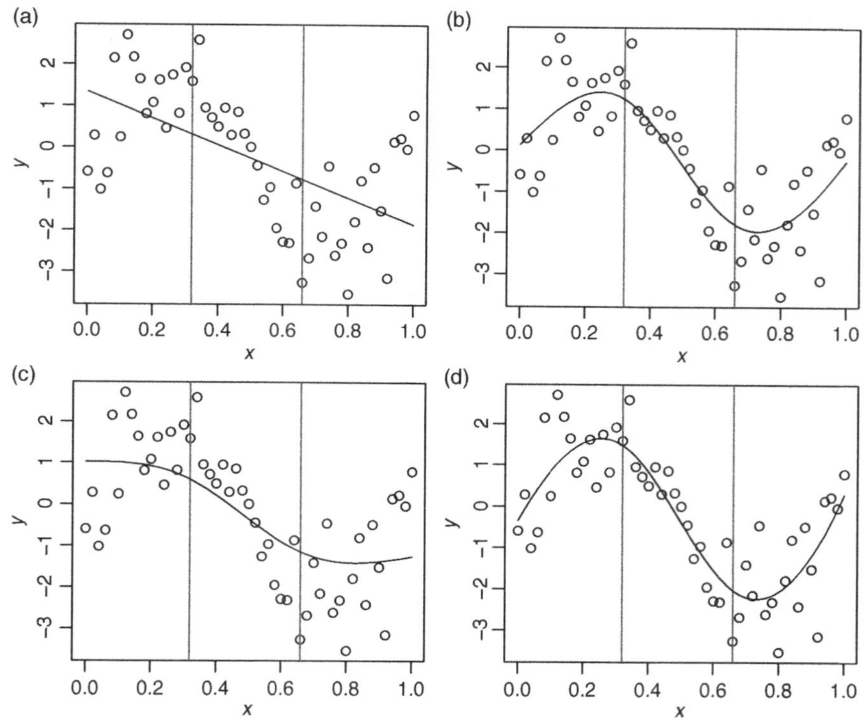

Figure 3.9 Illustration of smooth splines. The data used are the same as those of Figure 3.8. (a), (c) and (b) and (d) have degrees of freedom 2, 3, 4 and 5, respectively.

```
> require(stats)
> par(mfcol=c(2,2))
> xi <- c(0.32,0.66) # equally-spaced knots
> plot(x,y); abline(v=xi,col="red")
> m1 <- smooth.spline(x,y,df=2)
> lines(m1)
> plot(x,y); abline(v=xi,col="red")
> m2 <- smooth.spline(x,y,df=3)
> lines(m2)
> plot(x,y); abline(v=xi,col="red")
> m3 <- smooth.spline(x,y,df=4)
> lines(m3)
> plot(x,y); abline(v=xi,col="red")
> m4 <- smooth.spline(x,y,df=5)
> lines(m4)
```

3.4.1 Cubic and B-Splines

For a given set of knots $\{\xi_i | i = 1, \dots, K\}$, a flexible spline function is the *natural cubic spline*. It consists of cubic splines within intervals between knots and the additional constraints that the function is linear beyond the boundary knots. The additional constraints are achieved by requiring that the second and third derivatives of $f(\cdot)$ are zero at the boundary. Mathematically, the resulting smooth function can be written as

$$f(x) = \beta_0 + \beta_1 x + \beta_2 x^2 + \beta_3 x^3 + \sum_{j=1}^{K} \gamma_j (x - \xi_j)_+^3 \tag{3.17}$$

where, again, $(z)_+ = \max(0, z)$. However, the model in Equation (3.17) is not efficient in computing. It requires evaluation of the regressors at all points to the right of its knot. A more convenient form for estimation is to express the model using certain basis functions, each of which only requires evaluation in a small number of intervals. A well-known class of basis functions is the *B-splines*. de Boor (1978) discusses splines and B-splines in detail.

Also, for computational reason, we need to augment the knot sequence $\xi_1 < \xi_2 < \cdots < \xi_K$. Let $\xi_0 < \xi_1$ and $\xi_K < \xi_{K+1}$ be the two *boundary* knots. Typically, $[\xi_0, \xi_{K+1}]$ denotes the domain over which the spline is evaluated. Let M be the order of the spline function. Following Hastie, Tibshirani and Friedman (2001), we create an augmented knot sequence $\{\tau_j\}$ such that

- $\tau_1 \le \tau_2 \le \cdots \le \tau_M \le \xi_0$;
- $\tau_{i+M} = \xi_i$, for $i = 1, \dots, K$;
- $\tau_{K+1} \le \tau_{K+2} \le \cdots \le \tau_{K+2M}$.

The actual values of the newly added knots are arbitrary and they may assume the same values as the boundary knots ξ_0 and ξ_{K+1}, respectively. Let $B_{i,m}(x)$ be the ith B-spline basis function of order m for the knot sequence τ, where $m \le M$. One can define the basis functions recursively starting with the *Haar basis functions*:

$$B_{i,1}(x) = \begin{cases} 1 & \text{if } \tau_i \le x < \tau_{i+1} \\ 0 & \text{otherwise} \end{cases} \tag{3.18}$$

where $i = 1, \dots, K + 2M - 1$. Then,

$$B_{i,m}(x) = \frac{x - \tau_i}{\tau_{i+m-1} - \tau_i} B_{i,m-1}(x) + \frac{\tau_{i+m} - x}{\tau_{i+m} - \tau_{i+1}} B_{i+1,m-1}(x) \tag{3.19}$$

where $i = 1, \dots, K + 2M - m$. This recursion is referred to as the *Cox–de Boor recursion*. In practice, it is the convention to adopt $B_{i,1} = 0$ if $\tau_i = \tau_{i+1}$. By induction, we then have $B_{i,m} = 0$ if $\tau_i = \tau_{i+1} = \cdots = \tau_{i+m}$.

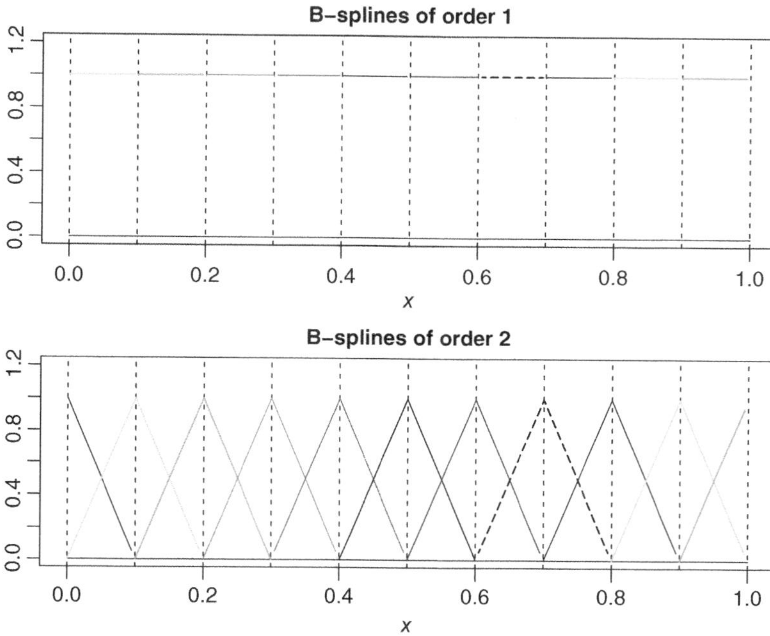

Figure 3.10 The sequence of B-splines of orders 1 and 2 with 10 knots equally spaced from 0 to 1.

From the recursion, it is easy to see that one only needs to evaluate $B_{i,m}(x)$ for $x \in [\tau_i, \tau_{i+m}]$. This local feature simplifies the computing in using the B-spline basis functions. Also, from the recursion, $\{B_{i,m}(x), i = 1, \ldots, K+4\}$ are the $K+4$ cubic B-spline basis functions for the knot sequence $\{\xi_j\}$. Figure 3.10 shows the sequences of B-splines of orders 1 and 2 with knots at the points 0.0, 0.1, …, 1.0 whereas Figure 3.11 shows the sequences of B-splines of orders 3 and 4 for the same knots. By selecting the degree M and the number of knots and their locations, one obtains a set of B-spline basis functions that can be used to approximate the function $f(X)$.

Example 3.4 An important model in finance is the Capital Asset Pricing Model (CAPM) relating excess returns of an asset to market excess returns. See, for instance, Sharpe (1964). Let R_t be the monthly excess return of General Motors stock from March 1967 to December 2008. Let M_t be the excess returns of the S&P composite index. The 3-month risk-free interest rate is used to construct the excess returns. All data are from the Center of Research for Security Prices (CRSP) of the University of Chicago. The CAPM model is

$$R_t = \alpha + \beta M_t + \epsilon_t \tag{3.20}$$

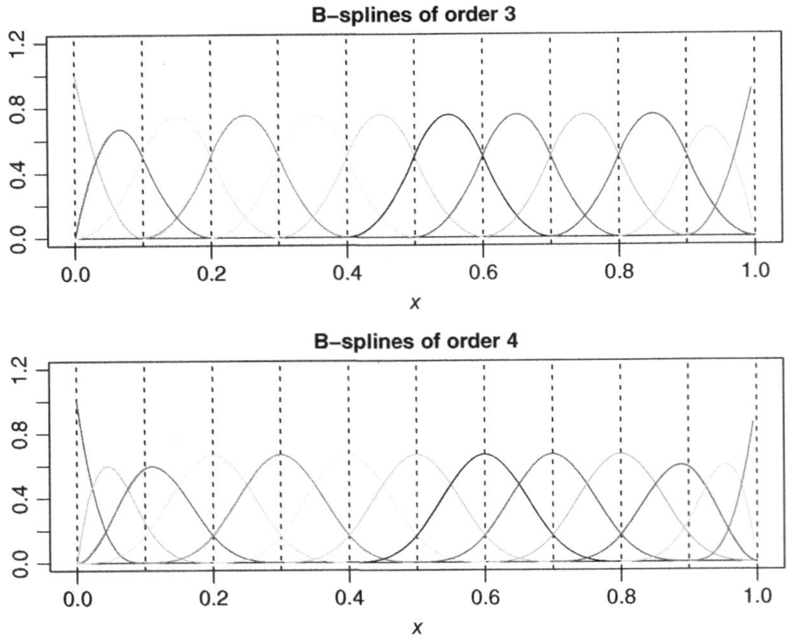

Figure 3.11 The sequence of B-splines of orders 3 and 4 with 10 knots equally spaced from 0 to 1.

and the coefficients are estimated by the ordinary least squares method. Figure 3.12 shows the scatter plot of R_t versus M_t and the fitted linear regression line (straight). Also shown in Figure 3.12 is a B-spline fit, the curve. The B-splines are degree 3 and default knot options. From the plot, the spline fit seems to fare better in this particular instance.

R demonstration: B-splines via the package `splines`.

```
> require(stats); require(splines)
> da <- read.table("m-gmsp6708.txt",header=T)
> da1 <- read.table("m-riskfree3m-6708.txt",header=T)
> rf <- da1[,4]/(100*12)
> gm <- da$GM-rf; sp <- da$SP-rf
> plot(sp,gm,xlab="Market excess returns",ylab='GM excess returns')
> m1 <- lm(gm~ sp)
> lines(sp,m1$fitted.values)
> m2 <- bs(sp,degree=3) ### B-splines of degree 3.
> m3 <- lm(gm~ m2)
> xx <- sort(sp,index=T)
> lines(sp[xx$ix],m3$fitted.values[xx$ix],col=2)
```

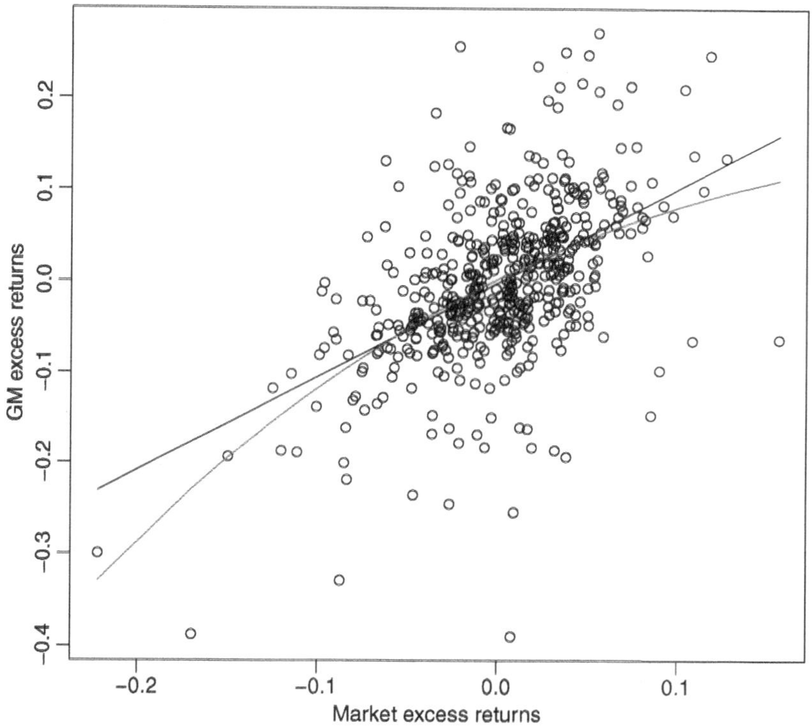

Figure 3.12 Scatter plot of monthly excess returns of General Motors stock versus the excess returns of the S&P market index from March 1967 to December 2008. The straight line and curve represent the linear regression and B-spline fit, respectively.

3.4.2 Smoothing Splines

In the discussion of spline functions, we simply use equally spaced knots over the domain of the X variable. In real applications, one needs to specify both the number of knots and their exact locations. Even though there are methods available in the literature to perform such selection, we shall not address them in this chapter. To mitigate this selection problem, we introduce the *smoothing-spline* method, which uses a maximal set of knots, but controls the complexity of the fit by certain regularization.

Consider all functions $f(x)$ with first two continuous derivatives. Smoothing spline method finds the one that minimizes the penalized residual sum of squares

$$\text{RSS}(f, \lambda) = \sum_{i=1}^{n} [y_i - f(x_i)]^2 + \lambda \int f''(s) ds, \tag{3.21}$$

where $\lambda \geq 0$ is the *smoothing parameter* and n denotes the sample size. The first term of Equation (3.21) measures the goodness of fit whereas the second term penalizes curvature in the function. If $\lambda = 0$, there is no penalty and any $f(x)$ that interpolates the data will suffice. If $\lambda = \infty$, then no second derivative is allowed and $f(x)$ is simply the least squares line. The idea of smoothing spline is to find a suitable penalty $\lambda \in (0, \infty)$ such that the resulting estimate $\hat{f}(x)$ works well in practice.

At first glance, Equation (3.21) involves functional space and seems hard to minimize. It turns out that it has a unique minimizer, which is a natural cubic spline with knots at the observed values $\{x_i | i = 1, \ldots, n\}$. The penalty term effectively puts constraints on the spline coefficients to reduce the degrees of freedom. See, for instance, Hastie et al. (2001, Chapter 5). Let $\{\xi_i | i = 1, \ldots, n\}$ be the knots (in increasing order). Define

$$N_1(x) = 1, \quad N_2(x) = x, \quad N_{k+2}(x) = d_k(x) - d_{k-1}(x), \tag{3.22}$$

where

$$d_k(x) = \frac{(x - \xi_k)_+ - (x - x_n)_+}{\xi_n - \xi_k}.$$

Then,

$$f(x) = \sum_{k=1}^{n} N_k(x)\beta_k,$$

and we have

$$\text{RSS}(f, \lambda) = (\boldsymbol{y} - \boldsymbol{N\beta})'(\boldsymbol{y} - \boldsymbol{N\beta}) + \lambda\boldsymbol{\beta}'\boldsymbol{\Omega}_n\boldsymbol{\beta}, \tag{3.23}$$

where $\boldsymbol{y}$ is the observed values of the y variable, $\boldsymbol{\beta}$ denotes the parameter vector, $\boldsymbol{N} = [N_{ij}]$ with $N_{ij} = N_j(x_i)$, and $\boldsymbol{\Omega}_n = [\Omega_{ij}]$ with $\Omega_{ij} = \int N_i''(s)N_j''(s)ds$. Equation (3.23) is in the form of the well-known ridge regression. Its solution is

$$\hat{\boldsymbol{\beta}} = (\boldsymbol{N}'\boldsymbol{N} + \lambda\boldsymbol{\Omega}_n)^{-1}\boldsymbol{N}'\boldsymbol{y} \tag{3.24}$$

and the fitted smoothing spline is

$$\hat{f}(x) = \sum_{i=1}^{n} N_i(x)\hat{\beta}_i.$$

Let $\boldsymbol{S}_\lambda = \boldsymbol{N}(\boldsymbol{N}'\boldsymbol{N} + \lambda\boldsymbol{\Omega}_n)^{-1}\boldsymbol{N}'$, which can be thought of as a generalized Hat-matrix of the linear least squares regression. We have $\hat{f} = \boldsymbol{S}_\lambda\boldsymbol{y}$. The smoothing spline estimator is thus a linear estimator, i.e. a weighted linear combination of $\boldsymbol{y}$. The effective degrees of freedom (df) of the smoothing spline is defined as

$$\text{df}_\lambda = \text{tr}(\boldsymbol{S}_\lambda), \tag{3.25}$$

which is the sum of the diagonal elements of $\boldsymbol{S}_\lambda$. The concept of degrees of freedom is useful and a widely used approach to parameterize the smoothing spline. Specifically, since df_λ is monotone in λ, one can fix λ by specifying df_λ.

In applications, the smoothing parameter λ, or equivalently the degrees of freedom df_λ, can be selected by cross-validation. When the sample size n is not too large, one can use the leave-one-out cross-validation. Here the criterion function is

$$\text{CV}(\hat{f}_\lambda) = \sum_{i=1}^{n}[y_i - \hat{f}_\lambda^{(-i)}(x_i)]^2 = \sum_{i=1}^{n}\left(\frac{y_i - \hat{f}_\lambda(x_i)}{1 - S_\lambda(i,i)}\right)^2 \tag{3.26}$$

where $\hat{f}_\lambda^{(-i)}(x_i)$ denotes the smoothing-spline fit with the ith observation omitted and $S_\lambda(i,i)$ is the (i,i)th element of $\boldsymbol{S}_\lambda$. The second equality of Equation (3.26) shows that, for a given λ, the CV can be computed using the original fitted values. There is no need to re-fit the model in computing the criterion. This is a nice result. One chooses the value of λ by the minimum value of $\text{CV}(\hat{f}_\lambda)$.

Example 3.5 Consider the growth rates of US quarterly real gross domestic product (GDP) from the second quarter of 1947 to second quarter of 2015 for 273 observations. Let x_t denote the GDP growth rate. The data are available from FRED. As discussed in Chapter 2, the x_t series exhibits certain nonlinear characteristics. Figure 3.13 shows the scatter plot of x_t versus x_{t-2}. Also shown are three fitted models via the smoothing spline method. The solid line denotes the fit with the smoothing parameter λ selected by the CV criterion in Equation (3.26). In this particular instance, the selected λ is around 1.34. The dashed line of Figure 3.13 is the fit with $\text{df}_\lambda = 5$, which corresponds to $\lambda = 1.18$. The dotted line, on the other hand, is the fit with $\text{df}_\lambda = 15$, which corresponds to $\lambda = 0.87$. From the plots, we see that the choice of $\text{df}_\lambda = 15$ is too large, resulting in a wiggly fit. Overall, the fits seem to confirm that some nonlinearity exists in the data.

R demonstration: Smoothing spline via the package `splines`.

```
da <- read.table("GDPC1.txt",header=T)
gdp <- diff(log(da$rgdp)); n <- length(gdp)
x <- gdp[1:(n-2)]; y <- gdp[3:n]
plot(x,y,xlab='gdp(t-2)',ylab='gdp(t)')
require(splines)
m1 <- smooth.spline(x,y,cv=T)  ## cross-validation
lines(m1)
names(m1)
 [1] "x"         "y"         "w"         "yin"       "data"      "lev"
 [7] "cv.crit"   "pen.crit"  "crit"      "df"        "spar"      "lambda"
[13] "iparms"    "fit"       "call"
```

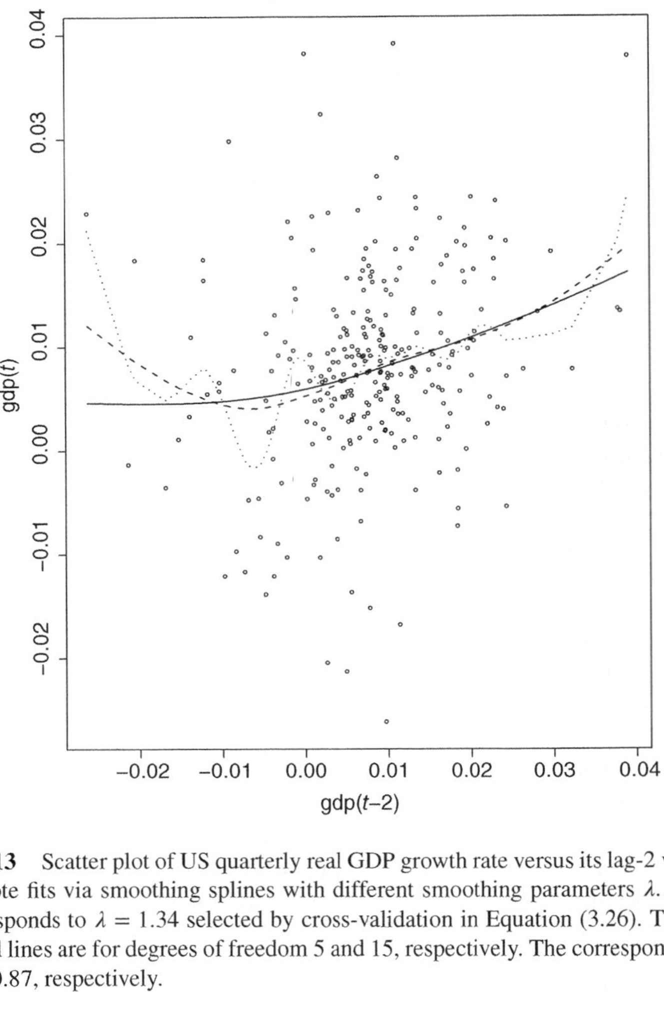

Figure 3.13 Scatter plot of US quarterly real GDP growth rate versus its lag-2 value. The lines denote fits via smoothing splines with different smoothing parameters λ. The solid line corresponds to $\lambda = 1.34$ selected by cross-validation in Equation (3.26). The dashed and dotted lines are for degrees of freedom 5 and 15, respectively. The corresponding λ are 1.18 and 0.87, respectively.

```
m1$spar
[1] 1.338832
m2 <- smooth.spline(x,y,df=5)
lines(m2,lty=2, lwd=2) #col="red")
m2$spar
[1] 1.179196
m3 <- smooth.spline(x,y,df=15)
lines(m3,lty=3, lwd=2) #col="blue")
m3$spar
[1] 0.8719036
```

Remark: There are other spline-based methods available in the literature. In particular, the thin plate splines of Wood (2003) have been widely used in fitting nonlinear regression models.

3.5 WAVELET SMOOTHING

In this section we consider another approach of using basis functions in time series analysis and smoothing. The tools are the *wavelets* and the approach is referred to as wavelet smoothing. There are several books that apply wavelet methods to time series analysis. See, for instance, Percival and Waldem (2000) and Nason (2008). Our discussion focuses on the approach of discrete wavelet transform.

3.5.1 Wavelets

Wavelet methods typically start with the selection of a scaling function, which is called *father wavelet*. Corresponding to the father wavelet, there is a basic wavelet function, which is used to generate a set of orthonormal basis functions. The basic wavelet is called the *mother wavelet* and the orthonormal basis functions are generated by *translations* and *dilations* of the mother wavelet. Mother wavelets must satisfy certain conditions such as absolute and square integrable and having vanishing moments. There are many choices available for the mother wavelet. See, for instance, Daubechies (1992). With a given set of wavelet basis functions, one can approximate the underlying *true* function via a linear combination of the basis functions. The main idea behind this is that any square-integrable function can be well approximated by wavelet basis functions. Data analysis can then be performed by processing the coefficients of the linear combination. In a nutshell, the process of wavelet data analysis can be shown as

Raw data	⇒	Wavelet decomposition	⇒	Thresholding
	⇒	Wavelet reconstruction	⇒	Processed data

Thus, the idea of wavelet analysis is similar to that of Fourier analysis. However, there are some important differences between Fourier analysis and wavelets. Fourier basis functions are localized in frequency but not in time. A small frequency change in Fourier transform will result in changes throughout the time domain. Wavelets are local in both frequency and time, and this difference provides advantages for wavelets in many applications. Another difference is that wavelets can provide a more compact way to approximate the underlying function by choosing a proper mother wavelet. On the other hand, Fourier analysis enjoys some advantages in estimating derivatives of the underlying function.

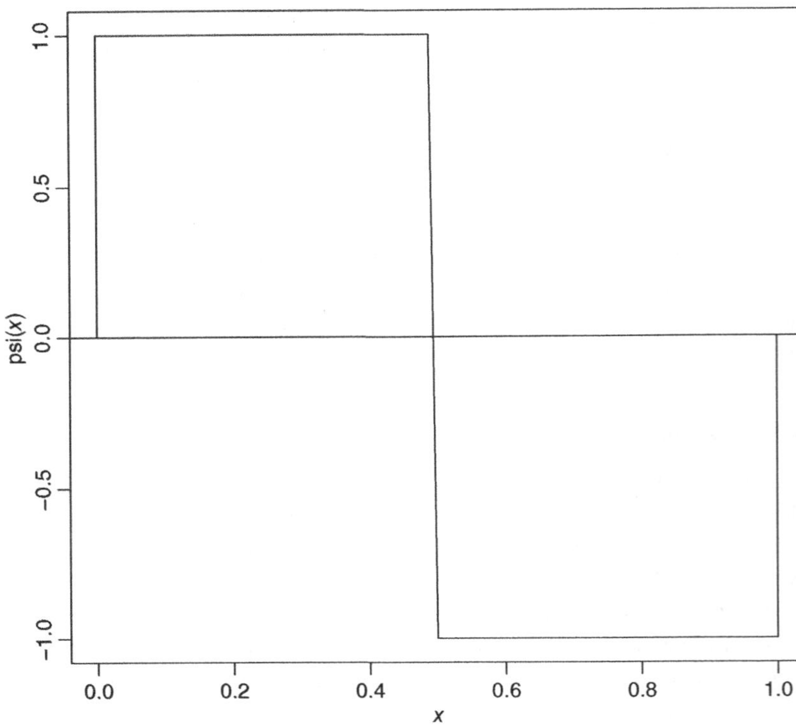

Figure 3.14 The Haar function.

To demonstrate wavelet analysis, we use the simplest *Haar* wavelet, defined by

$$\psi(x) = \begin{cases} 1 & \text{if } 0 \leq x < 0.5, \\ -1 & \text{if } 0.5 \leq x < 1. \end{cases} \tag{3.27}$$

Figure 3.14 shows the Haar function. To generate basis functions from the Haar wavelet, we use translations and dilations given by

$$\psi_{j,k}(x) = 2^{j/2}\psi(2^j x - k) \tag{3.28}$$

where k and j are integers. The translation is achieved by k and the dilation by powers of j. The constant $2^{j/2}$ is used so that the square-integral of $\psi_{j,k}(x)$ is 1. Figure 3.15 shows the plots of $\{\psi_{1,0}(x), \psi_{1,1}(x), \psi_{20,}(x), \psi_{2,1}(x), \psi_{2,2}(x), \psi_{2,3}(x)\}$ over the interval [0,1]. It is easy to see that the basis functions satisfy

$$\int_0^1 \psi_{j,k}^2(x)dx = 1, \quad \int_0^1 \psi_{j,k}(x)\psi_{u,v}(x)dx = 0$$

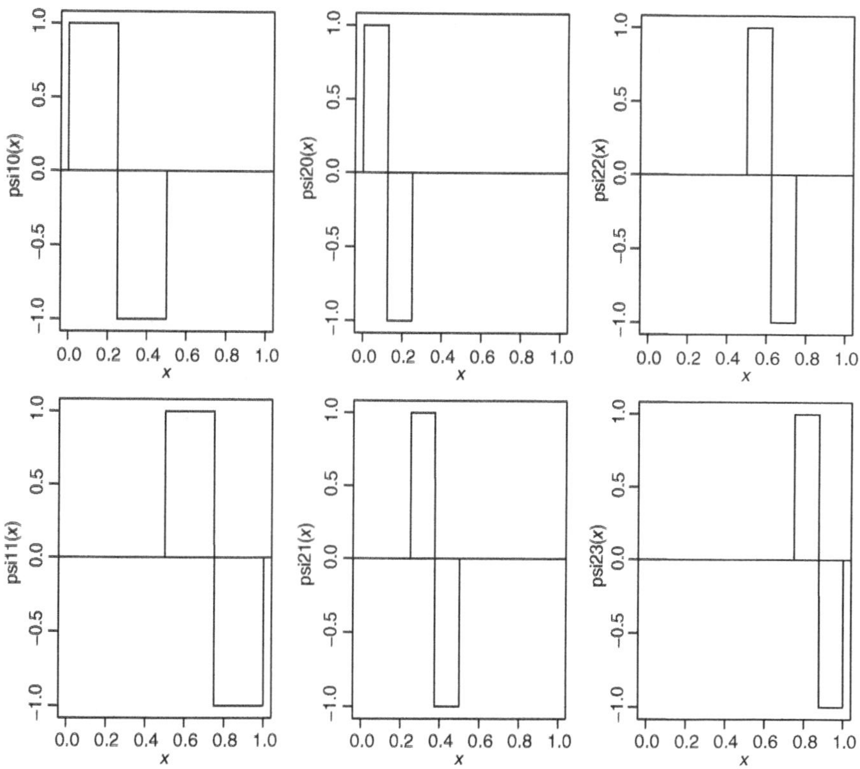

Figure 3.15 Some basis functions of Haar wavelets.

for $(j, k) \neq (u, v)$. Thus, the basis functions of the Haar wavelet are orthonormal functions. It is also known that any square-integrable function $f(x)$ can be written as a linear combination (possibly with infinite many terms) of the basis functions $\{\psi_{j,k}(x)\}$, where k and j are integers.

3.5.2 The Wavelet Transform

For the Haar wavelet, the father wavelet is

$$\phi(x) = \begin{cases} 1 & \text{if } 0 \leq x < 1, \\ 0 & \text{otherwise.} \end{cases} \tag{3.29}$$

It will become clear later that the father wavelet characterizes wavelet scale whereas the mother wavelet provides wavelet shape. One can start wavelet analysis by choosing a proper father wavelet. The choice often depends on the consideration of continuity, smoothness, and tail behavior for the data at hand. For a chosen father wavelet $\phi(x)$, the translation functions $\{\phi_{0,k}(x) = \phi(x-k)|k = 0, \pm1, \pm2, \cdots\}$ form an orthonormal basis for a reference space $V_0 \subset l^2(R)$, where $l^2(R)$ consists of all square-integrable functions on R. For the Haar wavelet $\phi(x)$, V_0 is formed by all piecewise constant functions with jumps at integers. Next, consider the dilations $\phi_{1,k}(x) = \sqrt{2}\phi(2x-k)$, where k is an integer. It is easy to see that $\{\phi_{1,k}(x)\}$ form an orthonormal basis for a space $V_1 \supset V_0$. For the Haar wavelet, V_1 consists of piecewise constant functions on intervals of length 0.5. In fact, we have a sequence of spaces $\cdots \supset V_2 \supset V_1 \supset V_0 \supset V_{-1} \supset \cdots$, where V_j is spanned by $\phi_{j,k}(x) = 2^{j/2}\phi(2^j x - k)$, where k is an integer. Mathematically, $\{V_j\}$ is dense in $l^2(R)$ and the intersection $\cap_j V_j = \emptyset$. Also, if $f(x) \in V_j$, then $f(2x) \in V_{j+1}$.

Next, we introduce the role played by the mother wavelet and its offspring. Since $V_0 \subset V_1$, any function in V_0 can be written as a linear combination of the basis of V_1. In particular, since the father wavelet (scaling function) is in V_0, we have

$$\phi(x) = \sum_k h_k \phi_{1,k}(x) = \sum_k h_k \sqrt{2}\phi(2x-k), \tag{3.30}$$

where the coefficient h_k is the inner product $h_k =< \phi(x), \sqrt{2}\phi(2x-k) >$. For the Haar scaling function, simple calculation shows

$$\phi(x) = \frac{\sqrt{2}}{2}\phi_{1,0}(x) + \frac{\sqrt{2}}{2}\phi_{1,1}(x) = \frac{\sqrt{2}}{2}[\sqrt{2}\phi(2x)] + \frac{\sqrt{2}}{2}[\sqrt{2}\phi(2x-1)].$$

Now, define

$$\psi(x) = \sum_k (-1)^k h_{-k+1}\phi_{1,k}(x) = \sum_k (-1)^k h_{-k+1}\sqrt{2}\phi(2x-k). \tag{3.31}$$

Readers will see immediately that we use the notation of the mother wavelet. This is indeed true. For the Haar wavelet, direct calculation shows that $\psi(x)$ of Equation (3.31) turns out to be precisely the mother wavelet of Equation (3.27), that is,

$$\begin{aligned}\psi(x) &= h_1\phi_{1,0}(x) + h_0\phi_{1,1}(x) = \frac{\sqrt{2}}{2}[\sqrt{2}\phi(2x)] - \frac{\sqrt{2}}{2}[\sqrt{2}\phi(2x-1)] \\ &= \phi(2x) - \phi(2x-1).\end{aligned}$$

Let W_0 be the orthogonal complement of V_0 to V_1. Then, we can express V_1 as $V_1 = V_0 \otimes W_0$. It is easy to see that the functions $\{\psi(x-k)\}$, where k is an integer, form an orthonormal basis for W_0. In general, let W_j be the orthogonal complement of V_j to V_{j+1}. We have $V_{j+1} = V_j \otimes W_j$ and $\{\psi_{j,k}(x) = 2^{j/2}\psi(2x-k)\}$ form a basis for W_j. From $\cup_j V_j = \cup_j W_j$, we see that $\{W_j\}$ is also dense in $l^2(R)$ and the family $\{\psi_{j,k}(x)\}$, where both j and k are integers, is a basis for $l^2(R)$.

Suppose that $f(x)$ is a function in V_J. By the recursion $V_{j+1} = V_j \otimes W_j$, we have

$$V_J = V_{J-1} \otimes W_{J-1} = V_{J-2} \otimes W_{J-2} \otimes W_{J-1} = V_0 \otimes W_0 \otimes W_1 \otimes \cdots \otimes W_{J-1}.$$

Consequently, $f(x)$ can be written as

$$f(x) = f_0(x) + \sum_{i=0}^{J-1} g_i(x),$$

where $f_0(x) \in V_0$ and $g_j(x) \in W_j$. In an application with T data points such that $T = 2^J$ for some integer J, V_J is as far as we can go. There are 2^j basis elements at the level j ($j = 1, \ldots, J-1$) and, adding up, we have a total of $2^J - 1$ elements in the spaces W_j and one in V_0. Since all spaces involved are orthogonal, all the basis functions are orthonormal. This type of structured orthonormal basis allows for an efficient *multiresolution analysis*.

The Haar wavelet is easy to understand, but is discontinuous and is too coarse for some applications. Many other wavelet functions have been proposed in the literature. Figure 3.16 shows examples of symmlet (symmlet-8) and Daubechies (d16) wavelets. These wavelets have certain nice features. For instance, a symmlet-p family has a support of $2p - 1$ consecutive intervals. The wider the support, the longer the wavelet has to go to zero. This in turn enables the wavelet to achieve additional smoothness. Also, the scaling function $\phi(x)$ of a symmlet-p wavelet has p vanishing moments, namely

$$\int \phi(x)x^j dx = 0, \quad j = 1, \ldots, p.$$

Consequently, any degree-p polynomial over the $T = 2^J$ time points can be reproduced exactly in the space V_0. This is an improvement over the Haar wavelet, which can only reproduce any constants in its V_0.

In general, for a given father wavelet $\phi(x)$, the transform operations are similar to those of the Haar basis:

1. If V_0 is spanned by translations $\phi(x - k)$, then $V_1 \supset V_0$ is spanned by $\phi_{1,k}(x) = \sqrt{2}\phi(2x - k)$ and $\phi(x) = \sum_k h_k \phi_{1,k}(x)$, where h_k are filter coefficients. For instance, for Daubechies wavelet with parameter 2 (d2), we have

$$h_0 = \frac{1+\sqrt{3}}{4\sqrt{2}}, \quad h_1 = \frac{3+\sqrt{3}}{4\sqrt{2}}, \quad h_2 = \frac{3-\sqrt{3}}{4\sqrt{2}}, \quad h_3 = \frac{1-\sqrt{3}}{4\sqrt{2}}.$$

2. W_0 is spanned by $\psi(x) = \sum_k g_k \phi_{1,k}(x)$, where the coefficients are $g_k = (-1)^k h_{1-k}$.

The sequence $\{h_k\}$ is called the *low pass* or *low band* filter whereas $\{g_k\}$ is the *high pass* or *high band* filter in the signal processing literature. The low-pass filter

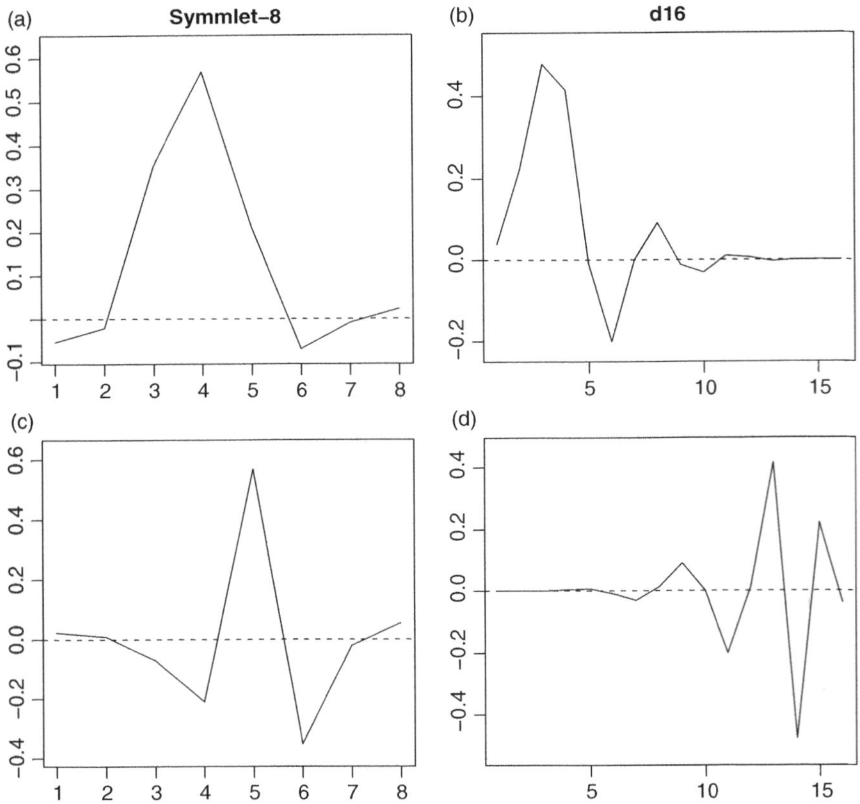

Figure 3.16 The father and mother basis functions of symmlet-8 and d16 wavelets. (a) and (b) are father wavelets and (a) and (c) are symmlet wavelets.

represents an *averaging* filter while the high-pass filter produces details. Thus, the wavelet coefficients correspond to details.

We now turn to data analysis. Let $\boldsymbol{x} = (x_1, \ldots, x_T)'$ be the vector of scalar time series, where, for simplicity, $T = 2^J$. Let $\boldsymbol{W}$ be the $T \times T$ matrix of the orthonormal wavelet basis functions evaluated at the T equally spaced observations. The vector $\tilde{\boldsymbol{x}} = \boldsymbol{W}'\boldsymbol{x}$ is called the *wavelet transform* of $\boldsymbol{x}$. Mathematically, $\tilde{\boldsymbol{x}}$ is the coefficient vector of full least squares regression and one can easily recover $\boldsymbol{x}$ from $\tilde{\boldsymbol{x}}$, namely $\boldsymbol{x} = \boldsymbol{W}\tilde{\boldsymbol{x}}$. To produce *smoothed* (or filtered) data, ideas of thresholding and regularization are useful. We discuss these ideas next.

3.5.3 Thresholding and Smoothing

Consider the vector of coefficients $\tilde{\boldsymbol{x}}$. The idea of thresholding is to keep the larger coefficients while constraining the smaller coefficients to zero. There are various

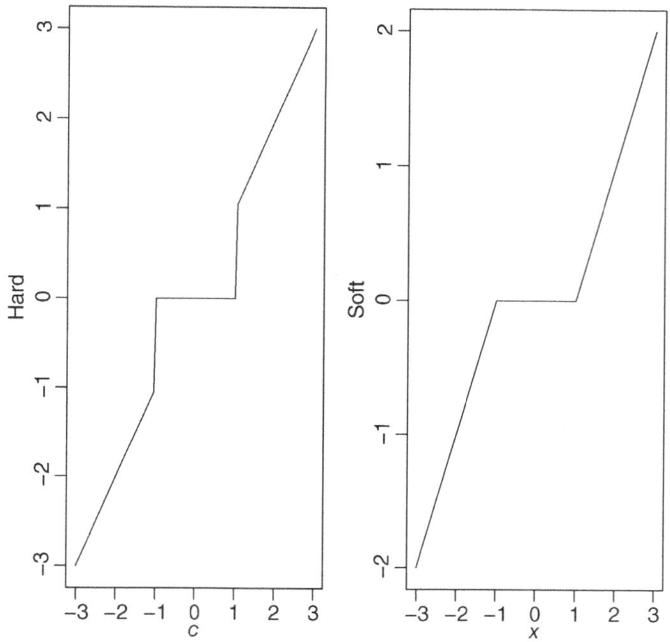

Figure 3.17 Illustrations of hard and soft thresholding with $\lambda = 1$.

types of thresholding available. The commonly used methods are *hard* and *soft* thresholding. For a pre-specified positive real number λ, the hard thresholding of a coefficient c is defined as

$$T_{hard}(c, \lambda) = cI(|c| > \lambda),$$

where $I(\cdot)$ denotes the indicator function. Thus, only large coefficients are kept. The soft thresholding is given by

$$T_{soft}(c, \lambda) = (c - sgn(c)\lambda)I(|c| > \lambda),$$

where $sgn(c)$ denotes the sign of c. Not only the small coefficients are set to zero, the large coefficients are reduced by the amount λ. Figure 3.17 shows hard and soft thresholding functions with $\lambda = 1$. Let $\boldsymbol{x}^*$ be the resulting coefficient vector after thresholding for a given λ. The smoothed data are then obtained by $\hat{\boldsymbol{x}} = \boldsymbol{W}\boldsymbol{x}^*$, which is the inverse wavelet transform. Donoho and Johnstone (1994) proposed a *universal* threshold based on properties of iid $N(0, 1)$ random variates, $\lambda = \hat{\sigma}\sqrt{2\ln(T)}$, where $\hat{\sigma}$ denotes an estimate of the standard deviation of the noise.

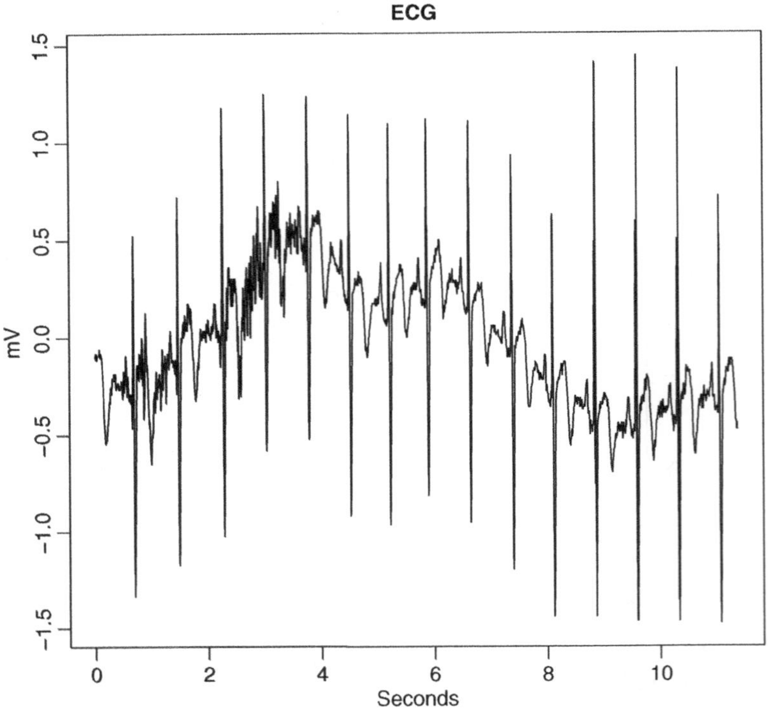

Figure 3.18 The electrocardiogram of a patient, taking 180 samplers per second for 2048 observations. The data are from the `wmtsa` package of R. See Percival and Walden (2000) for a description.

Another approach to obtain smoothed data is to apply regularization. Donoho and Johnstone (1994) considered a Stein unbiased risk estimation (*SURE shrinkage*):

$$\min_{\boldsymbol{\beta}} \|\boldsymbol{x} - \boldsymbol{W}\boldsymbol{\beta}\|^2 + 2\lambda\|\boldsymbol{\beta}\|_1, \tag{3.32}$$

which is the same as the well-known LASSO criterion for linear regression. Since $\boldsymbol{W}$ is orthonormal, a simple solution is

$$\hat{\beta}_j = sgn(\tilde{x}_j)(|\tilde{x}_j| - \lambda)_+, \tag{3.33}$$

where $\hat{\beta}_j$ is the jth element of $\hat{\boldsymbol{\beta}}$ and $\tilde{x}_j$ is the jth element of the wavelet transform $\tilde{\boldsymbol{x}}$. The inverse wavelet transform is then given by $\hat{\boldsymbol{x}} = \boldsymbol{W}\hat{\boldsymbol{\beta}}$.

Example 3.6 Figure 3.18 plots the electrocardiogram (ECG) data in the `wmtsa` package of R. Details of the data are given in Percival and Walden (2000, Section 4.10). The data consist of 2048 observations of electrocardiogram (ECG)

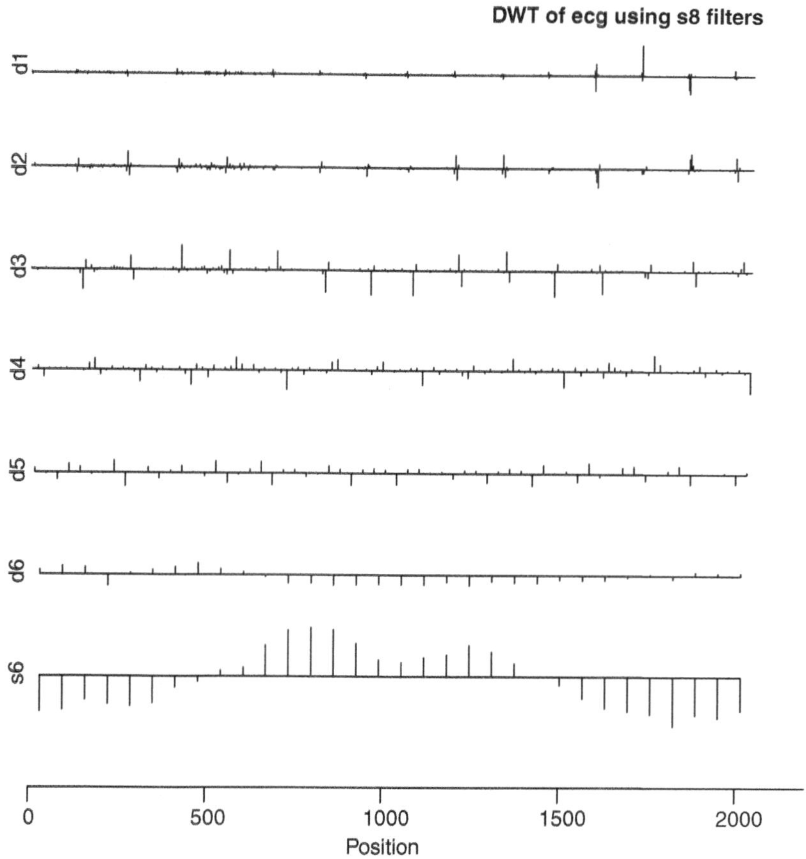

Figure 3.19 Discrete wavelet transform of the electrocardiogram data using six levels and the symmlet-8 wavelet.

measurements taken from a patient who occasionally experienced arrhythmia during the normal sinus rhythm. The measurements are in units of millivolts and collected at a rate of 180 samples per second. We use the data to demonstrate discrete wavelet transform with different wavelets and wavelet thresholding.

To begin, we apply discrete wavelet transform to the ECG data using both the symmlet-8 and Haar wavelets. Figures 3.19 and 3.20 show the results of discrete wavelet transform with $J = 6$ for symmlet-8 and Haar wavelet, respectively. We chose the level $J = 6$ for simplicity in deciphering the plots. Methods for selecting an optimal level are discussed in Percival and Walden (2000). From the plots, the differences in wavelet coefficients are visible. Note, for instance, that there are

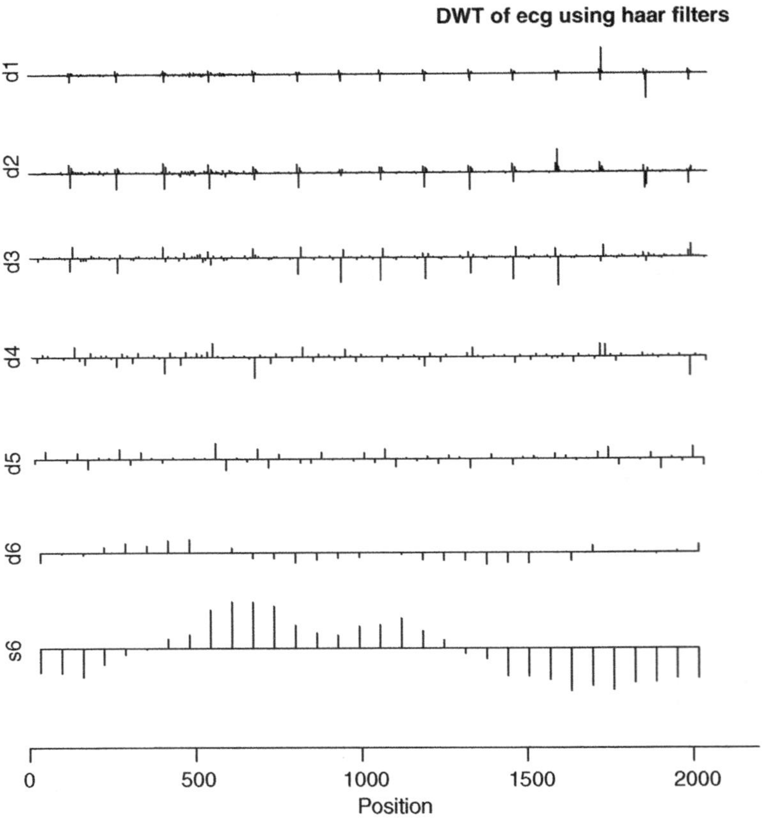

Figure 3.20 Discrete wavelet transform of the electrocardiogram data using six levels and the Haar wavelet.

more larger values in the `d1` plot of Figure 3.20. Figures 3.21 and 3.22 show the extended data analysis of the two discrete wavelet transforms of the ECG data. Extended data analysis is a graphical tool used to visually summarize the salient features of the discrete wavelet transform. The upper right plots of these two figures are the same as those of Figures 3.19 and 3.20. From the energy distribution of the two figures we see that the two wavelets work similarly, but also have some differences.

Next, we demonstrate the wavelet thresholding by applying both hard and soft thresholding with universal threshold to the `d1` coefficients of the discrete wavelet transform of the electrocardiogram data with symmlet-8 wavelet. Figure 3.23 shows the plots of empirical $d1$ and results of thresholding using the universal

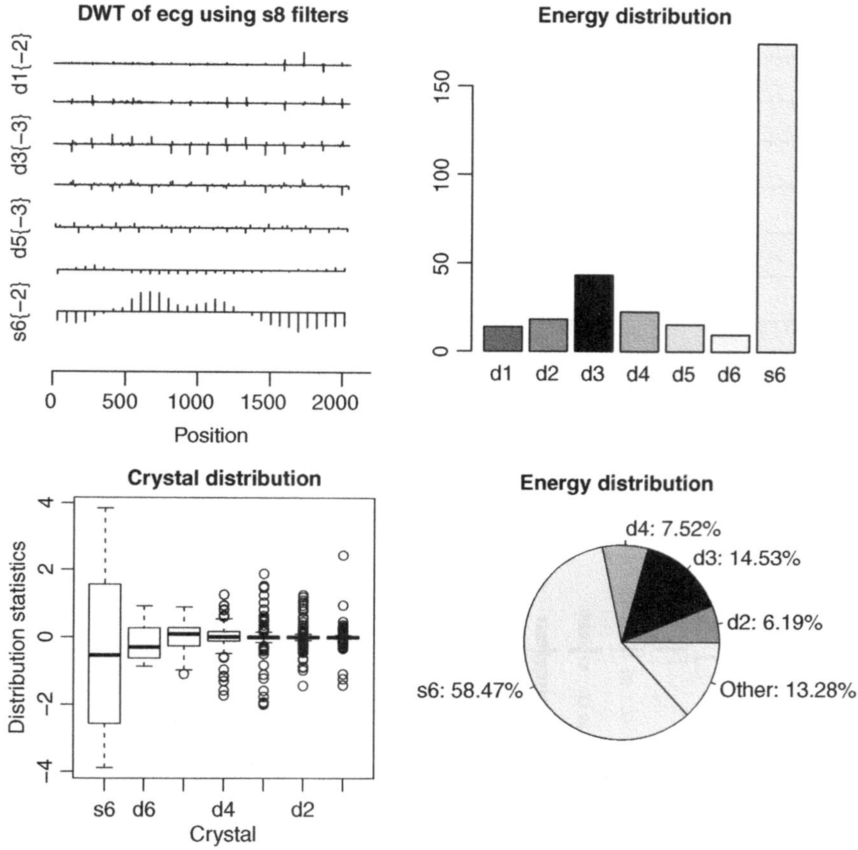

Figure 3.21 Extended data analysis of the discrete wavelet transform of electrocardiogram data using the symmlet-8 wavelet.

threshold. From the plots, the effects of thresholding are clearly seen. Finally, we apply the hard thresholding with universal threshold to the coefficients of `d1`, `d2`, `d3` and `d4` of the discrete wavelet transform with symmlet-8 wavelet, then use the resulting coefficients to perform inverse wavelet transform. Figure 3.24 shows the reconstructed electrocardiogram data (line), and the observed data (points). As expected, the reconstructed data are smoother.

Remark: There are several packages available in R for wavelet analysis, including `wmtsa`, `wavelets`, and `wavethresh`. We use `wmtsa` in the demonstration. The commands used are given below.

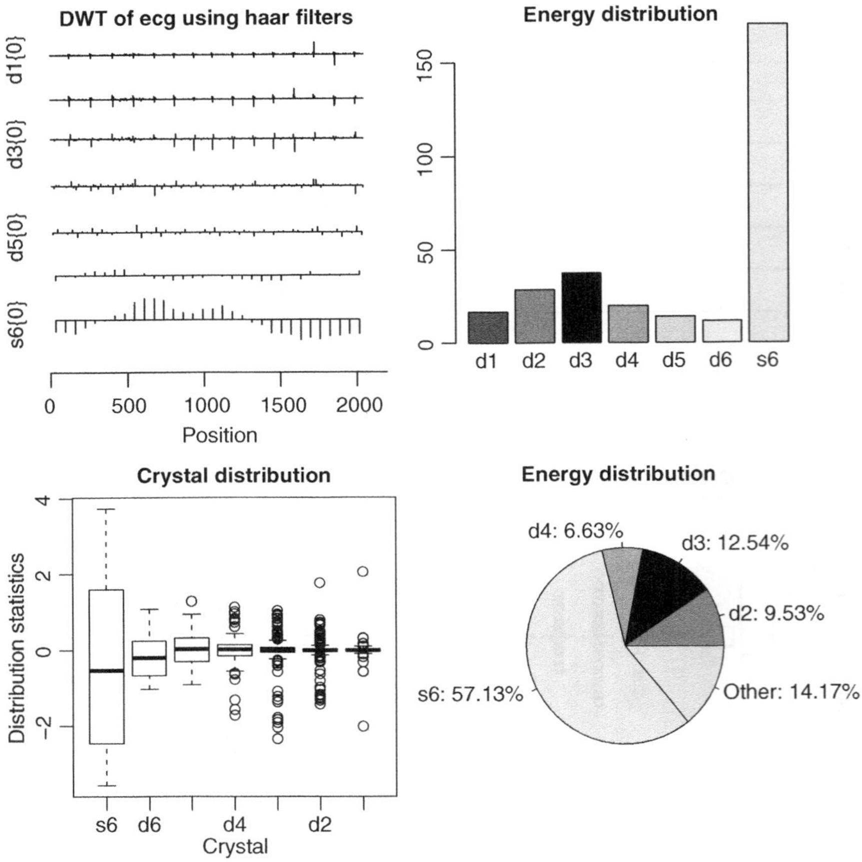

Figure 3.22 Extended data analysis of the discrete wavelet transform of electrocardiogram data using the Haar wavelet.

R demonstration: Wavelet transform via the `wmtsa` package.

```
> require(wmtsa)
> plot(ecg)
> m1 <- wavDWT(ecg,n.level=6)
> names(m1)
[1] "data"         "scales"       "series"       "dictionary" "shifted"
[6] "xform"
> plot(m1)
> eda.plot(m1)
> m2 <- wavDWT(ecg,n.level=6,wavelet="haar")
> plot(m2)
> eda.plot(m2)
```

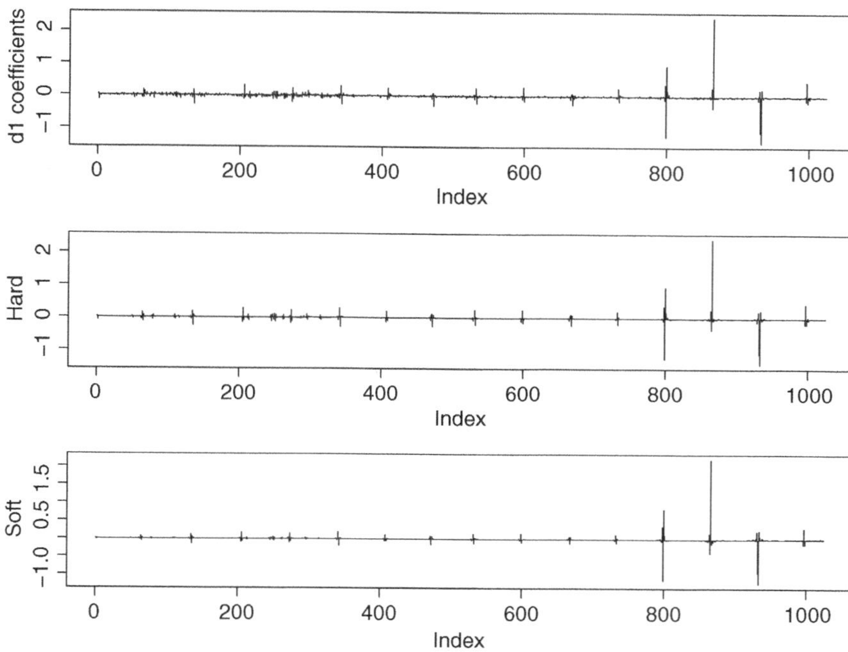

Figure 3.23 Hard and soft thresholding with universal threshold for the d1 coefficients of the discrete wavelet transform of the electrocardiogram data with symmlet-8 wavelet.

```
# Thresholding with "universal" threshold function
> d1 <- wavShrink(m1$data$d1) ## hard threshold
> d1a <- wavShrink(m1$data$d1,shrink.fun="soft")
> par(mfcol=c(3,1))
> plot(m1$data$d1,type="h",ylab="d1 coefficients")
> plot(d1,type="h",ylab='hard')
> plot(d1a,type="h",ylab='soft')
### Reconstruction
> d1 <- wavShrink(m1$data$d1)  ## Hard thresholding
> d2 <- wavShrink(m1$data$d2)
> d3 <- wavShrink(m1$data$d3)
> d4 <- wavShrink(m1$data$d4)
> m1copy <- m1
> m1copy$data$d1 <- d1
> m1copy$data$d2 <- d2
> m1copy$data$d3 <- d3
> m1copy$data$d4 <- d4
> ecg1 <- reconstruct(m1copy)
> ts.plot(ecg1,ylab='ecg',col="red")
> points(1:2048,as.numeric(ecg),cex=0.4)
```

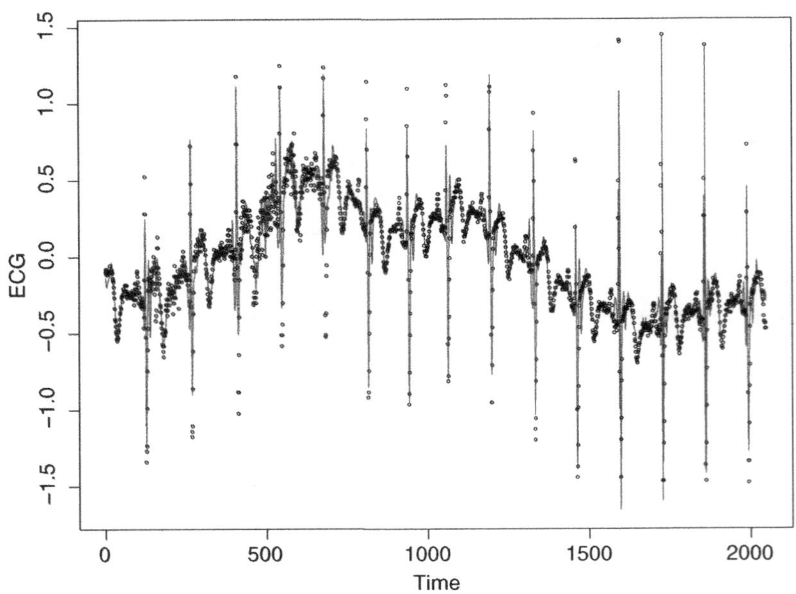

Figure 3.24 Observations and reconstructed values of electrocardiogram data. The reconstruction is done by the inverse wavelet transform after hard thresholding some of the coefficients of discrete wavelet transform of the observed data with the symmlet-8 wavelet.

3.6 NONLINEAR ADDITIVE MODELS

A zero-mean time series x_t follows a nonlinear additive autoregressive (NAAR) model of order p if it satisfies

$$x_t = \sum_{i=1}^{p} f_i(x_{t-i}) + a_t, \tag{3.34}$$

where $f_i(x_{t-i})$ is a smooth zero-mean function of x_{t-i} and $\{a_t\}$ is a sequence of independent and identically distributed random variables with mean zero and variance $0 < \sigma_a^2 < \infty$. This model was introduced by Chen and Tsay (1993). It is a special case of the general nonlinear AR(p) model $x_t = \mu + f(x_{t-1}, \ldots, x_{t-p}) + a_t$. See Equation (3.6). Except for the dynamic dependence, the model also belongs to the generalized additive models of Hastie and Tibshirani (1990).

For a given NAAR model, Chen and Tsay (1993) discussed two backfitting methods for model fitting. The two methods are the alternating conditional expectation (ACE) algorithm of Breiman and Friedman (1985) and the BRUTO algorithm of Hastie and Tibshirani (1990). These two algorithms are iterative, use nonparametric smoothing methods, and apply generalized cross-validation to select the smoothing parameters. They are called *backfitting procedures* since they are based on the fact that

$$f_i(x_{t-i}) = E\left[x_t - \sum_{k\neq i}^{p} f_k(x_{t-k})\right].$$

The smoothing methods used include smoothing splines, local polynomials, and the LOWESS procedures discussed in Section 3.3. Roughly speaking, for the NAAR model in Equation (3.34), one starts with some initial estimate $\hat{f}_i^{(0)}(x_{t-i})$ for each function $f_i(x_{t-i})$. To obtain the jth refined estimate of $f_i(x_{t-i})$, one considers the smoother

$$\hat{f}_i^{(j)}(x_{t-i}) = S_i\left[x_t - \sum_{k\neq i}^{p} \hat{f}_k^{(j-1)}(x_{t-k}) \middle| x_{t-i}\right],$$

where $S_i(\cdot)$ is smoother and the superscript (j) denotes the jth iteration. The smoother $S_i(\cdot)$ is typically represented by a symmetric matrix, as given in Section 3.4.2. The iteration cycles through $i = 1, \ldots, p$ for the jth iteration and continues until all individual fitted functions do not change. Basically, nonparametric estimation of the NAAR model in (3.34) involves the selection of p smoothing parameters via the generalized cross-validation, which is defined as

$$\mathrm{GCV}(\lambda_1, \ldots, \lambda_p) = \frac{\sum_{t=1}^{n}[x_t - \sum_{i=1}^{p}\hat{f}_{i,\lambda_i}(x_{t-i})]^2}{n[1 - \mathrm{tr}R(\lambda_1, \ldots, \lambda_p)/n]^2}, \tag{3.35}$$

where $R(\lambda_1, \ldots, \lambda_p)$ is the additive-fit operator for the given smoothing parameters $\lambda_1, \ldots, \lambda_p$ and $\hat{f}_{i,\lambda_i}(x_{t-i})$ denotes the fitted value of the function $f_i(x_{t-i})$. In practice, evaluating the denominator of Equation (3.35) requires intensive computation in order to optimize the p bandwidths simultaneously. See, for instance, the Newton method described in Gu and Wahba (1991).

To ease the computational burden, the BRUTO algorithm attempts to minimize the statistic

$$\mathrm{GCV}^B(\lambda_1, \ldots, \lambda_p) = \frac{\frac{1}{n}\sum_{t=1}^{n}[x_t - \sum_{i=1}^{p}\hat{f}_{i,\lambda_i}(x_{t-i})]^2}{\{1 - [1 + \sum_{i=1}^{p}(\mathrm{tr}S_i(\lambda_i) - 1)]/n\}^2}. \tag{3.36}$$

The use of $1 + \sum_{i=1}^{p}[\mathrm{tr}S_i(\lambda_i) - 1]$ in place of $\mathrm{tr}R(\lambda_1, \ldots, \lambda_p)$ reduces the computation from $O(n^3)$ to $O(n)$. Details and convergence of the BRUTO algorithm can be found in Hastie and Tibshirani (1990) and the references therein. A special feature of the criterion in Equation (3.36) is that the minimization is carried out one parameter at a time yet the objective function is global.

The ACE algorithm attempts to estimate nonparametric models and a suitable transformation of the dependent variable jointly. Specifically, the working model corresponding to the NAAR model in Equation (3.34) is

$$\theta(x_t) = \sum_{i=1}^{p} f_i(x_{t-i}) + a_t, \tag{3.37}$$

where $\theta(x_t)$ denotes a link function between the conditional expectation of x_t given the predictor $\sum_{i=1}^{p} f_i(x_{t-i})$ and the initial values $x_1, \dots, x_p$. The link function is assumed to be monotone, zero mean and unit variance for identifiablity. For the NAAR models, $\theta(x_t)$ is simply the identity function. The ACE algorithm for estimating the model in Equation (3.37) is as follows:

(i) Initialize: Let $\theta(x_t) = [x_t - E(x_t)]/\text{var}(x_t)^{1/2}$.

(ii) Fit an additive model to $\theta(x_t)$ to update the functions $f_1(x_{t-1}), \dots, f_p(x_{t-p})$.

(iii) Compute the expectation $\hat{\theta}(x_t) = E[\sum_{i=1}^{p} f_i(x_{t-i})|x_t]$ and standardize to obtain a new estimate $\theta(x_t) = \hat{\theta}(x_t)/\{\text{var}[\hat{\theta}(x_t)]\}^{1/2}$.

(iv) Alternate steps (ii) and (iii) until $E[\theta(x_t) - \sum_{i=1}^{p} f_i(x_{t-i})]^2$ does not change.

The NAAR model in Equation (3.34) can easily be extended to include some explanatory variables. For instance, let z_t denote an explanatory variable that might be helpful in understanding or predicting the time series x_t. A generalized nonlinear additive model then becomes

$$x_t = \mu + \sum_{i=1}^{p} f_i(x_{t-i}) + \sum_{j=0}^{s} g_j(z_{t-j}) + a_t, \tag{3.38}$$

where s is a nonnegative integer and $g_j(\cdot)$ are smooth functions similar to $f_i(\cdot)$. The estimation methods for NAAR models continue to apply to the model in Equation (3.38).

Example 3.7 Consider the weekly US regular gasoline prices, dollars per gallon, from January 7, 2008 to March 20, 2017 for 481 observations. It is known that the price of gasoline depends on the crude oil price so we also consider the weekly crude oil prices, dollars per barrel of West Texas Intermediate (WTI). The two price series are shown in Figure 3.25. In this analysis, we use the first-differenced price series, that is, we analyze the price changes. The sample size used is then $T = 480$. Let x_t be the weekly change series of the regular gasoline price and z_t be the change series of crude oil price. Figure 3.26 shows some scatter plots of x_t against lagged values of x_t and z_t. As expected, the plots exhibit dependence between x_t and the selected lagged variables.

If univariate time series models with exogenous variables are used, we obtain the model

$$x_t = 0.366x_{t-1} + 0.078x_{t-2} + 0.0094z_t + 0.0021z_{t-1} + e_t, \quad \hat{\sigma}_e = 0.0419, \tag{3.39}$$

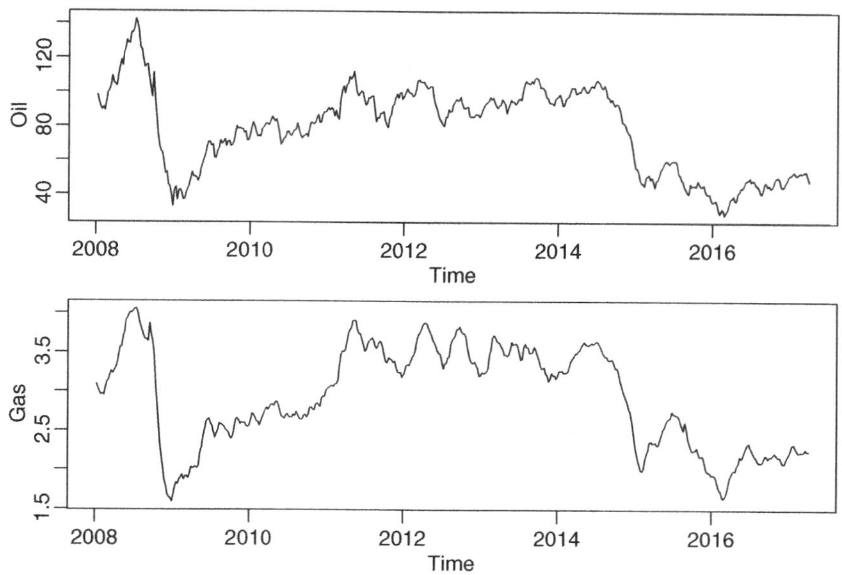

Figure 3.25 Time plots of weekly US regular gasoline prices and crude oil prices from January 7, 2008 to March 20, 2017.

where all coefficient estimates are statistically significant at the 5% level. Model checking indicates that this model fits the data reasonably well. On the other hand, the upper-right plot of Figure 3.26 shows some nonlinear relationship between x_t and z_t. This is understandable as the gasoline price tends to increase immediately following a jump in the crude oil price, but decline slowly after a drop in the oil price.

To explore the nonlinear relationship, we use a nonlinear additive model with explanatory variables x_{t-1}, x_{t-2}, z_t, and z_{t-1}. The estimation is carried out via the `gam` package of R by using smoothing splines with 4 degrees of freedom for each explanatory variable. The nonlinear additive model improves the fit. However, the associated analysis of variance for the nonparametric effects indicates that the explanatory variable x_{t-2} does not contribute significantly to the model. See below:

```
Anova for Nonparametric Effects
                Npar Df  Npar F      Pr(F)
s(pcgasm1, 4)         3 11.1235 4.623e-07 ***
s(pcgasm2, 4)         3  1.4769 0.2200580
s(pcoil, 4)           3  6.8333 0.0001634 ***
s(pcoilm1, 4)         3  4.4194 0.0044548 **
```

To simplify the model, we remove x_{t-2} and re-fit the nonlinear additive model. All three explanatory variables are now statistically significant at the usual 5% level.

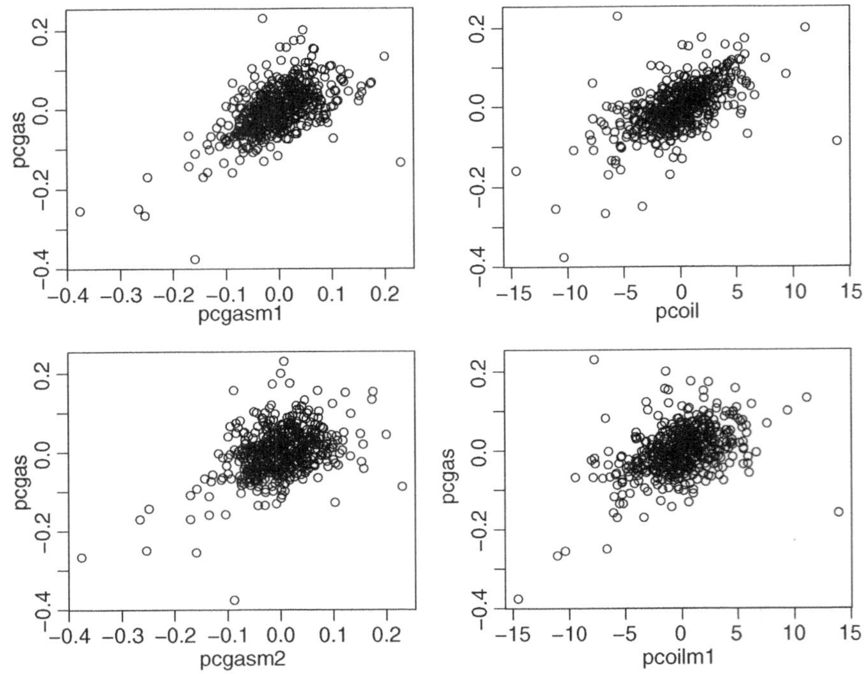

Figure 3.26 Selected scatter plots of price changes of regular gasoline and their lagged variables. The data are from January 7, 2008 to March 20, 2017.

For comparison, the residual standard error of the linear ARX model of Equation (3.39) is 0.0419 whereas that of the nonlinear additive model is 0.0390. Note that caution should be exercised when comparing residual sum of squares obtained from a nonparametric method, since it depends on the bandwidth. A small bandwidth often results in a small residual sum of squares. Comparing cross-validation criterion is often preferred. Figure 3.27 shows the smooth functions used in the nonlinear additive model. The plots also include the point-wise 95% confidence intervals of the smooth functions. From the plots, the nonlinear feature of the relationship is seen. Furthermore, we also use the analysis of variance to compare the linear model of Equation (3.39) with the fitted nonlinear additive model. The result is given below:

```
> anova(m2.lm,m2.gam)
Analysis of Variance Table

Model 1: pcgas ~ -1 + pcgasm1 + pcgasm2 + pcoil + pcoilm1
Model 2: pcgas ~ -1 + s(pcgasm1, 4) + s(pcoil, 4) + s(pcoilm1, 4)
```

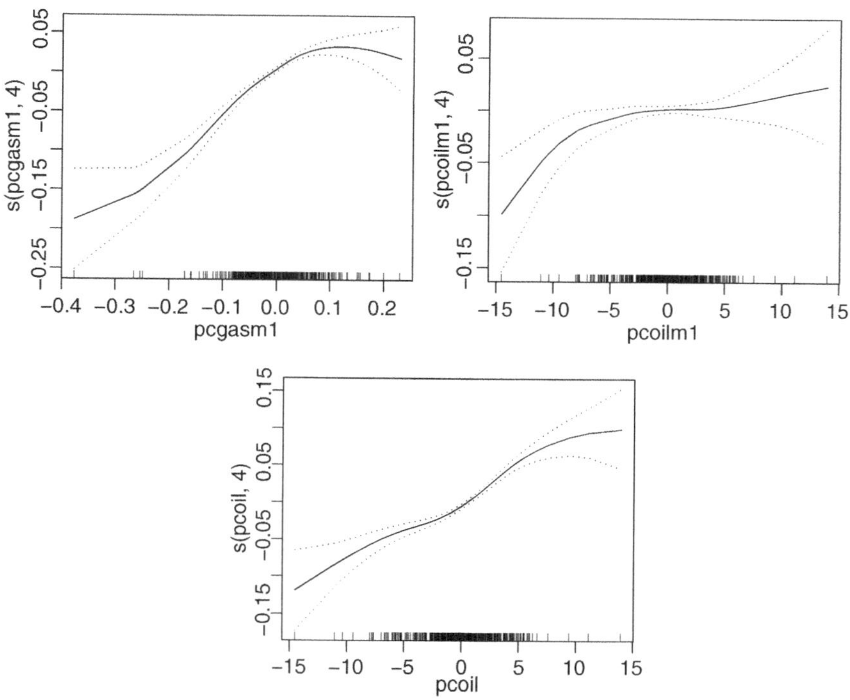

Figure 3.27 Smooth functions of explanatory variables resulted from fitting an additive model to the price change of the US weekly regular gasoline price. The dashed lines indicate point-wise 95% confidence intervals. The data are from January 7, 2008 to March 20, 2017.

```
Res.Df       RSS     Df Sum of Sq       F     Pr(>F)
1     474 0.83218
2     466 0.72499 8.0001    0.10719 8.6123 5.619e-11 ***
```

Clearly, the nonlinear model shows significant improvements over the linear ARX model. Finally, we use the `bruto` command of the `mda` package in R to confirm the fitted nonlinear additive model. The result of BTUTO fit is

```
> require(mda)
> m3.bruto <- bruto(X[,-1],X[,1])
> m3.bruto$type
pcgasm1  pcgasm2    pcoil  pcoilm1
  smooth excluded   smooth    linear
Levels: excluded linear smooth
```

The result confirms the removal of x_{t-2} and using smooth spline functions of x_{t-1} and z_t (pcoil).

R commands used in fitting the NAAR model via the gam package.

```
require(gam)
da <- read.csv("w-gasoil-pc.csv")
X <- data.frame(da[,-1])
m2.lm <- lm(pcgas~ -1+pcgasm1+pcgasm2+pcoil+pcoilm1,data=X)
m2 <- gam(pcgas~ -1+s(pcgasm1,4)+s(pcgasm2,4)+s(pcoil,4)+s(pcoilm1,4),data=X)
summary(m2)
m2.gam <- gam(pcgas~ -1+s(pcgasm1,4)+s(pcoil,4)+s(pcoilm1,4),data=X)
par(mfcol=c(2,2))
plot(m2.gas)
par(mfcol=c(1,1))
anova(m2.lm,m2.gam)
require(mda)
m3.bruto <- bruto(X[,-1],X[,1])
m3.bruto$type
```

3.7 INDEX MODEL AND SLICED INVERSE REGRESSION

Another class of nonlinear models useful in many scientific areas is the index models, especially the single index model. This model is closely related to dimension reduction and projection pursue in the statistical literature. Let x_t be the time series of interest and $z_t = (z_{1t}, \ldots, z_{kt})'$ be a k-dimensional vector of explanatory variables, the components of which may include lagged variables of x_t. A general index model can be written as

$$x_t = f(\boldsymbol{\beta}' z_t) + a_t, \tag{3.40}$$

where $\{a_t\}$ is a white noise series and independent of $\boldsymbol{\beta}' z_t$, $\boldsymbol{\beta}$ is a $k \times m$ full-rank matrix of real numbers with $m > 0$, and $f(\cdot)$ denotes a smooth function. Typically, we have $m \ll k$ so that x_t depends on m linear combinations of z_t rather than the k explanatory variables directly. The integer m is referred to as the effective dimension of the model or the number of indices, and the function $f(\cdot)$ is m-dimensional. In particular, if $m = 1$, then x_t depends on $v_t = \boldsymbol{\beta}' z_t$, which is a scalar, and the function $f(\cdot)$ is one-dimensional. The resulting model is called the *single index* model.

Strictly speaking, the $\boldsymbol{\beta}$ matrix of Equation (3.40) is not uniquely defined because $f(\cdot)$ is not precisely specified. This type of non-uniqueness does not cause any problems in applications because we can require that columns of $\boldsymbol{\beta}$ are normalized and orthogonal to each other so that $\boldsymbol{\beta}' \boldsymbol{\beta} = \boldsymbol{I}_m$, the $m \times m$ identity matrix. Since both $f(\cdot)$ and $\boldsymbol{\beta}$ are unknown, we need certain conditions to obtain consistent estimates of $\boldsymbol{\beta}$. To this end, many methods have been proposed in the literature, and one of these methods is the *sliced inverse regression* of Li (1991). A key concept of the index model is that the conditional distribution of x_t given z_t is the same as that of x_t given $\boldsymbol{\beta}' z_t$. Therefore, the subspace expanded by $\boldsymbol{\beta}' z_t$ is referred to as the central subspace, which is of dimension m, not k. Since this subspace of z_t is in the predictor space, it is referred to as an inverse subspace. Hence, the name *inverse regression* is used. A sufficient condition for the existence of a lower dimensional central inverse subspace is as follows. For any arbitrary non-zero vector $\boldsymbol{b} \in R^k$, there exists a constant c_0 and an m-dimensional constant vector $\boldsymbol{c} = (c_1, \dots, c_m)'$ such that

$$E(\boldsymbol{b}' z_t | \boldsymbol{\beta}' z_t) = c_0 + \boldsymbol{c}'(\boldsymbol{\beta}' z_t). \tag{3.41}$$

This condition essentially states that given $\boldsymbol{\beta}' z_t$, the expectation of any non-zero linear combination of z_t is in the subspace generated by $\boldsymbol{\beta}' z_t$. Under this condition Li (1991) established the following theorem.

Theorem 3.1 Consider the index model of Equation (3.40) and assume that the condition (3.41) holds. Then, the centered inverse regression curve $E(z_t \mid x_t) - E(z_t)$ is contained in the linear subspace spanned by $\boldsymbol{\Sigma}_z \boldsymbol{\beta}$, where $\boldsymbol{\Sigma}_z$ is the covariance matrix of z_t.

To make use of Theorem 3.1, one needs to estimate $\boldsymbol{\Sigma}_z$ and the covariance matrix of $E(z_t \mid x_t)$. The former is estimated by the sample covariance matrix of z_t. The latter, on the other hand, requires some creative thinking. Li (1991) provided a crude, but effective, estimate based on slices of z_t given x_t. This leads to the procedure of *sliced inverse regression*. To describe the procedure, we express the data as $(\boldsymbol{X}, \boldsymbol{Z})$, where $\boldsymbol{X} = (x_1, \dots, x_T)'$ and $\boldsymbol{Z}$ is a $T \times k$ design matrix such that the tth row is z_t'.

SIR procedure: Let H be a positive integer.

1. Sort the data $(\boldsymbol{X}, \boldsymbol{Z})$ by x_t (in increasing order).
2. Divide the data set into H non-overlapping slices as equally as possible. Denote the slices by $(\boldsymbol{X}_{(i)}, \boldsymbol{Z}_{(i)})$, $i = 1, \dots, H$.
3. For each slice, compute the sample mean of $\boldsymbol{Z}_{(i)}$. That is, compute the mean vector of the regressors of each slice.
   Let $\bar{\boldsymbol{Z}}_{(i)} = \frac{1}{n_i} \sum_{j=1}^{n_i} z_{(i),j}$, where n_i is the number of data points in the ith slice and $z_{(i),j}$ is the jth data point of the ith slice.
4. Compute the covariance matrix of the slice means of $\boldsymbol{Z}$ weighted by the size of the slice.

$$\hat{\boldsymbol{\Sigma}}_s = \frac{1}{T} \sum_{i=1}^{H} n_i (\bar{\boldsymbol{Z}}_{(i)} - \bar{\boldsymbol{Z}})(\bar{\boldsymbol{Z}}_{(i)} - \bar{\boldsymbol{Z}})',$$

where T is the sample size, $\bar{\mathbf{Z}}$ is the sample mean of $\mathbf{Z}$, and the subscript s is used to signify that $\mathbf{\Sigma}_s$ is formed using slices.

5. Find the SIR directions by performing the eigenvalue–eigenvector analysis of $\widehat{\mathbf{\Sigma}}_z^{-1}\widehat{\mathbf{\Sigma}}_s$, where $\widehat{\mathbf{\Sigma}}_z$ is the sample covariance matrix of z_t. That is, obtain
$$\widehat{\mathbf{\Sigma}}_s\widehat{\boldsymbol{\beta}}_i = \lambda_i\widehat{\mathbf{\Sigma}_z}\widehat{\boldsymbol{\beta}}_i, \quad i = 1, \dots, k,$$
where $\lambda_1 \geq \lambda_2 \geq \cdots \geq \lambda_k$. $\widehat{\boldsymbol{\beta}}_i$ is called the ith SIR direction.
6. Use each SIR direction to form a linear combination of z_t.
7. Estimate $f(\cdot)$ by smoothing x_t against $\widehat{\boldsymbol{\beta}}_i' z_t$.

The statistic $\widehat{\mathbf{\Sigma}}_s$ is used to estimate the covariance matrix of $E(z_t \mid x_t)$, and the last step can be done by one of the smoothing methods discussed in the chapter.

The eigenvalues λ_i of the SIR procedure can be used to find the dimension m of the central inverse subspace. Li (1991) also showed that, under the assumption that z_t are normally distributed, then $T(k-j)\bar{\lambda}_{k-j}$ follows a chi-square distribution with $(k-j)(H-j-1)$ degrees of freedom, where $\bar{\lambda}_{k-j}$ is the average of the $(k-j)$ smallest eigenvalues.

The number of slices H does not need to be large for the sliced inverse regression to perform well. Furthermore, there exist several extensions of the SIR method, including the principal Hessian directions of Li (1992) and Cook (1998), an adaptive approach by Xia et al. (2002), and the sliced average variance estimation (SAVE) of Shao et al. (2007). In applications, the `dr` package of R can be used to perform the SIR method and its extensions for dimension reduction.

Example 3.8 Consider again the weekly US regular gasoline prices and the crude oil prices of Example 3.7. Here we also use the price change series x_t and z_t, and our goal is to demonstrate the application of sliced inverse regression in exploring the dependence of x_t on its lagged values x_{t-1} and x_{t-2}, and on z_t and z_{t-1}. To this end, we have $z_t = (x_{t-1}, x_{t-2}, z_t, z_{t-1})'$.

Applying the `dr` command in the `dr` package, we obtain the following result:

```
> require(dr)
> da <- read.csv("w-gasoil-pc.csv")
> head(da)
  N  pcgas pcgasm1 pcgasm2 pcoil pcoilm1
1 1 -0.038  -0.050  -0.047 -2.10   -3.25
....
> X <- da[,-1]
```

```
> m1 <- dr(pcgas~ .,data=data.frame(X))
> summary(m1)
Call: dr(formula = pcgas ~ ., data = data.frame(X))

Method:  sir with 8 slices, n = 478.
Slice Sizes:
59 59 59 60 61 59 61 60

Estimated Basis Vectors for Central Subspace:
            Dir1      Dir2     Dir3      Dir4
pcgasm1 0.996244 -0.037141  0.65816  0.902187
pcgasm2 0.083244  0.999245 -0.75199 -0.431118
pcoil   0.023252 -0.011099 -0.02416 -0.002282
pcoilm1 0.005176 -0.002633  0.02744 -0.013846

              Dir1    Dir2     Dir3     Dir4
Eigenvalues 0.4983 0.03988 0.004562 0.001055
R^2(OLS|dr) 0.9975 0.99855 0.999395 1.000000

Large-sample Marginal Dimension Tests:
                 Stat df p.value
0D vs >= 1D 259.9249 28  0.0000
1D vs >= 2D  21.7489 18  0.2433
2D vs >= 3D   2.6851 10  0.9879
3D vs >= 4D   0.5043  4  0.9731
```

The default option of the `dr` command uses eight slices each with size approximately 60. The large-sample test indicates that the dimension of the effective central subspace is 1. The null hypothesis of zero dimension is clearly rejected with the p value being zero whereas that of one dimension cannot be rejected with p value 0.24. This conclusion can also be seen as the largest eigenvalue is 0.49, which explains about 99.75% of the variability. The second largest eigenvalue is only 0.040.

Let r_t denote the transformed data of the first principal direction shown by direction 1 of the SIR method. Figure 3.28 shows the scatter plot of x_t versus r_t. The solid line of the plot denotes the fitted values via the local smoothing method `lowess`. From the plot it is seen that there exists some minor nonlinearity between x_t and r_t. More importantly, the local smoothing suggests a slope change around $r_t \approx -0.2$. Therefore, we used a simple threshold model and obtained

$$x_t = \begin{cases} 0.366r_t + \epsilon_t, & \text{if } r_t \geq -0.2, \\ 0.462r_t + \epsilon_t, & \text{if } r_t < -0.2, \end{cases} \tag{3.42}$$

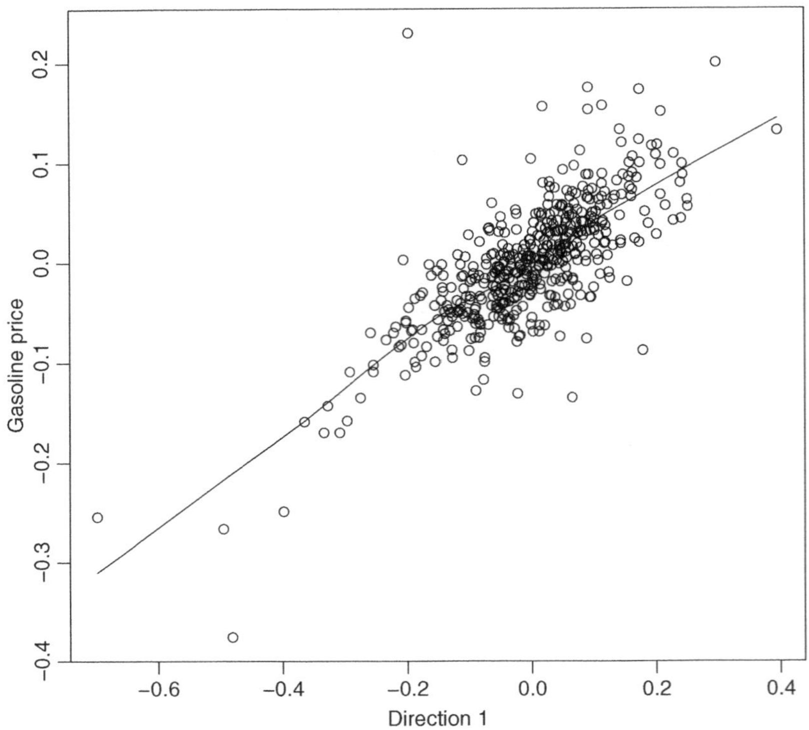

Figure 3.28 Scatter plot of changes in weekly regular gasoline price versus the first principal direction of the sliced inverse regression. The solid line denotes the local smoothing fit via `lowess`.

where the residual standard error is 0.042 and the two coefficient estimates are statistically significant at the 1% level. The Ljung–Box statistics of the residuals indicate that there exist no significant serial correlations in the residuals. Specifically, we have $Q(12) = 10.35$ with p value 0.59.

```
### Further Analysis
> TZ <- as.matrix(X[,-1])%*%m1$evectors
> idx <- c(1:478)[TZ[,1] < -0.2]
> R1 <- rep(0,478)
> R1[idx] <- 1
> LTZ <- TZ[,1]*R1
> m2 <- lm(X$pcgas~ -1+TZ[,1]+LTZ)
> summary(m2)

Call: lm(formula = X$pcgas ~ -1 + TZ[, 1] + LTZ)
```

```
Coefficients:
        Estimate Std. Error t value Pr(>|t|)
TZ[, 1]  0.36633    0.02017  18.165  < 2e-16 ***
LTZ      0.09613    0.03345   2.874  0.00424 **
---
Residual standard error: 0.04152 on 476 degrees of freedom
Multiple R-squared:  0.5697,    Adjusted R-squared:  0.5679

> Box.test(m2$residuals,lag=12,type='Ljung')
        Box-Ljung test
data:  m2$residuals
X-squared = 10.353, df = 12, p-value = 0.585
```

3.8 EXERCISES

3.1 Consider the half-hourly electricity demands of Mondays in Adelaide, Australia, from July 6, 1997 to March 31, 2007. The data are available from the `fds` package of R. See Hyndman (2016). There are 48 observations in a day and 508 weeks. The original data consist of demands from Sunday to Saturday, but we focus here on Monday only. Since the demands have an upward trend so we focus on $x_t = (1 - B^{48})d_t$ with d_t denoting the demand. Use the commands below to obtain the x_t series.

```
require(fds)
data(mondaydemand)
de <- c(mondaydemand$y)
xt <- diff(de,48)
```

(a) Perform discrete wavelet transform of x_t using six levels and the symmlet-8 wavelet. Show the plot and eda.plot, as in Example 3.6.

(b) Repeat part (a) using the Haar wavelet.

(c) Perform a reconstruction using hard thresholding, as in Example 3.6.

3.2 Consider the US monthly leading index from January 1982 to July 2017. The data can be obtained from FRED (code: USSLIND) and are seasonally adjusted. Let x_t denote the monthly leading index. Use lowess and local polynomials with degrees 0, 1, and 2 to explore the dependence of x_t on x_{t-6}. Show a scatter plot of x_t versus x_{t-6} with fitted lines of lowess and local polynomials. Is the relationship nonlinear? Why?

3.3 Consider again the US monthly leading index of question 2. This time use B-splines and smooth splines to explore the dependence of x_t on x_{t-6}. Show the scatter plot of x_t versus x_{t-6} with fitted lines via B-splines and smooth splines. Is the relationship nonlinear? Why?

3.4 Consider again the US monthly leading index of question 2.

(a) Fit a linear AR(6) model to x_t and refine the model by removing the estimates that are not statistically significant at the usual 5% level. Write down the fitted model.

(b) Fit a NAAR model

$$x_t = c_0 + f_1(x_{t-1}) + f_3(x_{t-3}) + f_4(x_{t-4}) + f_6(x_{t-6}) + \epsilon_t,$$

to the data, where $f_i(x_{t-i})$ is a spline function. You may use the `gam` package.

(c) Perform an analysis of variance to confirm that the NAAR model is significantly different from the linear AR(6) model.

3.5 Consider the US quarterly index of manufacturing real output from 1987.I to 2017.II. The data can be obtained from FRED (code: OUTMS). Let $x_t = \ln(O_t) - \ln(O_{t-1})$, where O_t is the real output index (2009=100).

(a) Build a linear AR model for x_t. Perform model checking and write down the fitted model.

(b) Fit the NAAR model

$$x_t = c_0 + f_1(x_{t-1}) + f_2(x_{t-2}) + \epsilon_t,$$

using the `gam` package. What are the estimates of the coefficients? Perform model checking. Is the model adequate? Why?

(c) Perform analysis of variance to confirm that the NAAR model contributes significantly over the linear AR model of part (a).

(d) Obtain the residual plots of the two fitted models in parts (a) and (b). Comments on the fit of the models.

REFERENCES

Bierens, H. J. (1983). Uniform consistency of kernel estimators of a regression function under generalized conditions. *Journal of the American Statistical Association* **78**: 699–707.

Bierens, H. J. (1987). Kernel estimators of regression functions in *Advances on Econometrics: Fifth World Congress*, ed. T. F. Bewley. Cambridge University Press, Cambridge.

Breiman, L. and Friedman, J. H. (1985). Estimating optimal transformation for multiple regression and correlation (with discussion). *Journal of the American Statistical Association* **80**: 580–619.

Chen, R. and Tsay, R. S. (1993). Nonlinear additive ARX models. *Journal of the American Statistical Association* **88**: 955–967.

Cleveland, W. S. (1979). Robust locally weighted regression and smoothing scatterplots. *Journal of the American Statistical Association* **74**: 829–836.

Collomb, G. and Härdle, W. (1986). Strong uniform convergence rates in robust nonparametric time series analysis and prediction: Kernel regression estimation from dependent observations. *Stochastic Processes and Their Applications* **23**: 77–89.

Cook, R. D. (1998). Principal Hessian directions revisited (with discussion). *Journal of the American Statistical Association* **93**: 84–100.

Daubechies, B. (1992). *Ten Lectures in Wavelets*. Society for Industrial and Applied Mathematics, Philadelphia.

de Boor, C. (1978). *A Practical Guide to Splines*. Springer-Verlag, New York.

Donoho, D. and Johnstone, I. (1994). Ideal spatial adaptation by wavelet shrinkage. *Biometrika* **81**: 425–455.

Epanechnikov, V. (1969). Nonparametric estimates of a multivariate probability density. *Theory of Probability and Its Applications* **14**: 153–158.

Fan, J. and Gijbels, I. (1996). *Local Polynomial Modeling and Its Applications*. Chapman and Hall, London.

Fan, J. and Yao, Q. (2003). *Nonlinear Time Series: Nonparametric and Parametric Methods*. Springer-Verlag, New York.

Gu, C. and Wahba, G. (1991). Minimizing GCV/GML scores with multiple smoothing parameters via the Newton method. *SIAM Journal of Scientific and Statistical Computing* **12**: 383–398.

Härdle, W. (1990). *Applied Nonparametric Regression*. Cambridge University Press, New York.

Härdle, W., Lütkepohl, H., and Chen, C. (1997). A review of nonparametric time series analysis. *International Statistical Review* **65**: 49–72.

Hastie, T. J. and Tibshirani, R. J. (1990). *Generalized Additive Models*. Chapmand and Hall, New York.

Hastie, T., Tibshirani, R. and Friedman, J. (2001). *The Elements of Statistical Learning: Data Mining, Inference, and Prediction*. Springer, New York.

Hyndman, R. J. (2016). Time Series Data Library. http://data.is/TSDLdemo. Accessed March, 2016.

Li, K. C. (1991). Sliced inverse regression for dimension reduction (with discussion). *Journal of the American Statistical Association* **86**: 316–342.

Li, K. C. (1992). On principal Hessian directions for data visualization and dimension reduction: Another application of Stein's lemma. *Journal of the American Statistical Association* **87**: 1025–1039.

Masry, E. and Tjφstheim, D. (1995). Nonparametric estimation and identification of nonlinear ARCH time series: Strong consistency and asymptotic normality. *Econometric Theory* **11**: 258–289.

Müller, H. G. (1988). *Nonparametric Regression Analysis of Longitudinal Data*. Lecture Notes in Statistics, 46. Springer-Verlag, Berlin.

Nadaraya, E. A. (1964). On estimating regression. *Theory of Probability and its Applications* **9**: 141–142.

Nason, G. P. (2008). *Wavelet Methods in Statistics with R*. Springer, New York.

Percival, D. B. and Walden, A. T. (2000). *Wavelet Methods for Time Series Analysis*. Cambridge University Press, Cambridge.

Priestley, M. B. and Chao, M. T. (1972). Nonparametric function fitting. *Journal of the Royal Statistical Society Series B* **34**: 384–392.

Robinson, P. (1983). Nonparametric estimation for time series models. *Journal of Time Series Analysis* **4**: 185–208.

Shao, Y., Cook, R. D. and Weisberg, S. (2007). Marginal tests with sliced average variance estimation. *Biometrika* **94**: 285–296.

Sharpe, W. (1964). Capital asset prices: A theory of market equilibrium under conditional risk. *Journal of Finance* **19**: 425–442.

Stone, C. J. (1977). Consistent nonparametric regression. *Annals of Statistics* **5**: 595–645.

Tong, H. (1990). *Non-Linear Time Series: A Dynamical System Approach*. Oxford University Press, Oxford.

Truong, Y. K. (1993). A nonparametric framework for time series analysis in *New Directions in Time Series Analysis*. Springer, New York.

Tweedie, R. L. (1975). Sufficient conditions for ergodicity and recurrence of Markov chain on a general state space. *Stochastic Processes and Their Applications* **3**: 385–403.

Watson, G. S. (1964). Smooth regression analysis. *Sankhya Series A* **26**: 359–372.

Wood, S. N. (2003). Thin plate regression splines. *Journal of the Royal Statistical Society, Series B* **65**: 95–114.

Xia, Y., Tong, H., Li, W.K., and Zhu, L. (2002). An adaptive estimation of dimension reduction space (with discussion). *Journal of the Royal Statistical Society, Series B* **64**: 363–410.

CHAPTER 4

NEURAL NETWORKS, DEEP LEARNING, AND TREE-BASED METHODS

In this chapter we introduce some recent developments in high-dimensional statistical analysis that are useful in time series analysis. The methods discussed include neural networks, deep learning, and regression trees. Our discussions focus on analysis of dynamically dependent data and their applications. Deep learning and neural networks have a long history with many successful applications and continue to attract lots of research interest. Readers are referred to Schmidhuber (2015) for an overview and to the online free book by M. Nielsen (2017) for detailed descriptions and computing codes.

4.1 NEURAL NETWORKS

Neural networks are semi-parametric statistical procedures that have evolved over time with the advancement in computational power and algorithms. They can be used in prediction or in classification. In this section, we shall only describe

Nonlinear Time Series Analysis, First Edition. Ruey S. Tsay and Rong Chen.

Companion website: www.wiley.com/go/tsay/nonlineartimeseries

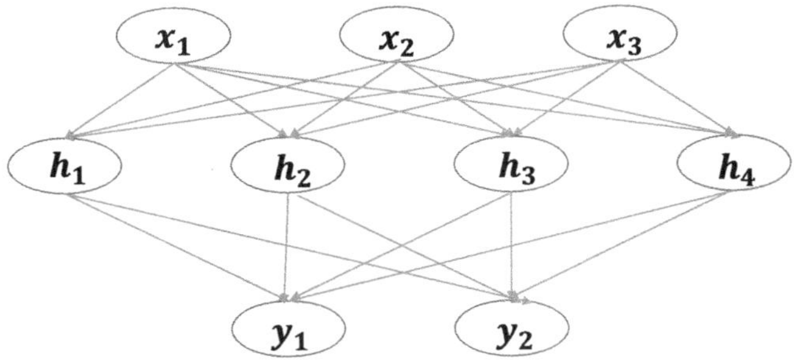

Figure 4.1 Schematic of a 3-4-2 feedforward neural network. The x nodes are input, the h nodes represent the hidden layer, and the y nodes are output.

the widely used *vanilla* feedforward networks. For general networks, readers are referred to textbooks on the topic. See, for instance, Hassoun (2003).

The concept of neural networks originates from models for the human brain, but it has evolved into a powerful statistical method for information processing. Some terminologies of the brain remain in use. Figure 4.1 shows a network diagram for a feedforward 3-4-2 network. The circles in the diagram are referred to as *nodes* or *neurons*, and the arrows signify the direction of information flow and the connection between nodes. In this particular network, there are three input variables, denoted by Xs, and two output variables, denoted by Ys. The network has a single hidden layer that has four nodes, denoted by hs. Each input variable is connected to every hidden node, which in turn is connected to every output variable. The name 3-4-2 is used to represent the structure of the network. The output variables Ys can be continuous or discrete. In classification applications, Y is discrete with y_i being the probability that the output belongs to the ith class. In real applications, multiple hidden layers may exist in a network. For simplicity, we only describe the case of a single hidden layer in our discussion. The same idea, however, applies to the general networks.

Suppose that $\boldsymbol{X} = (X_1, \ldots, X_p)'$ is the vector of p input variables, $\boldsymbol{H} = (H_1, \ldots, H_m)'$ is the vector of m hidden nodes, and $\boldsymbol{Y} = (Y_1, \ldots, Y_k)'$ is the vector of k output variables. Thus, we consider a $p - m - k$ feedforward network. An important feature of a given neural network is the way in which the information is processed from input nodes to the hidden nodes, then to the output nodes. Under the statistical framework used, we can write the network as

$$H_i = h(\alpha_{0i} + \boldsymbol{\alpha}_i'\boldsymbol{X}), \quad i = 1, \ldots, m, \tag{4.1}$$

$$L_j = \beta_{0j} + \boldsymbol{\beta}_j'\boldsymbol{H}, \quad j = 1, \ldots, k, \tag{4.2}$$

$$Y_j(\boldsymbol{X}) = g_j(\boldsymbol{L}), \quad j = 1, \ldots, k, \tag{4.3}$$

where $h(\cdot)$ and $g_j(\cdot)$ are *activation functions*, α_{0i} and β_{0j} are real parameters, $\boldsymbol{\alpha}_i = (\alpha_{1i}, \ldots, \alpha_{pi})'$ is a p-dimensional real vector, $\boldsymbol{\beta}_j = (\beta_{1j}, \ldots, \beta_{mj})'$ is an

m-dimensional real vector, and $\boldsymbol{L} = (L_1, \ldots, L_k)'$. The linear functions L_j are used as an intermediate step. The activation function $h(\cdot)$ is often chosen to be the *sigmoid* function (inverse of the logit function)

$$h(v) = \frac{1}{1 + e^{-v}}. \tag{4.4}$$

The choice of activation function $g_j(\cdot)$ depends on the output variables. If the output variables $\boldsymbol{Y}$ are continuous, then $g_j(\boldsymbol{L}) = L_j$, which is simply the identity function of the jth element of $\boldsymbol{L}$. On the other hand, if $\boldsymbol{Y}$ represents classification, then

$$g_j(\boldsymbol{L}) = \frac{e^{L_j}}{\sum_{v=1}^{k} e^{L_v}}, \tag{4.5}$$

which is the *softmax* function satisfying $0 \leq g_j(\cdot) \leq 1$ and $\sum_{j=1}^{k} g_j(\boldsymbol{L}) = 1$, that is, it is a multinomial model. In statistics, α_{ij} and β_{ij} of Equations (4.1)–(4.3) are unknown parameters. In neural networks, they are referred to as *weights*. For a given network, neural network modeling reduces to estimation of the unknown parameters provided that the activation functions are chosen. In this sense, neural networks belong to semi-parametric statistical models. In the special case that $k = 1$ and Y is binary, the activation function in (4.5) may reduce the heaviside function,

$$g(x) = \begin{cases} 0 & \text{if } x < 0, \\ 1 & \text{if } x \geq 0. \end{cases}$$

The resulting simple feedforward neural network is referred to as a *perceptron* in the literature.

Figure 4.2 shows the sigmoid function $h(\cdot)$ and two of its scaled versions $h(sv)$ with $s = 0.5$ and $s = 5$, respectively. From the plots, it is seen that for small v, $h(v)$ is approximately linear and $h(sv)$ approaches the 0-1 step function as $s \to \infty$. In applications, one can use $h(s(v - v_0))$ to shift the activation threshold from 0 to v_0.

Another common choice for the activation function is the *hyperbolic tangent*,

$$\tanh(x) = \frac{\sinh(x)}{\cosh(x)} = \frac{e^x - e^{-x}}{e^x + e^{-x}}. \tag{4.6}$$

Figure 4.3 shows the tanh function. A nice property of tanh function is $\frac{\partial \tanh(x)}{\partial x} = \text{sech}^2(x)$. In addition, the function has a nice Taylor series expansion and can be written as the fraction given below:

$$\begin{aligned} \tanh(x) &= x - \frac{1}{3}x^2 + \frac{2}{15}x^5 - \frac{17}{315}x^7 + \frac{62}{2835}x^9 - \cdots \\ &= \cfrac{x}{1 + \cfrac{x^2}{3 + \cfrac{x^2}{5 + \cdots}}} \end{aligned}$$

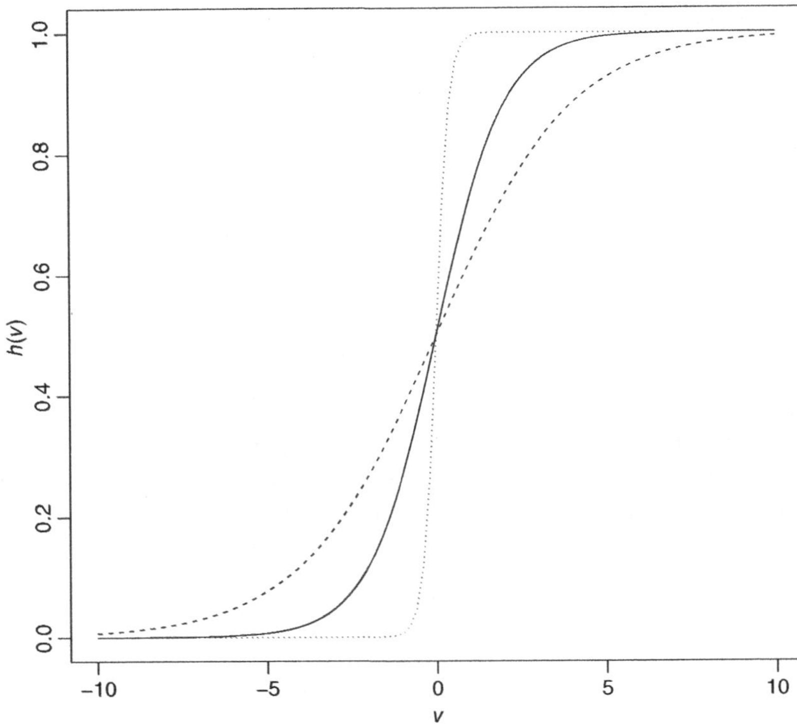

Figure 4.2 The activation function $h(v)$ of Equation (4.4): the solid line is $h(v)$, the dashed line is $h(0.5v)$, and the dotted line is $h(5v)$.

4.1.1 Estimation or Training of Neural Networks

Application of neural networks often divides the data into training and forecasting subsamples. The parameters (or weights) are estimated using data in the training subsample. The fitting is typically carried out by the *back propagation* method, which is a gradient descent procedure. For a given $p - m - k$ network, the weights consist of $\{\alpha_{0i}, \boldsymbol{\alpha}_i | i = 1, \dots, m\}$ with $m(p + 1)$ elements and $\{\beta_{0j}, \boldsymbol{\beta}_j | j = 1, \dots, k\}$ with $k(m + 1)$ elements. For ease in notation, we incorporate α_{0i} and β_{0j} into $\boldsymbol{\alpha}_i$ and $\boldsymbol{\beta}_j$, respectively, and define $\boldsymbol{h}_t = (h_{0t}, h_{1t}, \dots, h_{mt})'$, where $h_{0t} = 1$ and $h_{it} = h(\boldsymbol{\alpha}_i'\boldsymbol{x}_t)$ for $i > 0$, with the understanding that $\boldsymbol{x}_t$ includes 1 as its first element.

Let $\boldsymbol{\theta}$ be the collection of all weights. For continuous output variables, the objective function of model fitting is the sum of squared errors

$$\ell(\boldsymbol{\theta}) = \sum_{j=1}^{k} \sum_{t=1}^{N} [y_{jt} - Y_j(\boldsymbol{x}_t)]^2 \tag{4.7}$$

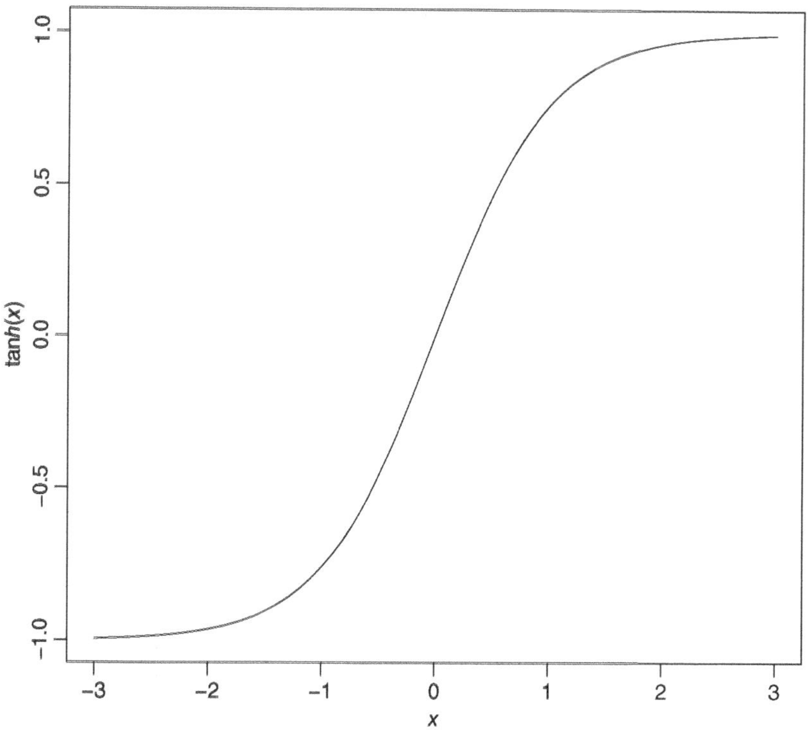

Figure 4.3 The hyperbolic tangent function.

where y_{jt} is the tth observation of the jth output variable, $\boldsymbol{x}_t$ denotes the tth observation of the input vector (including 1), and N denotes the sample size of the training subsample. For classification problems, we use the cross-entropy (or deviance) as the objective function,

$$\ell(\boldsymbol{\theta}) = -\sum_{j=1}^{k}\sum_{t=1}^{N} y_{jt} \log Y_j(\boldsymbol{x}_t), \tag{4.8}$$

and the classification rule is $\text{argmax}_j Y_j(\boldsymbol{x}_t)$.

In the following, we use the objective function in Equation (4.7) to describe the back propagation method. Rewrite the objective function as

$$\ell(\boldsymbol{\theta}) = \sum_{t=1}^{N} \ell_t, \quad \text{with} \quad \ell_t = \sum_{j=1}^{k} [y_{jt} - Y_j(\boldsymbol{x}_t)]^2.$$

Taking partial derivatives and using the chain rule, we have

$$\frac{\partial \ell_t}{\partial \beta_{ij}} = -2[y_{jt} - Y_j(\boldsymbol{x}_t)]g_j'(\boldsymbol{\beta}_j'\boldsymbol{h}_t)h_{it}, \quad i = 0, \ldots, m; \quad j = 1, \ldots, k, \tag{4.9}$$

$$\frac{\partial \ell_t}{\partial \alpha_{iv}} = -\sum_{j=1}^{k} 2[y_{jt} - Y_j(\boldsymbol{x}_t)]g_j'(\boldsymbol{\beta}_j'\boldsymbol{h}_t)\beta_{vj}h'(\boldsymbol{\alpha}_v'\boldsymbol{x}_t)x_{it}. \tag{4.10}$$

From Equations (4.9) and (4.10), a gradient decent update from the uth to the $(u+1)$th iteration assumes the form

$$\beta_{ij}^{(u+1)} = \beta_{ij}^{(u)} - \delta_u \sum_{t=1}^{N} \frac{\partial \ell_t}{\partial \beta_{ij}^{(u)}} \tag{4.11}$$

$$\alpha_{iv}^{(u+1)} = \alpha_{iv}^{(u)} - \delta_u \sum_{t=1}^{N} \frac{\partial \ell_t}{\partial \alpha_{iv}^{(u)}}, \tag{4.12}$$

where δ_u is the *learning rate*, which is usually taken to be a constant and, if necessary, can also be optimized by a line search that minimizes the error function at each update. On the other hand, δ_u should approach zero as $u \to \infty$ for online learning.

Rewrite Equations (4.9) and (4.10) as

$$\frac{\partial \ell_t}{\partial \beta_{ij}} = d_{jt}h_{it} \tag{4.13}$$

$$\frac{\partial \ell_t}{\partial \alpha_{iv}} = s_{vt}x_{it}. \tag{4.14}$$

The quantities d_{jt} and s_{vt} can be regarded as *errors* from the current model at the output and hidden layer nodes, respectively. From their definitions, these errors satisfy

$$s_{vt} = h'(\boldsymbol{\alpha}_v'\boldsymbol{x}_t)\sum_{j=1}^{k} \beta_{vj}d_{jt}, \tag{4.15}$$

which are the *back-propagation equations* from which the updates in Equations (4.11) and (4.12) can be carried out by a two-pass algorithm. Specifically, in the forward pass, the current weights are given and the predicted values $\widehat{Y}_j(\boldsymbol{x}_t)$ are computed from the model in Equations (4.1)–(4.3). In the backward pass, the errors d_{jt} are computed and then back-propagated via Equation (4.15) to give the errors s_{vt}. The two sets of errors are then used to compute the gradients for the updates in Equations (4.11) and (4.12). The back-propagation is relatively simple, but it may converge slowly in an application. In addition, some cares must be exercised in training neural networks. Readers are referred to Hastie et al. (2001, Chapter 11) for further discussions and examples of application.

The gradient decent updates in Equations (4.11) and (4.12) can be modified by adding a random disturbance term to overcome some difficulties or to improve efficiency in network training.

4.1.2 An Example

To demonstrate the application of neural networks, we consider the US weekly crude oil prices from January 3, 1986 to August 28, 2015 for 1548 observations. The data are available from the FRED (Federal Reserve Bank of St. Louis) and are the West Texas Intermediate prices, Cushing, Oklahoma. Let x_t denotes the weekly crude oil price at time index t. The time plot shows that x_t has an upward trend during the sample period so we employ the change series $y_t = x_t - x_{t-1}$ with sample size $T = 1547$. Figure 4.4 shows the time plot of y_t with the vertical line denoting the separation between modeling and forecasting subsamples.

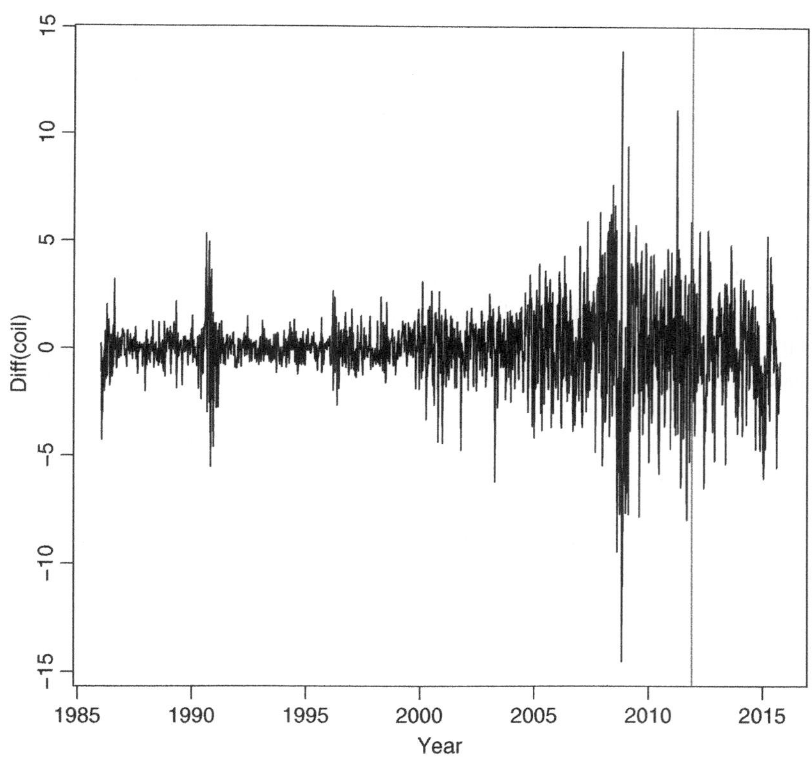

Figure 4.4 Time plot of the change series of weekly US crude oil prices, West Texas Intermediate, Cushing, Oklahoma, from January 1986 to August 2015. The vertical line separates the modeling and forecasting subsamples.

Model	RMSE	MAE
AR(0)	5.3331	1.7968
AR(8)	4.8930	1.7066
AR(13)	4.8658	1.6995
13-1-1 nnet	5.0744	1.7603
13-1-1 nnet	4.8488	1.6990
13-1-1 nnet	4.9945	1.7245
8-3-1 nnet	5.0210	1.7246

Table 4.1 Out-of-sample predictions of various models for the change series of US weekly crude oil prices from January 1986 to August 2015. The last 200 observations are used as the forecasting subsample. RMSE, root mean squared error; MAE, mean absolute errors; AR(0), sample mean of the modeling subsample used as prediction.

The forecasting subsample consists of the last 200 observations. From the plot, it is clear that the variability of the price changes y_t increased markedly after 2008.

For comparison, we use the root mean squared forecast errors (RMSE) and mean absolute forecast errors (MAE) to measure the performance of the model used in out-of-sample prediction. In addition, we use a linear AR(0) model (e.g., a white noise model) as a benchmark, and use two additional linear AR models. An AR(8) or AR(13) model fits the training data well. For the neural networks, we use the R package `nnet` in this example and employ $y_{t-1}, \dots, y_{t-13}$ as the input variables. Different numbers of nodes in the hidden layer are tried. Furthermore, we trained a given neural network multiple times to better understand its performance. Table 4.1 summarizes the out-of-sample performance of various models. From the table, the neural networks, in this particular case, produce similar forecasts as the linear AR models. As expected, the models used fare better than the benchmark AR(0).

R demonstration: Some commands used. The R command `NNsetting` of the `NTS` package is used to set up the input matrix and the output variable in both the training and forecasting subsamples for a time series.

```
da <- read.table('w-coilwti.txt',header=T)
yt <- diff(da[,4])
require(NTS)
m1 <- NNsetting(yt,nfore=200,lags=c(1:13))
names(m1)
X <- m1$X; y <- m1$y; predX <- m1$predX; predY <- m1$predY
m2 <- lm(y~-1+X)
yhat <- predY <- predX%*%matrix(m2$coefficients,13,1)
```

```
er1 <- yhat-predY
mean(er1^2); mean(abs(er1))
require(nnet)
m3 <- nnet(X,y,size=1,skip=T,linout=T,maxit=1000)
pm3 <- predict(m3,newdata=predX)
er2 <- pm3-predY
mean(er2^2); mean(abs(er2))
```

4.2 DEEP LEARNING

Deep learning is a large multilayer neural network. It makes use of the advances in optimization algorithms and computing power, and has become popular in recent years because we now have access to huge amounts of data and the computation power to train complicated neural networks. The word *deep* is often used to refer to the number of layers in a network. Thus, deep learning can be thought of as a multilayer perceptron network. To handle the complexity in data, neural networks have also been extended to include loops, adding feedback and memory to the networks over time. Such networks are called *recurrent neural networks*. The feedback and memory allow recurrent networks to learn and generalize across sequences of inputs. This in turn improves the power of neural networks. Therefore, deep learning can also be considered as a convolution neural network or long short-term memory recurrent neural network. Details are given below. In the literature, deep learning has been found useful in many applications, but, due to computational intensity, most of the current applications of deep learning are for classification (or labeled data). Interested readers are referred to Goodfellow et al. (2016) and Géron (2017) for further information.

The feedforward neural network discussed in Section 4.1 often encounters two main difficulties in application, namely *overfitting* and computing time. Overfitting can occur when we do not have enough data to train a complicated network. With recent developments in optimization methods, overfitting can be mitigated by the following techniques: (a) pruning or dropout, which randomly omits nodes from hidden layers during training, (b) weight decay, which uses ℓ_2 regularization to the parameters, and (c) sparsity, which applies ℓ_1 regularization to reduce the number of parameters. The pruning can reduce the correlations between the weights (i.e. parameters) of a network whereas ℓ_1 penalty is widely used in high-dimensional statistical inference.

The back-propagation procedure continues to apply in training deep learning networks. However, when the number of hidden layers is large and the network is recurrent, the gradient decent updates in Equations (4.11) and (4.12) can become inefficient. Consider, for example, the hyperbolic tangent activation function. The gradient is between -1 and 1 for each parameter. When the network is complicated, products of many small gradients can easily be close to zero, making the updating ineffective. This is referred to as the *vanishing gradient problem*. Hochreiter and Schmidhuber (1997) proposed the long short-term memory (LSTM) method to overcome this difficulty in network training. Specifically, the LSTM method augments the network with certain *forget* gates to prevent backpropagated errors from vanishing or exploding. Let $\boldsymbol{x}_t$ be the input vector and $\boldsymbol{y}_t$ be the output vector. In addition, let $\boldsymbol{s}_t$ be a state vector and $\boldsymbol{f}_t$, $\boldsymbol{i}_t$, and $\boldsymbol{u}_t$ be three gate vectors. Starting with $\boldsymbol{y}_0 = \boldsymbol{0}$ and $\boldsymbol{s}_0 = \boldsymbol{0}$ a traditional LSTM network can be written as

$$\begin{aligned} \boldsymbol{f}_t &= A_g(\boldsymbol{\beta}_f \boldsymbol{x}_t + \boldsymbol{\delta}_f \boldsymbol{y}_{t-1} + \boldsymbol{b}_f) \\ \boldsymbol{i}_t &= A_g(\boldsymbol{\beta}_i \boldsymbol{x}_t + \boldsymbol{\delta}_i \boldsymbol{y}_{t-1} + \boldsymbol{b}_i) \\ \boldsymbol{u}_t &= A_g(\boldsymbol{\beta}_u \boldsymbol{x}_t + \boldsymbol{\delta}_u \boldsymbol{y}_{t-1} + \boldsymbol{b}_u) \end{aligned} \tag{4.16}$$

$$\boldsymbol{s}_t = \boldsymbol{f}_t \circ \boldsymbol{s}_{t-1} + \boldsymbol{i}_t \circ A_s(\boldsymbol{\beta}_s \boldsymbol{x}_t + \boldsymbol{\delta}_s \boldsymbol{y}_{t-1} + \boldsymbol{b}_s) \tag{4.17}$$

$$\boldsymbol{y}_t = \boldsymbol{u}_t \circ A_y(\boldsymbol{s}_t) \tag{4.18}$$

where $\circ$ denotes the Hadamard product (element-wise product), $\boldsymbol{\beta}_j$ and $\boldsymbol{\delta}_j$ are parameter matrices for $j = f, i, u$ and s, and $\boldsymbol{b}_j$ are the associated parameter vectors. In the network, $A_j(\cdot)$ are activation functions for $j = g, s$ and y with $A_g(\cdot)$ being the sigmoid function, $A_s(\cdot)$ the hyperbolic tangent function, and $A_y(\cdot)$ either the hyperbolic tangent or the identity function.

A nice feature of the LSTM network is that it avoids the difficulty of vanishing gradient in network training. This is so because when the error values are back-propagated from the output, the error becomes trapped in the memory portion of the network. Thus, back-propagation can be used to train an LSTM network effectively.

4.2.1 Deep Belief Nets

In this section, we introduce the deep belief nets (DBN) to gain further insight concerning deep learning. In the 1980s, *Boltzmann machines* were introduced to learn arbitrary probability distributions over binary vectors. See, for instance, Hinton and Sejnowski (1986), among others. Let $\boldsymbol{X} = (x_1, \ldots, x_n)'$ denote a set of n nodes (or neurons), where each x_i is a binary variable. In a Boltzmann machine, all neurons are connected to each other. The probability that the ith neuron will assume the output 1 in the next step is given by

$$P(x_i^{(1)} = 1) = h\left(\frac{\sum_{j=1}^{n} w_{ij} x_j + b_i}{T}\right), \tag{4.19}$$

where the superscript (1) denotes the next step, b_i is the bias term (a constant term in statistics), $h(x)$ is the logistic activation function, w_{ij} is the connection weight between neurons i and j with $w_{ii} = 0$, and T denotes the *temperature*; the higher the temperature, the more random the output is, i.e. $P(x_i^{(1)} = 1) \to 0.5$ as $T \to \infty$. The nodes are divided into two groups, namely visible nodes and hidden nodes, with visible nodes consisting of input and output. The probability distribution is obtained when the machine is iterated many iterations until the probabilities no longer depend on the initial state, i.e. the probability distribution is a function of the weights and biases only. This is often achieved with the help of gradually decreasing the temperature T during the iterations. Clearly, the training of a Boltzmann machine can become difficult when the number of nodes n is large.

A *restricted Boltzmann machine* (RBM) is a Boltzmann machine in which there are no connections between visible nodes or between hidden nodes. The connections are between visible and hidden nodes. RBM is also known as *harmonium*. The restriction dramatically simplifies the computation or training of the machine. Let $\boldsymbol{v}$ and $\boldsymbol{h}$ denote the visible and hidden nodes of a Boltzmann machine, respectively. Using the idea of Gibbs sampling, the joint distribution of the nodes can be obtained by using the conditional distributions $p(\boldsymbol{h}|\boldsymbol{v})$ and $p(\boldsymbol{v} \mid \boldsymbol{h})$. The connections of RBM further imply that

$$p(\boldsymbol{h} \mid \boldsymbol{v}) = \prod_i p(h_i \mid \boldsymbol{v}) \quad \text{and} \quad p(\boldsymbol{v} \mid \boldsymbol{h}) = \prod_j p(v_j \mid \boldsymbol{h}).$$

Consequently, efficient block Gibbs sampling is possible, which alternates between sampling all of $\boldsymbol{h}$ simultaneously and sampling all of $\boldsymbol{v}$ simultaneously. For an introduction to Gibbs sampling, interested readers are referred to textbooks that contain Markov chain Monte Carlo methods. See, for instance, Tsay (2010, Chapter 12) and Robert and Casella (2004). We also include a short introduction in Chapter 8.

An efficient training algorithm was introduced by Carriera–Perpiñan and Hinton (2005) that marks a significant improvement. The algorithm is called *contrastive divergence*. In the literature of Boltzmann machine, the activation function is referred to as the *energy function*. With the separation of visible and hidden nodes, the energy function can be written as

$$p(\boldsymbol{h} \mid \boldsymbol{v}) \propto \exp[-E(\boldsymbol{v}, \boldsymbol{h})], \tag{4.20}$$

where

$$E(\boldsymbol{v}, \boldsymbol{h}) = -\boldsymbol{b}'\boldsymbol{v} - \boldsymbol{c}'\boldsymbol{h} - \boldsymbol{v}'\boldsymbol{W}\boldsymbol{h},$$

where $\boldsymbol{b}$ and $\boldsymbol{c}$ denote the vector of biases associated with $\boldsymbol{v}$ and $\boldsymbol{h}$, respectively, and $\boldsymbol{W} = [w_{ij}]$ is the weight matrix. This energy is linear so that we have

$$\frac{\partial}{\partial w_{ij}} E(\boldsymbol{v}, \boldsymbol{h}) = -v_i h_j, \tag{4.21}$$

where v_i is the ith node of $\boldsymbol{v}$ and h_j is the jth node of $\boldsymbol{h}$. The contrasive divergence works as follows. For a training $\boldsymbol{v}$, the algorithm starts by feeding it to the network so that the state of the hidden nodes can be computed by the activation function in Equation (4.19). With a given $\boldsymbol{h}$, one can also compute the state of the visible nodes by the same activation function (4.19). This results in a new vector $\tilde{\boldsymbol{v}}$. Next, one computes again the hidden nodes, resulting in a new vector $\tilde{\boldsymbol{h}}$. Making use of the derivative in Equation (4.21), the contrasive divergence weight updating becomes

$$w_{ij}^{(1)} = w_{ij} + \eta(\boldsymbol{v}'\boldsymbol{h} - \tilde{\boldsymbol{v}}'\tilde{\boldsymbol{h}}), \tag{4.22}$$

where η is the learning rate and again the superscript (1) denotes the next step in the training. A nice feature of this algorithm is that it simply goes forward, backward, and forward of the network. There is no need to wait for the network to reach an equilibrium in the iteration.

Several layers of RBMs can be stacked such that the hidden nodes of the first-layer RBM serves as the visible nodes of the second-layer RBM, and so on. Such a multilayer RBM network is called a *deep belief net* (DBN). The DBN becomes feasible because it is possible to train one layer of the network at a time using the contrasive divergence algorithm, starting with the first layer and then gradually moving up to the last layer. See Hinton et al. (2006).

4.2.2 Demonstration

There are several deep learning packages available in R. In this book, we use the packages `darch` by Drees et al. (2016) and `deepnet` by Xiao (2014). Details of the two packages can be found from their reference manuals available from R web page at https://www.r-project.org. These two packages apply deep learning to classification, and they are based on the RBM and DBN methodology. We consider two applications in this subsection.

Example 4.1 In this simple example, we use deep learning to model and predict the direction of the daily price movement of Amazon stock. The data period used is from January 3, 2007 to April 28, 2017. Let r_t be the daily log return of the Amazon stock, i.e. the first difference of its log prices. We define the direction of price movement as

$$y_t = \begin{cases} 1 & \text{if } r_t \geq 0, \\ 0 & \text{if } r_t < 0. \end{cases} \tag{4.23}$$

Since the Amazon price movement is likely to be affected by the movement of the financial market, we use the daily log returns of the S&P 500 index (tick symbol

GSPC) as an explanatory variable. We use the first 2000 data points as the training subsample and the remaining 598 observations as the forecasting subsample.

Let m_t be the daily log return of the S&P 500 index and d_t be the direction of S&P price movement. In this study, we use 12 input variables defined as

$$\boldsymbol{x}_t = (y_{t-1}, d_{t-1}, r_{t-1}, m_{t-1}, y_{t-2}, d_{t-2}, r_{t-2}, m_{t-2}, y_{t-3}, d_{t-3}, r_{t-3}, m_{t-3})'.$$

In other words, we use lags 1 to 3 of the past directions and past log returns as the input variables. As a benchmark for comparison, we use the logistic linear regression. Let $p_t = P(y_t = 1|F_{t-1})$ with F_{t-1} being the available information. The fitted logistic model in the training subsample is

$$\begin{aligned}\text{logit}(p_t) = 0.06 &- 0.26y_{t-1} + 0.16d_{t-1} + 0.80r_{t-1} - 9.06m_{t-1} + 0.07y_{t-2} \\ &+ 0.09d_{t-2} - 6.99r_{t-2} + 2.97m_{t-2} - 0.05y_{t-3} - 0.08d_{t-3} \\ &+ 2.21r_{t-3} - 2.64m_{t-3}\end{aligned} \tag{4.24}$$

where the coefficients of y_{t-1} and y_{t-2} are significant at the 5% level and that of m_{t-1} at the 10% level. If we classify the predicted price movement using the same procedure as Equation (4.23), then the prediction result of the fitted logistic linear regression is

Observed y_t	Predicted y_t	
	0	1
0	135	138
1	153	172

Therefore, the probability of correct prediction is $(135 + 172)/598 = 0.513$, which is slightly better than the random walk model.

Using the same input variables, we use `deepnet` to train several neural networks and use the fitted model to predict the price movement of Amazon stock. Figure 4.5 shows the time plot of the predicted probabilities of a 12-10-1 network. The plot looks reasonable. On the other hand, Figure 4.6 shows the predicted probabilities of a 12-5-5-1 network. The predicted probabilities are not centered around 0.5 so we modify the classification rule as

$$\hat{y}_t = \begin{cases} 1 & \text{if } p_t \leq \bar{p} \\ 0 & \text{if } p_t < \bar{p}, \end{cases}$$

where p_t denotes the fitted value of a neural network and $\bar{p}$ denotes the sample mean of p_t.

We also apply the `darch` package using the same input variables. Figure 4.7 shows the time plot of the predicted values of a 12-10-1 network via `darch`. The

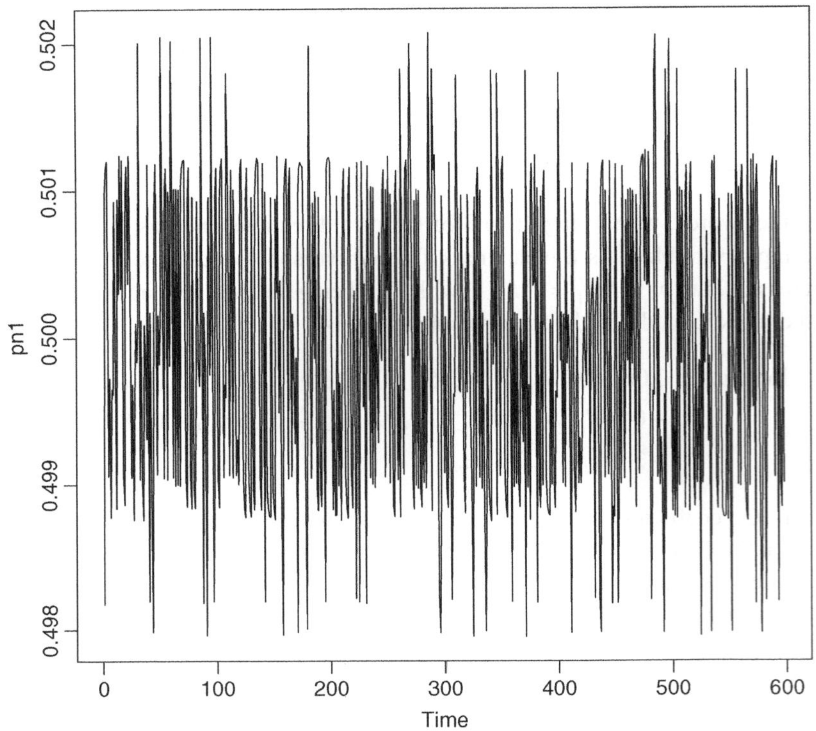

Figure 4.5 Time plot of the predicted values of a 12-10-1 neural network via `deepnet`.

mean of the predicted values is close to 0.5 so we used the classification procedure of Equation (4.23) to assess the prediction performance of the `darch` package.

Table 4.2 shows the prediction results of various neural networks using the deep learning packages. From the table it is seen that the deep learning methods perform similarly to the logistic linear regression in this simple example.

R demonstration: Deep learning with `deepnet` and `darch` packages.

```
> amzn <- read.csv("Amzn.csv")
> gspc <- read.csv("GSPC.csv")
> dim(amzn)
[1] 2599    7
> Gspc <- diff(log(as.numeric(gspc$Adj.Close)))
> damzn <- ifelse(Amzn >= 0,1,0)
> dgspc <- ifelse(Gspc >= 0,1,0)
> rtn <- cbind(Amzn,Gspc)
```

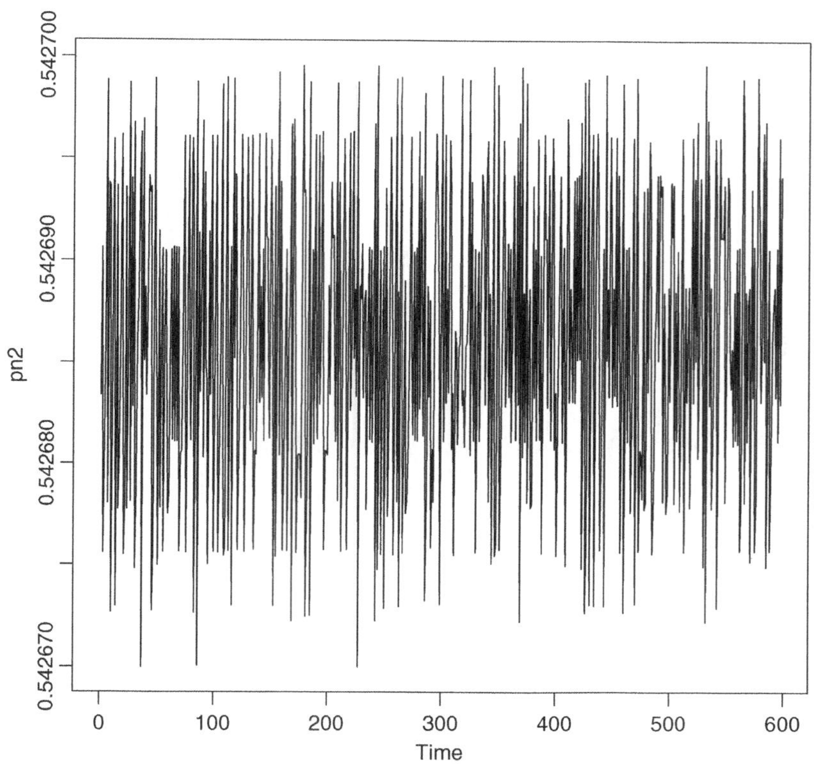

Figure 4.6 Time plot of the predicted values of a 12-5-5-1 neural network via the `deepnet` package.

```
> drtn <- cbind(damzn,dgspc)
> zt <- cbind(drtn,rtn)
> source("NNsetting.R")
> m1 <- NNsetting(zt,locY=1,nfore=598,lags=c(1:3))
> names(m1)
[1] "X"     "y"     "predX" "predY"
> X <- m1$X; y <- m1$y; predX <- m1$predX; predY <- m1$predY
> nameX <- c("dAmzn1","dspc1","Amzn1","spc1","dAmzn2","dspc2","Amzn2","spc2",
          "dAmzn3","dspc3","Amzn3","spc3")
> colnames(X) <- nameX
> colnames(predX) <- nameX

> require(deepnet)  ## deepnet package
> n1 <- nn.train(X,y,hidden=c(10))  ## Single hidden layer
> pn1 <- nn.predict(n1,predX)
> ts.plot(pn1)
```

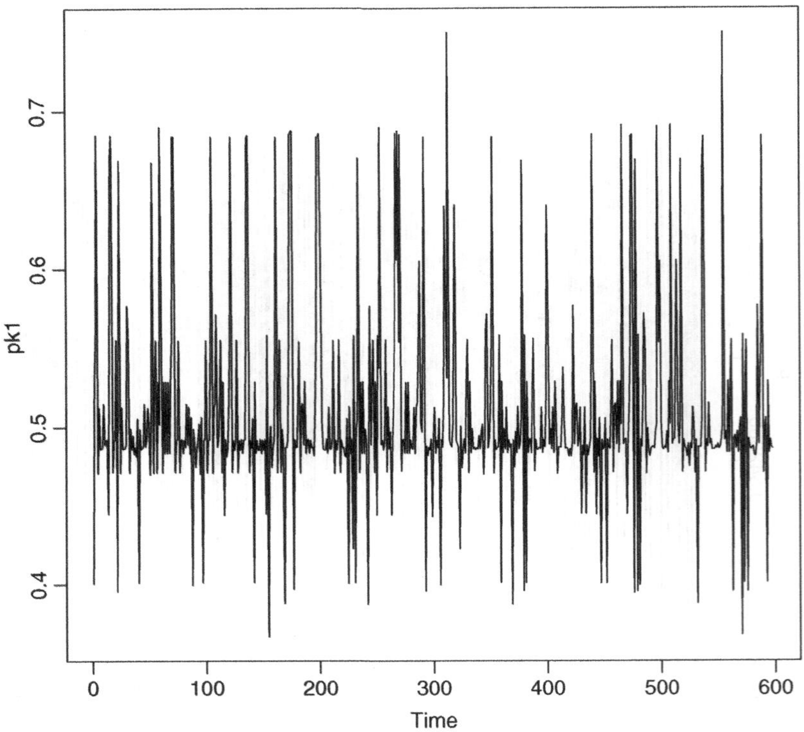

Figure 4.7 Time plot of the predicted values of a 12-10-1 neural network via the `darch` package.

```
> dpnl <- ifelse(pnl >= 0.5,1,0)
> table(predY,dpnl)
     dpnl
predY   0   1
    0 150 123
    1 160 165
#### the same result if mean is used.
> dpnl <- ifelse(pnl >= mean(pnl), 1,0)
> table(predY,dpnl)
     dpnl
predY   0   1
    0 150 123
    1 160 165
> n2 <- nn.train(X,y,hidden=c(5,5))  ## two hidden layers

> require(darch)     ## darch package
> dl <- darch(X,y,layers=10)     ## Single hidden layer
> pkl <- predict(dl,newdata=predX)
```

Methods	Two-way table		Probability of correct prediction
Logistic linear regression	135	138	(135+172)/598 = 0.513
	153	172	
deepnet 12-10-1	150	123	0.527
	160	165	
deepnet 12-100-1	154	119	0.518
	169	156	
deepnet 12-10-10-1	131	142	0.493
	161	164	
deepnet 12-10-10-10-1	137	136	0.492
	168	157	
darch 12-10-1	180	93	0.509
	201	124	
darch 12-100-1	198	75	0.503
	222	103	
darch 12-5-5-1	193	80	0.5
	219	106	

Table 4.2 Comparison of selected deep networks with logistic linear regression in predicting the direction of daily price movement of Amazon stock. The training sample is from January 3, 2007 to December 11, 2014 whereas the forecasting sample is from December 12, 2014 to April 28, 2017.

```
> ts.plot(pk1)
> pp1 <- ifelse(pk1 >= 0.5,1,0)
> table(predY,pp1)
     pp1
predY   0   1
    0 180  93
    1 201 124
> d3 <- darch(X,y,layers=c(12,5,5,1))  ## Two hidden layers of sizes 5 and 5.
```

Example 4.2 Modeling the trade-by-trade price changes is of particular interest in high-frequency financial data analysis. Many methods have been proposed in the literature, including the decomposition approach of Rydberg and Shephard (2003) and the ordered probit model of Hauseman et al. (1992). See Tsay (2013) for application of these methods. In this example, we apply deep learning as a modern approach to modeling the intraday price movements of a stock.

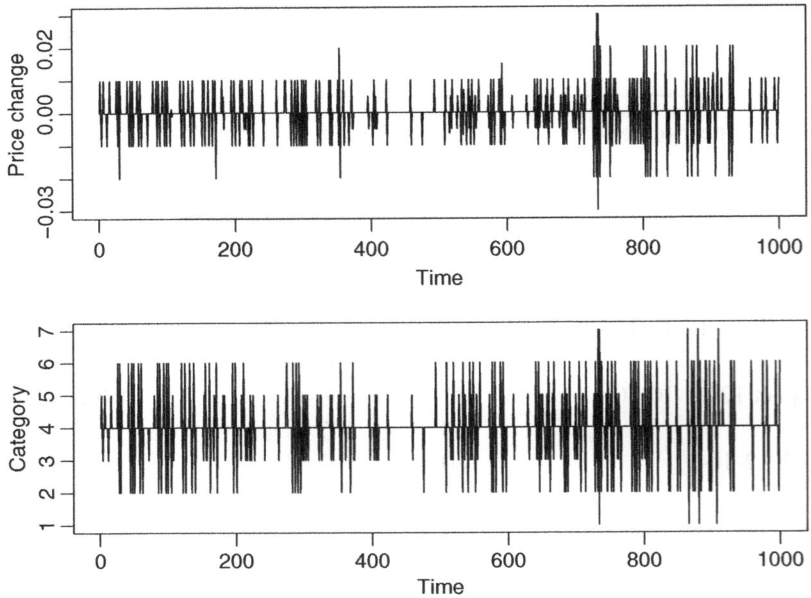

Figure 4.8 Time plots of trade-by-trade price changes and the associated categories of Walgreens stock on February 6, 2017. Only the last 1000 transactions during normal trading hours are shown.

Consider the trading of Walgreens stock on February 6, 2017. The data are available from the TAQ database of the New York Stock Exchange. Let y_i^* be the observed price change of the ith trade during the normal trading hours from 9:30am to 4:00pm, Eastern time. Since y_i^* tends to assume discrete values, we divide the price changes into seven categories, namely

$$< -0.02,\ [-0.02, -0.01),\ [-0.01, 0),\ 0,\ (0, 0.01],\ (0.01, 0.02],\ > 0.02,$$

where the unit is one US dollar. Let y_i be the category associated with y_i^*. Then, we have $y_i = 1$ if $y_i^* < -0.02$, $y_i = 2$ if $-0.02 \le y_i^* < -0.01$, and so on. Figure 4.8 shows the time plots of y_i^* and y_i of the last 1000 transactions on February 6, 2017. The discreteness of y_i^* is clearly seen. In addition, the plots show that y_i is a reasonable approximation of y_i^*.

The goal of our study is to predict y_i based on the observed intraday data. To this end, we also consider $d_i = t_i - t_{i-1}$ and $s_i = V_i/100$, where t_i is the trading time of the ith transaction measured in seconds starting from the midnight and V_i is the trading volume (number of shares) of the ith trade. Thus, d_i is the duration between transactions and s_i is a normalized size of the transaction.

With seven categories, we define six dummy variables for the price change. Specifically, let

$$x_{i,j} = \begin{cases} 1 & \text{if } y_i = j, \\ 0 & \text{if } y_i \neq j, \end{cases} \quad j = 2, \dots, 7. \tag{4.25}$$

Let $\boldsymbol{x}_i = (x_{i,2}, x_{i,3}, \dots, x_{i,7})'$ be the six-dimensional indicator variable of price change for the i-trade. In particular, $\boldsymbol{x}_i = \boldsymbol{0}$ implies that $y_i = 1$.

In our demonstration, we use the following 27 input variables

$$\boldsymbol{X}_i \equiv \{\boldsymbol{x}_{i-j},\ y^*_{i-j},\ d_{i-j},\ s_{i-j} | \, j = 1, 2, 3\},$$

and our goal is to obtain $P(y_i = j \mid \boldsymbol{X}_i)$. In other words, we like to predict y_i using selected prior transactions data.

On February 6, 2017, there were 29275 transactions available for the Walgreens stock. We use the first 27275 observations as the training subsample and reserve the last 2000 observations for out-of-sample prediction to perform model comparison. Besides several network models, we also used the ordered probit model as a benchmark. The fitted ordered probit model is shown in the R output below. In this particular instance, the ordered probit model does not fare well in prediction. As a matter of fact, the model predicts no price change for all of the last 2000 transactions. See the forecast tabulation in the R demonstration below.

For deep learning, we use the `darch` package and employ three models. They are 27-5-1, 27-10-1, and 27-10-5-1 networks. Figure 4.9 shows the training results of the 27-5-1 network. With default of 100 epochs, the best iteration appears to be 84. The predictions of the three networks for the last 2000 transactions are also shown in the R demonstration. In this particular instance, the 27-10-1 network appears to perform the best. In particular, it was able to correctly predict 3, 47, 1389, and 11 times for Categories 1, 3, 4, and 5, respectively.

In this example, we demonstrated that deep learning can be helpful in modeling trade-by-trade price changes of Walgreens stock. This is encouraging. On the other hand, we did not attempt to find the best deep learning model for the given data. This important question deserves careful investigation.

R demonstration: Ordered probit model and deep learning.

```
> da <- read.table("taq-wba-cpch-feb06-2017.txt",header=T)
> dim(da)
[1] 29275     4
### create dummy variables
> c2 <- c3 <- c4 <- c5 <- c6 <- c7 <- rep(0,29275)
> nT <- nrow(da)
> idx <- c(1:nT)[da[,1]==2]
```

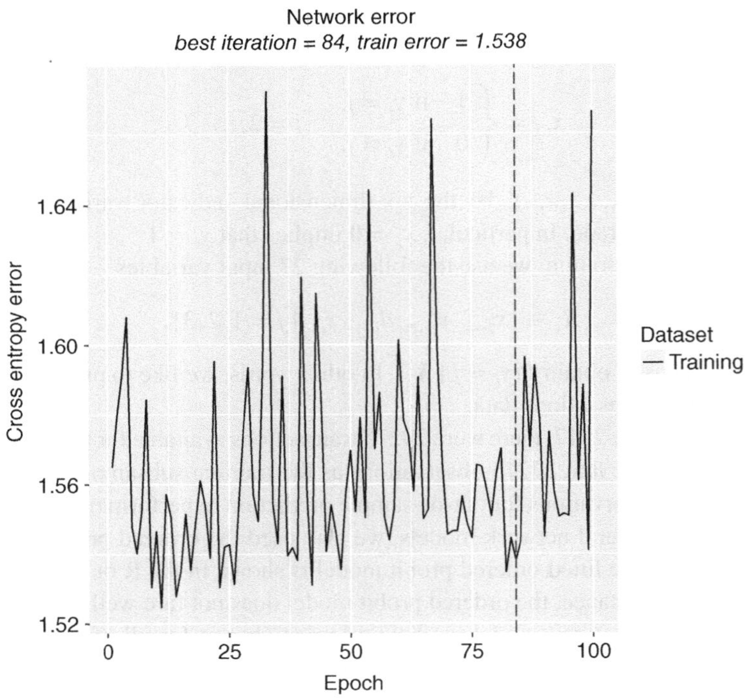

Figure 4.9 Fitting results of a 27-5-1 network via the `darch` package for the trade-by-trade price changes of Walgreens transactions data for February 6, 2017.

```
> c2[idx] <-1
> idx <- c(1:nT)[da[,1]==3]
> c3[idx] <-1
> idx <- c(1:nT)[da[,1]==4]
> c4[idx] <- 1
> idx <- c(1:nT)[da[,1]==5]
> c5[idx] <- 1
> idx <- c(1:nT)[da[,1]==6]
> c6[idx] <- 1
> idx <- c(1:nT)[da[,1]==7]
> c7[idx] <- 1
> Z <- cbind(da,c2,c3,c4,c5,c6,c7)
> source("NNsetting.R")
> n1 <- NNsetting(Z,nfore=2000,lags=c(1:3),include.lagY=F)
> X <- n1$X; y <- n1$y; predX <- n1$predX; predY <- n1$predY
> na <- c("pr","dur","size","c2","c3","c4","c5","c6","c7")
> na1 <- paste(na,"l1",sep=""); na2 <- paste(na,"l2",sep="")
> na3 <-paste(na,"l3",sep="")
```

```
> name <- c(na1,na2,na3)
> colnames(X) <- colnames(predX) <- name
> require(MASS)
> y <- as.factor(y)
> m1 <- polr(y~.,data=data.frame(X),method="probit")
> summary(m1)
Call: polr(formula = y ~ ., data = data.frame(X), method = "probit")
Coefficients:
              Value Std. Error  t value
prl1    -1.901e+01   2.004434  -9.4855
durl1    1.284e-02   0.002678   4.7971
sizel1   2.194e-03   0.001322   1.6596
c2l1    -2.955e-01   0.101196  -2.9200
c3l1    -6.422e-01   0.106859  -6.0097
c4l1    -1.005e+00   0.112055  -8.9728
c5l1    -1.300e+00   0.122176 -10.6376
c6l1    -1.610e+00   0.133076 -12.0973
c7l1    -1.759e+00   0.188524  -9.3279
prl2    -9.389e+00   1.999849  -4.6951
durl2    5.934e-03   0.002753   2.1552
sizel2   2.265e-03   0.001336   1.6956
c2l2    -3.550e-01   0.102485  -3.4640
c3l2    -4.644e-01   0.108436  -4.2830
c4l2    -6.333e-01   0.113046  -5.6022
c5l2    -6.913e-01   0.123311  -5.6063
c6l2    -9.001e-01   0.133976  -6.7186
c7l2    -1.218e+00   0.189065  -6.4434
prl3    -4.302e+00   1.847475  -2.3286
durl3   -8.912e-04   0.002770  -0.3218
sizel3  -3.752e-04   0.001043  -0.3597
c2l3    -2.086e-02   0.100412  -0.2077
c3l3    -9.264e-02   0.105887  -0.8749
c4l3    -1.311e-01   0.110385  -1.1876
c5l3    -1.790e-01   0.120459  -1.4858
c6l3    -2.369e-01   0.130089  -1.8208
c7l3    -3.802e-01   0.184299  -2.0629

Intercepts:
    Value    Std. Error t value
1|2  -4.3392   0.2111   -20.5586
2|3  -3.3606   0.2100   -16.0044
3|4  -2.8530   0.2098   -13.5971
4|5  -0.5979   0.2094    -2.8549
5|6  -0.0710   0.2093    -0.3391
6|7   0.8207   0.2091     3.9243

Residual Deviance: 54018.88
```

```
> pm1 <- predict(m1,newdata=predX,type="class")   ## prediction
> table(predY,pm1)
     pm1
predY    1    2    3    4    5    6    7
    1    0    0    0    4    0    0    0
    2    0    0    0  100    0    0    0
    3    0    0    0  180    0    0    0
    4    0    0    0 1437    0    0    0
    5    0    0    0  174    0    0    0
    6    0    0    0  100    0    0    0
    7    0    0    0    5    0    0    0

### Deep learning
> require(darch)
> y1 <- as.factor(y)
> nn1 <- darch(X,y1,layers=5)
> pnn1 <- predict(nn1,newdata=predX,type="raw")
> dim(pnn1)
[1] 2000    7
> x1 <- apply(pnn1,1,which.max)
> table(predY,x1)
     x1
predY    4    5
    1    4    0
    2   99    1
    3  171    9
    4 1420   17
    5  156   18
    6   99    1
    7    5    0
>
> nn2 <- darch(X,y1,layers=10)
> pnn2 <- predict(nn2,newdata=predX)
> x1 <- apply(pnn2,1,which.max)
> table(predY,x1)
     x1
predY    1    3    4    5
    1    3    0    1    0
    2    0    2   98    0
    3    0   47  129    4
    4    2   38 1389    8
    5    0   11  152   11
    6    0    5   95    0
    7    0    0    5    0
>
> nn3 <- darch(X,y1,layers=c(27,10,5,1))
> pnn3 <- predict(nn3,newdata=predX)
> x3 <- apply(pnn3,1,which.max)
```

```
> table(predY,x3)
      x3
predY    3    4    5
    1    0    4    0
    2    2   98    0
    3   46  123   11
    4   43 1367   27
    5    8  141   25
    6    3   95    2
    7    0    5    0
```

4.3 TREE-BASED METHODS

Tree-based methods have been widely used in the literature for prediction and classification, especially for independent observations. See, for instance, James et al. (2013). The methods include *regression tree, classification tree, bagging, random forests*, and *boosting*. The basic idea of tree-based statistical methods is stratifying or segmentation of the predictor space into multiple subregions, each of which contains relatively more homogeneous observations so that simple models can be used in these subregions. These methods are useful in understanding the basic structure embedded in the data and in interpretation. Although much of research into tree-based methods assumes independent data, the basic idea of the methods remains highly relevant for analysis of time series data. With some modifications on the procedure to account for the serial dependence in time series, tree-based methods can also be useful in nonlinear time series analysis. The goal of this section is to introduce some tree-based methods and to consider modifications to these methods in the presence of serial dependence. We shall also demonstrate the applications of tree-based methods in prediction.

4.3.1 Decision Trees

Binary trees are the building block of most tree-based statistical methods. By binary, we mean that each branch of the tree can only be divided into two subbranches. These trees are commonly referred to as *decision trees*.

4.3.1.1 Regression Tree We begin our discussion of decision trees with a simple example of the regression tree. Consider the quarterly growth rates, in percentages, of US real gross domestic product (GDP) from 1947.II to 2015.II for 273 observations. Figure 4.10 shows the time plot of the GDP growth rates. For a clear display of a fitted tree, we round the GDP growth rates to two decimal points.

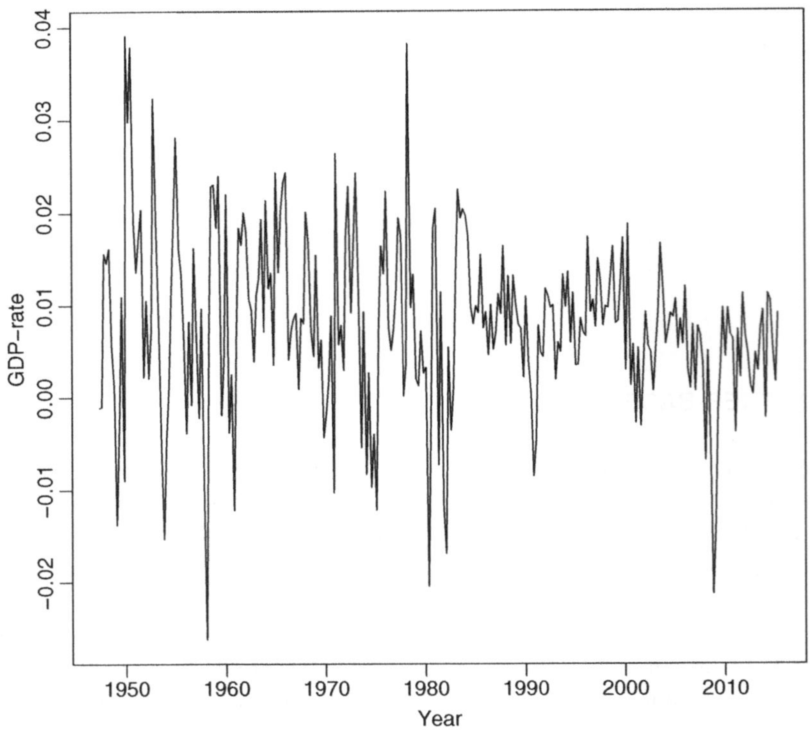

Figure 4.10 Time plot of the quarterly growth rates of US real gross domestic product from 1947.II to 2015.II.

Our goal is to predict the GDP growth rate using its own past three lagged values. Mathematically, one can think of the underlying model as

$$y_t = g(y_{t-1}, y_{t-2}, y_{t-3}) + a_t,$$

where y_t is the GDP growth rate at time t, a_t denotes the error term, and $g(\cdot)$ is a smooth function from R^3 to R. The decision tree represents a piecewise nonlinear approximation for $g(\cdot)$.

Figure 4.11 shows the output of a fitted regression tree for the GDP growth rates. From the plot, the tree is drawn upside down in the sense that the leaves are at the bottom of the tree. In this particular instance, the tree has 11 leaves, shown by the numerical values in the plot, i.e. −0.2450, 0.6529, 1.239, etc. Starting from the root at the top of the plot, the first split of the tree is determined by the lag-1 variable y_{t-1}, denoted by GDP1 in the plot. The left branch is for $y_{t-1} < 1.15$. The right branch (for $y_{t-1} \geq 1.15$) further splits into two subbranches determined by

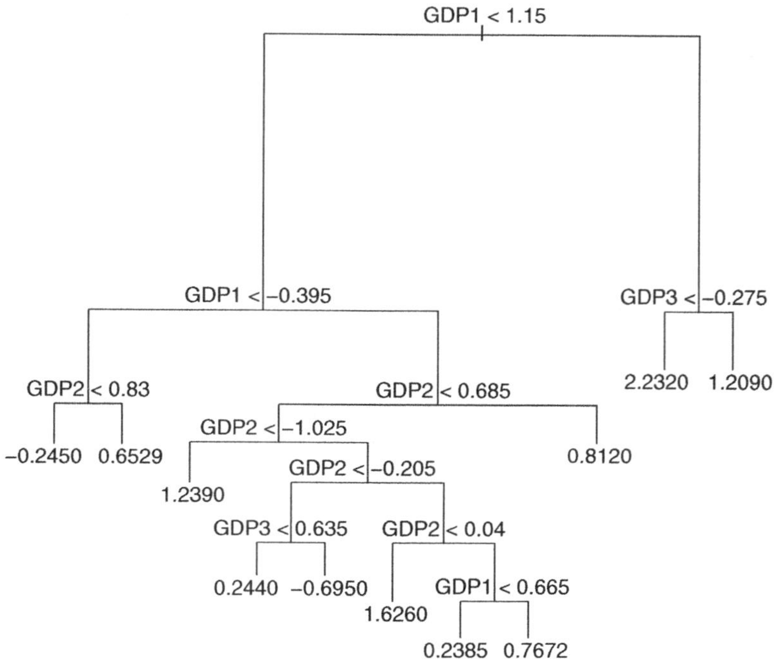

Figure 4.11 A regression tree for the quarterly growth rate of US real gross domestic product using three lagged values as explanatory variables. The data are from 1947.II to 2015.II.

the lagged variable y_{t-3}, denoted by GDP3 in the plot. The new left subbranch is for $y_{t-3} < -0.275$. The plot then shows two leaves, namely 2.232 and 1.209.

The first left branch of the tree is further divided into two subbranches by the lag-1 variable y_{t-1} with the new left subbranch satisfying $y_{t-1} < -0.395$. This new subbranch is further divided into two new sub-subbranches by the lag-2 variable y_{t-2} with the criterion $y_{t-2} < 0.83$, resulting in two leaves with values -0.2450 and 0.6529. The rest of the tree can be considered in a similar manner.

The tree in Figure 4.11 can be used in prediction. For instance, suppose that we have $(y_n, y_{n-1}, y_{n-2}) = (0.51, 1.05, 1.12)$ at the forecast origin $t = n$ and want to predict y_{n+1}. From the fitted tree, we see that the prediction is 0.812. Since the lag-1 value for $t = n + 1$ is $y_n = 0.51$, which is less than 1.15, so that y_{n+1} belongs to the left branch of the first split. Since $y_n \geq -0.395$, y_{n+1} then belongs to the right subbranch of the second split. Finally, the lag-2 value of y_{n+1} is $y_{n-1} = 1.05$, which is greater than 0.685, so that y_{n+1} belongs to the right subbranch at the third split. The prediction (or leaf) 0.812 is the sample mean for all data in the same subregion as y_{n+1}.

The tree in Figure 4.11 can be interpreted as follows. Among the three explanatory variables, the lag-1 GDP growth rate is the most important one in determining the GDP growth rate. This is not surprising given the dynamic dependence of the data. When the last quarter's growth rate is greater than or equal to 1.15%, then the lag-2 variable becomes irrelevant. The growth rate is then determined by the lag-3 variable. If $y_{t-3} < -0.275$, meaning that the US economy was in deep contraction, then the US economy was expected to grow faster at the rate of 2.23%. This phenomenon seems reasonable. It is well-known that consecutive negative GDP growths are rare. When the economy was in deep contraction at y_{t-3}, certain interventions were likely to take place resulting in a turn around in y_{t-1}. The effect of the intervention was likely to continue so that the growth rate y_t should be greater. Other branches of the tree might also have some nice interpretation.

As illustrated by the example of GDP growth rate, the regression tree divides the predictor space into non-overlapping subregions and uses the sample average of each subregion to produce prediction. Figure 4.12 shows the scatter plot of

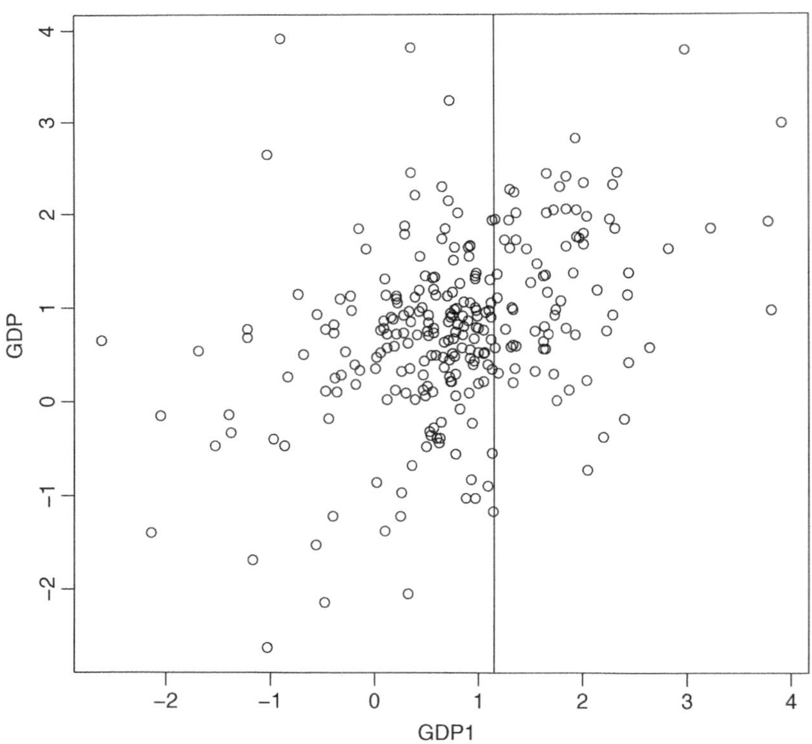

Figure 4.12 The scatter plot of US quarterly GDP growth rate versus the past quarter's growth rate. The vertical line is 1.15, and the data are from 1947.II to 2015.II.

y_t versus y_{t-1} of the GDP growth rates. The vertical line of the plot shows the division of y_t into two groups based on the values of y_{t-1} with the threshold 1.15. This threshold is determined by the following method. Let c be a real number, and let $R_1 = \{t \mid y_{t-1} < c\}$ and $R_2 = \{t \mid y_{t-1} \geq c\}$. Furthermore, let $\bar{y}_{1,c}$ and $\bar{y}_{2,c}$ be the sample means of y_t in R_1 and R_2, respectively. The sum of squared errors (SSE) corresponding to this given c is

$$SSE_1(c) = \sum_{t \in R_1} (y_t - \bar{y}_{1,c})^2 + \sum_{t \in R_2} (y_t - \bar{y}_{2,c})^2, \tag{4.26}$$

which is a function of c. The threshold 1.15 is then obtained by

$$1.15 = \text{argmin}_c SSE_1(c).$$

The choice of threshold 1.15 can be seen clearly from Figure 4.12. Of course, one performs such a division because

$$SSE_1(1.15) < SSE_0 \equiv \sum_{t=4}^{T} (y_t - \bar{y})^2,$$

where $T = 273$ is the sample size and $\bar{y}$ is the sample mean of y_t for $t = 4, \ldots, T$. The summation starts with $t = 4$ because we used y_{t-1}, y_{t-2}, and y_{t-3} as the explanatory variables.

In this particular example, we have three explanatory variables. The use of y_{t-1} to perform the first tree split actually involves variable selection. In theory, we also consider y_{t-2} and y_{t-3} to form two regions using the method stated in Equation (4.26). However, y_{t-1} is chosen because the resulting $SSE_1(1.15)$ is smaller than any binary partition based on either y_{t-2} or y_{t-3}.

Next, to grow the tree, we can consider R_1 and R_2 separately. Focus on the region R_2, which is the right branch of the first split. One can treat observations in R_2 as a new sample and apply the tree-growing process by selecting the best explanatory variable and the best threshold to form a new split. For the GDP growth rate example, R_2 is divided into $R_{21} = \{t \mid (y_{t-1} \geq 1.15) \cap (y_{t-3} < -0.275)\}$ and $R_{22} = \{t \mid (y_{t-1} \geq 1.15) \cap (y_{t-3} \geq -0.275)\}$. Turn to the region R_1, which is the left branch of the first split. One can also apply the tree-growing process. The decision to first split R_1 or R_2 is determined by the resulting sum of squared errors. One selects the second split so that the resulting tree, which has three branches, has the smallest sum of squared errors.

In theory, using the sum of squared errors as the objective function, one can continue to grow a binary tree by repeating the process of selecting the best predictor and the best threshold for splitting so as to minimize the objective function. However, a tree with many leaves often results in over-fitting, which in turn leads to suboptimal prediction. Some pruning is needed.

4.3.1.2 Tree Pruning Let $\mathcal{T}$ be a tree with $|\mathcal{T}|$ leaves. Let R_i be the ith region of the tree for $i = 1, \ldots, |\mathcal{T}|$. Furthermore, let y_t be the tth observation of the dependent variable and $\boldsymbol{x}_t$ be the corresponding vector of predictors. Let $\bar{y}_{R_i}$ be the sample mean of the dependent variable in region R_i, i.e.

$$\bar{y}_{R_i} = \frac{1}{|R_i|} \sum_t y_t I(\boldsymbol{x}_t \in R_i),$$

where $|R_i|$ denotes the number of data points in region R_i. The idea of tree pruning is to apply a penalty to a tree based on its complexity. Here the complexity of a tree is measured by the number of its leaves.

Suppose that the sample size is T so that the data are $(y_t, \boldsymbol{x}_t)$ for $t = 1, \ldots, T$. Similar to the selection of other statistical models discussed in the book, we use out-of-sample prediction to select a binary tree for the data. To this end, we divide the data into training and forecasting subsamples. Typically, for time series data the training sample consists of the first N data points and the forecasting subsample the last $T - N$ data points.

Use the training sample to grow a large tree, say $\mathcal{T}_0$, which is often determined by setting an upper bound for $|R_i|$, the number of data points in region R_i. For a given penalty α, there exists a subtree $\mathcal{T} \subset \mathcal{T}_0$ such that

$$\sum_{i=1}^{|\mathcal{T}|} \sum_{t, \boldsymbol{x}_t \in R_i} (y_t - \bar{y}_{R_i})^2 + \alpha |\mathcal{T}|, \tag{4.27}$$

is as small as possible. Clearly, Equation (4.27) shows that the penalty parameter α controls a trade-off between the complexity and goodness of fit of the subtree $\mathcal{T}$. Denote the subtree by $\mathcal{T}_\alpha$.

We then apply the selected subtree $\mathcal{T}_\alpha$ to produce predictions in the forecasting subsample and compute the mean squared forecast errors, which is also a function of α. Let α_0 be the value of α that gives the minimum mean squared forecast errors. Finally, we choose $\mathcal{T}_{\alpha_0}$ as the selected tree for y_t.

In the literature, cross-validation is often used to select the penalty parameter α in tree pruning. Such a procedure can still be used if y_t is assumed to be a linear AR process. Also, there is substantial research on Bayesian classification and regression tree (BCART) in the literature. See, for instance, Chipman et al. (1998). In the Bayesian framework, prior information is used to grow and prune a tree. Instead of using sample mean of the response variable in each subregion (leaf), linear regression models or other predictive models can also be used, similar to the piecewise linear approximations discussed in Chapter 3.

R demonstration: Regression tree.

```
> da <- read.table("GDPC1.txt",header=T)
> gdp <- diff(log(da$rgdp))
> tdx <- da$year + (da$mon/12)
> tdx <- tdx[-1]
> plot(tdx,gdp,xlab='year',ylab='gdp-rate',type='l')
> length(gdp)
[1] 273
> gdp <- round(gdp*100,2)
#
> X <- cbind(gdp[4:273],gdp[3:272],gdp[2:271],gdp[1:270])
> colnames(X) <- c("gdp","gdp1","gdp2","gdp3")
> require(tree)
> t1 <- tree(gdp~.,data=data.frame(X))
> summary(t1)
Regression tree:
tree(formula = gdp ~ ., data = data.frame(X))
Number of terminal nodes:  11
Residual mean deviance:  0.6654 = 172.3 / 259
Distribution of residuals:
    Min.   1st Qu.    Median      Mean  3rd Qu.     Max.
-3.27300 -0.46200 -0.03298  0.00000  0.51020  3.25700
> plot(t1)
> text(t1,pretty=0)
```

4.3.1.3 Classification Tree A classification tree is very similar to a regression tree, but it deals with a qualitative response variable instead of a quantitative one. As such, the same tree-building algorithm continues to apply. However, the categorical nature of the response variable requires using a different objective function. Suppose that the response variable y_t may assume K possible values, say $\{1, 2, \dots, K\}$. Denote the corresponding vector of explanatory variables by $\boldsymbol{x}_t$. Let R_i be the ith leaf (or region) of a tree. One can then estimate the probability of y_t belonging to Category k in R_i by

$$\hat{p}_{ik} = \frac{|R_{ik}|}{|R_i|}, \quad k = 1, \dots, K,$$

where $R_{ik} = \{j|(y_j = k) \cap (\boldsymbol{x}_j \in R_i)\}$ in the training sample and $|R_i|$ is the number of data points in R_i. For prediction, one may apply the simple majority rule, known as the most commonly occurring class, so that, given $\boldsymbol{x}_t \in R_i$

$$\hat{y}_t = k_0, \quad k_0 = \text{argmax}_k\{\hat{p}_{ik}\}.$$

The resulting classification error rate for y_t is then $e_t = 1 - \hat{p}_{i,k_0}$ provided that $\boldsymbol{x}_t \in R_i$. Experience shows that the classification error rate is not sufficiently sensitive for growing a classification tree. Two alternative measures are often used. They are the Gini index and the cross-entropy.

The *Gini index* is defined by

$$g_t = \sum_{k=1}^{K} \hat{p}_{ik}(1 - \hat{p}_{ik}), \tag{4.28}$$

which measures the total variance of the K categories. The Gini index takes a small positive value if all $\hat{p}_{ik}$ are close to either 0 or 1. Consequently, a small Gini index implies that R_i contains predominantly observations from a single category. In Equation (4.28), it is understood that $\hat{p}_{ik}$ depends on $\boldsymbol{x}_t$.

The second alternative measure is the *cross-entropy* defined by

$$c_t = -\sum_{k=1}^{K} \hat{p}_{ik} \log(\hat{p}_{ik}). \tag{4.29}$$

It is not hard to see that the cross-entropy $c_t \geq 0$ and it will take a small value if all probabilities $\hat{p}_{ik}$ are close to either 0 or 1. The latter property of the cross-entropy is similar to that of the Gini index. In applications, one may experiment with all three measurements to obtain a classification tree for prediction.

Example 4.3 Consider the daily measurement of $PM_{2.5}$ (fine particular matter with diameter less than 2.5 mm) of Shanghai, China, from January 1, 2013 to May 31, 2017. The original data are hourly, measured in units (μg/m^3), and can be downloaded from www.stateair.net, which is part of US Department of State air quality monitoring program. There were missing values in the hourly observations and we filled in those missing values by the previously available measurement. We then take the average of hourly data as the measurement of $PM_{2.5}$ for a given day. For simplicity in analysis, we removed the observation for February 29, 2016 so that each year has the same number of observations. Figure 4.13(a) shows the time plot of the $PM_{2.5}$ series whereas Figure 4.13(b) gives the sample autocorrelations of the series. From the plot and sample ACF, as expected, the series has an annual cycle.

Our goal here is to demonstrate the application of tree methods in prediction and to compare the results with other commonly used linear models. To this end, we reserve the last 365 observations as the forecasting subsample. Thus, there

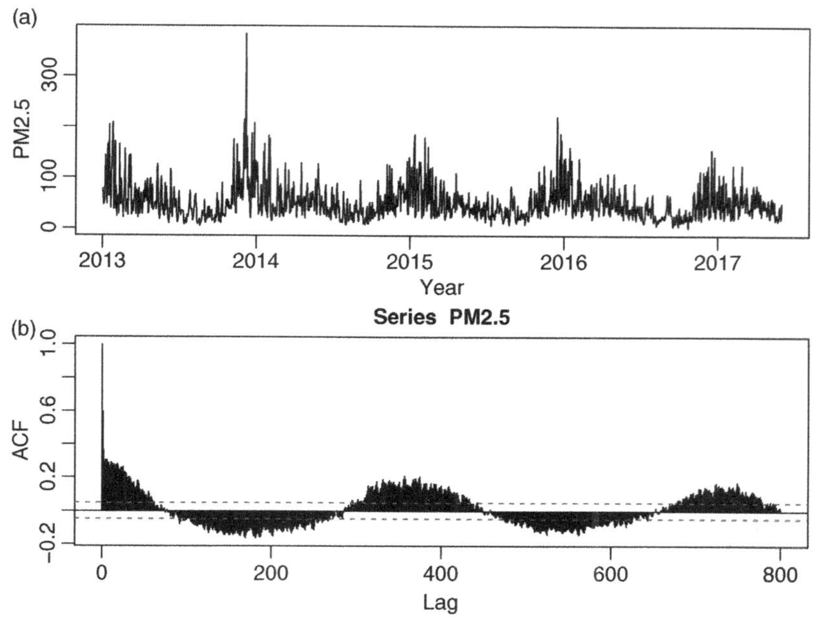

Figure 4.13 Time series of daily $PM_{2.5}$ of Shanghai, China, from January 1, 2013 to May 31, 2017: (a) the time plot of the series and (b) the sample autocorrelations of the data. The daily data are the average of hourly measurements.

are 1246 observations for training. Because of the annual cycle, we use the linear model

$$y_t = \beta_0 + \sum_{i=1}^{10} \beta_i y_{t-i} + \sum_{j=365}^{370} \beta_j y_{t-j} + e_t, \quad t = 371, \ldots, 1246, \tag{4.30}$$

as the benchmark model, where y_t denotes the $PM_{2.5}$ at time t, and e_t denotes the error term. This is a special AR(370) model. The fitted model is given in the R demonstration shown below. The R^2 of the model is 34.43%, and the model fits the series reasonably well. The Ljung–Box statistics of the residuals show $Q(400) = 370.16$ with p value 0.86. The p value would remain high even after adjusting the degrees of freedom of the chi-square distribution due to estimation. Thus, the residuals have no significant serial correlations.

Using the same explanatory variables, we apply the regression tree to the data. The resulting tree is shown in Figure 4.14. From the tree, it is seen that lag-1 variable y_{t-1} is the most important explanatory variable, and the seasonal lagged variables y_{t-370} and y_{t-367} are also important. But the residuals of the fitted tree show some minor serial correlations both at the regular and seasonal lags.

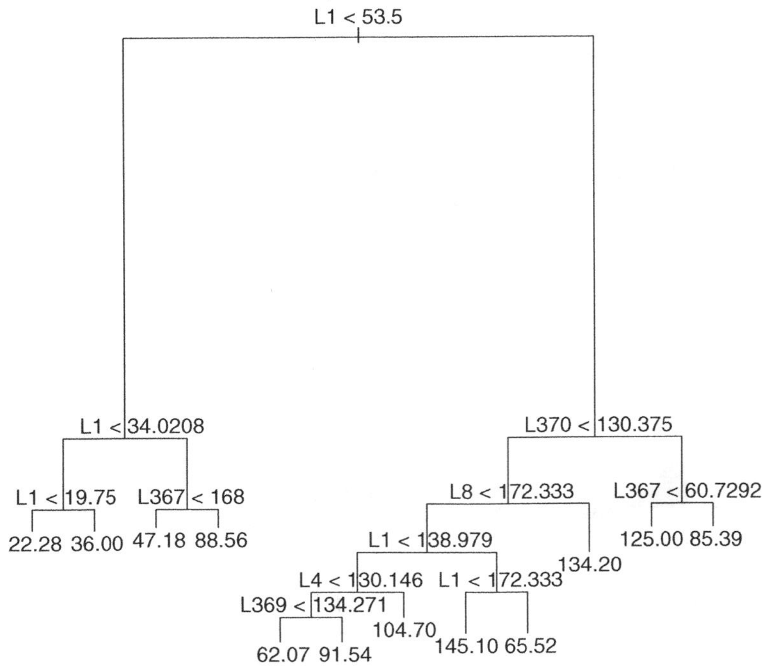

Figure 4.14 Regression tree for the daily $PM_{2.5}$ of Shanghai, China. The tree was built using the first 1246 observations as the training subsample.

The Ljung–Box statistics of the residuals give Q(400) = 566.59 with p value close to zero. This is not surprising as the fitted tree has 12 leaves, indicating that it only uses 12 values in prediction.

Turn to out-of-sample prediction. The linear AR(370) model in Equation (4.30) gives (MAE, RMSE) = (15.45, 20.19) whereas the fitted tree provides (MAE, RMSE) = (16.38, 22.27), where MAE and RMSE denote mean absolute error and root mean squared error, respectively. For the same reason as that stated in model checking, the performance of the regression tree in forecasting is not surprising either.

To demonstrate pruning, we apply the traditional cross-validation to the fitted tree. This can be justified as the model fitted is of the autoregressive type. Figure 4.15 shows the results of cross-validation. In this particular instance, the best tree seems to be of size 4. Figure 4.16 shows the prune tree for the Shanghai $PM_{2.5}$ during the training sample. Only the lag-1 and lag-370 variables are used in this simplified tree. Using the prune tree to perform out-of-sample prediction, we have (MAE,RMSE) = (17.12, 22.03).

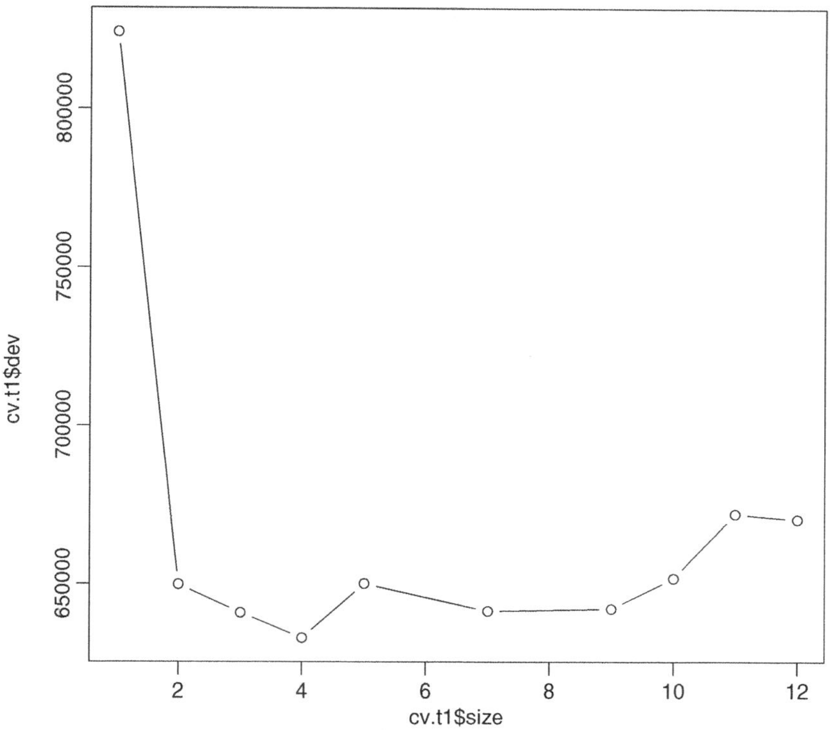

Figure 4.15 Results of cross-validation of regression tree for the daily $PM_{2.5}$ of Shanghai, China. The tree was built using the first 1246 observations as the training subsample.

R demonstration: Tree.

```
> PM2.5 <- scan(file="d-Shanghai-1317.txt")
> tdx <- c(1:length(PM2.5))/365+2013
> par(mfcol=c(2,1))
> plot(tdx,PM2.5,xlab='year',ylab='PM2.5',type='l')
> acf(PM2.5,lag=800)
> m1 <- NNsetting(PM2.5,nfore=365,lags=c(1:10,365:370))
> names(m1)
[1] "X"      "y"      "predX" "predY"
> X <- m1$X; y <- m1$y; predX <- m1$predX; predY <- m1$predY
> n1 <- paste("L",1:10,sep="")
> n2 <- paste("L",365:370,sep="")
> colnames(X) <- colnames(predX) <- c(n1,n2)
> m2 <- lm(y~.,data=data.frame(X))
> summary(m2)
Call:  lm(formula = y ~ ., data = data.frame(X))
```

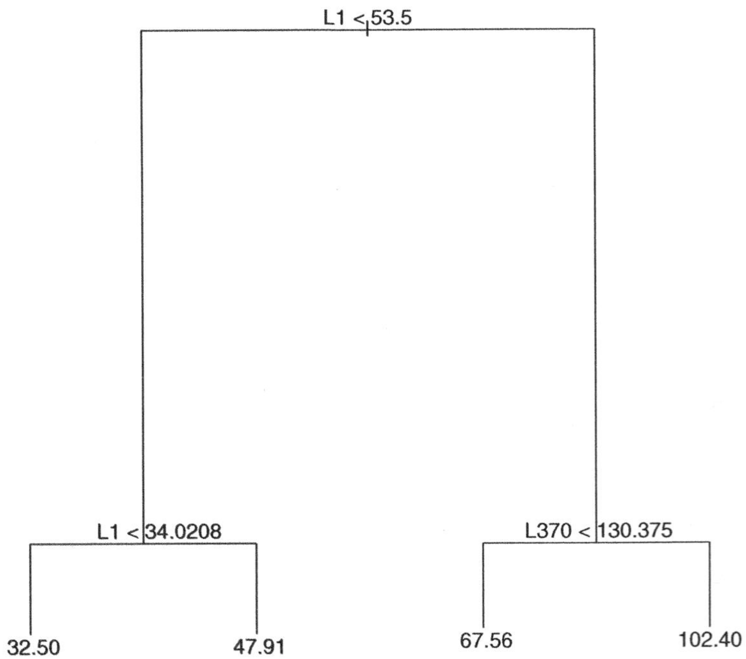

Figure 4.16 Prune regression tree for the daily $PM_{2.5}$ of Shanghai, China. The tree was built using the first 1246 observations as the training subsample and prune by cross-validation.

```
Coefficients:
             Estimate Std. Error t value Pr(>|t|)
(Intercept) 11.330124   2.605231   4.349 1.53e-05 ***
L1           0.487490   0.033792  14.426  < 2e-16 ***
L2          -0.046739   0.037458  -1.248  0.21246
L3          -0.062058   0.037249  -1.666  0.09607 .
L4           0.026367   0.037273   0.707  0.47951
L5           0.082690   0.037183   2.224  0.02641 *
L6           0.006062   0.037240   0.163  0.87074
L7          -0.005810   0.037323  -0.156  0.87632
L8           0.106191   0.037370   2.842  0.00459 **
L9          -0.107141   0.037513  -2.856  0.00439 **
L10          0.110018   0.033664   3.268  0.00113 **
L365        -0.019086   0.029047  -0.657  0.51131
L366         0.035823   0.032708   1.095  0.27372
L367         0.051650   0.032712   1.579  0.11472
L368        -0.011674   0.032646  -0.358  0.72074
L369         0.051792   0.032573   1.590  0.11219
L370         0.058908   0.028975   2.033  0.04236 *
```

```
---
Residual standard error: 25.01 on 859 degrees of freedom
Multiple R-squared:  0.3443,    Adjusted R-squared:  0.3321
F-statistic: 28.19 on 16 and 859 DF,  p-value: < 2.2e-16
> acf(m2$residuals,lag=400)
> Box.test(m2$residuals,lag=400,type="Ljung")
        Box-Ljung test
data:  m2$residuals
X-squared = 370.16, df = 400, p-value = 0.8552
> pm2 <- predict(m2,newdata=data.frame(predX))
> er <- predY-pm2
> mean(abs(er))
[1] 15.44751
> sqrt(mean(er^2))
[1] 20.18687

> require(tree)
> t1 <- tree(y~.,data=data.frame(X))
> par(mfcol=c(1,1))
> plot(t1)
> text(t1,pretty=0)
> pt1 <- predict(t1,newdata=data.frame(predX))
> er3 <- pt1-predY
> mean(abs(er3))
[1] 16.3767
> sqrt(mean(er3^2))
[1] 22.2729
### Compute residuals for model checking
> pp1 <- predict(t1,newdata=data.frame(X))
> resi <- y-pp1
> Box.test(resi,lag=400,type="Ljung")
        Box-Ljung test
data:  resi
X-squared = 566.59, df = 400, p-value = 7.614e-08
### Pruning tree
> cv.t1 <- cv.tree(t1)
> plot(cv.t1$size,cv.t1$dev,type="b")
> prune.t1 <- prune.tree(t1,best=4)
> plot(prune.t1)
> text(prune.t1,pretty=0)
> prun <- predict(prune.t1,newdata=data.frame(predX))
> per <- predY-prun
> mean(abs(per))
[1] 17.11805
> sqrt(mean(per^2))
[1] 22.03309
```

Example 4.4 Consider again the daily $PM_{2.5}$ of Shanghai, but classify the measurements into four categories as

$$(0, 50],\ (50, 100],\ (100, 150],\ (150, \infty].$$

The classification is similar to that used in Stateair.net and the categories are referred to as good, moderate, sensitive, and unhealthy, respectively. On the Stateair.net, the unhealthy category is further divided into four classes, but the maximum of the dependent variable during the sample period is only about 221 so it suffices to use four categories.

We use the same explanatory variables as those of Equation (4.30). Figure 4.17 shows the fitted tree for the $PM_{2.5}$ series in the training sample. Again, the lag-1 variable remains the most important determining factor in predicting $PM_{2.5}$, but some seasonal lagged variables are also relevant. Applying the fitted tree in Figure 4.17 to the forecasting subsample, we see that the success rate is approximately 71.8%, but the model fails to predict the occurrence of Categories 3 and 4.

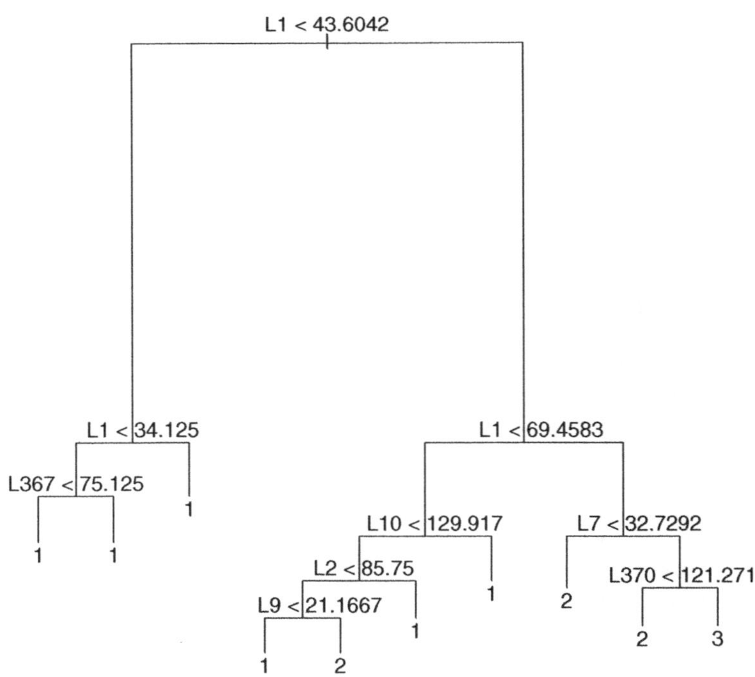

Figure 4.17 Classification tree for the daily $PM_{2.5}$ of Shanghai, China. The tree was built using the first 1246 observations as the training subsample. The $PM_{2.5}$ was divided into four categories, namely (0,50], (50,100], (100,150], (150,∞].

R demonstration: Classification tree.

```
> cy <- rep(1,876)
> idx <- c(1:876)[y > 50]
> cy[idx] <- 2
> idx <- c(1:876)[y > 100]
> cy[idx] <- 3
> idx <- c(1:876)[y > 150]
> cy[idx] <- 4
> cpY <- rep(1,365)
> jdx <- c(1:365)[predY > 50]
> cpY[jdx] <- 2
> jdx <- c(1:365)[predY > 100]
> cpY[jdx] <- 3
> jdx <- c(1:365)[predY > 150]
> cpY[jdx] <- 4
> cy <- as.factor(cy); cpY <- as.factor(cpY)
> t2 <- tree(cy~.,data=data.frame(X))
> plot(t2)
> text(t2,pretty=0)
> pt2 <- predict(t2,newdata=data.frame(predX),type="class")
> table(cpY,pt2)
   pt2
cpY   1   2   3   4
  1 207  54   0   0
  2  33  55   2   0
  3   4   9   0   0
  4   0   1   0   0
> (207+55)/365
[1] 0.7178082
```

4.3.1.4 Bagging Bagging stands for *Bootstrap Aggregation.* It is a general and powerful method for reducing the variance of a statistical model. Bagging is motivated by the idea that averaging reduces variance. For instance, the sample mean (a simple average) of an iid sample reduces the variance from σ^2 to σ^2/T, where T is the sample size and σ^2 is the variance of the underlying random variable. Consider the decision trees of the previous section. The prediction of a regression tree tends to have a large root mean squared error (RMSE) because the tree only uses a relatively small number of leaves in prediction. If one can take the average of predictions of many trees, one could reduce the forecast errors (or RMSE). A question then arises as how to obtain multiple regression trees. The *bootstrap* method comes to the rescue because one can draw many bootstrap samples from the training sample and use each bootstrapped sample to obtain a regression tree.

Consider the simple one-step ahead prediction at the forecast origin $t = n$. Let $\hat{y}_{n+1,m}$ be the prediction obtained from the mth bootstrapped sample, where $m = 1, \ldots, M$, with M being a sufficiently large positive integer. Then, the simple average

$$\hat{y}_{n+1,bag} = \frac{1}{M} \sum_{m=1}^{M} \hat{y}_{n+1,m}$$

is called the bagged prediction. This process is referred to as *bagging*. For the classification tree, one can replace the averaging by the *majority votes* among the M bootstrapped predictions to achieve the effect of bagging.

In practice, the trees of the bootstrapped samples are grown deep and are not pruned. The idea here is to keep the bias of an individual tree small. Of course, for time series data, one needs to apply proper methods to draw the bootstrap samples. There are many methods available in the literature, including block bootstrap, wild bootstrap, and frequency domain bootstrap. See, for instance, Lahiri (2003) and the references therein. For an autoregressive-type of time series, one can apply simple random sampling with replacement to the rows of the predictor matrix to obtain bootstrap samples. While this approach may not be ideal to draw a bootstrap sample, the independence of the error terms and the simplicity of the procedure provide some justifications.

A byproduct of bagging is the *out-of-bag error estimation*. As mentioned before, bagging fits trees to the bootstrapped samples. For the linear regression setup, the default setting is to use approximately two-thirds of the observations in the training sample. The remaining observations not used to fit a given bagged tree are referred to as the *out-of-bag* (OOB) observations. We can use the fitted tree to predict the response variable of every OOB observation. This will yield roughly $M/3$ predictions for each observation in the training sample. Using a simple average of those predictions to obtain a single prediction, one can compute the OOB mean squared error for the bagged model. For time series data, one can use the OOB predictions to compute the (pseudo predictive) residuals of a bagged tree. The residuals can then be used for model checking. Strictly speaking, the fitted values are averages of out-of-sample predictions over many bagged trees, so pseudo predictive residuals might be a more appropriate term.

Example 4.5 Again, consider the daily $PM_{2.5}$ of Shanghai used in Example 4.3. We now apply bagging to the series. For illustrative purposes, we use $M = 500$ and 1000 bootstrap iterations. Once a bagged tree is built, we applied it to obtain out-of-sample predictions of the 365 observations in the forecasting sample. The resulting (MAE, RMSE) are (15.72, 20.37) and (15.69, 20.32), respectively, for $M = 500$ and 1000. As expected, the results show improvements over the single tree of Example 4.3. They also indicate that the number of bootstrap iterations is not critical so long as it is sufficiently large. In our application, the improvements from 500 iterations to 1000 iterations are relatively small. Finally, the results of bagging are close to those of the linear model in Equation (4.30). To check the bagged trees, we use the average of the fitted values of the tree with 1000 iterations to compute the residuals. The Ljung–Box statistics of the residuals gives $Q(400) = 397.21$ with p value 0.53, indicating that the bagged tree describes the dynamic dependence of Shanghai $PM_{2.5}$ reasonably well. Details of using bagging are given in the R demonstration.

R demonstration: Bagging is a special case of random forests with the subcommand `mtry` being the number of explanatory variables. The R package `randomForest` is used in the demonstration.

```
> require(randomForest)
> dim(X)
[1] 876  16
> bag1 <- randomForest(y~.,data=data.frame(X),mtry=16,importance=TRUE)
> bag1
Call: randomForest(formula = y~.,data=data.frame(X), mtry=16,importance=TRUE)
               Type of random forest: regression
                     Number of trees: 500
No. of variables tried at each split: 16

          Mean of squared residuals: 655.6926
                    % Var explained: 29.93
> pbag1 <- predict(bag1,newdata=data.frame(predX))
> er <- pbag1-predY
> mean(abs(er))
[1] 15.72161
> sqrt(mean(er^2))
[1] 20.36633
>
> set.seed(1)
> bag2 <- randomForest(y~.,data=data.frame(X),mtry=16,ntree=1000)
> bag2
Call: randomForest(formula=y ~ ., data=data.frame(X),mtry=16,ntree=1000)
               Type of random forest: regression
                     Number of trees: 1000
No. of variables tried at each split: 16

          Mean of squared residuals: 649.4435
                    % Var explained: 30.6
> pbag2 <- predict(bag2,newdata=data.frame(predX))
> er2 <- pbag2-predY
> mean(abs(er2))
[1] 15.688
> sqrt(mean(er2^2))
[1] 20.32263
> names(bag2)
```

```
 [1] "call"            "type"          "predicted"     "mse"
 [5] "rsq"             "oob.times"     "importance"    "importanceSD"
 [9] "localImportance" "proximity"     "ntree"         "mtry"
[13] "forest"          "coefs"         "y"             "test"
[17] "inbag"           "terms"
> resi <- y - bag2$predicted
> ts.plot(resi)
> acf(resi,lag=400)
> Box.test(resi,lag=400,type="Ljung")
        Box-Ljung test
data:  resi
X-squared = 397.21, df = 400, p-value = 0.53
```

4.3.2 Random Forests

Random forests are bagging with a small tweak to reduce the correlations between trees. Like bagging, random forests also build many trees from the bootstrapped training samples, but they modify the tree-building process. Suppose that there are p explanatory variables with p being large. Let $m \leq p$ be a positive integer. Typically, $m < p$ such that $m \approx \sqrt{p}$. To build a decision tree for a given bootstrapped training sample in random forecasts, at each time a split is considered, a random sample of m predictors is chosen from the p explanatory variables to serve as split candidates. Thus, the split is only allowed to use one of those selected m predictors. A fresh random sample of m predictors is drawn at each split.

Consequently, in a random forecast, trees are built using a randomly chosen set of predictors as split candidates at each split. The rationale for using such a tree-building process is to avoid the possibility that all trees in a random forest are dominated by a single strong predictor. Consider the special case in which one predictor is very strong and other predictors are moderate, then the strong predictor will be chosen to perform the first split for all trees in the forest under bagging. The predictions of the resulting trees are likely to be highly correlated. The high correlations, in turn, reduce the number of effective bootstrap iterations, which seems to be in contradiction with the fundamental concept of bagging.

A word of caution is in order for applying random forests to time series data. Due to the dynamic dependence, strong predictors are likely to occur in a time series, and one often needs those predictors in a model for it to provide good fit. Consider, for example, the daily $PM_{2.5}$ series of Shanghai used in Section 4.3.1.4. The lag-1 and lag-370 values are strong explanatory variables as the series has regular and seasonal dependence. It seems important to include these two explanatory variables in a decision tree. This consideration seems to go against the idea of random forests. A careful study of applying random forests to time series data is needed. This is particularly so when the time series involved has strong serial dependence.

Example 4.6 Again, consider the daily $PM_{2.5}$ of Shanghai of Example 4.3. We also reserve the last 365 observations for out-of-sample prediction. Let y_t be the $PM_{2.5}$ at time t. Two sets of explanatory variables are used. First, $\boldsymbol{X}_{1t} = (y_{t-1}, \ldots, y_{t-10}; y_{t-365}, \ldots, y_{t-370})'$ and $\boldsymbol{X}_{2t} = (y_{t-1}, \ldots, y_{t-15}; y_{t-365}, \ldots, y_{t-374})'$. Here $\boldsymbol{X}_{1t}$ and $\boldsymbol{X}_{2t}$ have 16 and 25 predictors, respectively. We apply random forests to the data and compute the mean absolute error and root mean squared error of the out-of-sample predictions. The results are given below:

Model	MAE	RMSE
$\boldsymbol{X}_{1t}$, mtry = 4, ntree = 500	15.77	20.10
$\boldsymbol{X}_{1t}$, mtry = 4, ntree = 1000	15.64	19.97
$\boldsymbol{X}_{1t}$, mtry = 8, ntree = 500	15.64	20.06
$\boldsymbol{X}_{2t}$, mtry = 5, ntree = 500	15.68	19.93
$\boldsymbol{X}_{2t}$, mtry = 5, ntree = 1000	15.66	19.94
$\boldsymbol{X}_{2t}$, mtry = 8, ntree = 500	15.43	19.82

where "mtry" denotes the number of explanatory variables used for each split and "ntree" is the number of trees used.

From the results, it is seen that random forests provide slight improvements over bagging and some improvements over the linear model in Equation (4.30). Also, the results of random forests are not sensitive to the number of trees used.

R demonstration: Random forests via the package `randomForest`.

```
> da <- read.table("d-Shanghai-1317.txt")
> m1 <- NNsetting(da[,1],nfore=365,lags=c(1:10,365:370))
> X <- m1$X; y <- m1$y; predX <- m1$predX; predY <- m1$predY
> n1 <- paste("L",1:10,sep="")
> n2 <- paste("L",365:370,sep="")
> colnames(X) <- colnames(predX) <- c(n1,n2)
> require(randomForest)
> rf2 <- randomForest(y~.,data=data.frame(X),mtry=4,ntree=1000)
> rf2
Call: randomForest(formula=y~.,data=data.frame(X),mtry=4,ntree=1000)
               Type of random forest: regression
                     Number of trees: 1000
No. of variables tried at each split: 4

          Mean of squared residuals: 638.2131
                    % Var explained: 31.8
```

```
> prf2 <- predict(rf2,newdata=data.frame(predX))
> er2 <- prf2-predY
> mean(abs(er2))
[1] 15.64032
> sqrt(mean(er2^2))
[1] 19.97458
######## add more predictors
> m2 <- NNsetting(da[,1],nfore=365,lags=c(1:15,365:374))
> X <- m2$X; y <- m2$y; predX <- m2$predX; predY <- m2$predY
> n1 <- paste("L",1:15,sep="")
> n2 <- paste("L",365:374,sep="")
> colnames(X) <- colnames(predX) <- c(n1,n2)
> dim(X)
[1] 872  25
> rf4 <- randomForest(y~.,data=data.frame(X),mtry=5)
> rf4
Call: randomForest(formula = y ~ ., data = data.frame(X), mtry= 5)
               Type of random forest: regression
                     Number of trees: 500
No. of variables tried at each split: 5

          Mean of squared residuals: 634.4673
                    % Var explained: 32.42
> prf4 <- predict(rf4,newdata=data.frame(predX))
> er4 <- prf4-predY
> mean(abs(er4))
[1] 15.68707
> sqrt(mean(er4^2))
[1] 19.93318
```

4.4 EXERCISES

4.1 Consider the transactions data of Walgreens stock on February 6, 2017. Apply the classification tree to the data to model the price movement using the same training and forecasting subsamples as those of Example 4.2 of the chapter. Compare the prediction accuracy with that of deep learning.

4.2 Use the methods of neural network, deep learning, and classification tree to predict the $PM_{2.5}$ series of Beijing. You may use the same lagged explanatory variables as those of the Shanghai data in Example 4.3. For deep learning, classify the $PM_{2.5}$ into six categories, namely

$$(0,50],\ (50,100],\ (100,150],\ (150,200],\ (200,300],\ (300,\infty].$$

Use the last 365 observations as the forecasting subsample.

4.3 Consider again the Beijing data. Use bagging with 300 bootstrap iterations to predict the $PM_{2.5}$ of Beijing. Again, reserve the last 365 observations as the forecasting subsample. Repeat the bagging analysis with 600 bootstrap iterations. Are there serial correlations in the residuals of the bagged tree with 600 iterations? Why?

4.4 Consider again the daily $PM_{2.5}$ series of Beijing. Reserve the last 365 observations for out-of-sample prediction. Apply random forests to obtain predictions. Use the same lagged values as in Example 4.6 as the predictors. Try several random forests and comment on their performance.

4.5 Consider the US monthly unemployment rates from January 1948 for September 2017 for 837 observations. The data can be downloaded from FRED (code: UNRATE) and are seasonally adjusted. The series has strong serial correlations. The goal of this exercise is to explore the effect of strong serial dependence on the performance of bagging and random forests. To this end, use the first 700 observations as the modeling subsample and reserve the last 137 observations as the forecasting subsample. Let $\boldsymbol{X}_t = (x_{t-1}, \dots, x_{t-24})'$ be the predictors, where x_t is the unemployment rate at time t.

(a) Apply bagging with 1000 bootstrap iterations to the modeling subsample and compute the mean absolute forecast error and the root mean squared forecast errors of the out-of-sample predictions.

(b) Apply random forests with `mtry = 5` and `ntree=500` and compute the out-of-sample prediction measures as part (a).

(c) Repeat part (b) with `mtry = 8` and `btree=500`.

(d) Compare and comment on the forecasting accuracy of parts (a) to (c). In particular, what is the effect of strong serial correlations on bagging and random forests, if any?

REFERENCES

Carreira-Perpiñan, M. A. and Hinton, G. E. (2005). On contrastive divergence learning. In *Proceedings of the Tenth International Workshop on Artificial Intelligence and Statistics* (AISTATS'05), eds. R. G. Cowell and Z. Ghahramani, pp. 30–40. Society for Artificial Intelligence and Statistics.

Chipman, H. A., George, E. I., and McCulloch, R. E. (1998). Bayesian CART model search (with discussion). *Journal of the American Statistical Association* **93**: 935–960.

Drees, M., J. Rueckert, C. M. Friedrich, G. Hinton, R. Salakhutdinov, and C. E. Rasmussen (2016). *darch*: Deep Architures and Restricted Boltzmann Machines. R package.

Géron, A. (2017). *Hands-on Machine Learning with SciKit learn and TensorFlow: Concepts, Tools, and Techniques to Build Intelligent Systems.* O'Reilly Media, Sebastopol, CA.

Goodfellow, I., Dengio, Y. and Courville, A. (2016). *Deep Learning*. MIT Press, Boston.

Hassoun, M. H. (2003). *Fundamentals of Artificial Neural Networks.* A Bradford Book. MIT Press, Boston.

Hastie, T., Tibshirani, R. and Froedman, J. (2001). *The Elements of Statistical Learning: Data Mining, Inference, and Prediction*. Springer, New York.

Hauseman, J., Lo, A. and MacKinlay, C. (1992). An ordered probit analysis of transaction stock prices. *Journal of Financial Economics* **31**: 319–379.

Hinton, G. E. and Sejnowski, T. J. (1986). Learning and relearning in Boltzmann machines. In *Parallel Distributed Processing*, eds. D. E. Rumelhart and J. L. McClelland, Vol. 1, Chapter 7, pp. 282–317. MIT Press, Cambridge, MA.

Hinton, G. E., Osindero, S., and Teh, Y. (2006). A fast learning algorithm for deep belief nets. *Neural Computation*, **18**: 1527–1154.

Hochreiter, S. and Schmidhuber, J. (1997). Long short-term memory. *Neural Computation* **9**: 1735–1780.

James, G., Witten, D., Hastie, T., and Tibshirani, R. (2013). *An Introduction to Statistical Learning with Applications in R.* Springer, New York.

Lahiri, S. N. (2003). *Resampling Methods for Dependent Data.* Springer, New York.

Nielsen, M. (2017). *Neural Networks and Deep Learning*. A free online book at http://neuralnetworksanddeeplearning.com.

Robert, C. P. and Casella, G. (2004). *Monte Carlo Statistical Methods*, 2nd edition. Springer, New York.

Rydberg, T. H. and Shephard, N. (2003). Dynamics of trade-by-trade price movements decomposition and models. *Journal of Financial Econometrics* **1**: 2–25.

Schmdhuber, J. (2015). Deep learning in neural networks: An overview. *Neural Networks*, **61**: 85–117.

Tsay, R. S. (2010). *Analysis of Financial Time Series*, 3rd edition. John Wiley, Hoboken, NJ.

Tsay, R. S. (2013). *An Introduction to Analysis of Financial Data with R.* John Wiley, Hoboken, NJ.

Xiao, R. (2014). *deepnet*: Deep learning toolkit in R. R package.

CHAPTER 5

ANALYSIS OF NON-GAUSSIAN TIME SERIES

In this chapter we study the analysis of non-Gaussian time series. The series considered consist of discrete-valued data, functional data, and financial data, with the latter being serially uncorrelated but dependent. We focus on parametric models except for functional data. For discrete-valued time series, we discuss two modeling approaches. The first modeling approach employs the generalized autoregressive moving-average (GARMA) models and can be thought of as an extension of the generalized linear models (GLM) of McCullagh and Nelder (1989) to serially dependent data. This general class of models was considered by Benjamin et al. (2003). For empirical analysis, we use the R package `glarma` of Dunsmuir et al. (2015), where GLARMA stands for the generalized linear autoregressive moving-average models. We also introduce a recent extension of the model by Zheng et al. (2015), the martingalized GARMA (M-GARMA) model, which addresses some of the shortcomings of the GARMA models. The second modeling approach employs the autoregressive conditional mean models, which are closely related to the autoregressive conditional duration (ACD)

Nonlinear Time Series Analysis, First Edition. Ruey S. Tsay and Rong Chen.

Companion website: www.wiley.com/go/tsay/nonlineartimeseries

models of Engle and Russell (1998). See Heinen (2003) for the autoregressive conditional Poisson (ACP) model and Hautsch (2012) for applications of the model in financial econometrics. There are other approaches to modeling time series of counts. See Jung and Tremayne (2011) for a comprehensive review.

For financial data, we introduce some commonly used volatility models. Our discussion is brief, and we refer interested readers to books on analysis of financial time series, e.g. Tsay (2010) and Ruppert and Matteson (2015, Chapter 14). The R packages `fGarch` and `rugarch` are useful for analyzing asset return series. Finally, for functional data, we discuss some simple autoregressive models, including convolution models, for modeling and prediction. Nonparametric spline methods are used in estimation.

5.1 GENERALIZED LINEAR TIME SERIES MODELS

Non-Gaussian time series analysis has a long history, especially for count data. See the references in Jung and Tremayne (2011), Dunsmuir and Scott (2015), and Weiss (2017). Following Cox (1981), statistical models for discrete-valued time series are classified into *observation-driven* and *parameter-driven* models. It turns out that the observation-driven models are often easier to estimate and we shall focus on such models in this section.

Benjamin et al. (2003) proposed a broad class of generalized autoregressive moving-average (GARMA) models for modeling non-Gaussian time series. Many existing models in the literature are special cases of the GARMA models. Let y_t be a (non-Gaussian) time series and $\boldsymbol{x}_t$ be a finite-dimensional vector of explanatory variables. Denote by $\boldsymbol{F}_t = \sigma\{y_{t-1}, y_{t-2}, \dots ; \boldsymbol{x}_t, \boldsymbol{x}_{t-1}, \dots\}$ the σ-field generated by the past values of y_t and the current and past values of $\boldsymbol{x}_t$. The GARMA(p, q) model of Benjamin et al. (2003) postulates that the conditional distribution of y_t given $\boldsymbol{F}_t$ is a member of the exponential family distributions via some link function similar to those of the generalized linear models of McCullagh and Nelder (1989). Specifically, the conditional density function of y_t given $\boldsymbol{F}_t$ is

$$f(y_t \mid \boldsymbol{F}_t) = \exp\left\{\frac{y_t\vartheta_t - b(\vartheta_t)}{\varphi} + a(y_t, \varphi)\right\}, \tag{5.1}$$

where ϑ_t and φ are the canonical and scale parameters and the functions $b(\cdot)$ and $a(\cdot)$ define the specific member of the exponential family of interest. The conditional expectation and variance of y_t given $\boldsymbol{F}_t$ are, respectively,

$$E(y_t \mid \boldsymbol{F}_{t-1}) = \mu_t = b'(\vartheta_t), \quad \text{Var}(y_t \mid \boldsymbol{F}_{t-1}) = \varphi b''(\vartheta_t),$$

where $b'(\vartheta_t)$ and $b''(\vartheta_t)$ denote the first and second derivatives of $b(\cdot)$ with respect to its argument. Furthermore, a link function $g(\cdot)$ is used to relate the conditional expectation μ_t to the canonical parameter so that $g(\mu_t) = \vartheta_t$. For simplicity, Benjamin et al. (2003) and Dunsmuir and Scott (2015) use the form

$$\vartheta_t = \boldsymbol{x}_t'\boldsymbol{\beta} + w_t, \tag{5.2}$$

where w_t is a dynamically dependent variable and $\boldsymbol{\beta}$ denotes the coefficient vector of explanatory variables. If $w_t = 0$ in Equation (5.2), then the model reduces to the conventional GLM of McCullagh and Nelder (1989) and the time series y_t is conditionally uncorrelated given $\boldsymbol{x}_t$.

Benjamin et al. (2003) consider a general form for w_t as

$$w_t \equiv g(\mu_t) = \sum_{j=1}^{p} \phi_j A(y_{t-j}, \boldsymbol{x}_{t-j}, \boldsymbol{\beta}) + \sum_{j=1}^{q} \delta_j M(y_{t-j}, \mu_{t-j}), \tag{5.3}$$

where p and q are non-negative integers, $\boldsymbol{\phi} = (\phi_1, \ldots, \phi_p)'$ and $\boldsymbol{\delta} = (\delta_1, \cdots, \delta_q)'$ are the autoregressive and moving-average parameters, and $A(\cdot)$ and $M(\cdot)$ are autoregressive and moving-average components. The authors suggest several possibilities for the choice of $M(\cdot)$, including deviance residuals, Pearson residuals, residuals on the original scale $y_t - \mu_t$, and residuals on the predictor scale $g(y_t) - \vartheta_t$.

The ARMA formulation of Equation (5.3) is too general for practical use. Benjamin et al. (2003) focus on the simple form

$$\vartheta_t = \boldsymbol{x}_t'\boldsymbol{\beta} + \sum_{j=1}^{p} [g(y_{t-j}) - \boldsymbol{x}_{t-j}'\boldsymbol{\beta}] + \sum_{j=1}^{q} \theta_j [g(y_{t-j}) - \vartheta_{t-j}]. \tag{5.4}$$

Let $\epsilon_t = g(y_t) - \vartheta_t = g(y_t) - g(\mu_t)$ be the deviation of $g(y_t)$ from $g(\mu_t)$. By adding $g(y_t) - \vartheta_t$ to both sides of Equation (5.4), we obtain

$$g(y_t) = \boldsymbol{x}_t'\boldsymbol{\beta} + \sum_{j=1}^{p} [g(y_{t-j}) - \boldsymbol{x}_{t-j}'\boldsymbol{\beta}] + \epsilon_t + \sum_{j=1}^{q} \theta_j \epsilon_{t-j}. \tag{5.5}$$

Thus, under the GARMA framework, the transformed series $g(y_t)$ assumes the formulation of an ARMAX model commonly used in time series analysis. In this sense, GARMA models are models on the predictor scale, not the original scale.

However, there is a key difference between the ARMA and GARMA models. The error ϵ_t of the GARMA model in Equation (5.5) is not necessarily a white noise series. In particular, for a nonlinear function $g(\cdot)$, $E(\epsilon_t \mid \boldsymbol{F}_{t-1})$ might be nonzero. If the link function $g(\cdot)$ is the identity function, then it is easy to show, via iterated expectations, that $\{\epsilon_t\}$ consists of a sequence of zero-mean and serially uncorrelated random variates. The distributions of ϵ_t, however, might be time dependent. The martingalized GARMA models discussed in the next section are proposed to overcome this difficulty.

The ARMAX formulation of Equation (5.5) is helpful in understanding the GARMA models. On the other hand, because the observed response y_t enters the equation after the transformation by the link function, care must be exercised in some applications. In fact, an arbitrary threshold for observations is often needed for the model to work. Consider, for instance, the Poisson response with log link, i.e. $g(y_t) = \log(y_t)$. Then, $g(y_t)$ is not defined when $y_t = 0$. In many applications of count data, $y_t = 0$ is common. One may even encounter zero-inflated count data.

Remark: The acronym GARMA is also used by Woodward et al. (2016) to denote the Gegenbauer ARMA process, which involves a $(1 - 2uB + B^2)$ factor in the autoregressive polynomial, where $|u| < 1$ and B is the backshift operator. The R package `tswge` associated with Woodward et al. (2016) can be used to estimate Gegenbauer ARMA models, but not the GARMA models discussed here.

5.1.1 Count Data and GLARMA Models

Count data are perhaps the most commonly encountered non-Gaussian data in time series analysis. See, for example, Davis et al. (2003) and Davis and Wu (2009), among others. Dunsmuir et al. (2015) and Dunsmuir and Scott (2015) consider the generalized linear autoregressive moving-average (GLARMA) models for count data. These models are special cases of the GARMA models with some simplifying assumptions. The conditional distribution of GLARMA models is

$$f(y_t \mid F_t) = \exp\left\{y_t\vartheta_t - g_t b(\vartheta_t) + c_t\right\}, \tag{5.6}$$

where g_t and c_t are sequences of constants possibly depending on the observations y_t. Compared with Equation (5.1), the scale parameter $\varphi = 1$ under the GLARMA model, but the function $b(\cdot)$ is scaled by a factor g_t. This is a restricted form of the exponential family. The conditional means and variances of the responses of Equation (5.6) are $\mu_t \equiv E(y_t \mid F_t) = g_t b'(\vartheta_t)$ and $\sigma_t^2 \equiv \text{Var}(y_t \mid F_t) = g_t b''(\vartheta_t)$. Note that the negative binomial distribution is not a member of the exponential family when the scale parameter is not fixed. The canonical parameter employed by the `glarma` package of Dunsmuir et al. (2015) is defined as

$$\vartheta_t = \boldsymbol{x}_t'\boldsymbol{\beta} + O_t + w_t, \tag{5.7}$$

where O_t is an offset term, which provides an added flexibility for parameter interpretation. For example, for Poisson data such as disease counts, O_t is the logarithm of population, which results in a model for the incidence per unit of population. There is no coefficient parameter associated with O_t.

Function	Poisson	Binomial	Negative binomial
$f(y_t \mid \boldsymbol{F}_t)$	$\frac{\mu_t^{y_t} e^{-\mu_t}}{y_t!}$	$\binom{m_t}{y_t} \pi_t^{y_t}(1-\pi_t)^{m_t-y_t}$	Equation (5.8)
a_t	1	m_t	–
c_t	$-\log(y_t!)$	$\log\binom{m_t}{y_t}$	–
$b(\vartheta_t)$	e^{ϑ_t}	$\log(1+e^{\vartheta_t})$	–
μ_t	e^{ϑ_t}	$m_t\pi_t$	e^{ϑ_t}
σ_t^2	μ_t	$m_t\pi_t(1-\pi_t)$	$\mu_t + (\mu_t^2/\alpha^2)$
Link	log	logit	log

Table 5.1 Distributions and the associated link functions supported by the `glarma` package.

The response distributions used in the `glarma` package are Poisson and binomial, both of which can be put in the form of Equation (5.6), and the negative binomial distribution with density function given by

$$f(y_t \mid \boldsymbol{F}_t, \alpha) = \frac{\Gamma(\alpha + y_t)}{\Gamma(\alpha)\Gamma(y_t + 1)} \left[\frac{\alpha}{\alpha + \mu_t}\right]^{\alpha} \left[\frac{\mu_t}{\alpha + \mu_t}\right]^{y_t}, \tag{5.8}$$

where $\mu_t = \exp(\vartheta_t)$. This cannot be put into the one parameter exponential family form when α is unknown. In practice, negative binomial distribution is useful in the presence of over-dispersion. Clearly, negative binomial density converges to the Poisson density as $\alpha \to \infty$. Table 5.1 provides a summary of the conditional distributions and associated link functions employed by the `glarma` package. Further details of the discrete distributions used for count data are given in the appendix to this chapter, including the double Poisson distribution of Efron (1986).

With Equation (5.2), the serial dependence of GLARMA models is governed by

$$w_t = \sum_{i=1}^{p} \phi_i(w_{t-i} + e_{t-i}) + \sum_{j=1}^{q} \theta_j e_{t-j} \tag{5.9}$$

where $e_t = (y_t - \mu_t)/v_t$ for some scaling sequence $\{v_t\}$. Three specifications of v_t are supported by the `glarma` package:

- $v_t = \sigma_t$, the standard deviation of the predictive distribution given in Table 5.1. The resulting e_t is called the Pearson residual.

- $v_t = \sigma_t^2$, the conditional variance. This follows the suggestion of Creal et al. (2008). The resulting e_t is called the score-type residual.
- $v_t = 1$, no scaling is used and e_t is called the identity residual.

An advantage of using the model in Equation (5.9) is that the residuals can be calculated recursively. However, some discussions are in order because the parametrization of Equation (5.9) is unconventional. First, the model represents a fixed function as it contains no random component. This is so because w_t is known once $\boldsymbol{F}_t$ is given. Second, let $q_* = \max\{p, q\}$. The model can be rewritten as

$$w_t = \sum_{i=1}^{p} \phi_i w_{t-i} + \sum_{j=1}^{q_*} \tilde{\theta}_j e_{t-i} \tag{5.10}$$

where the coefficients $\tilde{\theta}_j$ are given below: (a) If $p \leq q$, then

$$\tilde{\theta}_j = \begin{cases} \theta_j + \phi_j & \text{for } j = 1, \ldots, p \\ \theta_j & \text{for } j = p+1, \ldots, q, \end{cases}$$

and (b) if $p > q$, then

$$\tilde{\theta}_j = \begin{cases} \theta_j + \phi_j & \text{for } j = 1, \ldots, q \\ \phi_j & \text{for } j = q+1, \ldots, p. \end{cases}$$

If we assume that $w_t = 0$ and $e_t = 0$ for $t \leq 0$, then $w_t = 0$ for all t if and only if $\tilde{\theta}_j = 0$ for $j = 1, \ldots, q_*$. In this case, there is no serial dependence in the GLARMA model. This is equivalent to $\phi_j = -\theta_j$ for $j = 1, \ldots, p$ and $\theta_j = 0$ for $j = p+1, \ldots, q_*$. Consequently, under the null hypothesis of no serial dependence, some nuisance parameters may exist and cannot be estimated. In practice, this has implications for the convergence in estimation. Readers are referred to Dunsmuir and Scott (2015) for further discussions concerning parameter identifiability. Finally, stochastic properties of GLARMA models can be studied by leveraging the ARMA formulation of the model. For instance, sufficient conditions for the existence of the first two moments of y_t are (a) all zeros of the autoregressive polynomial $\phi(B) = 1 - \sum_{i=1}^{p} \phi_i B^i$ are outside the unit circle, (b) the explanatory variables $\boldsymbol{x}_t$ satisfy that $0 < \sigma_t < \infty$, and (c) the recursions in Equation (5.9) are initialized with $w_t = e_t = 0$ for $t \leq 0$.

There is no simple approach currently available to specify a GLARMA model for a given time series of count data. In some applications, preliminary analysis via traditional time series methods is helpful. We demonstrate such an application via an example. For a well-specified GLARMA model, one can estimate the parameters via the maximum likelihood method because the log likelihood function can be calculated recursively.

Column	Variable	Description
1	Count	Daily asthma counts
2	Intercept	Vector of 1s
3	Sunday	Dummy variable for Sundays
4	Monday	Dummy variable for Mondays
5	CosAnnual	$\cos(2\pi t/365)$, annual cosine function
6	SinAnnual	$\sin(2\pi t/365)$, annual sine function
7	H7	Scaled, lagged, and smoothed humidity
8	NO2max	Maximum daily nitrogen oxide
9-16	T1.1990–T2.1993	Smooth curves for school terms in each year

Table 5.2 Variables used in the asthma studies.

Example 5.1 Consider the asthma data of Davis et al. (2003), which are available from the `glarma` package. Let y_t be the daily number of asthma cases presenting at the emergency department of a hospital in the south west region of Sydney, Australia. The data are part of a study concerning the relationship between atmospheric pollution and asthma cases. Table 5.2 provides a description of the variables in the data set. There are 16 variables (including the intercept) and 1461 observations. Figure 5.1 shows the time-plot of the daily asthma counts.

We start with some preliminary analyses to gain insights into the data. Let $z_t = \sqrt{y_t}$ and treat z_t as a real-valued series. The goal of the project is to seek the relationship between atmospheric pollution and asthma cases so we start with a linear regression model and obtain the result below:

```
>require(glarma)
> data("Asthma",package="glarma")
> y <- Asthma[,1]
> X <- as.matrix(Asthma[,-1])
> z <- sqrt(y)
> m1 <- lm(z~-1+.,data=data.frame(X))
> summary(m1)
Call: lm(formula = z ~ -1 + ., data = data.frame(X))

Coefficients:
           Estimate Std. Error t value Pr(>|t|)
Intercept  1.19440    0.05177  23.072  < 2e-16 ***
Sunday     0.12873    0.04855   2.652 0.008095 **
Monday     0.18631    0.04776   3.901 0.000100 ***
CosAnnual -0.16244    0.03022  -5.376 8.88e-08 ***
```

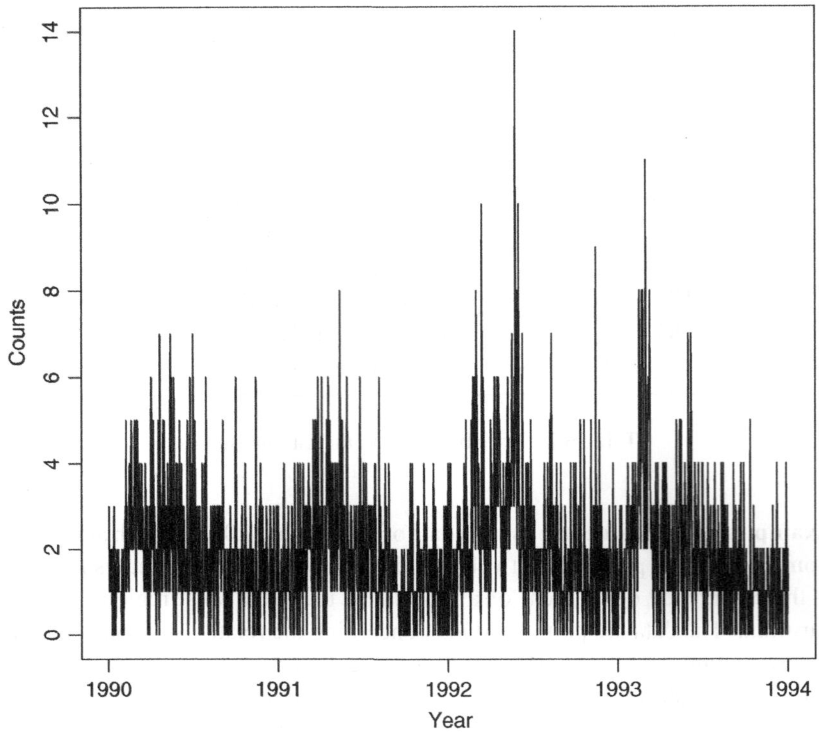

Figure 5.1 Time plot of the daily asthma counts in a hospital in the south west region of Sydney, Australia from 1990 to 1993.

```
SinAnnual   0.13169     0.03272    4.025 5.99e-05 ***
H7          0.10311     0.04667    2.210 0.027292 *
NO2max     -0.08138     0.02761   -2.947 0.003256 **
T1.1990     0.20539     0.05374    3.822 0.000138 ***
T2.1990     0.11551     0.05321    2.171 0.030124 *
T1.1991     0.05616     0.05481    1.025 0.305715
T2.1991     0.15497     0.05457    2.840 0.004579 **
T1.1992     0.21848     0.05383    4.059 5.19e-05 ***
T2.1992     0.28513     0.05315    5.364 9.46e-08 ***
T1.1993     0.37351     0.05393    6.926 6.48e-12 ***
T2.1993     0.06352     0.05461    1.163 0.244928
---
Residual standard error: 0.6296 on 1446 degrees of freedom
Multiple R-squared:  0.7977,    Adjusted R-squared:  0.7956
F-statistic:    380 on 15 and 1446 DF,  p-value: < 2.2e-16
```

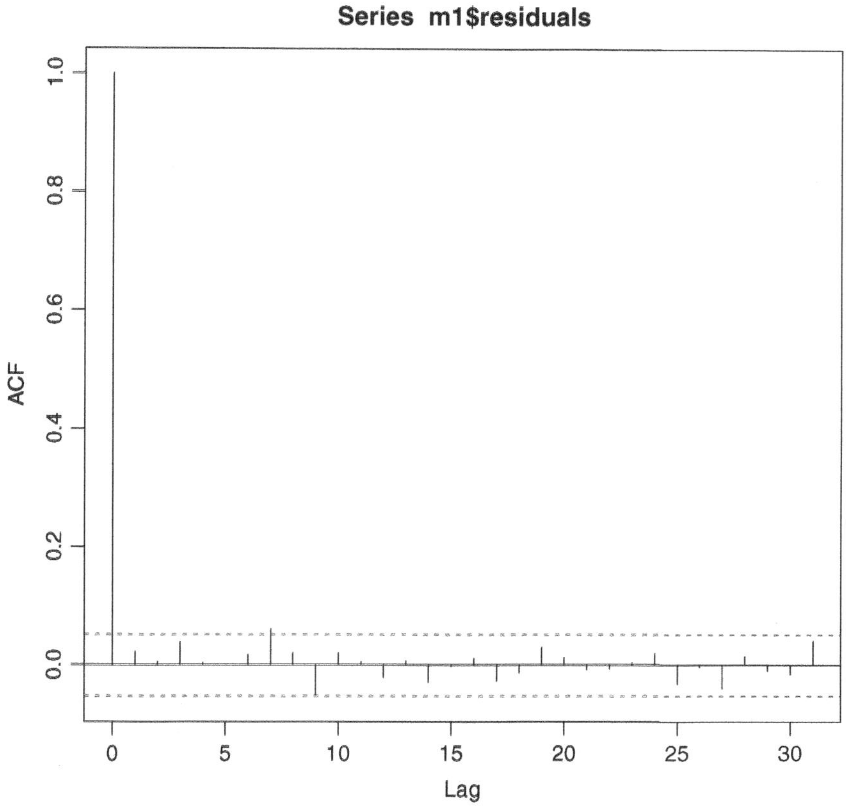

Figure 5.2 Residual autocorrelation function of a linear regression model for the square root series of asthma counts.

The regression result shows that all explanatory variables but T1.1991 and T2.1993 are statistically significant at the 5% level, indicating that air pollution variables are indeed affecting asthma counts. Figure 5.2 shows the sample autocorrelations of the residuals of the linear regression model. From the plot, it seems that there exists some minor serial correlation at lag-7. This is not surprising as people tend to have different activities on weekdays and weekends.

To refine the model, we employ a linear regression model with time series errors. See Tsay (2010). The result becomes

```
> m2 <-arima(z,seasonal=list(order=c(0,0,1),period=7),xreg=X,include.mean=F)
> m2
Call:arima(x=z,seasonal=list(order=c(0,0,1),period=7),xreg=X,include.mean=F)
```

```
Coefficients:
         sma1  Intercept  Sunday  Monday  CosAnnual  SinAnnual      H7   NO2max
       0.0648     1.1899  0.1296  0.1870    -0.1607     0.1314  0.1009  -0.0789
s.e.   0.0269     0.0524  0.0512  0.0505     0.0315     0.0344  0.0470   0.0277
       T1.1990  T2.1990  T1.1991  T2.1991  T1.1992  T2.1992  T1.1993  T2.1993
        0.2076    0.116   0.0561   0.1565   0.2189   0.2866   0.3731   0.0621
s.e.    0.0565    0.056   0.0575   0.0573   0.0566   0.0559   0.0567   0.0573

sigma^2 estimated as 0.3908:  log likelihood = -1386.74,  aic = 2807.48
> Box.test(m2$residuals,lag=400,type="Ljung")
        Box-Ljung test
data:  m2$residuals
X-squared = 332.02, df = 400, p-value = 0.9943
```

Model checking shows that there exist no significant serial correlations in the residuals. The Ljung–Box statistics of the residuals give Q(400) = 333, which is not significant even after adjusting for the degrees of freedom. The residual ACF is shown in Figure 5.3(b). The MA coefficient at lag-7 is small, but statistically

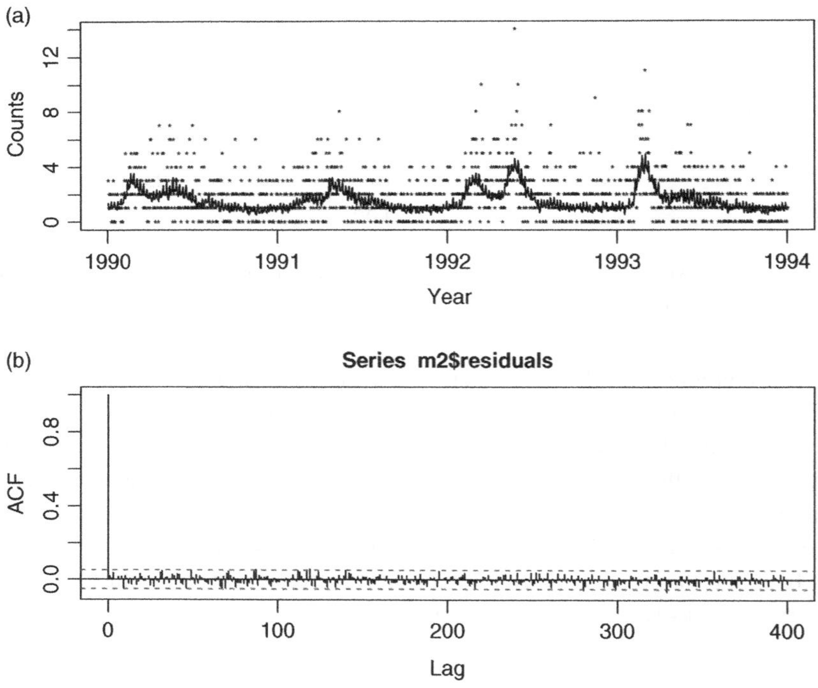

Figure 5.3 Results of a linear regression model with time series errors for the square root series of asthma counts: (a) fitted values and (b) residual autocorrelation functions.

significant at the 5% level. The fitted model confirms the relationship between pollution measurements and asthma counts. Figure 5.3(a) shows the observed y_t and the squares of fitted values.

We now turn to modeling y_t directly using the GLARMA models. Based on information obtained from the linear regression analysis and allowing for possible over-dispersion, we use a GLARMA model with negative binomial distribution. We use Pearson residuals and the Newton–Raphson method in the estimation. The fitted model is

```
m4 <- glarma(y,X,thetaLags=7,type="NegBin",method="NR",alphaInit=0,
    maxit=100,grad=1e-6,residuals="Pearson")
> summary(m4)
Call: glarma(y = y,X=X,type="NegBin", method="NR", residuals = "Pearson",
    thetaLags = 7, alphaInit = 0, maxit = 100, grad = 1e-06)

Negative Binomial Parameter:
      Estimate Std.Error z-ratio Pr(>|z|)
alpha    37.19     25.44   1.462    0.144

GLARMA Coefficients:
        Estimate Std.Error z-ratio Pr(>|z|)
theta_7  0.04392   0.01936   2.269   0.0233 *

Linear Model Coefficients:
           Estimate Std.Error z-ratio Pr(>|z|)
Intercept   0.58397   0.06331   9.225  < 2e-16 ***
Sunday      0.19455   0.05760   3.377 0.000732 ***
Monday      0.22999   0.05642   4.076 4.57e-05 ***
CosAnnual  -0.21450   0.03965  -5.410 6.31e-08 ***
SinAnnual   0.17728   0.04153   4.269 1.96e-05 ***
H7          0.16843   0.05634   2.990 0.002794 **
NO2max     -0.10404   0.03392  -3.067 0.002163 **
T1.1990     0.19903   0.05845   3.405 0.000662 ***
T2.1990     0.13087   0.05897   2.219 0.026477 *
T1.1991     0.08587   0.06746   1.273 0.203058
T2.1991     0.17082   0.05951   2.871 0.004096 **
T1.1992     0.25276   0.05669   4.459 8.24e-06 ***
T2.1992     0.30572   0.05103   5.991 2.08e-09 ***
T1.1993     0.43607   0.05233   8.334  < 2e-16 ***
T2.1993     0.11412   0.06269   1.821 0.068679 .

    Null deviance: 1989.9  on 1460  degrees of freedom
Residual deviance: 1442.6  on 1444  degrees of freedom
AIC: 4873.511
```

```
Number of Newton Raphson iterations: 6

LRT and Wald Test:
Alternative hypothesis: model is a GLARMA process
Null hypothesis: model is a GLM with the same regression structure
          Statistic p-value
LR Test       7.047 0.00794 **
Wald Test     5.147 0.02329 *
---
> par(mfcol=c(2,2))
> plot(m4,which=c(1,2,3,5),titles=c("","","","PIT for glarma(NegBin)"))
```

From the output, the MA coefficient is small, but statistically significant. The estimate of α is relatively large, indicating that a Poisson distribution might be sufficient. The likelihood ratio test and the Wald test indicate that a GLARMA model is needed. Figure 5.4 shows the model checking of the fitted GLARMA model.

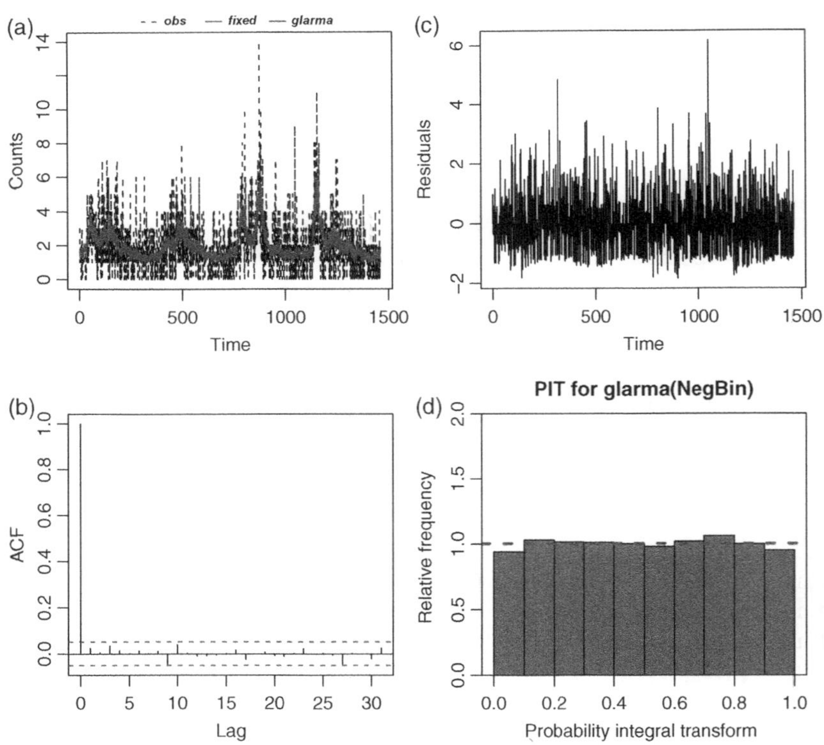

Figure 5.4 Model checking of a fitted GLARMA model for the asthma counts: (a) the fitted values, (b) the residual autocorrelations, (c) the residual plot, and (d) the probability integral transformation.

The plots indicate that the model is adequate. Overall, the model also confirms the effect of pollution measurements on the asthma counts.

Remark: There are other R packages that can fit GARMA models. See, for instance, `VGAM` and `gamlss.util`. Also, the package `tscount` of R can handle time series of count data.

5.2 AUTOREGRESSIVE CONDITIONAL MEAN MODELS

A highly related class of models for time series of count data is the autoregressive conditional mean (ACM) models. Motivated by the autoregressive conditional duration (ACD) models of Engle and Russell (1998), Heinen (2003) proposes the autoregressive conditional Poisson (ACP) models. Since then, various extensions of ACM models have been developed. Again, let y_t be the time series of interest and $\boldsymbol{F}_t = \sigma\{y_{t-1}, y_{t-2}, \ldots; \boldsymbol{x}_t, \boldsymbol{x}_{t-1}, \ldots\}$. The conditional mean models postulate

$$E(y_t \mid \boldsymbol{F}_t) = \exp(\boldsymbol{x}_t'\boldsymbol{\beta})\lambda_t, \tag{5.11}$$

and λ_t follows the model

$$\lambda_t = \omega + \sum_{i=1}^{p} \alpha_i \left[\frac{y_{t-i}}{\exp(\boldsymbol{x}_{t-i}'\boldsymbol{\beta})}\right] + \sum_{j=1}^{q} \gamma_j \lambda_{t-j}, \tag{5.12}$$

where p and q are nonnegative integers, $\omega > 0$, and α_i and γ_j are parameters satisfying certain conditions so that λ_t is positive and finite. The conditional distribution $p(y_t \mid \boldsymbol{F}_t)$ is Poisson, negative binomial or double Poisson. Descriptions of these three distributions are given in the appendix. The negative binomial distribution allows for over-dispersion whereas the double Poisson can handle both over- and under-dispersion.

Properties of the conditional mean models have been derived in the literature. For example, assuming that there exist no explanatory variables, Heinen (2003) showed that the ACP process with λ_t in Equation (5.12) is stationary if $\omega > 0$ and $0 \leq \sum_{i=1}^{p} \alpha_i + \sum_{j=1}^{q} \gamma_j < 1$. The unconditional mean of the process is given by

$$E(y_t) \equiv \mu = \frac{\omega}{1 - \sum_{i=1}^{p} \alpha_i - \sum_{j=1}^{q} \gamma_j}.$$

In addition, for the ACP(1,1) model, the unconditional covariance function is given by

$$\operatorname{Cov}(y_t, y_{t-h}) = \begin{cases} \frac{[1-(\alpha_1+\gamma_1)^2+\alpha_1^2]\mu}{1-(\alpha_1+\gamma_1)^2}, & h = 0 \\ \frac{\alpha_1[1-\gamma_1(\alpha_1+\gamma_1)](\alpha_1+\gamma_1)^{h-1}\mu}{1-(\alpha_1+\gamma_1)^2}, & h \geq 1. \end{cases} \tag{5.13}$$

Rewriting the variance of the ACP(1,1) process as

$$\mathrm{Var}(y_t) = \mu \left[1 + \frac{\alpha_1^2}{1 - (\alpha_1 + \gamma_1)^2} \right],$$

we see that $\mathrm{Var}(y_t) > E(y_t)$ provided that $\alpha_1 \neq 0$, implying that the dynamic dependence of the ACP(1,1) model introduces over-dispersion in the observed data even though the innovations are Poisson. This property is similar to that of a GARCH(1,1) model with Gaussian innovations for volatility. The maximum likelihood method is often used to estimate autoregressive conditional mean models. The likelihood function can be calculated recursively similar to that of the standard time series analysis. The statistic $r_t = y_t - \exp(\boldsymbol{x}_t \widehat{\boldsymbol{\beta}})\hat{\lambda}_t$ is then used as residual for model checking. An alternative specification for the ACM models is to modify Equation (5.11) so that $E(y_t \mid \boldsymbol{F}_t) = \exp(\boldsymbol{x}_t' \boldsymbol{\beta}) \exp(\tilde{\lambda}_t)$. In this case, the latent process $\tilde{\lambda}_t$ may assume negative values.

Example 5.2 Consider the trading activity on the New York Stock Exchange of Glatfelter stock with tick symbol `GLT` from January 3 to March 31, 2005 for 61 trading days. Glatfelter Company is a manufacturer of specialty papers and fiber-based engineered materials located at York, PA. It is classified as a small cap stock for which the trading intensity is of interest to high-frequency traders. Let y_t be the number of transactions within a 5-minute interval, and the series has been used by Jung et al. (2011) and Tsay (2014a). Following Jung et al. (2011), we ignore the trading within the first 15 minutes each day so that there are 75 observations in a given trading day and the sample size is 4575.

Figure 5.5 shows that there exists a diurnal pattern in the GLT trading activities. To handle the diurnal pattern, we used eight indicator variables for the first and last four 5-minute intervals of the trading hours. Specifically, $\{x_{1t}, \ldots, x_{4t}\}$ are the dummy variables for time intervals (9:45,9:50], ..., (9:55,10:00], respectively, and $\{x_{5t}, \ldots, x_{8t}\}$ are the dummy variables for time intervals (15:55,16:00], ..., (15:40,15:45], respectively, where all times are Eastern. We apply the ACM(1,1) model to the series and the results of maximum likelihood estimation are given in Table 5.3. From the table, it is clear that the model fits the data well, the ACM(1,1) parameters are highly significant, and the diurnal pattern is mainly due to increased trading at the market closing. The latter is not surprising as do not consider the trading in the first 15 minutes each day. Finally, from the estimate of θ for the negative binomial and double Poisson distributions, the data show evidence of over-dispersion.

R demonstration: The command `ACMx` of the `NTS` package is used, and the command `hfDummy` creates dummy variables for trading intervals with subcommand

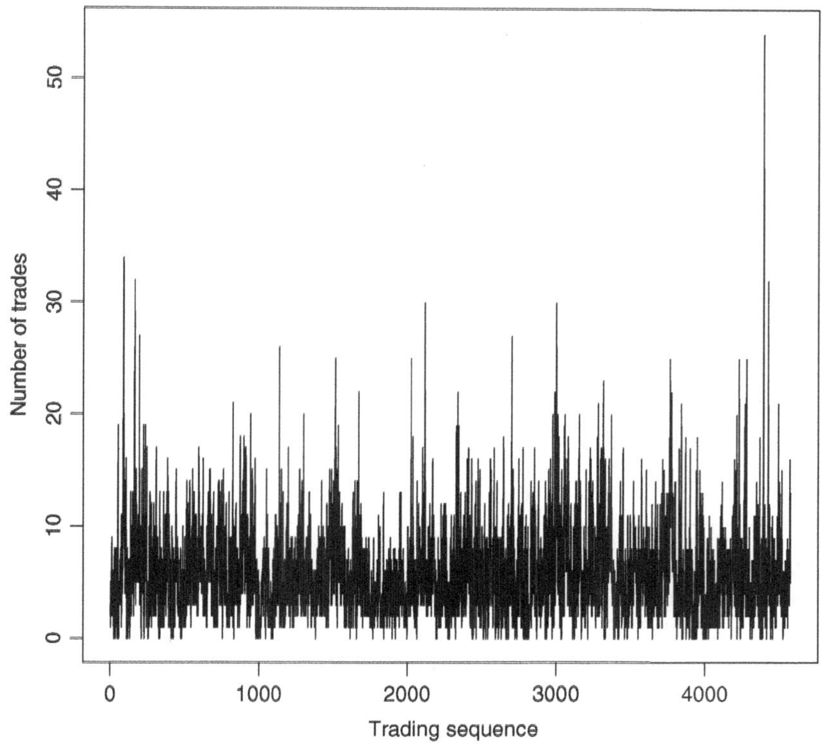

Figure 5.5 Time plot of the number of transactions in 5-minute intervals of Glatfelter stock from January 3 to March 31, 2005. The trading within the first 15 minutes of a trading day is ignored.

`Fopen` denoting the number of intervals from market open, `Tend` the number of intervals to the market close, and `skipmin` denotes the number of minutes omitted from the opening.

```
> da <- read.table("taq-JBES2011.txt",header=T)
> y <- da$GLT
> require(NTS)
> X <- hfDummy(5,Fopen=4,Tend=4,days=61,skipmin=15)
> X <- as.matrix(X)
> colnames(X) <- paste("X",1:8,sep="")
> m1 <- ACMx(y,X=X)   ## Output omitted. Use default order (1,1).
> names(m1)
> acf(m1$sresi,lag=230)
> m2 <- ACMx(y,X=X,cond.dist="nb")
> m3 <- ACMx(y,X=X,cond.dist="dp")
```

Parameter	Poisson	Neg-binomial	Double Poisson
β_1	−0.020(0.049)	0.006(0.077)	−0.020(0.069)
β_2	−0.033(0.049)	−0.018(0.077)	−0.034(0.070)
β_3	−0.109(0.052)	−0.112(0.078)	−0.107(0.073)
β_4	0.110(0.048)	0.103(0.076)	0.107(0.067)
β_5	0.599(0.038)	0.609(0.070)	0.594(0.054)
β_6	0.169(0.047)	0.186(0.075)	0.165(0.066)
β_7	0.138(0.048)	0.143(0.076)	0.134(0.068)
β_8	0.128(0.049)	0.126(0.076)	0.128(0.068)
ω	0.503(0.048)	0.483(0.068)	0.535(0.071)
α_1	0.173(0.009)	0.171(0.013)	0.171(0.012)
γ_1	0.735(0.015)	0.741(0.022)	0.735(0.022)
θ		4.913(0.198)	0.497(0.009)
Q(230)	92.15(0.09)	91.45(0.10)	93.62(0.07)

Table 5.3 Maximum likelihood estimates of an autoregressive conditional mean model of order (1,1) for the number of trades in 5-minute intervals of Glatfelter stock from January 1 to March 31, 2005. The sample size is 4575. The numbers in parentheses for the estimates are standard errors and that for $Q(m)$ statistics is p value.

5.3 MARTINGALIZED GARMA MODELS

Zheng et al. (2015) proposed the martingalized GARMA models to mitigate some shortcomings of Equation (5.5). Specifically, the authors introduce an additional transformation of the time series y_t so that the error terms of the ARMAX formulation of a GARMA model become a martingale difference sequence. The new model is called a martingalized GARMA model or M-GARMA model for short. To this end, Zheng et al. (2015) assumed that the conditional probability mass function of y_t given $\boldsymbol{F}_t$ can be parameterized as

$$p(y_t \mid \boldsymbol{F}_t) = f(y_t \mid \mu_t, \boldsymbol{\varphi}) \tag{5.14}$$

where $\boldsymbol{\varphi}$ denotes the collection of all time-invariant parameters. This assumption says that all past information concerning y_t is summarized in μ_t. Furthermore, there exist a function $h(y_t)$ and the link function $g_\varphi(\mu_t) = E[h(y_t) \mid \boldsymbol{F}_t]$ so that Equation (5.4) becomes

$$g_\varphi(\mu_t) = \boldsymbol{x}_t'\boldsymbol{\beta} + \sum_{i=1}^{p} \phi_i[h(y_{t-i}) - \boldsymbol{x}_{t-i}'\boldsymbol{\beta}] + \sum_{j=1}^{q} \theta_j[h(y_{t-j}) - g_\varphi(\mu_{t-j})]. \tag{5.15}$$

Distribution	Density	$E(y_t \mid \boldsymbol{F}_{t-1})$	$\mathrm{Var}(y_t \mid \boldsymbol{F}_{t-1})$
Lognormal	$\log N(\log(\mu_t) - \sigma^2/2, \sigma^2)$	μ_t	$(e^{\sigma^2} - 1)\mu_t^2$
Gamma	$\Gamma(c\mu_t^d, c\mu_t^{d-1})$	μ_t	μ_t^{2-d}/c
Inv-gamma	Inv-$\Gamma(c\mu_t^d + 1, c\mu_t^d)$	μ_t	$c\mu_t^{1+d}/(c\mu_t^{d-1} - 1)$
Weibull	$\mathrm{Weibull}(k, \mu_t/\Gamma(1 + k^{-1}))$	μ_t	$\mu_t^2 \left[\frac{\Gamma(1+2k^{-1})}{\Gamma^2(1+k^{-1})} - 1\right]$
Beta	$\mathrm{Beta}(\tau\mu_t, \tau(1 - \mu_t))$	μ_t	$\frac{\mu_t(1-\mu_t)}{1+\tau}$
Poisson	$\mathrm{Poisson}(\mu_t)$	μ_t	μ_t
	$g_\varphi(\mu_t)$	$h(y_t)$	$\mathrm{Var}[h(y_t) \mid \boldsymbol{F}_{t-1}]$
Lognormal	$\log(\mu_t) - \frac{\sigma^2}{2}$	$\log(y_t)$	σ^2
Gamma	$\psi(c\mu_t^d) - \log(\mu_t^{d-1}) - \log(c)$	$\log(y_t)$	$\psi_1(c\mu_t^2)$
Inv-gamma	$\log(c\mu_t^d) - \psi(c\mu_t^{d-1} + 1)$	$\log(y_t)$	$\psi_1(c\mu_t^{d-1} + 1)$
Weibull	$\approx \log(\mu_t) - \frac{1}{2}\left[\frac{\Gamma(1+2k^{-1})}{\Gamma^2(1+k^{-1})} - 1\right]$	$\log(y_t)$	$\approx \frac{\Gamma(1+2k^{-1})}{\Gamma^2(1+k^{-1})} - 1$
Beta	$\psi(\tau\mu_t) - \psi(\tau(1 - \mu_t))$	$\log\left(\frac{y_t}{1-y_t}\right)$	$\psi_1(\tau\mu_t) + \psi_1(\tau(1 - \mu_t))$
Poisson	$\approx \sqrt{\mu_t}$	$\sqrt{y_t}$	$\approx \frac{1}{4}$

Table 5.4 Some commonly used conditional distributions, their recommended y-link functions, the corresponding link functions, and the first two moments, where $\boldsymbol{F}_{t-1}$ denotes the lagged values of y_t. $\psi(\cdot)$ and $\psi_1(\cdot)$ denote the gigamma and trigamma functions, respectively. See Zheng et al. (2015).

By adding $h(y_t) - g_\varphi(\mu_t)$ to both sides of Equation (5.15), we have

$$h(y_t) = \boldsymbol{x}_t'\boldsymbol{\beta} + \sum_{i=1}^{p} \phi_i[h(y_{t-i}) - \boldsymbol{x}_{t-i}'\boldsymbol{\beta}] + \epsilon_t + \sum_{j=1}^{q} \theta_j\epsilon_{t-j}, \tag{5.16}$$

where $\epsilon_t = h(y_t) - g_\varphi(\mu_t)$. By the construction of $h(\cdot)$ and $g_\varphi(\cdot)$, it is easy to show, via iterated expectations, that (a) $E(\epsilon_t) = E[E(h(y_t) - g_\varphi(\mu_t) \mid \boldsymbol{F}_t)] = 0$ and (b) $E(\epsilon_t\epsilon_{t-j}) = E[\epsilon_{t-j}E(\epsilon_t \mid \boldsymbol{F}_t)] = 0$ for $j > 0$. Therefore, $\{\epsilon_t\}$ is a martingale difference sequence. The function $h(\cdot)$ is referred to as the *y-link function* and $g_\varphi(\cdot)$ is the *link function*.

From the discussion, it is seen that the link function $h(\cdot)$ is used to transform the y_t series into a linear process. Zheng et al. (2015) recommended that one uses a y-link function that does not involve any unknown parameter. In this case, the orders p and q and the ARMA parameters of Equation (5.16) can be obtained by studying $h(y_t)$. The link function $g_\varphi(\cdot)$, on the other hand, may contain unknown parameters. The authors also suggested ways to select a proper transformation $h(\cdot)$. Table 5.4 provides some commonly used conditional distributions and their recommended y-link functions; see Zheng et al. (2015).

To gain insights into the M-GARMA model, we rewrite Equation (5.15) as

$$\eta_t = v(\boldsymbol{x}_t, \boldsymbol{\beta}) + \sum_{i=1}^{m} \tilde{\phi}_i h(y_{t-i}) + \sum_{j=1}^{m} \tilde{\theta}_j \eta_{t-j}, \tag{5.17}$$

where $\eta_t = g_\varphi(\mu_t)$, $v(\boldsymbol{x}_t, \boldsymbol{\beta})$ is a function of $\boldsymbol{\mathcal{F}}_t$ and $\boldsymbol{\beta}$ in the presence of $\boldsymbol{x}_t$ and is a constant otherwise, $m = \max\{p, q\}$, $\tilde{\phi}_i = \phi_i + \theta_i$ for $i = 1, \dots, m$, $\tilde{\theta}_j = -\theta_j$ for $j = 1, \dots, q$ with $\phi_i = 0$ for $i > p$ and $\theta_j = 0$ for $i > q$. Equation (5.17) shows that η_t is a linear predictor of $h(y_t)$, which is a linear combination of past transformed responses $h(y_{t-1}), \dots, h(y_{t-p})$ and the past predictors $\eta_{t-1}, \dots, \eta_{t-q}$. This fixed function framework is similar to that of the generalized autoregressive conditional heteroscedastic (GARCH) models for volatility to be discussed in the next section.

Some stochastic properties of M-GARMA models are derived in Zheng et al. (2015) using the *geometric drift condition* of Meyn and Tweedie (2009). The authors consider two estimation methods for the model. The first method uses Gaussian pseudo likelihood and the second method is the maximum likelihood estimation. Finally, they also demonstrate two applications of the M-GARMA models.

5.4 VOLATILITY MODELS

Let y_t be the return (or log return) of an asset at time t. The volatility of the asset is defined as the conditional standard deviation of y_t given the information $\boldsymbol{\mathcal{F}}_{t-1} = \sigma\{y_{t-1}, y_{t-2}, \dots,\}$. In finance, volatility plays an important role because it serves as a measurement of risk and is widely used in asset pricing. Figure 5.6 shows the daily log returns of the SPDR S&P 500 ETF (SPY) from January 3, 2007 to December 31, 2016. This exchange traded fund (ETF) follows the S&P 500 index, which is commonly used to represent the US financial market. From the plot, it is seen that the returns exhibit periods of high variability. This phenomenon is called *volatility clustering* in finance. Selected summary statistics of the SPY log returns are

Mean	Standard error	Median	Skewness	Excess kurtosis
0.0003	0.0131	0.0006	−0.099	12.93

Clearly, the log returns have high excess kurtosis and center around zero. The plot and summary statistics show the stylized characteristics of asset returns, namely zero mean, heavy tails, and volatility clustering.

The goal of volatility modeling is then to consider statistical models that can adequately describe the observed characteristics of asset returns. Two general

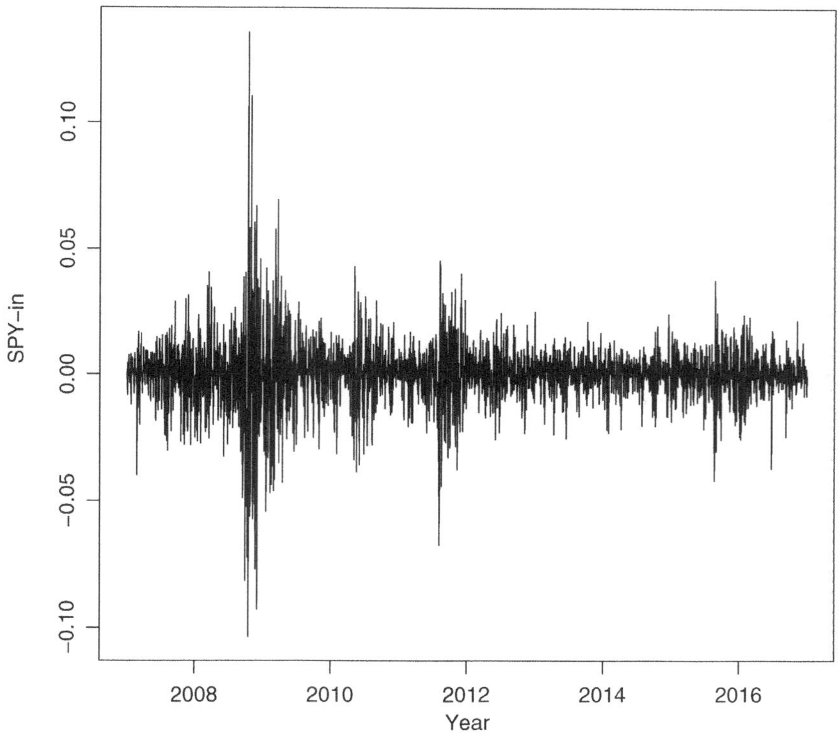

Figure 5.6 Time plot of daily log returns of the S&P 500 index (SPY) from January 3, 2007 to December 31, 2016.

approaches have been developed in the literature to volatility modeling. The first approach employs a fixed function to govern the time evolution of conditional variance, and the second approach uses a stochastic equation. See, for instance, Tsay (2010, Chapter 3), among others. In this chapter, we adopt the first approach and briefly introduce the class of generalized autoregressive conditional heteroscedastic (GARCH) models of Bollerslev (1986) and their variants.

An asset return can be decomposed as

$$y_t = E(y_t \mid \boldsymbol{F}_{t-1}) + a_t \equiv \mu_t + a_t, \tag{5.18}$$

where μ_t is the conditional expectation and a_t is the deviation of y_t from μ_t. In Tsay (2010), μ_t is called the predictable component and a_t the unpredictable component of an asset return. A general form for μ_t is

$$\mu_t = \boldsymbol{x}'_{t-1}\boldsymbol{\delta} + \phi_0 + \sum_{i=1}^{p} \phi_i y_{t-1} + \sum_{j=1}^{q} \theta_j a_{t-j}, \tag{5.19}$$

where $\boldsymbol{x}_{t-1}$ denotes a vector of available explanatory variables, $\boldsymbol{\delta}$ is the coefficient vector, ϕ_0 is a constant, and ϕ_i and θ_j are the usual autoregressive and moving-average parameters, respectively. In most applications, μ_t is simply a constant, especially for high-frequency returns. The unpredictable part of asset returns is written as

$$a_t = \sigma_t \epsilon_t, \quad \epsilon_t \sim_{iid} D(0, 1), \tag{5.20}$$

where $\sigma_t^2 = \mathrm{Var}(y_t \mid \boldsymbol{F}_{t-1}) = E[(y_t - \mu_t)^2 \mid \boldsymbol{F}_{t-1}]$ and ϵ_t is a sequence of independent and identically distributed random variates with mean zero and variance 1, denoted by $D(0, 1)$. The distribution of ϵ_t is Gaussian, standardized Student-t with v degrees of freedom, or the generalized error distribution. In some cases, ϵ_t assumes a skew distribution such as a skew normal or Student-t. Again, see Tsay (2010, Chapter 3) for further details, among others.

The most commonly used GARCH model is in the form

$$\sigma_t^2 = \omega + \alpha a_{t-1}^2 + \beta \sigma_{t-1}^2, \tag{5.21}$$

where $\omega > 0, 0 \le \alpha, \beta < 1$ and $\alpha + \beta < 1$. The model in Equation (5.21) is called a GARCH(1,1) model in the literature and has been widely studied. It is easy to show that

$$E(a_t^2) = \mathrm{Var}(a_t) = \frac{\omega}{1 - \alpha - \beta}, \quad \text{if} \quad \alpha + \beta < 1,$$

$$\frac{E(a_t^4)}{[E(a_t^2)]^2} = \frac{3[1 - (\alpha + \beta)^2]}{1 - (\alpha + \beta)^2 - 2\alpha^2} \ge 3, \quad \text{if} \quad 1 - 2\alpha^2 - (\alpha + \beta)^2 > 0.$$

The second result was obtained assuming $\epsilon_t \sim N(0, 1)$, and it implies that a_t has heavy tails under the GARCH(1,1) model even though ϵ_t is Gaussian. The formula also shows that the heavy tails induced by the GARCH dynamics mainly come from the parameter α. Let $\eta_t = a_t^2 - \sigma_t^2$ be the deviation of a_t^2 from its conditional expectation. The model in Equation (5.21) can be rewritten as

$$a_t^2 = \omega + (\alpha + \beta) a_{t-1}^2 - \beta \eta_{t-1}, \tag{5.22}$$

which is in the form of an ARMA model. The statistic η_t satisfies (a) $E(\eta_t) = 0$ and (b) $\mathrm{Cov}(\eta_t, \eta_{t-j}) = 0$ for $j > 0$. However, η_t is not necessarily identically distributed. The ARMA formulation in Equation (5.22) is helpful in understanding properties of GARCH(1,1) models. For instance, for a GARCH(1,1) process to be weakly stationary, we need $0 < \alpha + \beta < 1$. Also, by leveraging the properties of an ARMA process, we have $E(a_t^2) = \omega/(1 - \alpha - \beta)$.

From Equation (5.21), we see that a large σ_{t-1}^2 or a_{t-1}^2 leads to a large σ_t^2, implying that a volatile period at $t - 1$ tends to be followed by another volatile period at time t. Therefore, the GARCH model is capable of producing volatility clusterings observed in practice.

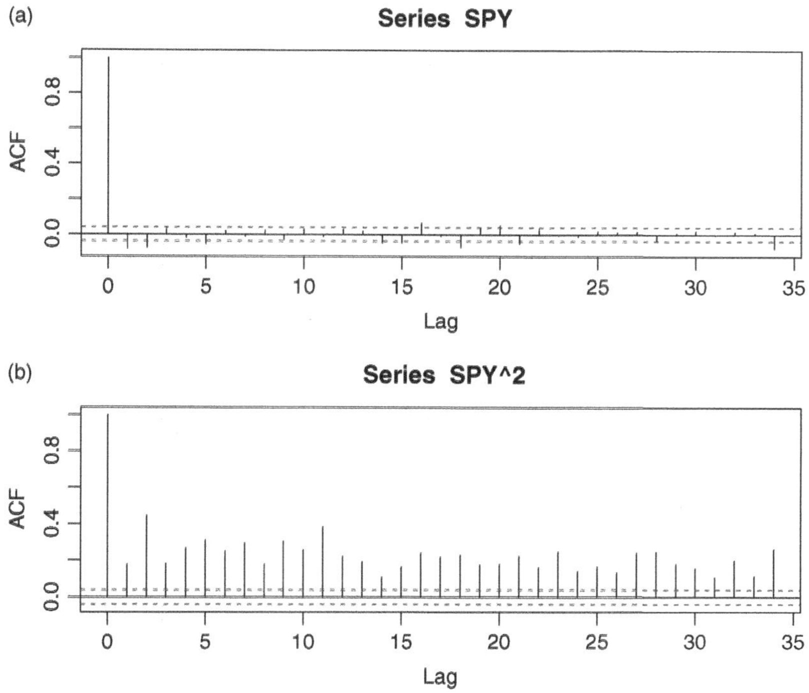

Figure 5.7 Sample autocorrelation functions of the daily log returns of (a) SPY and (b) their squared series. The sample period is from January 3, 2007 to December 31, 2016. The upper panel shows the ACF of the log return series.

GARCH models can be estimated by either the quasi maximum likelihood method or the conditional maximum likelihood method. Asymptotic properties of the estimators have been established in the literature. See, for example, Tsay (2014b) and the references therein. There are several R packages available to estimate volatility models. For GARCH models and their variants, both `fGarch` and `urgarch` packages work well.

Example 5.3 Consider the daily log returns of SPY shown in Figure 5.6. Denote the log return by y_t. Similar to most asset return series, y_t has weak serial correlations, but y_t^2 has strong serial dependence. See Figure 5.7 for the sample ACF of y_t and y_t^2. Therefore, we employ the model

$$y_t = \mu + a_t, \quad a_t = \sigma_t \epsilon_t \tag{5.23}$$

$$\epsilon_t \sim D(0, 1)$$
$$\sigma_t^2 = \omega + \alpha a_{t-1}^2 + \beta \sigma_{t-1}^2, \tag{5.24}$$

Parameter	Gaussian	Standardized Student-t	Skew standardized Student-t
$\mu \times 10^4$	6.84(1.68)	9.03(1.51)	6.44(1.60)
$\omega \times 10^6$	3.49(0.55)	2.48(0.62)	2.30(0.59)
α	0.128(0.014)	0.146(0.019)	0.141(0.018)
β	0.846(0.015)	0.849(0.017)	0.851(0.016)
v	–	5.16(0.58)	5.64(0.68)
Skew	–	–	0.888(0.024)
$Q(20)$	26.75(0.14)	27.12(0.13)	26.93(0.14)
$Q^*(20)$	20.30(0.44)	19.70(0.48)	19.39(0.50)

Table 5.5 Results of maximum likelihood estimation of a GARCH(1,1) model for the daily log returns of SPY. The sample period is from January 3, 2007 to December 31, 2016, and $Q(20)$ and $Q^*(20)$ are the Ljung–Box statistics of residuals and squared residuals, respectively. The numbers in parentheses are standard errors for estimates and p values for test statistics.

where $D(0, 1)$ denotes $N(0, 1)$, standardized Student-t with v degrees of freedom, or skew standardized Student-t with v degrees of freedom. Table 5.5 summarizes the results of maximum likelihood estimation of the model in Equations (5.23)–(5.24). R commands and details of the fitted models are given below. From the table and the R output, we make the following observations. First, the GARCH(1,1) model fits the data reasonably well, as shown by the Ljung–Box statistics of the residuals and squared residuals. Second, as expected, the SPY log returns have heavy tails. The normality assumption of the Gaussian model is clearly rejected and the fitted degrees of freedom for the standardized Student-t distributions are around 5. Third, the innovations of the fitted GARCH(1,1) model are skewed. The null hypothesis of symmetry is rejected because the test statistic is $t = (0.888 - 1)/0.024 \approx -4.67$, which is highly significant compared with its asymptotic N(0,1) reference distribution. Finally, the fitted GARCH(1,1) model with skew standardized Student-t distribution is used to produce one-step to five-step ahead predictions. In this particular instance, the unconditional standard error of the log returns is approximately 0.0168 ($\approx \sqrt{2.298 \times 10^{-6}/(1 - 0.1405 - 0.8514)}$). This seems to be far away from the five-step ahead prediction of 0.00652. However, as the forecast horizon increases, the volatility prediction would approach 0.0168. For instance, the 100-step ahead volatility prediction is 0.0131 for the fitted model.

Example 5.3 demonstrates a typical volatility modeling of asset returns. It confirms that most asset returns are not normally distributed. They tend to have heavy tails and, in some cases, they are not symmetrically distributed around the expected

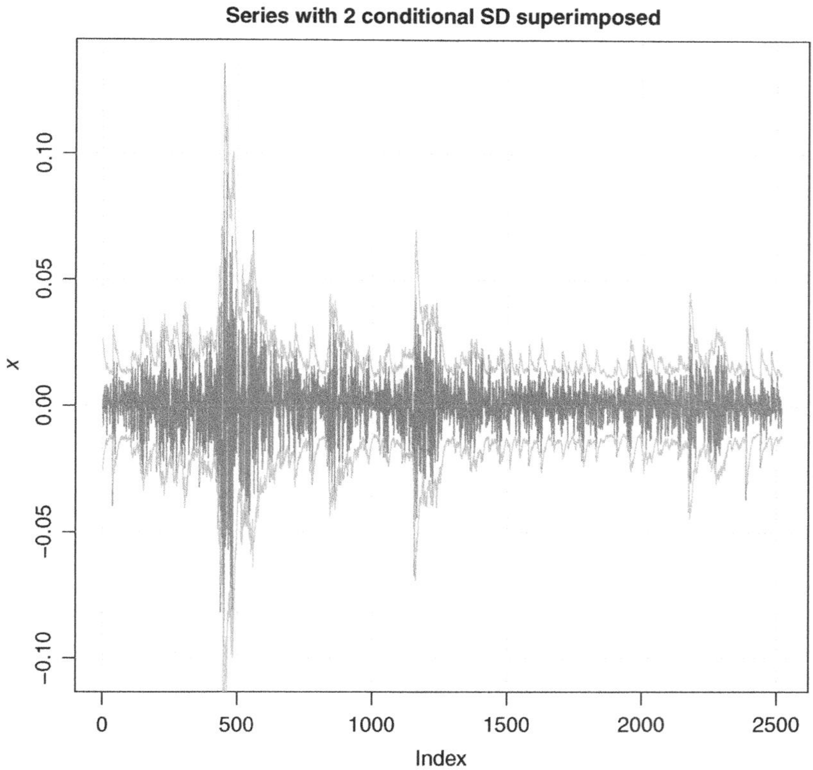

Figure 5.8 Time plot of daily log returns and point-wise 95% confidence intervals for the SPY. The sample period is from January 3, 2007 to December 31, 2016. The volatilities are obtained from the GARCH(1,1) model with skew innovations of Table 5.5.

return. Figure 5.8 shows the log return series and the point-wise 95% confidence band based on the fitted GARCH(1,1) model with Gaussian innovations. Similar plots can be obtained for other fitted models. Figure 5.9 shows the QQ-plots of the residuals of a fitted GARCH(1,1) model with standardized Student-t and skew standardized Student-t. The QQ-plot of the model with skew innovations fares better than that with symmetric innovations.

R demonstration: GARCH modeling with various innovation distributions.

```
> require(fGarch)
> m1 <- garchFit(~garch(1,1),data=spy,trace=F)
> summary(m1)
Title: GARCH Modelling
```

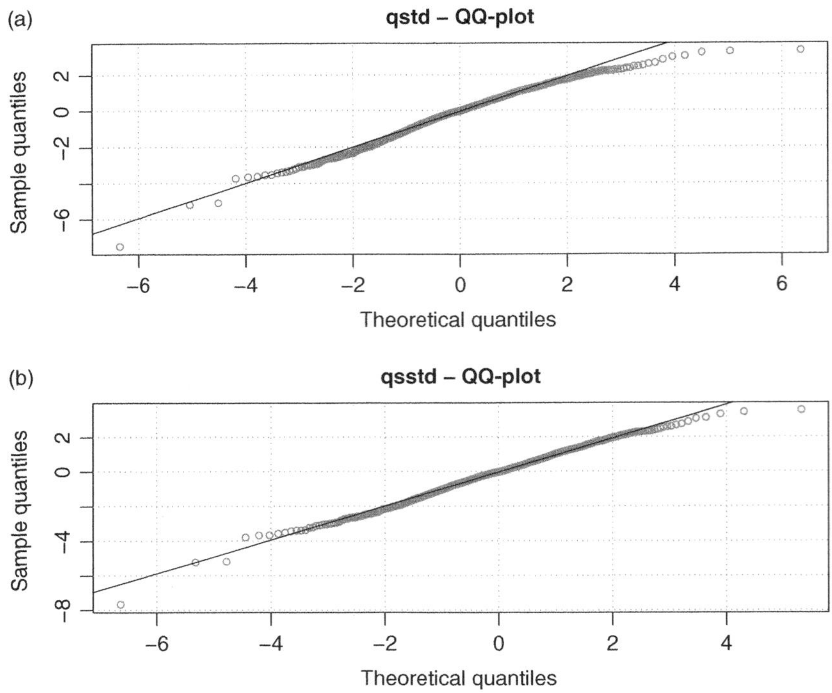

Figure 5.9 QQ plots of fitted GARCH(1,1) models with (a) standardized Student-t and (b) skew standardized Student-t innovations. The models are given in Table 5.5.

```
Call: garchFit(formula = ~garch(1, 1), data = spy, trace=F)
Conditional Distribution: norm
Error Analysis:
        Estimate  Std. Error  t value Pr(>|t|)
mu      6.842e-04  1.681e-04    4.070 4.70e-05 ***
omega   3.494e-06  5.469e-07    6.388 1.68e-10 ***
alpha1  1.275e-01  1.403e-02    9.086  < 2e-16 ***
beta1   8.464e-01  1.463e-02   57.856  < 2e-16 ***
---
Standardised Residuals Tests:
                                Statistic p-Value
 Jarque-Bera Test   R    Chi^2  557.7797  0
 Ljung-Box Test     R    Q(10)  16.93149  0.07589384
 Ljung-Box Test     R    Q(20)  26.75075  0.1424233
 Ljung-Box Test     R^2  Q(10)  14.30636  0.1594707
 Ljung-Box Test     R^2  Q(20)  20.29979  0.4393226
> plot(m1)
```

```
> m2 <- garchFit(~garch(1,1),data=spy,cond.dist="std",trace=F)
> summary(m2)
Title: GARCH Modelling
Call: garchFit(formula=~garch(1,1),data=spy,cond.dist="std",trace=F)
Conditional Distribution: std
Error Analysis:
        Estimate  Std. Error  t value Pr(>|t|)
mu      9.031e-04   1.509e-04    5.986 2.16e-09 ***
omega   2.477e-06   6.248e-07    3.965 7.34e-05 ***
alpha1  1.456e-01   1.915e-02    7.601 2.95e-14 ***
beta1   8.490e-01   1.681e-02   50.499  < 2e-16 ***
shape   5.160e+00   5.813e-01    8.876  < 2e-16 ***
---
Standardised Residuals Tests:
                                Statistic p-Value
 Ljung-Box Test     R    Q(10)  17.45736  0.06483688
 Ljung-Box Test     R    Q(20)  27.11551  0.1320453
 Ljung-Box Test     R^2  Q(10)  10.25488  0.4184234
 Ljung-Box Test     R^2  Q(20)  19.70427  0.4765604
 LM Arch Test       R    TR^2   11.08226  0.5218827

Information Criterion Statistics:
      AIC       BIC       SIC      HQIC
-6.430537 -6.418958 -6.430545 -6.426335

> m3 <- garchFit(~garch(1,1),data=spy,cond.dist="sstd",trace=F)
> summary(m3)
Title: GARCH Modelling
Call:garchFit(formula=~garch(1,1),data=spy,cond.dist="sstd",trace=F)
Conditional Distribution: sstd
Error Analysis:
        Estimate  Std. Error  t value Pr(>|t|)
mu      6.441e-04   1.603e-04    4.017 5.90e-05 ***
omega   2.298e-06   5.858e-07    3.922 8.77e-05 ***
alpha1  1.405e-01   1.790e-02    7.848 4.22e-15 ***
beta1   8.514e-01   1.628e-02   52.310  < 2e-16 ***
skew    8.882e-01   2.352e-02   37.760  < 2e-16 ***
shape   5.637e+00   6.844e-01    8.236 2.22e-16 ***
---
Standardised Residuals Tests:
                                Statistic p-Value
 Ljung-Box Test     R    Q(10)  17.46512  0.06468508
 Ljung-Box Test     R    Q(20)  26.92805  0.1373005
```

```
 Ljung-Box Test      R^2  Q(10)   9.8221     0.4562374
 Ljung-Box Test      R^2  Q(20)   19.38767   0.4967704

Information Criterion Statistics:
      AIC        BIC        SIC       HQIC
-6.437849 -6.423954 -6.437861 -6.432807
> predict(m3,5)
  meanForecast    meanError standardDeviation
1 0.0006440697 0.005880318       0.005880318
2 0.0006440697 0.006049519       0.006049519
3 0.0006440697 0.006212803       0.006212803
4 0.0006440697 0.006370635       0.006370635
5 0.0006440697 0.006523421       0.006523421
```

A weakness of the GARCH models is that they do not distinguish past negative returns from the positive ones. On the other hand, large past negative returns tend to have a greater impact on the volatility than their positive counterparts. This is known as the *leverage effect* in finance. To overcome this weakness, threshold GARCH (TGARCH) or GJR models have been proposed. See Glosten at al. (1993) and Zakoian (1994), among others. A TGARCH(1,1) volatility model can be written as

$$\sigma_t^2 = \omega + (\alpha + \gamma N_{t-1})a_{t-1}^2 + \beta\sigma_{t-1}^2, \tag{5.25}$$

where $\omega > 0, 0 \le \alpha, \beta < 1, 0 < \alpha + \gamma + \beta < 1$, and N_{t-1} is given by

$$N_{t-1} = \begin{cases} 0 & \text{if } a_{t-1} \ge 0, \\ 1 & \text{if } a_{t-1} < 0. \end{cases}$$

The parameter γ is referred to as the leverage effect and it is expected to be positive. Rewriting the model as

$$\sigma_t^2 = \begin{cases} \omega + \alpha a_{t-1}^2 + \beta\sigma_{t-1}^2 & \text{if } a_{t-1} \ge 0, \\ \omega + (\alpha + \gamma)a_{t-1}^2 + \beta\sigma_{t-1}^2 & \text{if } a_{t-1} < 0, \end{cases}$$

we see that the model is in the form of a threshold model discussed in Chapter 2 with a_{t-1} serving as the threshold variable and 0 being the threshold.

For the daily log returns of SPY of Example 5.3, the fitted TGARCH(1,1) model is

$$\begin{aligned} y_t &= 2.46 \times 10^{-4} + a_t, \quad a_t = \sigma_t\epsilon_t, \\ \epsilon_t &\sim t_{6.08}^*(0.846), \\ \sigma_t^2 &= 2.89 \times 10^{-6} + (0.07 + 1.0)a_{t-1}^2 + 0.849\sigma_{t-1}^2, \end{aligned}$$

where $t_v^*(\xi)$ denotes a skew standardized Student-t distribution with v degrees of freedom and skew parameter ξ. Model checking shows that the model fits the data well; see the R output below. The estimated leverage effect is statistically significant at the 5% level, and the AIC of the model is lower than that without the leverage parameter in Table 5.5.

R demonstration: TGARCH model.

```
> m4 <- garchFit(~garch(1,1),data=spy,leverage=T,trace=F,cond.dist="sstd")
> summary(m4)
Title: GARCH Modelling
Call: garchFit(formula=~garch(1,1),data spy,cond.dist="sstd",leverage=T,trace=F)
Conditional Distribution: sstd

Error Analysis:
        Estimate  Std. Error  t value Pr(>|t|)
mu      2.457e-04  1.571e-04    1.563    0.118
omega   2.890e-06  5.379e-07    5.372 7.80e-08 ***
alpha1  6.951e-02  1.646e-02    4.223 2.41e-05 ***
gamma1  1.000e+00  2.018e-01    4.956 7.21e-07 ***
beta1   8.492e-01  1.446e-02   58.713  < 2e-16 ***
skew    8.463e-01  2.330e-02   36.321  < 2e-16 ***
shape   6.083e+00  7.768e-01    7.831 4.88e-15 ***
---
Standardised Residuals Tests:
                                Statistic p-Value
 Ljung-Box Test     R    Q(10)  14.5514   0.1492931
 Ljung-Box Test     R    Q(20)  23.74802  0.2535727
 Ljung-Box Test     R^2  Q(10)  13.71561  0.1863633
 Ljung-Box Test     R^2  Q(20)  21.74093  0.3547326
 LM Arch Test       R    TR^2   16.33507  0.1763667

Information Criterion Statistics:
      AIC       BIC       SIC      HQIC
-6.484754 -6.468543 -6.484769 -6.478871
```

Another commonly used volatility model with leverage effect is the exponential GARCH (or EGARCH) model of Nelson (1991). The volatility equation of a simple EGARCH(1,1) model can be written as

$$\log(\sigma_t^2) = \omega + \alpha_1 \epsilon_{t-1} + \gamma_1[|\epsilon_{t-1}| - E(|\epsilon_{t-1}|)] + \beta_1 \log(\sigma_{t-1}^2), \qquad (5.26)$$

where $\epsilon_{t-1} = a_{t-1}/\sigma_{t-1}$ and

$$E(|\epsilon_{t-1}|) = \begin{cases} \sqrt{2/\pi} & \text{for } N(0,1) \\ \frac{2\Gamma((v+1)/2)\sqrt{v-2}}{\Gamma(v/2)\sqrt{\pi}} & \text{for } t_v^*. \end{cases}$$

From Equation (5.26), we see that the coefficients of ϵ_{t-1} are $(\alpha_1 + \gamma_1)$ and $(\alpha_1 - \gamma_1)$, respectively, for positive and negative ϵ_{t-1}. Taking into consideration the sign of ϵ_{t-1}, α_1 draws the difference between positive and negative ϵ_{t-1}. Therefore, α_1 of Equation (5.26) is called the leverage parameter. The EGARCH model can be estimated using the `rugarch` package. See the demonstration below for the SPY log return series. As expected, the leverage effect is statistically significant.

R demonstration: EGARCH model via the `rugarch` package. Output edited.

```
> require(rugarch)
>spec4 <- ugarchspec(variance.model=list(model="eGARCH"),mean.model=
list(armaOrder=c(0,0)),distribution.model="sstd")
> m6 <- ugarchfit(data=spy,spec=spec4)
> m6
*---------------------------------*
*          GARCH Model Fit        *
*---------------------------------*
Conditional Variance Dynamics
-----------------------------------
GARCH Model      : eGARCH(1,1)
Mean Model       : ARFIMA(0,0,0)
Distribution     : sstd

Optimal Parameters
------------------------------------
        Estimate  Std. Error   t value Pr(>|t|)
mu      0.000197    0.000295   0.66883 0.503602
omega  -0.262853    0.054392  -4.83257 0.000001
alpha1 -0.214794    0.025768  -8.33573 0.000000
beta1   0.971598    0.004886 198.83937 0.000000
gamma1  0.141275    0.007465  18.92445 0.000000
skew    0.832645    0.013163  63.25713 0.000000
shape   6.198791    1.636482   3.78788 0.000152

Information Criteria
------------------------------------
Akaike  -6.4939, Bayes    -6.4777
Shibata   -6.4939, Hannan-Quinn -6.4880

Weighted Ljung-Box Test on Standardized Residuals
------------------------------------
                        statistic p-value
Lag[1]                      5.963 0.01461
Lag[2*(p+q)+(p+q)-1][2]     6.437 0.01683
Lag[4*(p+q)+(p+q)-1][5]     7.626 0.03647
```

```
d.o.f=0
H0 : No serial correlation

Weighted Ljung-Box Test on Standardized Squared Residuals
------------------------------------
                          statistic p-value
Lag[1]                        4.246 0.03934
Lag[2*(p+q)+(p+q)-1][5]       5.234 0.13547
Lag[4*(p+q)+(p+q)-1][9]       6.351 0.25986
d.o.f=2
```

Finally, stochastic volatility models can be estimated by state space models (see Chapters 6 and 8).

5.5 FUNCTIONAL TIME SERIES

In recent years, functional data analysis has attracted much research interest; see, for example, Ramsay et al. (2009). In time series analysis, Bosq (2000) proposed the functional autoregressive processes, and Liu et al. (2016) and Kowal et al. (2017) developed various functional autoregressive models. In applications, many dynamically dependent data can be treated as functional time series and handled accordingly. Besse et al. (2000) studied forecasting functional climatic variations and Chen and Li (2017) considered an adaptive functional autoregressive model to predict electricity price curves. In this section, we briefly introduce some functional autoregressive models and discuss their properties.

To begin, we consider an example of functional time series. Figure 5.10 shows the time series plots of electricity demand on Sundays in Adelaide, Australia, starting from July 6, 1997. Each line consists of 48 observations representing the electricity demand of each 30-minute time interval starting from midnight. The plot has 508 lines and is not easy to comprehend. For ease of presentation, Figure 5.11 shows the demand plots of the first 10 Sundays. In the figure, we added $50(i-1)$ to the actual demands for the ith Sunday so that the lines show certain separation. The original data set has 3556 days from July 6, 1997 to March 31, 2007 organized by days in a week. The data are obtained from Shang and Hyndman (2011) and have been analyzed by several authors including Tsay (2016). Here we only show a small portion of the data and treat each plot as an electricity demand curve for a given Sunday. Each curve consists of 48 observations and exhibits certain diurnal pattern caused by temperature and human activities. For instance, the electricity demand was low around midnight when most people were sleeping and the temperature was low. In this particular instance, the goal of functional time series analysis is to study the time evolution of the demand curves.

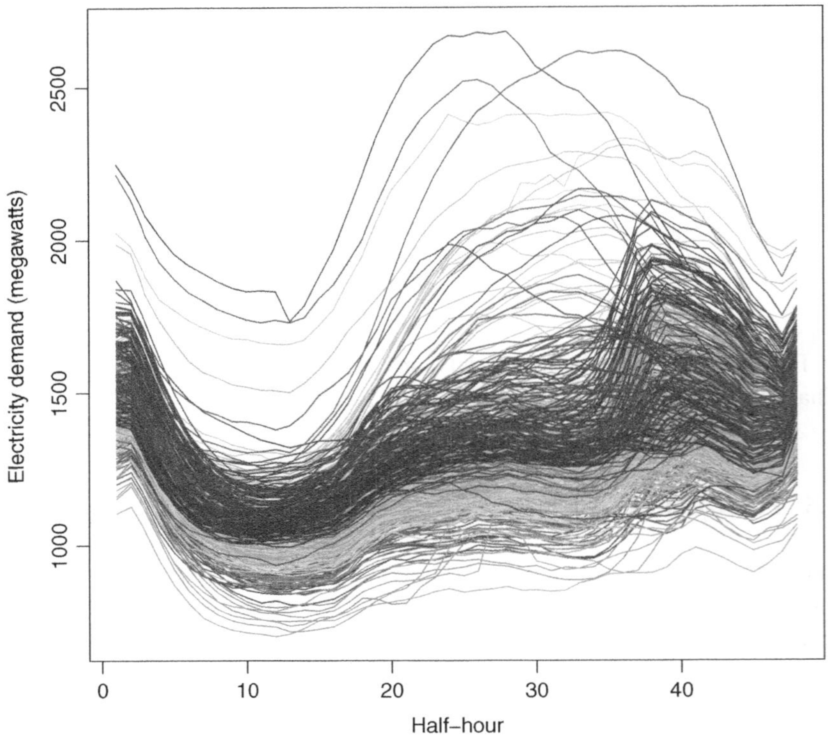

Figure 5.10 Time series plots of electricity demand on Sundays in Adelaide, Australia, starting from July 6, 1997. The demands are for 30-minute time intervals and there are 508 Sundays.

As illustrated by the electricity demand curves, we focus on the scalar functional time series. Let D be a compact subset of the real line. For instance, $D = [0, 1]$. We consider the functional space $L_2(D, \mathcal{B}(D), \lambda)$, where $\mathcal{B}$ denotes the Borel σ-field, λ is the Lebesgue measure, and $L_2(D) = \{f : \int_D f^2(x)dx < \infty\}$ is the set of all continuous squared-integrable real-valued functions on D. Let $X_t(s)$ be a functional time series at time t in $L_2(D)$, that is, $y_t(s) : D \to R$, where $s \in D$ and $\int_D y_t^2(s)ds < \infty$. Functional time series analysis is concerned with properties and time evolution of $\{y_t(s)\}$. In applications, $y_t(s)$ is only observed at some finite discrete points in D.

Most of the concepts of time series analysis continue to apply to the functional time series $y_t(s)$. For example, the series $\{y_t(s)|s \in D, t = 1, 2, \dots\}$ is weakly stationary if (a) $E[y_t(s)] = \mu(s)$ and (b) $\text{Cov}[y_{t+h}(s_1), y_{m+h}(s_2)] = \text{Cov}[y_t(s_1), y_m(s_2)] = \omega(|t-m|, |s_1 - s_2|)$, where h is an arbitrary integer, $s_i \in D$, and $\omega(.,.)$ denotes

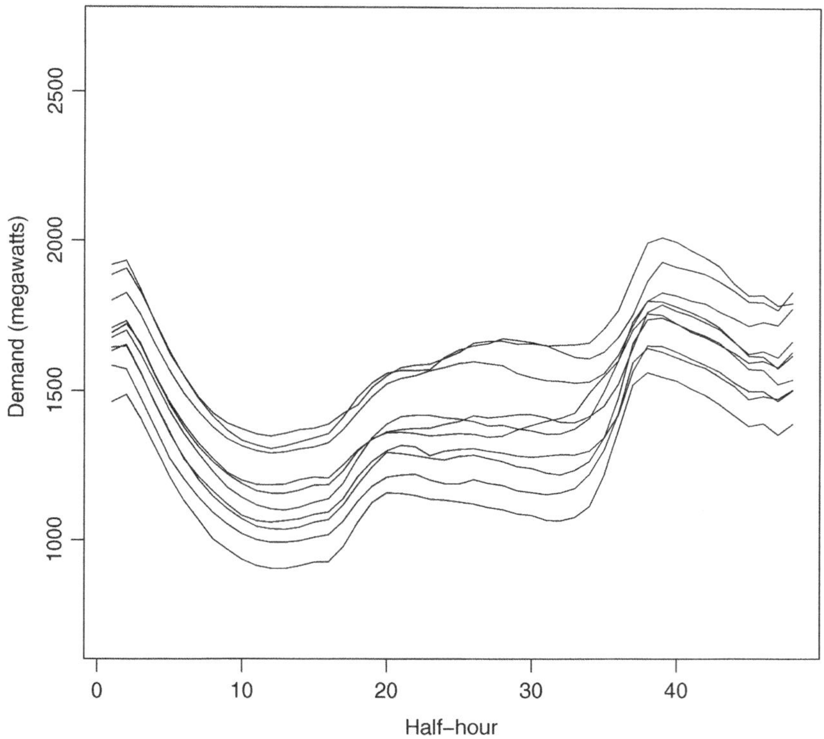

Figure 5.11 Time series plots of electricity demand on 10 Sundays in Adelaide, Australia, starting from July 6, 1997. The demands are for 30-minute time intervals. In the plot, $50(i-1)$ were added to the actual demands for the ith Sunday to aid presentation.

a well-defined function of its arguments. In other words, $y_t(s)$ is weakly stationary if its first two moments are time-invariant and space-invariant. The covariance function $\omega(\ell_1, \ell_2)$ of a stationary functional time series only depends on the time lags that separate the series and the distance between the two positions in D. In general, $\omega(\ell_1, \ell_2) \to 0$ sufficiently fast as $\ell_1 \to \infty$.

The most commonly used functional autoregressive (FAR) model for $y_t(s)$ is the FAR model of order 1, i.e. FAR(1), which assumes the form

$$y_t(s) - \mu(s) = \Phi(y_{t-1} - \mu) + \epsilon_t(s),$$

where $\mu(s)$ is a time-invariant function in D, $\{\epsilon_t(s)\}$ are independent white-noise processes in D such as a sequence of independent Gaussian processes with mean zero and $E(|\epsilon_t(s)|^2) < \infty$, and $\Phi(\cdot)$ is a linear operator on $L_2(D)$. See Bosq (2000).

Often one further employs an integral function for $\Phi(\cdot)$ and rewrites the FAR(1) model as

$$y_t(s) - \mu(s) = \int \phi(s,u)[y_{t-1}(u) - \mu(u)]du + \epsilon_t(s), \quad s \in D \tag{5.27}$$

where $\phi(s,u)$ is a smooth squared-integrable function on R (or a subset of R determined by the space D). In the literature, one often assumes that the employed smooth function satisfies certain Lipschitz conditions.

5.5.1 Convolution FAR models

Following Liu et al. (2016), we assume $D = [0, 1]$ and consider the convolution FAR (CFAR) models. The smooth functions employed belong to the following classes:

$$\text{Lip}^h[-1,1] = \{f : [-1,1] \to R, |f(x+\delta) - f(x)| \le M\delta^h, M < \infty, 0 < h \le 1\},$$
$$\text{Lip}_2^{\xi}[-1,1] = \{f \in C^r[-1,1] : f^{(r)} \in \text{Lip}^h[-1,1], \xi > 1, r = \lfloor \xi \rfloor, h = \xi - r\}.$$

If $f \in \text{Lip}_2^{\xi}[-1,1]$, then ξ is called the moduli of smoothness of $f(\cdot)$. These Lipschitz conditions are used to derive properties of the estimators of functional AR models.

A zero-mean CFAR(1) model then becomes

$$y_t(s) = \int_0^1 \phi_1(s-u) y_{t-1}(u) du + \epsilon_t(s), \quad s \in [0,1], \tag{5.28}$$

where $\phi_1 \in L_2[-1,1]$ is called the convolution function and $\{\epsilon_t\}$ are independent and identically distributed Ornstein–Uhlenbeck (O-U) processes on [0,1] satisfying the stochastic differential equation $d\epsilon_t(s) = -\rho\epsilon_t(s)ds + \sigma dW_s$, with $\rho > 0$, $\sigma > 0$ and W_s being a Wiener process.

The use of O-U processes in Equation (5.28) is for simplicity as the processes are stationary with well-defined covariance structure. Specifically, we have

$$\epsilon_t(s_1) \sim N\left(0, \frac{\sigma^2}{2\rho}\right), \quad \text{corr}(\epsilon_t(s_1), \epsilon_t(s_2)) = e^{-\rho|s_1 - s_2|}, \ s_1, s_2 \in [0,1].$$

Other Gaussian processes can also be used. If $\phi_1(\cdot)$ of Equation (5.28) is continuous, then $y_t(s)$ is continuous, but not differentiable. However, the deterministic component

$$f_t(s) = \int_0^1 \phi_1(s-u) y_{t-1}(u) du,$$

is differentiable.

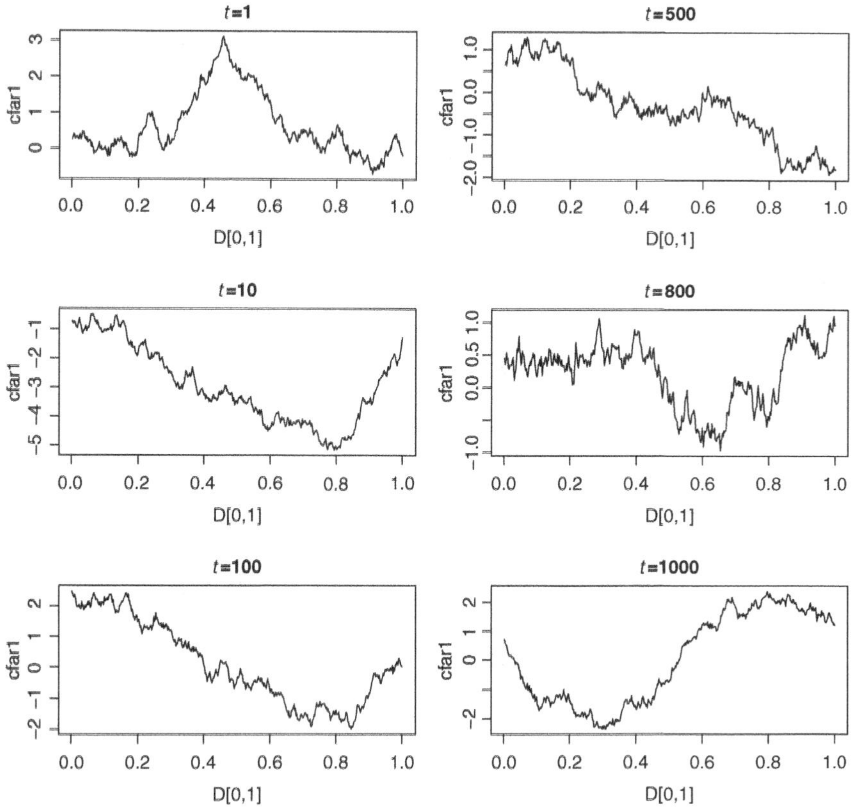

Figure 5.12 Selected time series plots of a simulated CFAR(1) process with $\rho = 2, \sigma = 1$, T=1000, and N=500, where T is the sample size and N is the number of equally spaced points in [0,1]. The $\phi_1(\cdot)$ function used is the probability density function of the standard Gaussian distribution.

To demonstrate, we generate a CFAR(1) time series with $\rho = 2, \sigma = 1, T$=1000, and $N = 500$, where T denotes the sample size and N is the number of equally spaced points in [0,1]. The $\phi_1(\cdot)$ function used is the probability density function of the standard Gaussian distribution. Figure 5.12 shows the time plots of selected $\{y_t(s)\}$. With $\rho = 2$ and $N = 500$, the variance and lag-1 serial correlations of the O-U processes used are 0.25 and 0.996, respectively. Thus, the innovation processes used are stationary but close to the non-stationary boundary. It is then not surprising to see that the $y_t(s)$ series can vary markedly when t increases.

Properties of CFAR(1) models have been studied by Liu et al. (2016). In particular, a sufficient condition for the weak stationarity of $y_t(s)$ in Equation (5.28) is that

$$k = \sup_{0\leq s\leq 1}\left(\int_0^1 \phi_1^2(s-u)du\right)^{1/2} < 1. \tag{5.29}$$

Clearly, if $\|\phi_1\|_2^2 < 1$, then the stationarity condition in Equation (5.29) holds and the resulting $y_t(s)$ process is weakly stationary. For a stationary $y_t(s)$, define $\psi_1(s,u) = \phi_1(s-u)$ and

$$\psi_\ell(s,u) = \int_0^1 \psi_{\ell-1}(s,v)\phi_1(v-u)dv, \quad \ell \geq 2.$$

Then, by repeated substitutions, we have

$$y_t(s) = \epsilon_t(s) + \sum_{\ell=1}^{\infty}\int_0^1 \psi_\ell(s,u)\epsilon_{t-\ell}(u)du. \tag{5.30}$$

This is a functional moving-average representation of a stationary CFAR(1) process of Equation (5.28).

Finally, the CFAR(1) model can be generalized easily to CFAR(p) models. Again, for simplicity, we assume the mean function of the process is zero. The functional time series $y_t(s)$ on [0,1] is a (zero-mean) CFAR(p) process if

$$y_t(s) = \sum_{i=1}^{p}\int_0^1 \phi_i(s-u)y_{t-i}(u)du + \epsilon_t(s), \quad s \in [0,1], \tag{5.31}$$

where p is a positive integer, $\phi_i \in L_2[-1,1]$ is the lag-i convolution function, and $\epsilon_t(s)$ is an O-U process as before. Let

$$k_i = \sup_{0\leq s\leq 1}\left(\int_0^1 \phi_i^2(s-u)du\right)^{1/2},$$

for $i = 1,\dots,p$. Then, a sufficient condition for the weak stationarity of the CFAR(p) model in Equation (5.31) is that all the roots of the characteristic function

$$1 - k_1 z - \cdots - k_p z^p = 0$$

are outside the unit circle. This is a generalization of the scalar AR(p) time series. For a weakly stationary CFAR(p) process, define

$$\psi_1(s,u) = \phi_1(s-u),$$

$$\psi_\ell(s,u) = \sum_{i=1}^{\ell-1}\int_0^1 \phi_i(s-v)\psi_{\ell-i}(v,u)dv + \phi_\ell(s,u), \quad \ell = 2,\dots,p,$$

$$\psi_\ell(s,u) = \sum_{i=1}^{p} \int_0^1 \phi_i(s-v)\psi_{\ell-i}(v,u)dv, \quad \ell = p+1, \ldots .$$

Then, the CFAR(p) process has an MA(∞) representation

$$y_t(s) = \epsilon_t(s) + \sum_{\ell=1}^{\infty} \int_0^1 \psi_\ell(s,u)\epsilon_{t-\ell}(u)du.$$

5.5.2 Estimation of CFAR Models

Consider the CFAR(p) model in Equation (5.31). Under the smoothness assumption, one can use linear interpolation to evaluate $y_t(s)$ at any given point $s_1 \in [0,1]$. This implies that for functional time series analysis the observed data can be irregularly spaced and the series may have different numbers of observed values at different time points. Assume that the sample size is T. For estimation, we consider $y_t(s)$ at $s_j = j/N$, where $j = 0, 1, \ldots, N$ and N is a pre-specified positive integer. Let $\boldsymbol{s} = (s_0, s_1, \ldots, s_N)'$ be the grid points for $y_t(s)$. If $y_t(s_j)$ is not observed and $s_{n-1} \le s_j \le s_n$, let

$$\tilde{y}_t(s_j) = \frac{(s_n - s_j)y_t(s_{n-1}) + (s_j - s_{n-1})y_t(s_n)}{1/N}.$$

We use $\tilde{y}_t(s_j)$ as a proxy for $y_t(s_j)$. Since the convolution functions $\phi_i(\cdot)$ are unknown, we use B-splines to provide approximations. Specifically, we employ

$$\phi_i(\cdot) \approx \tilde{\phi}_{k,i} = \sum_{j=1}^{k} \tilde{\beta}_{k,i,j} B_{k,j}(\cdot), \quad i = 1, \ldots, p, \tag{5.32}$$

where $\{B_{k,j}(\cdot) | j = 1, \ldots, k\}$ are uniform cubic B-spline basis functions with k degrees of freedom. See Chapter 3 for some information concerning spline basis functions.

Using linearly interpolated values, if needed, and B-spline approximations, we obtain approximate residuals

$$\tilde{\epsilon}_{t,n} \equiv \epsilon_t(s_n) \approx y_t(s_n) - \sum_{i=1}^{p}\sum_{j=1}^{k} \tilde{\beta}_{k,i,j} \int_0^1 B_{k,j}(s_n - u)\tilde{y}_{t-i}(u)du. \tag{5.33}$$

Let $\tilde{\boldsymbol{\epsilon}}_t = (\tilde{\epsilon}_{t,0}, \tilde{\epsilon}_{t,1}, \ldots, \tilde{\epsilon}_{t,N})'$. We can put the approximate residuals in a matrix form

$$\tilde{\boldsymbol{\epsilon}}_t = y_t(\boldsymbol{s}) - \boldsymbol{M}_t \tilde{\boldsymbol{\beta}}_k,$$

where $\boldsymbol{M}_t = (\boldsymbol{M}_{t,1}, \dots, \boldsymbol{M}_{t,p})$, $\boldsymbol{M}_{t,i}$ is an $(N+1) \times k$ matrix with entries $(\boldsymbol{M}_{t,i})_{nj} = \int_0^1 B_{k,j}(s_n - u)\tilde{y}_{t-i}(u)du$, and $\tilde{\boldsymbol{\beta}}_k$ is a $pk \times 1$ vector with entries $\tilde{\boldsymbol{\beta}}_k = (\tilde{\boldsymbol{\beta}}'_{k,1}, \cdots, \tilde{\boldsymbol{\beta}}'_{k,p})'$, and $(\boldsymbol{\beta}_{k,i})_j = \tilde{\beta}_{k,i,j}$. Here we use $(\boldsymbol{A})_{ij}$ to denote the (i,j)th element of the matrix $\boldsymbol{A}$ and $(\boldsymbol{b})_j$ the jth element of the vector $\boldsymbol{b}$.

Since $B_{k,j}(\cdot)$ are known functions and fixed, $\boldsymbol{M}_t$ is known given the observations. Furthermore, under the O-U process, the equally spaced grid points s_n imply that the random process $\{\epsilon_t(s_n) | 0 \leq n \leq N\}$ follows an AR(1) model with AR coefficient $e^{-\rho/N}$ and covariance matrix $\boldsymbol{\Sigma}$ with $(\boldsymbol{\Sigma})_{ij} = e^{-\rho|i-j|/N}\sigma^2/(2\rho)$. Therefore, the parameters $\boldsymbol{\beta} = \{\beta_{ij} | i = 1, \dots, p; j = 1, \dots, k\}, \sigma^2$, and ρ can be estimated by maximizing the approximated log-likelihood function

$$S_{k,T,N}(\boldsymbol{\beta}, \sigma^2, \rho) = -\frac{(N+1)(T-p)}{2} \ln\left(\frac{\pi\sigma^2}{\rho}\right) - \frac{N(T-p)}{2} \ln\left(1 - \frac{1}{e^{2\rho/N}}\right)$$
$$-\frac{1}{2} \sum_{t=p+1}^{T} \epsilon'_{\beta,t} \boldsymbol{\Sigma}^{-1} \epsilon_{\beta,t}, \tag{5.34}$$

where $\epsilon_{\beta,t} = y_t(s) - \boldsymbol{M}_t\boldsymbol{\beta}$. Let $\varphi = e^{-\rho/N}$, $\omega = \sigma^2/(2\rho)$, $\boldsymbol{\Sigma}_0 = 2\rho\boldsymbol{\Sigma}/\sigma^2$, and $e(\boldsymbol{\beta}, \varphi) = \sum_{t=p+1}^{T} \epsilon'_{\beta,t}\boldsymbol{\Sigma}_0^{-1}\epsilon_{\beta,t}$, where $\boldsymbol{\Sigma}_0$ is the correlation matrix of $\epsilon_t(\boldsymbol{s})$. The objective function of Equation (5.34) can be written as

$$S_{k,T,N}(\boldsymbol{\beta}, \varphi, \omega) = -\frac{(N+1)(T-p)}{2} \ln(2\pi\omega)$$
$$-\frac{N(T-p)}{2} \ln(1-\varphi^2) - \frac{e(\boldsymbol{\beta}, \varphi)}{2\omega}. \tag{5.35}$$

Given $\boldsymbol{\beta}$ and φ, the maximizer of ω of Equation (5.35) is

$$\hat{\omega} = \frac{e(\boldsymbol{\beta}, \varphi)}{(N+1)(T-p)}. \tag{5.36}$$

Hence, we have

$$\max_{\boldsymbol{\beta},\varphi,\omega} S_{k,T,N}(\boldsymbol{\beta}, \varphi, \omega) = \max_{\boldsymbol{\beta},\varphi} \left[-\frac{(N+1)(T-p)}{2} \ln e(\boldsymbol{\beta}, \varphi) \right.$$
$$\left. - \frac{N(T-p)}{2} \ln(1-\varphi^2) + c \right]$$
$$= \max_{\varphi} \left[-\frac{(N+1)(T-p)}{2} \ln e(\widehat{\boldsymbol{\beta}}(\varphi), \varphi) \right.$$
$$\left. - \frac{N(T-p)}{2} \ln(1-\varphi^2) + c \right], \tag{5.37}$$

where c is a constant and

$$\hat{\boldsymbol{\beta}}(\varphi) = \arg\min_{\boldsymbol{\beta}} e(\boldsymbol{\beta}, \varphi)$$
$$= \left(\sum_{t=p+1}^{T} \boldsymbol{M}_t' \boldsymbol{\Sigma}^{-1} \boldsymbol{M}_t \right)^{-1} \left(\sum_{t=p+1}^{T} \boldsymbol{M}_t' \boldsymbol{\Sigma}^{-1} y_t(s) \right). \tag{5.38}$$

Equation (5.37) is a one-dimensional optimization problem and can be easily solved to obtain $\hat{\varphi}$. Consequently, via Equations (5.36) and (5.38), we can obtain estimates for the unknown parameters σ^2, ρ, and $\boldsymbol{\beta}$. The convolution function $\phi_i(\cdot)$ can then be estimated by $\hat{\phi}_{k,i}(\cdot) = \sum_{j=1}^{k} \hat{\beta}_{k,i,j} B_{k,j}(\cdot)$.

5.5.3 Fitted Values and Approximate Residuals

With fitted convolution functions, one can obtain the fitted value and an updated smooth value of $y_t(s)$ for any given point $s \in [0, 1]$. Specifically, for $s \in (s_n, s_{n+1})$, let

$$\tilde{f}_t(s) = \sum_{i=1}^{p} \int_0^1 \hat{\phi}_{k,i}(s-u) \tilde{y}_{t-i}(u) du, \tag{5.39}$$

$$\tilde{y}_t^*(s) = \tilde{f}_t(s) + (e^{-\hat{\rho}(v_1)}, e^{-\hat{\rho}(v_2)}) \hat{\boldsymbol{\Sigma}}_0^{-1} \begin{bmatrix} y_t(s_n) - \tilde{f}_t(s_n) \\ y_t(s_{n+1}) - \tilde{f}_t(s_{n+1}) \end{bmatrix}, \tag{5.40}$$

where $v_1 = s - s_n$ and $v_2 = s_{n+1} - s$, and $\hat{\boldsymbol{\Sigma}}_0$ is the estimated correlation matrix between $\epsilon_t(s_n)$ and $\epsilon_t(s_{n+1})$ with off-diagonal elements being $e^{-\hat{\rho}/N}$. If $s = s_n$, then $\tilde{y}_t^*(s) = y_t(s)$, the observed value. For ease of reference, we refer to $\tilde{f}_t(s)$ as the fitted value of $y_t(s)$ and $\hat{\epsilon}_t(s_n) = y_t(s_n) - \tilde{f}_t(s_n)$ as an approximate residual. Equation (5.40) is similar to that of Kriging in the spatial statistics.

5.5.4 Prediction

Given a fitted CFAR(p) model and the data $\{y_t(s) | t = 1, \ldots, T\}$, one can obtain the minimum one-step ahead squared-error prediction as

$$\hat{y}_{T+1}(s) = \sum_{i=1}^{p} \int_0^1 \hat{\phi}_{k,i}(s-u) \tilde{y}_{T+1-i}^*(u) du, \tag{5.41}$$

where $\tilde{y}_{T+1-i}^*(\cdot)$ is the fitted process of Equation (5.40).

5.5.5 Asymptotic Properties

Asymptotic properties of the estimators of CFAR models discussed in Section 5.5.2 have been studied by Liu et al. (2016). The authors employ the sieve framework of Grenander (1981) and establish certain results as N (the number of points in [0,1]), T (the sample size), and k (the degrees of B-splines) go to infinity at certain rates. In general, sieve methods are concerned with parameter estimation in an infinite-dimensional space $\boldsymbol{\Theta}$ for which optimization of the objective function cannot be directly solved. Instead, one optimizes the objective function over an increasing sequence of subspaces $\{\boldsymbol{\Theta}_n, n = 1, 2, \ldots\}$. The subspaces are called sieve spaces and the associated sequence of estimators are called sieve estimators. Under certain conditions such as the sieve subspace $\boldsymbol{\Theta}_n$ converges to $\boldsymbol{\Theta}$ as n increases, sieve estimators are shown to be consistent. For further information on sieve estimation, see Chen (2008) and Halberstam and Richert (2013). For the CFAR models, Liu et al. (2016) obtained the limiting distributions of the estimators of B-spline coefficients assuming that $\phi_i(\cdot) \in \text{Lip}_2^{\xi}[-1, 1]$ with $\xi > 1$ as $T \to \infty$ and $N^2/T \to \infty$, provided that σ^2 and ρ are given. Interested readers are referred to Liu et al. (2016) for details and proofs.

Finally, Liu et al. (2016) also proposed a method for determining the order k of the B-splines and a F-test for selecting the order of a CFAR model. Out-of-sample prediction can also be used to select the CFAR order.

5.5.6 Application

To demonstrate analysis of functional time series, consider the electricity demand on Sundays shown in Figure 5.10. Also available are the half-hour temperature data at Adelaide Airport. See Shang and Hyndman (2011). Figure 5.13 shows the time plots of the temperature series. From the plots, there exist some outlying or missing values in the temperature series. Therefore, we used simple interpolation to fill in the 11 outlying observations. This is a very small portion of the data (11 out of 24,384) and we do not expect the interpolation has any material effect on the analysis. Since the effect of temperature on electricity demand is nonlinear, we use the linear regression

$$Y_t = 1255.9 + 0.924T_t + 68.12(T_t - 25)_+ + y_t, \quad t = 1, \ldots, 24384, \quad (5.42)$$

where Y_t denotes the observed electricity demand, T_t is the temperature $(T_t - 25)_+ = \max\{0, T_t - 25\}$, and y_t is the error term. The coefficients are estimated by the ordinary least squares method and they are statistically significantly at the 1% level. In what follows, we treat the residual series y_t of Equation (5.42) as the electricity demand. Next, to simplify the modeling process, we remove the

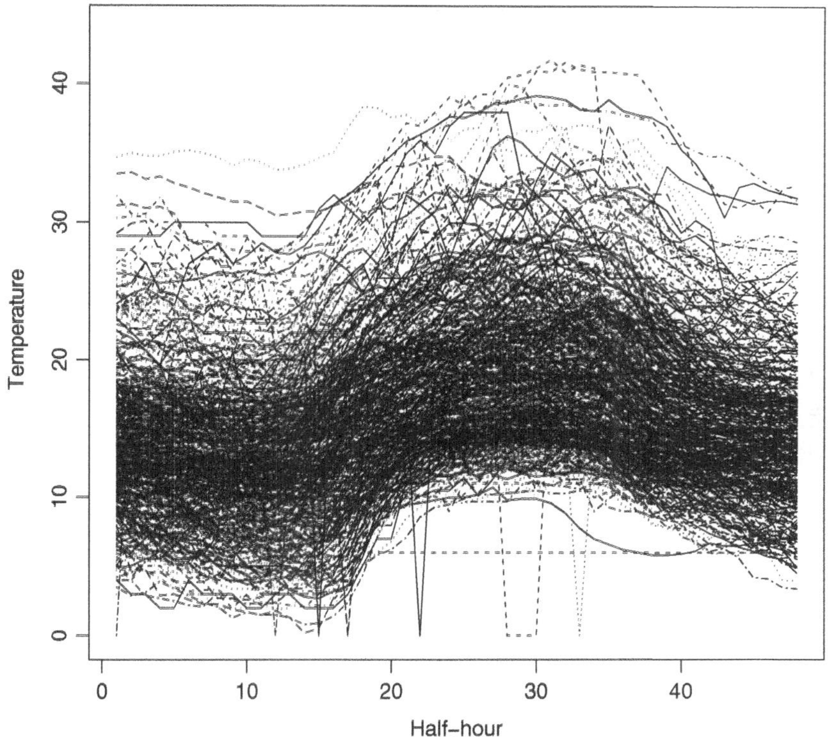

Figure 5.13 Time series plots of temperature for Sundays at Adelaide Airport, Australia, starting from July 6, 1997. The temperatures (in °C) are for 30-minute time intervals and there are 508 Sundays.

mean demand from the y_t series. Specifically, for each half-hour interval, we estimate the mean $\mu_t = E(y_t)$ using the sample mean over the 508 weeks. Figure 5.14 shows the mean function.

Let $\tilde{y}_t = y_t - \hat{\mu}_t$ be the mean-adjusted electricity demand, where $\hat{\mu}_t$ denotes the sample mean of y_t over the 508 weeks (with t fixed). Applying the F-test of Liu et al. (2015) to select the order of CFAR models with $k = 6$ for the B-splines, we obtain

Order	0	1	2	3	4	5	6
F-statistic	188.2	22.03	4.81	2.73	1.45	1.00	1.09
p value	0	0	0	0.01	0.19	0.42	0.37

Therefore, a CFAR(4) model is chosen for the $\tilde{y}_t$ series. Figure 5.15 shows the plots

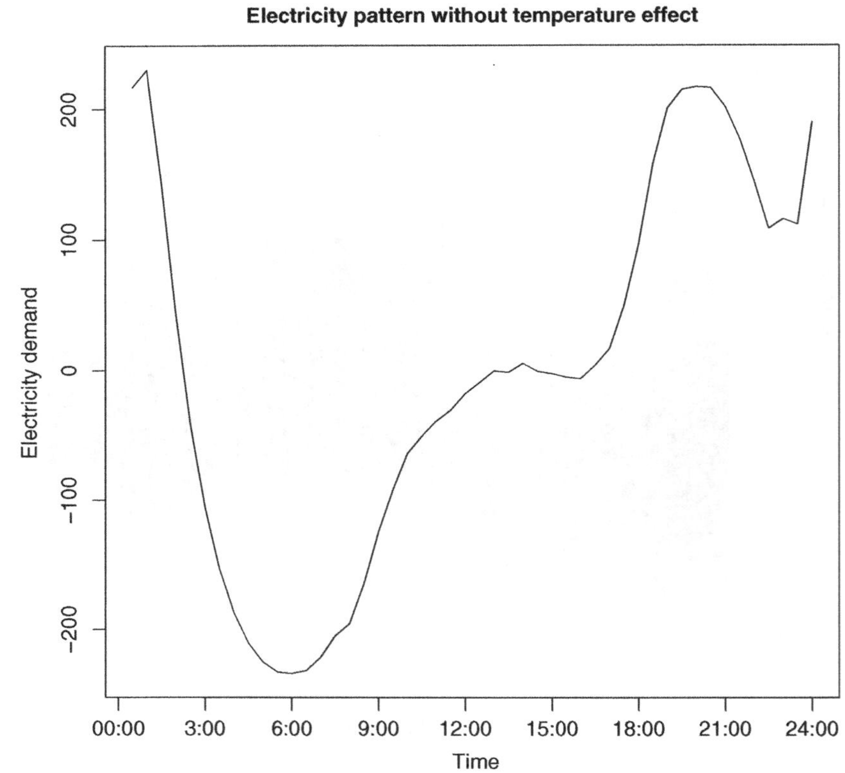

Figure 5.14 The sample mean function of temperature-adjusted electricity demand on Sundays.

of estimated coefficient functions. The estimated parameters of the O-U innovation process are $\hat{\rho} = 2.08$ and $\hat{\sigma} = 208.9$. From Figure 5.15, the convolution function of all four coefficient functions puts more weights around $x = 0$. This is reasonable. On the other hand, $\hat{\phi}_1$ and $\hat{\phi}_3$ have higher weights for negative x whereas $\hat{\phi}_2$ and $\hat{\phi}_4$ have heavier weights for positive x. Model checking shows that the fitted CFAR(4) model provides a reasonable fit to $\tilde{y}_t$. Finally, Figure 5.16 shows the one-step ahead prediction function for the electricity demand at the forecast origin $t = 508$. This predictive function is obtained by adding $\hat{\mu}_t$ to $\tilde{y}_{508}(1)$, where $\tilde{y}_{508}(1)$ denotes the one-step ahead prediction of the fitted CFAR(4) model.

R demonstration: Commands used in CFAR modeling.

```
require(fds)
data(sundaydemand)
```

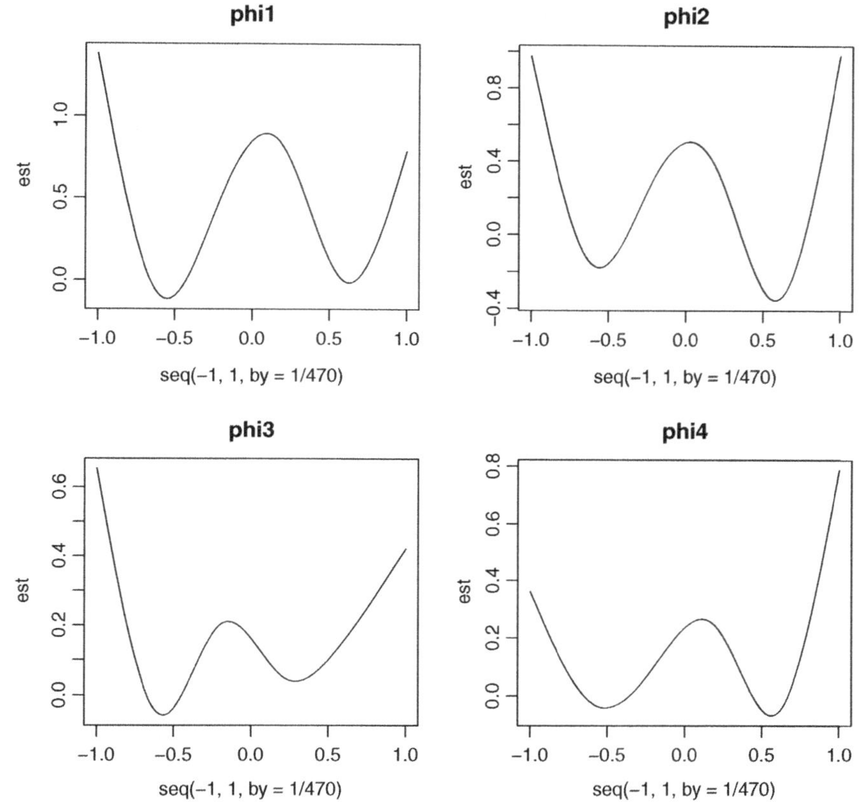

Figure 5.15 The estimated coefficient functions of a CFAR(4) model for the temperature-adjusted electricity demand on Sundays.

```
electric <- sundaydemand$y
dim(electric)  ## 48 by 508
ele=as.matrix(electric)
test=matrix(ele,48*508,1)
par(mfrow=c(1,1))
plot(test[1:(24*150)],type='l')
plot(test,type='l')

data(sundaytempairport)
temp=sundaytempairport$y
### fill in the missing values.
temp[1,90] <- (temp[1,89]+temp[1,91])/2
temp[12,6] <- (temp[12,5]+temp[12,7])/2
temp[14,269] <- (temp[14,268]+temp[14,270])/2
```

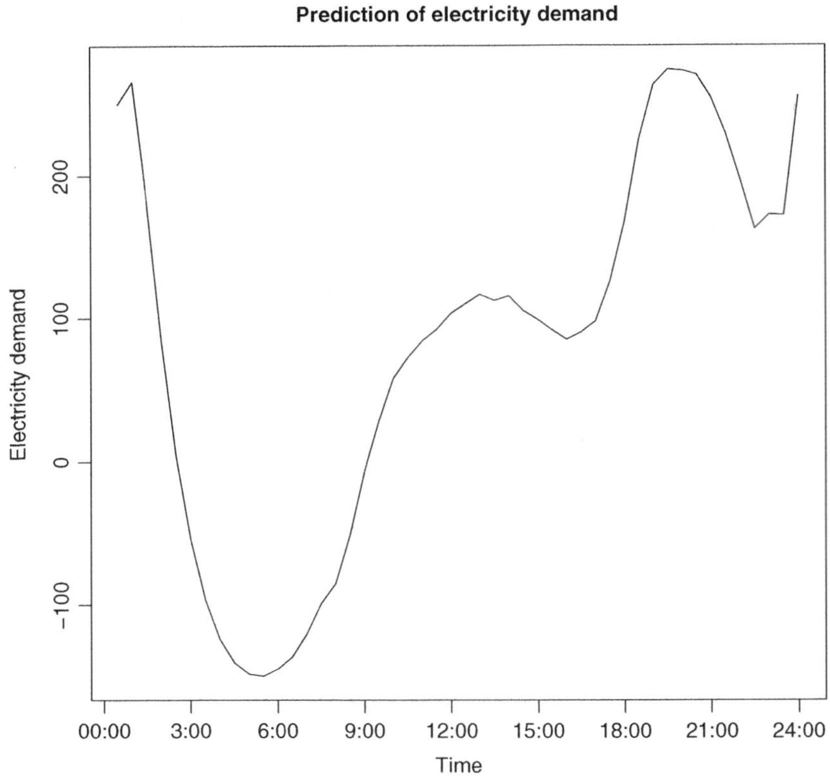

Figure 5.16 Time plot of the one-step ahead prediction for electricity demand on Sundays. The forecast origin is $T = 508$.

```
temp[15,79] <- (temp[15,78]+temp[15,80])/2
temp[17,11] <- (temp[17,10]+temp[17,12])/2
temp[17,46] <- (temp[17,45]+temp[17,47])/2
temp[22,37] <- (temp[22,36]+temp[22,38])/2
temp[28,92] <- (temp[28,91]+temp[28,93])/2
temp[29,92] <- (temp[29,91]+temp[29,93])/2
temp[30,92] <- (temp[30,91]+temp[30,93])/2
temp[33,9] <- (temp[33,8]+temp[33,10])/2

n.row=dim(temp)[1]
n.col=dim(temp)[2]
dep=matrix(electric,n.row*n.col,1)
temp=as.matrix(temp)
indep=matrix(temp,n.row*n.col,1)
```

```
indep2=indep
for(i in 1:(n.row*n.col)){
        indep2[i]=max(0,indep[i]-25)
}
lm3=lm(dep~indep+indep2)
summary(lm3)

data=t(matrix(lm3$residuals,n.row,n.col))
t=dim(data)[1]
mean=apply(data,2,mean)
plot(mean,type='l')

###############
### Average removed
data_re=data- matrix(rep(mean,each=t),t, dim(data)[2])
#pdf('plot.pdf')
par(mfrow=c(1,3))
plot(data[1,],type='l',ylim=c(min(data)*1.1,max(data)*1.1),
main='Original data',xlab='time',ylab='temperature',xaxt="n")
axis(1,at=seq(0,48,by=6),labels=c("00:00","3:00","6:00","9:00","12:00",
 "15:00","18:00","21:00","24:00"))
for(i in 2:508){
        points(data[i,],type='l',col=grey(i/508))
}
plot(mean,type='l',main='Temperature pattern',
     xlab='time',ylab='temperature',xaxt="n")
axis(1,at=seq(0,48,by=6),labels=c("00:00","3:00","6:00","9:00","12:00",
 "15:00","18:00","21:00","24:00"))
plot(data_re[1,],type='l',ylim=c(min(data_re)*1.1,max(data_re)*1.1),
main='Average removed',xlab='time',ylab='temperature',xaxt="n")
axis(1,at=seq(0,48,by=6),labels=c("00:00","3:00","6:00","9:00","12:00",
 "15:00","18:00","21:00","24:00"))
for(i in 2:508){
        points(data_re[i,],type='l',col=grey(i/508))
}
#dev.off()

par(mfrow=c(1,1))
#pdf('mean.pdf')
plot(mean,type='l',main='Electricity pattern without temperature effect',
    xlab='time',ylab='Electricity demand',xaxt="n")
axis(1,at=seq(0,48,by=6),labels=c("00:00","3:00","6:00","9:00","12:00",
 "15:00","18:00","21:00","24:00"))
#dev.off()
```

```
require(NTS)
grid=470
df_b=5
p.max=8
F_test(data_re,p.max,df_b,grid) ###CFAR(4)

#### Estimation
p=4
est=est_cfar(data_re,p,6,grid)
### end of addition

#pdf("phi.pdf")
par(mfrow=c(2,2))
plot(seq(-1,1,by=1/470),est$phi_func[1,],type='l',main='phi1',
xlab='x',ylab=",ylim=c(min(est$phi_func)*1.1,max(est$phi_func)*1.1))
plot(seq(-1,1,by=1/470),est$phi_func[2,],type='l',main='phi2',
xlab='x',ylab=",ylim=c(min(est$phi_func)*1.1,max(est$phi_func)*1.1))
plot(seq(-1,1,by=1/470),est$phi_func[3,],type='l',main='phi3',
xlab='x',ylab=",ylim=c(min(est$phi_func)*1.1,max(est$phi_func)*1.1))
plot(seq(-1,1,by=1/470),est$phi_func[4,],type='l',main='phi4',
xlab='x',ylab=",ylim=c(min(est$phi_func)*1.1,max(est$phi_func)*1.1))
#dev.off()

f=data_re
n=dim(f)[2]-1
t=dim(f)[1]
grid=470
index=seq(1,grid+1,by=grid/n)
pred=p_cfar(est,f,1)
pred_ele=pred[index]+mean
#pdf("prediction.pdf")
plot(pred_ele,type='l',main="Prediction of Electricity Demand",
xlab='Time', ylab='Electricity demand',xaxt="n")
axis(1,at=seq(0,48,by=6),labels=c("00:00","3:00","6:00","9:00","12:00",
 "15:00","18:00","21:00","24:00"))
#dev.off()
```

APPENDIX: DISCRETE DISTRIBUTIONS FOR COUNT DATA

Four probability distributions are often used in modeling discrete-valued time series: the binomial distribution, the Poisson distribution, the negative binomial distribution, and the double Poisson distribution. Let $\mathcal{F}_{t-1}$ be the information available at time $t - 1$. The probability mass functions (pmf) and the first two moments of these distributions are given below.

1. Binomial: $y_t \mid \mathcal{F}_{t-1} \sim B(m_t, p_t)$ with pmf

$$B(m_t, p_t) = \binom{m_t}{y_t} p_t^{y_t}(1-p_t)^{m_t-y_t}, \quad 0 < p_t < 1.$$

The mean and variance of the distribution are $E(y_t \mid \mathcal{F}_{t-1}) = m_t p_t$ and $\text{Var}(y_t \mid \mathcal{F}_{t-1}) = m_t p_t(1-p_t)$.

2. Poisson: $y_t \mid \mathcal{F}_{t-1} \sim Po(y_t \mid \mu_t)$ with pmf

$$Po(y_t \mid \mu_t) = \frac{\exp(-\mu_t)\mu_t^{y_t}}{y_t!}, \quad \mu_t > 0.$$

The first two moments are $E(y_t \mid \mathcal{F}_{t-1}) = \text{Var}(y_t \mid \mathcal{F}_{t-1}) = \mu_t$.

3. Negative binomial: $y_t \mid \mathcal{F}_{t-1} \sim NB(\theta, \mu_t)$ with pmf

$$NB(y_t \mid \theta, \mu_t) = \frac{\Gamma(y_t+\theta)}{y_t!\Gamma(\theta)}\left(\frac{\theta}{\mu_t+\theta}\right)^{\theta}\left(\frac{\mu_t}{\mu_t+\theta}\right)^{y_t}, \quad \theta > 0.$$

Here $E(y_t \mid \mathcal{F}_{t-1}) = \mu_t$ and $\text{Var}(y_t \mid \mathcal{F}_{t-1}) = \mu_t + \frac{\mu_t^2}{\theta}$. The distribution allows for over-dispersion with $\theta > 0$.

4. Double Poisson of Efron (1986) with pmf

$$\begin{aligned} DP(y_t \mid \theta, \mu_t) &= c(\theta, \mu_t)\theta^{1/2}[Po(y_t \mid \mu_t)]^{\theta}[Po(y_t \mid y_t)]^{1-\theta} \\ &= c(\theta, \mu_t)\theta^{1/2}e^{-\theta\mu_t}\left(\frac{e^{-y_t}y_t^{y_t}}{y_t!}\right)\left(\frac{e\mu_t}{y_t}\right)^{\theta y_t}, \quad \theta > 0, \end{aligned}$$

where $c(\theta, \mu_t)$ serves as the normalization factor and can be approximated by

$$\frac{1}{c(\theta, \mu_t)} \approx 1 + \frac{1-\theta}{12\mu_t\theta}\left(1 + \frac{1}{\mu_t\theta}\right).$$

Here $E(y_t \mid \mathcal{F}_{t-1}) = \mu_t$ and $\text{Var}(y_t \mid \mathcal{F}_{t-1}) \approx \mu_t/\theta$, and the distribution allows for over- and under-dispersion.

5.6 EXERCISES

5.1 Consider the autoregressive conditional Poisson model of order (1,1). Assume that there exist no explanatory variables. Derive the result of Equation (5.13).

5.2 Consider the number of transactions in 5-minute intervals for the Wausau Paper Corporation (WPP) from January 3 to March 31, 2005. See Example 5.2 for further information on the data. The data are given in the file `taq-JBES2011.txt`. Fit an ACM(1,1) model to the series using Poisson, negative binomial, and double Poisson distributions. Write down the fitted models and perform model checking.

5.3 Consider the number of transactions in 5-minute intervals for the Empire District Electric (EDE) stock from January 3 to March 31, 2005. See Example 5.2 for further information on the high-frequency financial data. The data and explanatory variables are in the file `taq-edeX.txt`. Fit a GLARNA model with subcommand `thetaLags=c(1:12)` and negative binomial distribution to the series. Perform model checking and write down the fitted model.

5.4 Consider the EDE transaction series of Problem 3. Delete the `intercept` from the explanatory variables, then fit an ACM(1,1) model with negative binomial distribution. Perform model checking and write down the fitted model. Compare the ACM and glarma models. Which model do you prefer? Why?

5.5 Consider the daily log returns of Amazon stock from January 3, 2007 to December 31, 2016. The simple returns of the stock obtained from the Center for Research in Security Prices, University of Chicago are in the file `d-amzn0716.txt`. Find a GARCH model for the series. Perform model checking and write down the fitted model. Obtain one- to five-step ahead predictions of the log return and its volatility.

5.6 Consider the daily log returns of Amazon stock in the prior question. Obtain a threshold GARCH model for the series. Perform model checking and write down the fitted model. Is the leverage effect statistically significant at the 5% level? Why?

5.7 Consider the electricity demand and temperature on Saturdays in Adelaide, Australia from July 6, 1997 for 508 weeks. The data are available from the R package `fds`.

(a) Show the time plots of electricity demand and temperature for the half-hour intervals.

(b) Remove the effect of temperature on electricity demand using a similar linear regression model to that of Equation (5.42).

(c) Let y_t be the temperature-adjusted electricity demand. Obtain a time plot of the mean demand function.

(d) Let $\tilde{y}_t$ be the mean-adjusted series of y_t. Build a CFAR model for $\tilde{y}_t$.

(e) Obtain the one-step ahead prediction for the demand function at the forecast origin $T = 508$.

REFERENCES

Benjamin, M. A., Rigby, R. A., and Stasinopoulos, D. M. (2003). Generalized autoregressive moving average models. *Journal of the American Statistical Association* **98**: 214–223.

Besse, P. C., Cardot, H. and Stephenson, D. B. (2000). Autoregressive forecasting of some functional climatic variations. *Scandinavian Journal of Statistics*, **27**: 673-687.

Bollerslev, T. (1986). Generalized autoregressive conditional heterskedasticity. *Journal of Econometrics* **31**: 307–327.

Bosq, D. (2000). *Linear Processes in Function Spaces: Theory and Applications.* Springer Science & Business Media, Berlin.

Chen, X. (2008). Large sample sieve estimation of semi-nonparametric models. In *Handbooks of Econometrics*, Volume 6B, pp. 5549–5632.

Chen, Y. and Li, B. (2017). An adaptive functional autoregressive forecast model to predict electricity price curves. *Journal of Business & Economic Statistics* **35**: 371–388.

Cox, D. (1981). Statistical analysis of time series: some recent developments. *Scandinavian Journal of Statistics* **8**: 93–115.

Creal, D., Koopman, S., and Lucas, A. (2008). A general framework for observation-driven time-varying parameter models. Discussion paper, Tinbergen Institute.

Davis, R. A. and Wu, R. (2009). A negative binomial model for time series of counts. *Biometrika* **96**: 735–749.

Davis, R. A., Dunsmuir, W. T. M., and Streett, S. B. (2003). Observation-driven models for Poisson counts. *Biometrika* **90**: 777–790.

Dunsmuir, W. T. M. and Scott, D. J. (2015). The glarma package for observation-driven time series regression of counts. *Journal of Statistical Software* **67**: 1–36.

Dunsmuir, W. T. M., Li, C., and Scott, D. J. (2015). `glarma`: *Generalized Linear Autoregressive Moving-Average Models.* R package.

Efron, B. (1986). Double exponential families and their use in generalized linear models. *Journal of the American Statistical Association* **81**: 709–721.

Engle, R. F. and Russell, J. R. (1998). Autoregressive conditional duration: a new model for irregularly spaced transaction data. *Econometrica* **66**: 1127–1162.

Glosten, L. R., Jagannathan, R., and Runkle, D. E. (1993). On the relation between the expected value and the volatility of nominal excess return on stocks. *Journal of Finance* **48**: 1779–1801.

Grenander, U. (1981). *Abstract Inference.* Wiley, New York.

Halberstam, H. and Richert, H. E. (2013). *Sieve Methods.* Dover Publications, New York.

Hautsch, N. (2012). *Econometrics of Financial High-Frequency Data*. Springer-Verlag, Berlin, Heidelberg.

Heinen, A. (2003). Modelling time series count data: an autoregressive conditional Poisson model. Available at SSRN: https://ssrn.com/abstract=1117187 or http://dx.doi.org/10.2139/ssrn.1117187.

Jung, R. C. and Tremayne, A. R. (2011). Useful models for time series of counts or simply wrong ones? *Advances in Statistical Analysis* **95**: 59–91.

Jung, R. C., Liesenfeld, R., and Richard, J. F. (2011). Dynamic factor models for multivariate count data: an application to stock-market trading activity. *Journal of Business & Economic Statistics* **29**: 73–85.

Kowal, D. R., Matteson, D. S., and Ruppert, D. (in press). Functional autoregression for sparsely sampled data. *Journal of Business & Economic Statistics*.

Liu, X., Xiao, H., and Chen, R. (2016). Convolutional autoregressive models for functional time series. *Journal of Econometrics* **194**: 263–282.

McCullagh, P. and Nelder, J. (1989). *Generalized Linear Models*. Chapman and Hall, London.

Meyn, S. and Tweedie, R. (2009). *Markov Chains and Stochastic Stability*, 2nd edition. Cambridge University Press, Cambridge.

Nelson, D. B. (1991). Conditional heteroscedasticity in asset returns: a new approach. *Econometrika* 59: 347–370.

Ramsay, J. B. (1969). Tests for specification errors in classical linear least squares regression analysis. *Journal of the Royal Statistical Society Series B* **31**: 350–371.

Ruppert, D. and Matteson, D. S. (2015). *Statistics and Data Analysis for Financial Engineering with R examples*, 2nd edition. Springer, New York.

Shang, H. L. and Hyndman, R. J. (2011). *FDS: Functional Data Sets Package in R*. R Development Core Team, Vienna.

Tsay, R. S. (2010). *Analysis of Financial Time Series*, 3rd edition. John Wiley, Hoboken, NJ.

Tsay, R. S. (2014a). Dynamic models for multivariate time series of count data. *Proceedings of the Business & Economics Section*. American Statistical Association.

Tsay, R. S. (2014b). *Multivariate Time Series Analysis with R and Financial Applications*. John Wiley, Hoboken, NJ.

Tsay, R. S. (2016). Some methods for analyzing big dependent data. *Journal of Business & Economic Statistics* **34**: 673–688.

Weiss, C. (2017). *An Introduction to Discrete-Valued Time Series*. Wiley, Hoboken, NJ.

Woodward, W. A., Gray, H. L., and Elliott, A. C. (2016). *Applied Time Series Analysis with R*, 2nd edition. Chapman & Hall, CRC Press, Boca Raton, FL.

Zakoian, J. M. (1994). Threshold heteroscedastic models. *Journal of Economic Dynamics and Control* **18**: 931–955.

Zheng, T., Xiao, H., and Chen, R. (2015). Generalized ARMA models with martingale difference errors. *Journal of Econometrics* **189**: 492–506.

CHAPTER 6

STATE SPACE MODELS

State space models (SSMs) provide a general framework for studying stochastic processes, especially when the data might be contaminated by measurement errors. Empirical time series often consist of indirect observations of an unobserved dynamic process that is of interest to the investigators. For example, the observed location of an airplane on a radar screen is a noisy version of the true location of the plane, and is only part of the complex dynamics of the airplane's movement, which includes the speed and acceleration of the plane in addition to its location. Cellphone signals are a modulated version of a sequence of digital signals sent through yet another dynamic process, the time-varying communication channel, that is to be decoded. Observed economic indicators such as the gross domestic product (GDP), unemployment rate, and non-farm payroll are often noisy and indirect observations of the true underlying economic processes of an economy. The series of equity option prices at different strike prices and expiration dates reflects indirectly the dynamics of volatility and risk neutral density of the asset. There are

Nonlinear Time Series Analysis, First Edition. Ruey S. Tsay and Rong Chen.

Companion website: www.wiley.com/go/tsay/nonlineartimeseries

many more. We provide later a list of real examples to demonstrate the applications of SSMs in Section 6.2. As a matter of fact, many of the models discussed in the previous chapters are special cases of SSMs. In Section 6.3, we focus on linear Gaussian models and briefly introduce the Kalman filter and Kalman smoothing. Four examples are given to demonstrate the applications of the linear Gaussian state space model.

6.1 A GENERAL MODEL AND STATISTICAL INFERENCE

A *generalized SSM* is in the form

$$\text{State equation:} \qquad x_t = h_t(\boldsymbol{x}_{t-1}, \varepsilon_t) \quad \text{or} \quad x_t \sim q_t(\cdot \mid \boldsymbol{x}_{t-1}) \tag{6.1}$$

$$\text{Observation equation:} \qquad y_t = g_t(\boldsymbol{x}_t, e_t) \quad \text{or} \quad y_t \sim f_t(\cdot \mid \boldsymbol{x}_t), \tag{6.2}$$

where x_t is the (unobservable) state variable, y_t is the observation, and $\boldsymbol{x}_j = (x_1, \ldots, x_j)'$ denotes the entire history of the state x_t before and at time j, a positive integer. Let $\boldsymbol{y}_t = (y_1, \ldots, y_t)'$ denote the entire past sequence of the observations y_t before and at time t. Both the state variable x_t and the observation y_t can be multi-dimensional. The underlying states evolve (uncontrolled) through the function $h_t(\cdot)$ and the state innovation ε_t, with a known distribution. Equivalently, x_t evolves through the conditional distribution $q_t(\cdot)$. The function $h_t(\cdot)$ and the distribution of ε_t (or equivalently the conditional distribution $q_t(\cdot)$) are assumed to be known, with possibly unknown parameters. On the other hand, the information about the underlying states is observed indirectly through y_t via the function $g_t(\cdot)$ and with observational noise e_t. Equivalently, y_t depends on $\boldsymbol{x}_t$ through the conditional distribution $f_t(\cdot)$. Again, the function $g_t(\cdot)$ and the distribution of e_t are assumed to be known, with possibly unknown parameters. It is often assumed that the state innovation ε_t and the observation noise e_t are independent. This independence condition can be relaxed, but care must be exercised in statistical inference when the condition does not hold.

For the general SSM in Equations (6.1) and (6.2), the state transition is not necessarily Markovian and the current state may depend on the entire history of the state trajectory, and in certain cases on the past observations as well. Similarly, the observation at time t may also depend on the entire state trajectory up to time t (inclusive).

A *Markovian SSM* assumes that $h_t(\boldsymbol{x}_{t-1}, \varepsilon_t)$ only depends on x_{t-1} and $g_t(\boldsymbol{x}_t, e_t)$ only depends on x_t, that is,

$$\text{State equation:} \qquad x_t = h_t(x_{t-1}, \varepsilon_t) \quad \text{or} \quad x_t \sim q_t(\cdot \mid x_{t-1}) \tag{6.3}$$

$$\text{Observation equation:} \qquad y_t = g_t(x_t, e_t) \quad \text{or} \quad y_t \sim f_t(\cdot \mid x_t). \tag{6.4}$$

Hence x_t forms a first-order Markov chain, and y_t is directly related to the current state x_t only, though indirectly y_t is related to all the past values of x_t.

Since a Markovian structure creates significant simplification in notation, modeling, and statistical inference, it is the preferred formulation whenever possible. For example, if $h_t(\cdot)$ and $g_t(\cdot)$ in Equations (6.1) and (6.2) only involve a finite number of lagged variables $x_{t-1}, \ldots, x_{t-p}$, then one can expand the dimension of the state variable by letting $x_t^* = (x_t, x_{t-1}, \ldots, x_{t-p+1})$ and reconfigure h_t and g_t so that $x_t^* = h_t^*(x_{t-1}^*, \varepsilon_t)$ and $y_t = g_t^*(x_t^*, e_t)$. Then the system becomes a Markovian SSM. Any input series is implicitly incorporated in the time-varying functions $h_t(\cdot)$ and $g_t(\cdot)$ in the above equations. In the following, unless specifically stated, we shall focus on the Markovian SSM.

Graphically, the following diagram depicts the structure of the Markovian SSM in Equations (6.3) and (6.4)

$$
\begin{array}{ccccccc}
\cdots & & y_t & & y_{t+1} & & \cdots \\
 & & \uparrow & & \uparrow & & \\
\cdots & \longrightarrow & x_t & \rightarrow & x_{t+1} & \longrightarrow & \cdots
\end{array}
$$

At any given time t, statistical inference on the states $x_1, \ldots, x_t$, given the observations $y_1, \ldots, y_t$ up to time t, can be obtained through the posterior distribution

$$
\begin{aligned}
p(x_1, \ldots, x_t \mid \boldsymbol{y}_t) &\propto p(x_1, \ldots, x_t, y_1, \ldots, y_t) \\
&\propto \prod_{i=1}^{t} p(y_i \mid x_1, \ldots, x_i, y_1, \ldots, y_{i-1}) p(x_i \mid x_1, \ldots, x_{i-1}, y_1, \ldots, y_{i-1}) \\
&\propto \prod_{i=1}^{t} f_i(y_i \mid x_i) q_i(x_i \mid x_{i-1}).
\end{aligned} \tag{6.5}
$$

In practice, there are generally four main statistical inference objectives associated with such a system:

(i) *Filtering*: Obtain the marginal posterior distribution $p(x_t \mid \boldsymbol{y}_t)$ of the current state x_t given the observations up to time t, and estimate the current state (termed as a concurrent estimation)

$$E[\varphi(x_t) \mid \boldsymbol{y}_t],$$

for some integrable function $\varphi(\cdot)$, given the observations up to time t. Various functions $\varphi(\cdot)$ can be used for different problems, including the (conditional) mean (with $\varphi(s) = s$), higher-order moments (with $\varphi(s) = s^k$), tail probability (with $\phi(s) = I(s > c)$), expected shortfall (with $\varphi(s) = sI(s > c)$), and many others.

(ii) *Prediction*: Obtain the marginal posterior distribution $p(x_{t+1} \mid y_1, \ldots, y_t)$ of the future state x_{t+1} given the observations up to time t and make prediction

$$E[\varphi(x_{t+1}) \mid \boldsymbol{y}_t],$$

for some integrable function $\varphi(\cdot)$.

(iii) *Smoothing*: Obtain the posterior distribution $p(x_1, \ldots, x_{t-1} \mid y_1, \ldots, y_t)$ of the past states $x_1, \ldots, x_{t-1}$ given the observations up to time t. A special case is the delayed estimation problem, making inference on x_{t-d} at time t, given all the available observations $y_1, \ldots, y_t$. Another special case of interest is to estimate the most likely path (state trajectory) $x_1, \ldots, x_t$ that maximizes $p(x_1, \ldots, x_t \mid \boldsymbol{y}_t)$.

(iv) *Likelihood and parameter estimation*: Let θ be the collection of all unknown parameters in the model, including the parameters in the functions $h_t(\cdot)$ and $g_t(\cdot)$ and the parameters associated with the distributions of the state innovation ε_t and the observational noise e_t. The likelihood function, given all observations $\boldsymbol{y}_T = \{y_t, t = 1, \ldots, T\}$ is

$$L(\theta) = p(\boldsymbol{y}_T \mid \theta) = \int p(y_1, \ldots, y_T, x_1, \ldots, x_T \mid \theta) dx_1 \ldots dx_T. \quad (6.6)$$

Here $L(\theta)$ can also be viewed as the normalizing constant of Equation (6.5).

A different formulation is

$$L(\theta) = p(y_1, \ldots, y_T \mid \theta) = \prod_{t=1}^{T} p(y_t \mid \boldsymbol{y}_{t-1}, \theta), \quad (6.7)$$

where

$$\begin{aligned} p(y_t \mid \boldsymbol{y}_{t-1}, \theta) &= \int p(y_t \mid x_t, \boldsymbol{y}_{t-1}, \theta) p(x_t \mid \boldsymbol{y}_{t-1}, \theta) dx_t \\ &= \int p(y_t \mid x_t, \theta) p(x_t \mid \boldsymbol{y}_{t-1}, \theta) dx_t \\ &= \int f_t(y_t \mid x_t, \theta) p(x_t \mid \boldsymbol{y}_{t-1}, \theta) dx_t. \end{aligned}$$

If we can obtain $p(x_t \mid \boldsymbol{y}_{t-1}, \theta)$, the predictive density, recursively and are able to evaluate the integral, then the likelihood function can be evaluated. We shall introduce some approaches to achieve this objective for both the linear Gaussian case and the nonlinear non-Gaussian case.

Maximum likelihood estimation of θ can then be carried out by maximizing Equations (6.6) or (6.7).

In some engineering applications such as target tracking and wireless communication, Objectives (i) and (ii) are often carried out on-line in real time, with a system without unknown parameters. In such applications, it is standard to use iterative procedures that can quickly update the system when a new observation becomes available. A special case of Objective (iii), the delayed estimation of $E[\varphi(x_{t-d}) \mid y_1, \dots, y_t]$, is also often carried out on-line.

Objectives (iii) and (iv) are often encountered in time series applications in which the entire observed sequence $y_1, \dots, y_T$ is available and one is interested in estimating the underlying state $\boldsymbol{x}_T$ and the unknown parameters θ based on all available data. Signal extraction and seasonal adjustment are two such applications.

6.2 SELECTED EXAMPLES

SSMs have been widely used in many scientific areas. In this section we present a small fraction of potential applications of the model. These examples demonstrate, in detail, how SSMs can be used. They also provide various approaches to formulating a specific problem into a Markovian SSM. Here we concentrate on the construction of an SSM for a given problem. Analysis of the models, including filtering, prediction, and estimation, will be discussed in latter sections. Some of the examples considered have been studied in the previous chapters.

6.2.1 Linear Time Series Models

The classical ARIMA model of Box and Jenkins (1976) is widely used in linear time series analysis. It can be represented by a linear and Gaussian SSM. In fact, the SSM formulation is the most convenient and efficient representation of ARIMA models for the maximum likelihood estimation of the parameters. Specifically, a zero-mean ARMA(p, q) model assumes the form

$$y_t = \phi_1 y_{t-1} + \dots + \phi_p y_{t-p} + \varepsilon_t + \theta_1 \varepsilon_{t-1} + \dots + \theta_q \varepsilon_{t-q}, \tag{6.8}$$

where $\varepsilon_t \sim_{iid} N(0, \sigma^2)$ and p and q are nonnegative integers. Using the back-shift operator B such that $By_t = y_{t-1}$, the model can be rewritten as

$$\phi(B) y_t = \theta(B) \varepsilon_t,$$

where $\phi(B) = 1 - \phi_1 B - \dots - \phi_p B^p$ and $\theta(B) = 1 + \theta_1 B + \dots + \theta_q B^q$ are the AR and MA polynomials, respectively. Let z_t be a process satisfying $\phi(B) z_t = \varepsilon_t$, then $y_t = \phi^{-1}(B)\theta(B)\varepsilon_t = \theta(B) z_t$. If we treat $x_t = (z_t, \dots, z_{t-p+1})'$ as the unobserved state variable, and assume that $q < p$, then the ARMA model (6.8) can be rewritten as an SSM:

$$\begin{aligned} \text{State equation} \quad x_t &= \boldsymbol{H} x_{t-1} + \boldsymbol{W} \varepsilon_t, \\ \text{Observation equation} \quad y_t &= \boldsymbol{G} x_t, \end{aligned}$$

where

$$\boldsymbol{H} = \begin{bmatrix} \phi_1 & \phi_2 & \cdots & \phi_{p-1} & \phi_p \\ 1 & 0 & \cdots & 0 & 0 \\ 0 & 1 & \cdots & 0 & 0 \\ \vdots & \vdots & \ddots & \vdots & \vdots \\ 0 & 0 & \cdots & 1 & 0 \end{bmatrix}, \tag{6.9}$$

and $\boldsymbol{W} = (1, 0, \ldots, 0)'$ and $\boldsymbol{G} = (1, \theta_1, \theta_2, \ldots, \theta_{p-1})$ are p-dimensional vectors with $\theta_j = 0$ for $j > q$, and $\varepsilon_t \sim N(0, \sigma^2)$. In this setting there is no observation noise. Occasionally this may cause certain instability in the inference procedures (filtering and smoothing in Sections 6.3.1 and 6.3.3) since zero observational noise removes a guarantee of the invertibility of a matrix that needs to be inverted in (6.29).

Shumway and Stoffer (2006) proposed an alternative representation. Assume $p \geq q$ with $\theta_j = 0$ for $j > q$. They rewrite model (6.8) in the following form.

$$\begin{aligned} \text{State equation} \quad x_{t+1} &= \boldsymbol{H} x_t + \boldsymbol{W} \varepsilon_t, \\ \text{Observation equation} \quad y_t &= \boldsymbol{G} x_t + \boldsymbol{V} \varepsilon_t, \end{aligned}$$

where

$$\boldsymbol{H} = \begin{bmatrix} \phi_1 & 1 & 0 & \cdots & 0 \\ \phi_2 & 0 & 1 & \cdots & 0 \\ \vdots & \vdots & \vdots & \ddots & \vdots \\ \phi_{p-1} & 0 & 0 & \cdots & 1 \\ \phi_p & 0 & 0 & \cdots & 0 \end{bmatrix}, \quad \boldsymbol{W} = \begin{bmatrix} \theta_1 + \phi_1 \\ \vdots \\ \theta_q + \phi_q \\ \phi_{q+1} \\ \vdots \\ \phi_p \end{bmatrix},$$

and $\boldsymbol{G} = (1, 0, \ldots, 0)$ and $\boldsymbol{V} = 1$. Note that in this representation, the innovation ε_t appears in both the state and observation equations. Additional care is needed in its inference procedures. Also, in this format the elements of the state vector x_t can be obtained iteratively as

$$x_{i,t} = \sum_{j=i}^{p} \phi_j y_{t+i-1-j} + \sum_{j=i}^{p} \theta_j \epsilon_{t+i-1-j}, \quad i = 1, \ldots, p. \tag{6.10}$$

There are other ways to represent an ARMA model as an SSM, including those in Harvey (1993), Aoki (2013), and Akaike (1975). See Tsay (2010, Chapter 11) for further details.

6.2.2 Time Series With Observational Noises

Traditional time series analysis often assumes that the process is observed accurately. However, in practice this is seldom the case. For example, all economic indicators are observed with errors due to imperfect sampling and survey, and all physical measurements, such as level of river flow, pollution indexes and others, are also observed with errors due to the limited precision of the measuring instruments and the random noise in the environment. One should then treat the underlying true process as the unobserved state, and the observations are its noisy realizations. A linear AR(p) process z_t with observational noises can be written as an SSM:

$$\begin{aligned}\text{State equation} \quad x_t &= \boldsymbol{H}x_{t-1} + \boldsymbol{W}\varepsilon_t,\\ \text{Observation equation} \quad y_t &= \boldsymbol{G}x_t + e_t,\end{aligned}$$

where the state variable is $x_t = (z_t, z_{t-1}, \ldots, z_{t-p+1})'$, $\boldsymbol{H}$ is given in (6.9), $\boldsymbol{W}' = \boldsymbol{G} = (1, 0, \ldots, 0)$, and e_t denotes the measurement error. Here z_t follows an AR(p) model and $y_t = z_t + e_t$ is its noisy observation.

In fact, most of the parametric time series models with observed noises can be written as SSMs. For example, an exponential AR model of order 1 with observational noises can be written as an SSM:

$$\begin{aligned}x_t &= \exp\{a + b|x_{t-1} - c|\}x_{t-1} + \varepsilon_t,\\ y_t &= x_t + e_t.\end{aligned}$$

See Haggan and Ozaki (1981) and Tong (1990) for exponential AR models.

A special SSM that has been widely used is the *local level structural model.* This model assumes that the underlying state follows a simple random walk and the observation is a noisy version of the state. Specifically,

$$\begin{aligned}\text{State equation} \quad x_t &= x_{t-1} + \varepsilon_t,\\ \text{Observation equation} \quad y_t &= x_t + e_t.\end{aligned}$$

One can also include (random) acceleration terms to create a local trend component. For example,

$$\begin{aligned}\text{State equation} \quad c_t &= c_{t-1} + u_t + \varepsilon_{1t},\\ u_t &= u_{t-1} + \varepsilon_{2t},\\ \text{Observation equation} \quad y_t &= c_t + e_t.\end{aligned}$$

Here $x_t = (c_t, u_t)$ is the state variable.

The model can be further extended to include seasonal effects (*local trend and seasonal model*) in the form of

$$c_t = c_{t-1} + \varepsilon_{1t}; \quad u_t = u_{t-1} + \varepsilon_{2t}; \quad s_t = s_{t-d} + \varepsilon_{3t},$$

with the observation $y_t = c_t + u_t + s_t + e_t$. Here d denotes the seasonality (or seasonal period) of y_t, e.g. 12 for monthly data and 4 for quarterly series. The three noise series ε_{it} are often assumed to be mutually independent. To write this seasonal model into a Markovian state space representation, let the state variable be $x_t = (c_t, u_t, s_t, \ldots, s_{t-(d-1)})$. Then we have

$$\begin{aligned} \text{State equation} \quad & x_t = \boldsymbol{H}x_{t-1} + \boldsymbol{W}w_t, \\ \text{Observation equation} \quad & y_t = \boldsymbol{G}x_t + \boldsymbol{V}v_t, \end{aligned}$$

where $\boldsymbol{H}$ and $\boldsymbol{W}$ are $(d+2) \times (d+2)$ and $(d+2) \times 3$ matrices given by

$$\boldsymbol{H} = \begin{bmatrix} 1 & 0 & 0 & 0 & 0 & \cdots & 0 & 0 \\ 0 & 1 & 0 & 0 & 0 & \cdots & 0 & 0 \\ 0 & 0 & 0 & 0 & 0 & \cdots & 0 & 1 \\ 0 & 0 & 1 & 0 & 0 & \cdots & 0 & 0 \\ 0 & 0 & 0 & 1 & 0 & \cdots & 0 & 0 \\ \vdots & \vdots & \vdots & \vdots & \vdots & \ddots & \vdots & \vdots \\ 0 & 0 & 0 & 0 & 0 & \cdots & 1 & 0 \end{bmatrix}, \quad \boldsymbol{W} = \begin{bmatrix} \sigma_1 & 0 & 0 \\ 0 & \sigma_2 & 0 \\ 0 & 0 & \sigma_3 \\ 0 & 0 & 0 \\ \vdots & \vdots & \vdots \\ 0 & 0 & 0 \\ 0 & 0 & 0 \end{bmatrix},$$

$\boldsymbol{G} = [1, 1, 1, 0, \ldots, 0]$, and $\boldsymbol{V} = \sigma_e$. The noises are $w_t \sim N(0, \boldsymbol{I}_{3\times3})$ and $v_t \sim N(0, 1)$. The noises are assumed to be independent.

Such local structural models have been widely used in analyzing economic time series. See Harvey (1989), West and Harrison (1989), and Durbin and Koopman (2001). Tsay (2005, Chapter 11) used the local trend model to analyze realized volatility (the observation) while assuming that the underlying true volatility follows a random walk.

6.2.3 Time-varying Coefficient Models

In addition to the time series y_t, suppose that there exists a set of observed predictors $\{u_{1t}, \ldots, u_{dt}\}$. Assume that y_t depends on the predictors linearly but with time-varying coefficients. See, for example, the time-varying coefficient models in Chapter 2. In this case, a useful approach is to employ an SSM and treat the coefficients as the underlying state, with a certain structure. For example, if

$$y_t = \sum_{i=1}^{d} \beta_{it} u_{it} + e_t, \quad \beta_{it} = \beta_{i,t-1} + \epsilon_{it}, \quad i = 1, \ldots, d,$$

then we have

$$\begin{aligned} \text{State equation} \quad & x_t = x_{t-1} + \varepsilon_t, \quad i = 1, \ldots, d, \\ \text{Observation equation} \quad & y_t = \boldsymbol{G}_t x_t + e_t, \end{aligned}$$

where $G_t = (u_{1t}, \ldots, u_{dt})$ and $x_t = (\beta_{1t}, \ldots, \beta_{dt})'$. The state equation can be extended to include other time series models if needed.

The time-varying coefficient model is flexible and widely applicable. For example, it has been used to model the time-varying behavior of the capital asset pricing model (CAPM). The CAPM of Sharpe (1964) and Fama (1970) is one of the fundamental models of asset pricing in finance. It assumes that

$$y_t = \alpha + \beta M_t + e_t,$$

where y_t is the excess return (over the risk-free interest rate) of an asset (stock, stock portfolio, or other type of assets) and M_t is the excess return of the "market", often represented by a stock index (such as the S&P 500 index). Since the characteristics of the underlying asset change over time, the model can be extended to allow the coefficients (α, β) to evolve slowly, resulting in the following SSM:

$$\begin{aligned} \text{State equation} \quad \alpha_t &= \alpha_{t-1} + \varepsilon_{0,t}, \\ \beta_t &= \beta_{t-1} + \varepsilon_{1,t}, \\ \text{Observation equation} \quad y_t &= \alpha_t + \beta_t M_t + e_t, \end{aligned}$$

where the state variable $x_t = (\alpha_t, \beta_t)'$.

The model has also been extended to include the Fama–French factor model,

$$\begin{aligned} \text{State equation} \quad x_t &= x_{t-1} + \varepsilon_t \\ \text{Observation equation} \quad y_t &= F_t x_t + e_t, \end{aligned}$$

where F_t is a $1 \times k$ matrix consisting of the Fama–French factors (such as market, size factor, book-to-equity ratio factor, and momentum factor). See Fama and French (1993, 1996). A variety of the time-varying coefficient CAPM and Fama–French factor models have been studied by Adrian and Franzoni (2009), Faff et al. (2000), and Gonzalez-Rivera (1997). We show an example of building and estimating such a model in Section 6.3.6.

Note that if one includes the lagged variables y_{t-j} $(j > 0)$ as part of the covariates series $(u_{1,t}, \ldots, u_{d,t})$, the model becomes the time-varying autoregressive model discussed in Chapter 2.

6.2.4 Target Tracking

Designing a sophisticated target tracking algorithm is an important task for both civilian and military surveillance systems, particularly when a radar, sonar, or optical, chemical, or nuclear material sensor is operated in a cluttered and noisy environment, or sensors produce unconventional observations, or innovations are non-Gaussian. See Bar-Shalom and Fortmann (1988). SSMs are often used for tracking problems since the characteristics of the target (location, speed, motion status, etc.) are often unobserved and change when the target moves. The motion characteristics can be summarized as the state variable and a motion model is often available as the state equation in an SSM. Observations from sensors are then linked to the underlying state in an observation equation, under certain assumptions on the sensors and observational noises. We briefly describe some of the tracking applications.

1. Single target in a clean environment: Consider a single target moving in a straight line with random (Gaussian) acceleration. One observes only the location of the target with a Gaussian observational error $e_t \sim N(0, r^2)$. The unobserved state variable is $x_t = (d_t, s_t)'$, where d_t is the true location (or distance to a fixed point) at time t and s_t is the speed at time t. Let ΔT be the time duration between two consecutive observations and we assume that the target maintains a random but constant acceleration between two consecutive observations, i.e. $a_t = w_t/\Delta T$, where w_t is the total acceleration within the period and is assumed to be $N(0, q^2)$. Hence the motion model consists of

$$\begin{aligned} d_t &= d_{t-1} + s_{t-1}\Delta T + 0.5\Delta T w_t, \quad w_t \sim N(0, q^2), \\ s_t &= s_{t-1} + w_t. \end{aligned}$$

Using $x_t = (d_t, s_t)'$ as the state variable, the system can be written as

$$\begin{aligned} \text{State equation} \quad & x_t = \boldsymbol{H}x_{t-1} + \boldsymbol{W}\varepsilon_t \\ \text{Observation equation} \quad & y_t = \boldsymbol{G}x_t + e_t, \end{aligned}$$

where

$$\boldsymbol{H} = \begin{bmatrix} 1 & \Delta T \\ 0 & 1 \end{bmatrix}, \quad \boldsymbol{W} = \begin{bmatrix} 0.5\Delta Tq \\ 1 \end{bmatrix},$$

and $\boldsymbol{G} = [1, 0]$, $\varepsilon_t \sim N(0, 1)$ and $e_t \sim N(0, r^2)$. This tracking system is linear and Gaussian. The objective of tracking is to estimate the true location and speed x_t based on all the observations up to time t, that is, we are interested in $p(x_t \mid y_1, \dots, y_t)$.

The model can easily be extended to track a target moving in a three-dimensional space such as an airplane or missile based on observations received from a radar system by expanding the dimensions.

2. Tracking a one-dimensional target in clutter: Here we continue with the prior one-dimensional target tracking example. Suppose at time t we observe total m_t signals $\boldsymbol{y}_t = (y_{t1}, \dots, y_{tm_t})$, where

$$m_t \sim \text{Bernoulli}(p_d) + \text{Poisson}(\lambda)$$

Among the m_t signals, the true signal $d_t + e_t$ has probability p_d to be observed and the rest are all false signals, such as deceiving objects, electro-magnetic interferences, and others. These false signals are distributed uniformly and independently in the detection region Ω. To set up an SSM for the problem, we include an indicator I_t in the state variable. This indicator shows which of the signals is the true signal $d_t + e_t$, that is, $I_t = i$ if the ith observation is the true signal and $I_t = 0$ if the true signal is not received at time t. Hence the state variable becomes $x_t = (d_t, s_t, I_t)$. We have:
State equations (motion model)

$$\begin{aligned} d_t &= d_{t-1} + s_{t-1}\Delta T + 0.5\Delta T w_t, \\ s_t &= s_{t-1} + w_t, \\ P(I_t = 0) &= 1 - p_d, \quad \text{and} \quad P(I_t = i) = p_d/m_t, \quad i = 1, \ldots, m_t. \end{aligned}$$

Observation equations

$$\begin{aligned} y_{ti} &= d_t + e_t, && \text{if } i = I_t, \\ y_{ti} &\sim \text{Unif}[\Omega] && \text{if } i \neq I_t. \end{aligned}$$

Figure 6.1 shows the observed signals through time, from a simulation of the above system with $\Delta T = 1, p_d = 0.95$, $\text{Var}(w_t) = 0.1^2$, $\text{Var}(e_t) = 0.5^2$, and a fixed number signals $m_t = 50$ in the detection range $[-80, 80]$. The horizontal axis is the time. Each column of dots represents the observed signals at time t on a one-dimensional line (in the y axis).

3. Tracking a maneuvering target in a clean environment: Consider a single target moving in a space with random (Gaussian) acceleration and unknown maneuvering. For simplicity, we discuss the case where the target is moving in a one-dimensional space. Extension to the three-dimensional setting is trivial.

If one assumes that there are three possible modes of maneuvering (fast, slow, and no maneuvering), each corresponding to an additional random acceleration with large, small, and zero variances, respectively, then we can build the following SSM using a maneuvering indicator $I_t \in \{2, 1, 0\}$. Let u_t be the maneuvering acceleration, $u_t \sim N(0, \sigma_i^2)$ if $I_t = i, i = 0, 1, 2$, where $\sigma_2 > \sigma_1$ and $\sigma_0 = 0$. Let the state variable be $X_t = (d_t, s_t, I_t) \equiv (x_t', I_t)$, where d_t and s_t are the one-dimensional location and speed, respectively. Let y_t be the one-dimensional location observation with noise. Assuming I_t follows a Markov chain with transition probability matrix $Q(i, j)$, we have

$$\begin{aligned} x_t &= \boldsymbol{H} x_{t-1} + \boldsymbol{W}_{I_t} \epsilon_t \\ P(I_t \mid I_{t-1}) &= Q(I_{t-1}, I_t) \\ y_t &= \boldsymbol{G} x_t + e_t, \end{aligned}$$

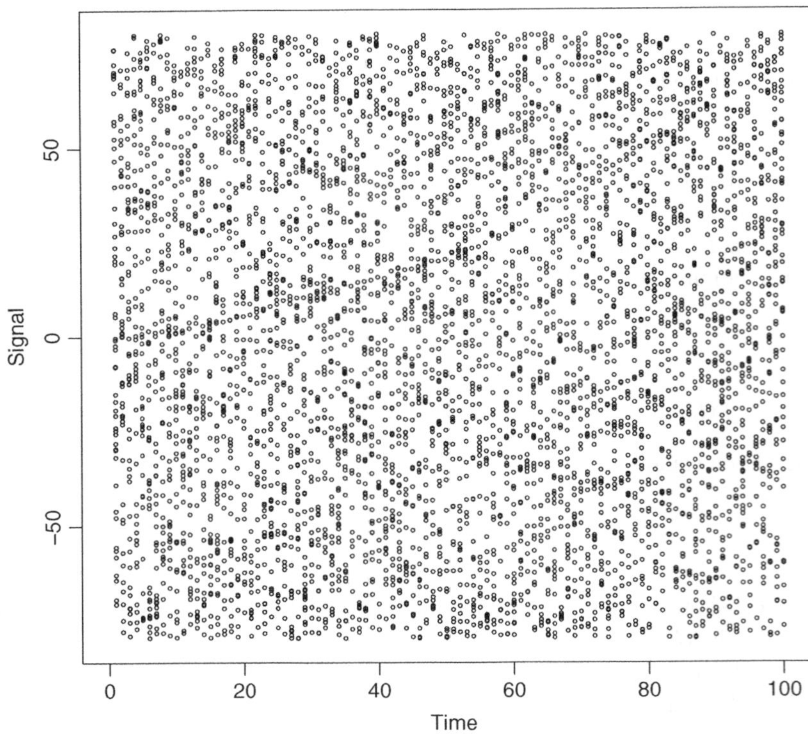

Figure 6.1 Observations of single target in a cluttered environment.

where $\boldsymbol{H}$ and $\boldsymbol{G}$ are defined as before in Case 1,

$$\boldsymbol{W}_i = \begin{bmatrix} 0.5\Delta T\sqrt{q^2+\sigma_i^2} \\ \sqrt{q^2+\sigma_i^2} \end{bmatrix},$$

for $i = 0, 1, 2$, $\epsilon_t \sim N(0, 1)$ and $e_t \sim N(0, r^2)$.

4. Passive sonar: Tracking an airplane with a radar or tracking a ship or submarine with a sonar is based on emitting a radio or sound signal and receiving its reflection from the target. By timing the round trip of the signal, the distance of the target can be measured. However, the emitted signals can be detected by the target being tracked, hence revealing the existence of the tracking unit. For concealment of a submarine but maintaining the tracking capability, passive sonar is often used. Without emitting any sound wave, a passive sonar quietly detects any sound signals generated by the motion of the target ship or submarine. However, without emitting an active signal, there is no reflection, hence the distance of the source cannot be directly observed. In this case, only the direction (or bearing) of the target (related to the detector) is observed, with error. With multiple detectors (e.g., multiple ships equipped with passive sonar devices and with constant communications, or multiple passive sonar devices installed on different locations on a single ship, or if the detector is moving (and hopefully without being detected by the target), the location of the target can be tracked, similar to the triangulation algorithm. However, such a problem becomes highly nonlinear.

Suppose the target is moving in a two-dimensional plane and there are two stationary detectors located on the same plane at $\eta_i = (\eta_{i1}, \eta_{i2})$ $(i = 1, 2)$ corresponding to a Cartesian coordinate. At each time t the observations consist of two angles (bearings) ϕ_{it} $(i = 1, 2)$ of the target related to the detectors, with noise. Suppose the target is moving with random acceleration as previously discussed. The state variable is then $x_t = (d_{1t}, d_{2t}, s_{1t}, s_{2t})'$, with the same state equations (in two dimensions) as above. The observation equation can be written as

$$\phi_{it} = \text{arc tan}\left(\frac{d_{2t} - \eta_{i2}}{d_{1t} - \eta_{i1}}\right) + e_{it} \quad \text{for} \quad i = 1, 2.$$

We restrict ϕ_{it} in $[-\pi/2, \pi/2]$.

Figure 6.2 depicts the observations in relation to the detectors and the target. This is a highly nonlinear SSM and has been studied in the literature by Kronhamn (1998), Arulampalam et al. (2004), and Peach (1995) among many others. We show an implementation of the tracking with bearing-only signals in Example 7.1.

5. Mobile network for nuclear material surveillance: In Grelaud et al. (2018), a mobile sensor network is proposed to detect nuclear material in large cities to prevent terrorist attacks using dirty bombs or portable nuclear weapons. A mobile sensor network is formed by installing inexpensive nuclear sensors on

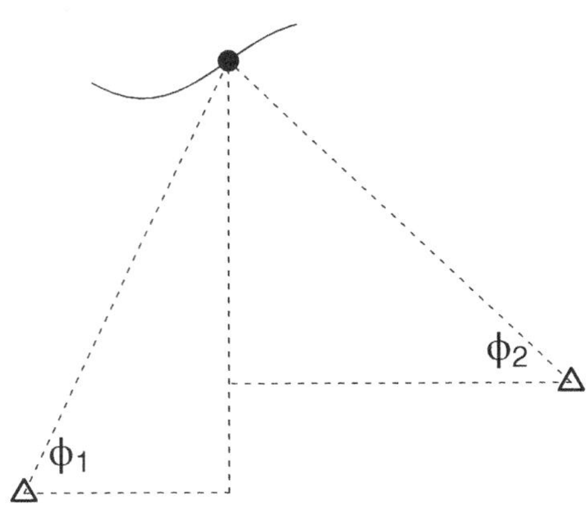

Figure 6.2 Illustration of passive sonar observation.

many constantly moving taxi cabs and police vehicles equipped with a global positioning system (GPS). The sensors would send binary signals (positive or negative) to a command center for on-line surveillance and detection in real time. This is of relatively lower cost than a fixed sensor network because inexpensive and less accurate sensors can be used due to the mobility of the sensors. It is also difficult to be tempered, while a fixed sensor network can easily be tempered to create blind spots in the network.

The problem is slightly different from the standard tracking problem due to the large number of binary signals received, with both false positive and false negative errors. Also the motion model needs to be adjusted since the target is moving within a road system in a large city.

Suppose a sensor has a detection range r for a specific source (target). This range is an unknown parameter as it depends on the source energy. Given r, denote S as the distance indicator, $S = 1$ if $d < r$ and $S = 0$, otherwise. This serves as the true signal if a perfect sensor is a distance d away from the source.

We assume that the sensors have detection errors, with sensor false-positive rate $\gamma = P(D = 1 \mid S = 0)$ and false-negative rate $\delta = P(D = 0 \mid S = 1)$, where D is the binary signal the sensor sends to the control center. Let $x_t = (\eta_t, s_t)'$ be the state variable that consists of the motion characteristics of the target (location and speed). The state equation is then a motion model for the source, possibly based on a city map. The observations at time t include the locations and the emitted signals of the n mobile sensors deployed at time t, $\{(y_{it}, D_{it}), i = 1, \ldots, n\}$. Let $d_{it} = ||y_{it} - \eta_t||$ be the distance between the ith mobile sensor and the current source location η_t and let $S_{it} = I(d_{it} < r)$. Then the observation equation is the likelihood

$$p(y_{1t}, \ldots, y_{nt}, D_{1t}, \ldots, D_{nt} \mid s_t) = (1-\delta)^{n_{1t}}\delta^{n_{2t}}\gamma^{n_{3t}}(1-\gamma)^{n_{4t}},$$

where $n_{1t} = \sum_{i=1}^n D_{it}S_{it}$ is the number of true positives, $n_{2t} = \sum_{i=1}^n (1 - D_{it})(1 - S_{it})$ is the number of true negatives, $n_{3t} = \sum_{i=1}^n D_{it}(1 - S_{it})$ is the number of false positives, and $n_{4t} = \sum_{i=1}^n (1 - D_{it})S_{it}$ is the number of false negatives, given the source location is at η_t. They can be calculated based on the observations and x_t.

Figure 6.3 depicts a set of observations, assuming all vehicles are on the streets, mapped by the grid. The shaded area is the detection area surrounding the target. The open circles are the locations of the sensors sending negative signals (those in the shaded area are false-negative signals). The black circles are the locations of the sensors sending positive signals (those outside the shaded area are false-positive signals).

Other tracking problems: Other tracking problems formulated as SSMs includes tracking multiple targets in Shumway and Stoffer (1991), Avitzour (1995), Hue et al. (2002), and Orton and Fitzgerald (2002); tracking and discriminating multiple targets in Srivastava et al. (2001); two-dimensional tracking (tanks, cars) with

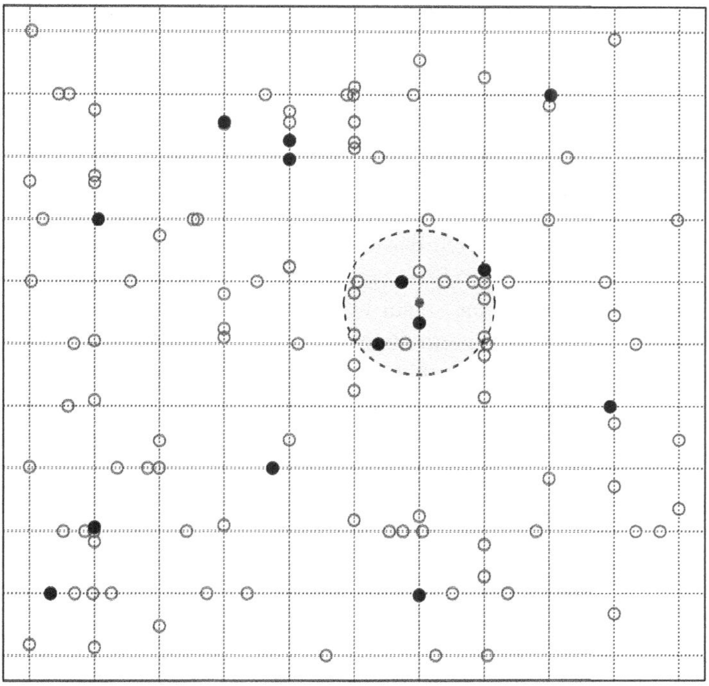

Figure 6.3 Illustration of mobile sensor observations.

radar in Irwin et al. (2002); two-dimensional tracking (cars, cellular phones) in a cellular network, with or without the assistance of a map in Liu et al. (1998); tracking and guidance of Salmond and Gordon (2001); two-dimensional tracking with GPS in Gustaffson et al. (2002); tracking with passive sonar in Paris and Jauffret (2004); tracking with a sequence of images (computer vision) in Isard and Blake (1996), Blake et al. (1998, 2001), Torma and Szepesvari (2003), Bruno et al. (2007), and Bruno and Moura (2000); audio tracking in Nickel et al. (2007) and Itohara et al (2011); mobile robot localization in Dellaert et al. (1999) and Fox et al. (1999, 2001). Tutorials and overviews of applications of SSM in target tracking can be found in Arulamapalam et al. (2002), Ristic et al. (2004), and Haug (2012).

6.2.5 Signal Processing in Communications

Signal processing is one of the major scientific areas in which SSM is commonly used. A sequence of information signals (along with other information) sent from a sender must be extracted by the receiver from a sequence of observations that have been coded, modulated, and transmitted through the transmission media and observed with noises. There is a vast amount of literature on using SSMs for signal processing and digital communications, including the books by Chitode (2009), Proakis (2001), and Kung (1988). An overview of applications of SSMs in signal processing can be found in Wang et al. (2002), Djuric et al. (2003), and Djuric (2001). The special issue on Monte Carlo methods for statistical signal processing of *IEEE Transactions on Signal Processing* (Volume 50, 2002) contains many excellent examples.

In this section, we present an example. Many mobile communication channels can be modeled as Rayleigh flat-fading channels, in which the digital signal $d_t \in \{-1, 1\}$ is modulated by a fading coefficient α_t and transmitted. The receiver receives a noisy version of the signal $y_t = \alpha_t d_t + e_t$. The fading coefficient α_t is assumed to follow the Butterworth filter of order $r = 3$, i.e. an ARMA(3, 3) model with roots close to the unit circle (hence slow moving). Based on our discussion of representing an ARMA(p, q) model with an SSM in Section 6.2.1, we can write the ARMA(3,3) model as $x_t = \boldsymbol{H}x_{t-1} + \boldsymbol{W}w_t$ and $\alpha_t = \boldsymbol{G}x_t$. Hence, the whole system can be written as

$$\begin{array}{ll}\text{State equations} & \left\{\begin{array}{lcl} x_t & = & \boldsymbol{H}x_{t-1} + \boldsymbol{W}w_t \\ d_t & \sim & p(\cdot \mid d_{t-1}) \end{array}\right., \\ \text{Observation equation} & y_t = \boldsymbol{G}x_t d_t + \boldsymbol{V}v_t,\end{array}$$

where (x_t', d_t) is the joint state variable, x_t is a four-dimensional complex variable, $d_t \in \{-1, 1\}$, and w_t and v_t are one-dimensional complex random variables. The state innovation w_t is often assumed to follow a complex Gaussian distribution, and the observational noise v_t follows either a Gaussian distribution or a mixture of Gaussian distributions. The signals d_t are either iid or have a Markovian structure. Figure 6.4 shows a graphical illustration of the model. Figure 6.5 shows a simulated example of the fading coefficients α_t on the left and the observed y_t on the right.

The objective then is to extract the digital signals $d_1, \ldots, d_t$ transmitted over such channels, given the observed sequence $y_1, \ldots, y_t$, in real time. Note that the problem suffers the *phase ambiguity* problem. Since

$$p(\boldsymbol{\alpha}_t, \boldsymbol{d}_t \mid \boldsymbol{y}_t) = p(-\boldsymbol{\alpha}_t, -\boldsymbol{d}_t \mid \boldsymbol{y}_t),$$

it is not possible to distinguish $(\boldsymbol{\alpha}_t, \boldsymbol{d}_t)$ from $(-\boldsymbol{\alpha}_t, -\boldsymbol{d}_t)$ given only $\boldsymbol{y}_t$.

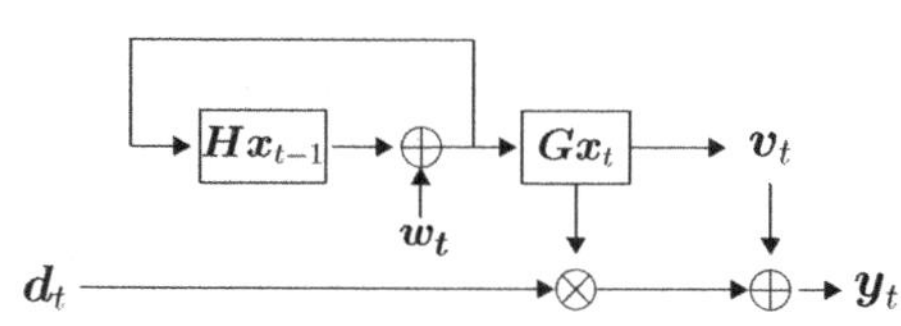

Figure 6.4 Illustration of a fading channel.

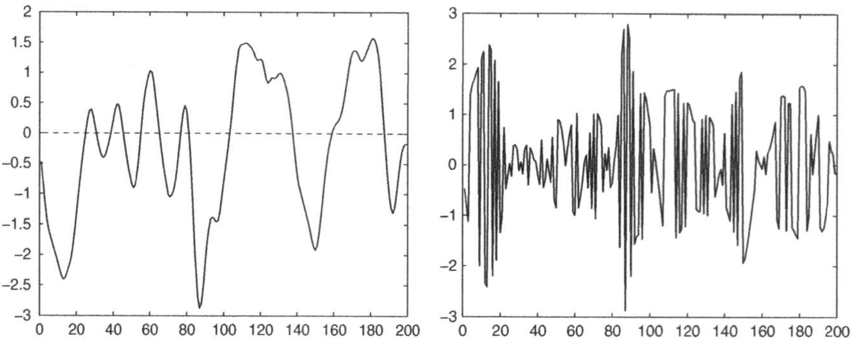

Figure 6.5 Example of fading channel signals.

To resolve the phase ambiguity, differential coding is often utilized. Specifically, if the information sequence is $d_1, \ldots, d_t$, then one transmits the sequence $d_1^*, \ldots, d_t^*$, where $d_{t-1}^* d_t^* = d_t$, $d_0^* = 1$. Hence, on the receiver side, it is only required to estimate $d_t = d_{t-1}^* d_t^*$, which does not have phase ambiguity.

A standard detector is the *differential detector*:

$$\hat{d}_t = sign(y_t y_{t-1}) = sign(\alpha_t \alpha_{t-1} d_t + \alpha_t d_t^* e_{t-1} + \alpha_{t-1} d_{t-1}^* e_t + e_{t-1} e_t).$$

When α_t changes slowly without altering its sign often, $\alpha_t \alpha_{t-1}$ is positive most of the time. When the noise is small, $\hat{d}_t$ should be accurate for most time points. However, whenever α_t changes its sign, the detector tends to make an error, no matter how small the noise e_t is. This results in an error floor. As the signal to noise ratio goes to ∞, the probability of making a wrong detection goes to the frequency at which α_t changes its sign. This frequency is a property of the channel and does not vanish even if the observational noise e_t vanishes. Figure 6.6 shows the bit-error rate of the differential decoder in a simulated example. It flats out as the signal to noise ratio increases. In the figure the known channel bound is the bit-error rate if the fading coefficient α_t is given and known. The "genie-bound" is the bit-error rate using a genie-assisted estimated sequence of fading coefficients $\hat{\alpha}_t$, which is obtained as if a known information sequence d_t^{**} is transmitted through the exact channel, being modulated with the exact sequence α_t, and its corresponding y_t^{**} were received and used to estimate α_t. An objective is to construct a procedure that is able to break the error floor of the differential decoder and hopefully with performance approaching the genie-bound. Chen et al. (2000) and Wang et al. (2002) used an SSM approach for this problem and were able to design a procedure with performance close to the genie-bound, shown in Figure 6.6. The solid line with circles is the bit-error rate using a concurrent estimator: using the received observations up to time t to estimate the transmitted bit d_t at time t. The procedure

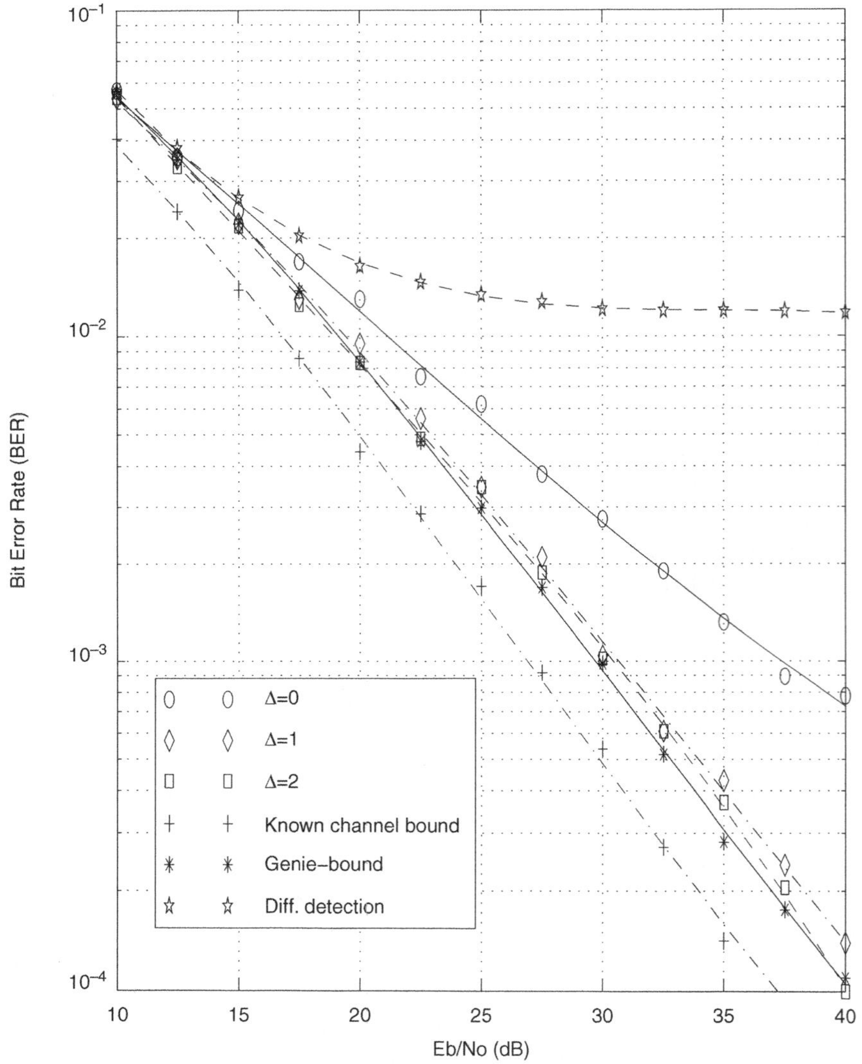

Figure 6.6 Bit-error rate of the illustrative example of fading channel signals.

is able to outperform the differential decoder. The $\Delta = 1$ and $\Delta = 2$ lines are the performance using the same SSM approach, except allowing a slight time delay of 1 and 2 periods, respectively, that is, we estimate d_t at time $t + 1$ (or $t + 2$), with observations up to time $t + 1$ ($t + 2$). The delay method is discussed later in Section 8.3.

6.2.6 Dynamic Factor Models

Let $\{y_{it}\}$, $i = 1, \ldots, N$ and $t = 1, \ldots T$ be a set of N time series with length T. A linear dynamic factor model (DFM) for $\{y_{it}\}$ assumes the form

$$y_{it} = \boldsymbol{\lambda}_{i0}\boldsymbol{\eta}_t + \boldsymbol{\lambda}_{i1}\boldsymbol{\eta}_{t-1} + \cdots + \boldsymbol{\lambda}_{iq}\boldsymbol{\eta}_{t-q} + \varepsilon_{it}, \quad i = 1, \ldots, N, \tag{6.11}$$

$$\boldsymbol{\eta}_t = \sum_{k=1}^{p} \boldsymbol{\Phi}_k \boldsymbol{\eta}_{t-k} + \boldsymbol{e}_t, \tag{6.12}$$

where q is a nonnegative integer and $\boldsymbol{\eta}_t = (\eta_{1t}, \ldots, \eta_{rt})'$ is a r-dimensional time series of unobserved factors satisfying $r \ll N$. The row vector $\boldsymbol{\lambda}_{ij}$ is called factor loading of the ith series on the jth lagged values of the factors. The collection of $\{\boldsymbol{\lambda}_{ij} | i = 1, \ldots, N; j = 1, \ldots, q\}$ is called the loading matrix.

The idiosyncratic noise ε_{it} is assumed to be uncorrelated with the factors $\boldsymbol{\eta}_t$ in all leads and lags. In addition, ε_{it} is also assumed to be uncorrelated with each other in all leads and lags, though it may possess some autocorrelations. Some research allows for weak cross-correlations between ε_{it} values. See Chamberlain (1983) and Chamberlain and Rothschild (1983). Equation (6.12) assumes that the factor follows a r-dimensional linear vector AR(p) model, with coefficient matrices $\{\boldsymbol{\Phi}_k, k = 1, \ldots, p\}$. The noise process $\boldsymbol{e}_t$ is assumed to be a white noise series and uncorrelated with ε_{it}.

To see the model more clearly, consider the simplest DFM:

$$y_{it} = \lambda_i \eta_t + \varepsilon_{it}, \quad i = 1, \ldots, N,$$
$$\eta_t = \phi \eta_{t-1} + e_t.$$

Here all co-movements of the time series y_{it} are driven by the common factor η_t, a latent process following an AR(1) model. The factor loading λ_i provides a link between the common factor and the ith series y_{it}. Clearly, this is an SSM with η_t being the unobserved state variable.

The DFM was first proposed by Geweke (1977). It gained popularity with Sargent and Sims (1977), who showed that two dynamic factors could explain a large fraction of the variance of many important US quarterly macroeconomic variables. In Forni et al. (2002), the DFM is extended to the generalized DFM that allows infinite order dynamics and nonorthogonal idiosyncratic components. Extensive surveys on the theory, methods, and applications of the DFM can be found in Stock and Watson (2006, 2012) and Bai and Ng (2008).

A typical analysis using the DFM starts with the estimation of the common factors (or more precisely the space spanned by the common factors) and the determination of the number of factors r. See Bai and Ng (2002), and Hallin and Liška (2007), among others. The order of the vector AR (VAR) model for the common factors needs to be determined. When the dimension N is small and the noise is assumed to be Gaussian, maximum likelihood estimators, along with a model selection criterion for determining the number of common factors r in (6.11) and the VAR order p in (6.12), can be obtained.

However, this approach encounters some difficulties when the number of time series N becomes large, for there are many parameters to estimate and the relationship between the variables becomes complex. If N is large, principle component analysis (PCA) and its variants are often used to estimate the common factors. The least squares method is then used to estimate the loading matrix and the VAR model. One can then use the SSM representation of the DFM to refine the estimates of common factors by treating the estimated parameters of the second steps as given. See, for instance, Forni et al. (2009) and Doz (2007). We consider an example of estimating the DFM in Example 6.4.

6.2.7 Functional and Distributional Time Series

Analysis of functional time series was discussed in Chapter 5 using nonparametric approaches. In this chapter we consider a different approach via a structural model formulation under the SSM framework. Let D be the domain of a functional time series $y_t(\cdot)$, i.e. $y_t(s) : D \to R$ for every time point t. The time series $\{y_t(\cdot), t = 1, \ldots, \}$ is said *to be driven by a dynamic process* if, for any fixed t, the function $y_t(\cdot)$ can be written as

$$y_t(s) = f_t(s; \theta_t) + \varepsilon_t(s), \quad s \in D, \tag{6.13}$$

where the function $f_t(\cdot)$ is known up to some parameter θ_t, θ_t follows a dynamic process over time, and $\varepsilon_t(s)$ is a noise process defined on D with $E[\varepsilon_t(s)] = 0$ for all $s \in D$. Often it is further assumed that $\varepsilon_{t_1}(\cdot)$ and $\varepsilon_{t_2}(\cdot)$ are independent for $t_1 \neq t_2$, but some dependence between $\varepsilon_t(s_1)$ and $\varepsilon_t(s_2)$ is allowed for $s_1, s_2 \in D$. In model (6.13), the dependence of the process $y_t(\cdot)$ is completely characterized by the parameter process θ_t and the noise process $\varepsilon_t(\cdot)$. The process $\{\theta_t\}$ is called the driving process. This is similar to the *parameter driving models* in modeling non-Gaussian time series of Chapter 5. In most applications, $y_t(\cdot)$ is only observed at a finite number of locations in D. In the following, we assume that for each time t, a set of observations $\{y_t(s_{ti}), i = 1, \ldots, m_t\}$ is available and they satisfy $y_t(s_{ti}) = f_t(s_{ti}, \theta_t) + \varepsilon_t(s_{ti})$.

The model can be written as an SSM:

$$\theta_t = h(\theta_{t-1}, \ldots, \theta_{t-p}, e_t, \ldots, e_{t-q}, \boldsymbol{\beta}), \tag{6.14}$$

$$y_t(s_{ti}) = f(s_{ti}, \theta_t) + \varepsilon_t(s_{ti}), \quad i = 1, \ldots, m_t, \tag{6.15}$$

where $h(\cdot)$ is a known function with unknown parameters $\boldsymbol{\beta}$, p and q are non-negative integers, and e_t is a sequence of scalar or vector white noises. For example, one can use a VAR(p) model for (6.14).

When the functional time series $\{y_t\}$ is driven by a finite dimensional process θ_t, it in fact assumes a hierarchical model with functional observations and dynamic latent processes. This is similar to the DFM, with a nonlinear loading relationship and the driving process as the dynamic factors. The difference is that the loading relationship is completely specified here whereas the loading matrices are unknown in a DFM.

For distributional time series, let $\pi_t(\cdot)$ be a sequence of distributions, indexed by time t. An objective of analyzing such time series is to study the dependence between the distributions and to make predictions of future distributions. Suppose at time t we observe independent data $y_{ti} \sim \pi_t, i = 1, \ldots, m_t$. Similar to the case of functional time series, if the distributions π_t belong to a parametric family $\pi_t(y) = \pi(y; \theta_t)$ and θ_t forms a dynamic process, then the distributional time series $\{\pi_t\}$ is driven by a dynamic process.

For distributional time series, similar to Equations (6.14) and (6.15), we have

$$\begin{aligned} \theta_t &= h(\theta_{t-1}, \ldots, \theta_{t-p}, e_t, \ldots, e_{t-q}, \boldsymbol{\beta}), \\ y_{ti} &\sim \pi(\,\cdot\,; \theta_t), \quad i = 1, \ldots, m_t, \end{aligned} \tag{6.16}$$

where $h(\cdot)$ presents a time series model with unknown parameters $\boldsymbol{\beta}$ and e_t is a scalar or vector white noise series. Distributional time series have many applications. For example, the simplest version of the popular stochastic volatility model in finance of Hull and White (1987) and Jacquier et al. (1994) takes the form $y_t \sim N(0, \sigma_t^2)$, where $\log(\sigma_t^2) = \beta_0 + \beta_1 \log(\sigma_{t-1}^2) + e_t$. This is a restricted case of model (6.16), with one observation per time period. As another example, assume that an underlying time series η_t is observed multiple times with noises, e.g. $y_{ti} \sim N(\eta_t, \sigma_t^2)$. Here we have the flexibility of adding certain structure for σ_t^2 such as that in the stochastic volatility models, or heteroscedasticity in the form $\sigma_t^2 = \sigma^2\eta_t^2$.

In what follows we consider two specific examples.

Example A: Yield curve of interest rates. An important topic in finance is the dynamic behavior of the term structure of interest rates (yield curve) of government bonds. The yield curve is the functional relationship between interest rate and its time to maturity. It plays a crucial role in fixed-income portfolio management, risk management, and many other finance and economic decisions. Modeling yield curves has been studied extensively in finance and economics. See, for instance, Vasicek (1977), Cox et al. (1985), Hull and White (1990), Heath et al. (1992), and Duffie and Kan (1998). However, most of the research papers focus on estimating the yield curve at a fixed time point instead of considering the dynamical evolution of the curve over time. These studies do not address the important issue of prediction, which is important for both derivative pricing and risk

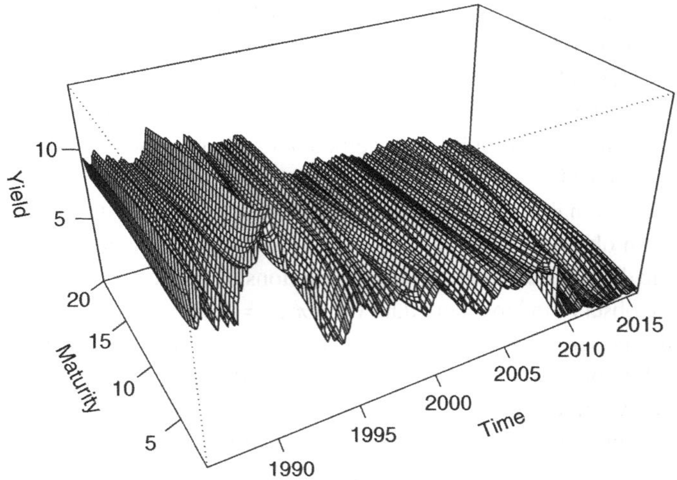

Figure 6.7 UK monthly yield curve.

management. Figure 6.7 shows the smoothed monthly yield curves from January 1985 to December 2015 of the UK nominal Government liability curve, obtained from http://www.bankofengland.co.uk/statistics/Pages/yieldcurve/archive.aspx. Clearly the curves (as a function of maturity) change over time with a nontrivial pattern. Diebold and Li (2006) proposed a dynamic Nelson–Siegel curve to study the dynamic behavior of the yield curves. The yield curve of Nelson and Siegel (1987) assumes the form

$$y_t(s) = \beta_{1t} + \beta_{2t}\left(\frac{1 - e^{-\lambda_t s}}{\lambda_t s}\right) + \beta_{3t}\left(\frac{1 - e^{-\lambda_t s}}{\lambda_t s} - e^{-\lambda_t s}\right) + \varepsilon_t(s),$$

where s is the time to maturity and $y_t(s)$ is the yield at time t for a bond with maturity s. The noise $\varepsilon_t(s)$ is assumed to be independent for different s. The Nelson–Siegel yield curve is a parametric family of functions that is flexible and can adopt various shapes by changing the parameters $(\beta_1, \beta_2, \beta_3)$. Figure 6.8 shows four realizations of the Nelson–Siegel curve. The increasing curve in Figure 6.8(a) is typical, reflecting that under the normal economic conditions the interest rate of a bond should increase with the time to maturity. However, in practice the yield curve can be inverted or humped, reflecting abnormal economic conditions such as recessions or financial crises.

With a fixed $\lambda_t = \lambda$, Diebold and Li (2006) assumed that the time-varying coefficients $(\beta_{0t}, \beta_{1t}, \beta_{2t})$ follow a linear VAR(1) process. Forecasting of a d-period ahead yield curve can then be carried out using the predicted values

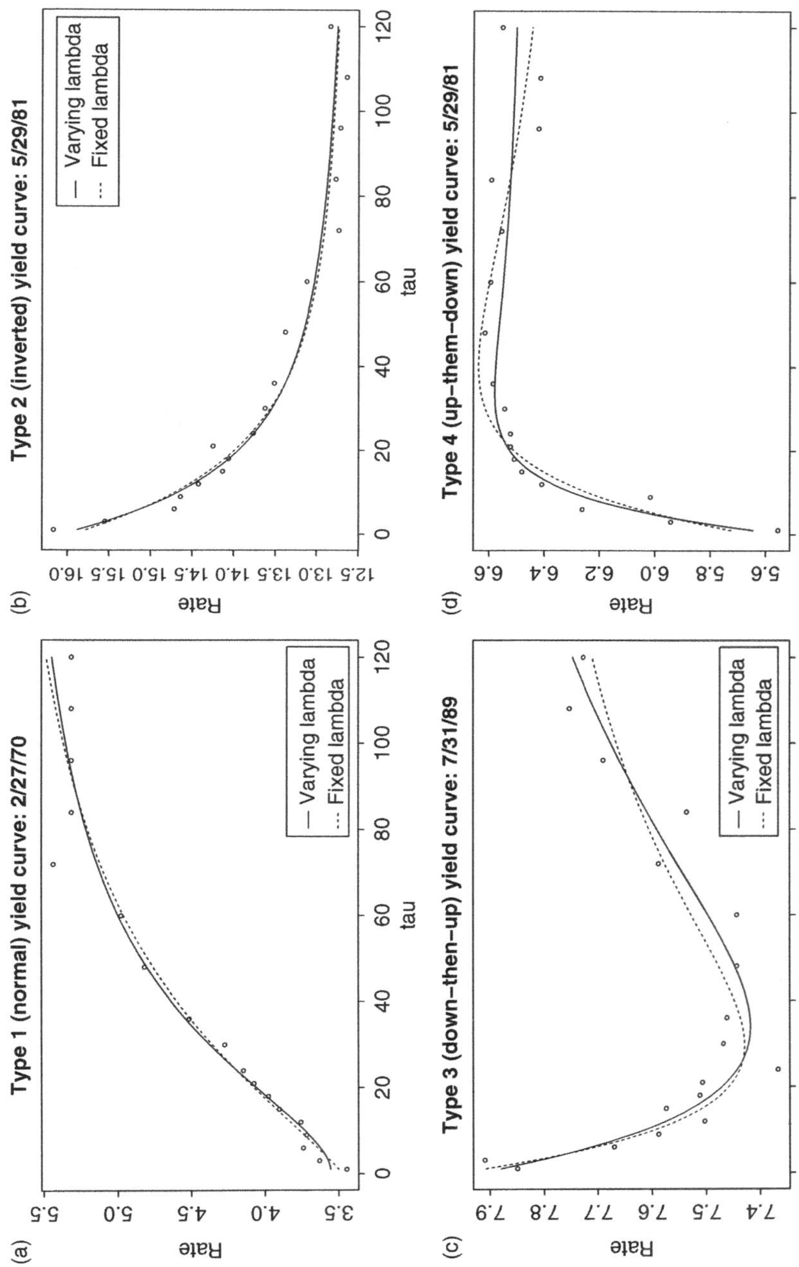

Figure 6.8 Four realizations of the Nelson–Siegel curve.

$(\hat{\beta}_{0(t+d)}, \hat{\beta}_{1(t+d)}, \hat{\beta}_{2(t+d)})$. Their approach falls exactly under the framework of functional time series driven by a finite dimensional process.

When λ_t is fixed, the model is also a time-varying coefficient model of Section 6.2.3. We prefer to treat the yield curve as a function and analyze it accordingly. The functional value at locations other than where data are observed can be estimated from the model. If λ_t varies over time, then the model becomes a nonlinear SSM.

Example B: Cross-sectional stock returns. Expected returns of equity have always been the focus of investment and finance. Since a stock index only reflects the central location (mean) of the entire ensemble of stocks in the market, it provides important but limited information on the financial market. On the other hand, the cross-sectional distribution of all stock returns provides an accurate description of the market. Some studies of cross-sectional return distribution can be found in Lillo and Mantegna (2000a,b), Campbell et al. (2001), Hamilton and Lin (1996), Schwert (1989), and Tsay (2016).

Figure 6.9 shows the empirical density functions of daily log returns of US stocks in 2012 and 2013. The density functions are obtained by Gaussian kernel estimation of all stock returns and the plots are arranged so that the density

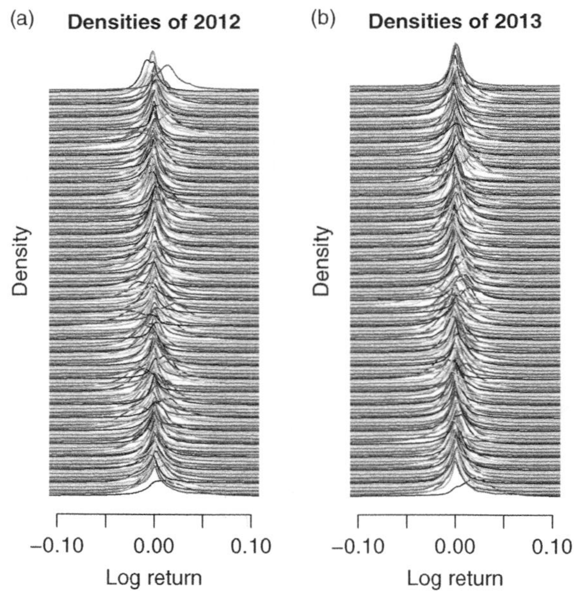

Figure 6.9 Empirical density functions of cross-sectional daily log returns of US stocks in (a) 2012 and (b) 2013. The density function of the first trading day is at the bottom of the plot. The plots are from Tsay (2016).

function of the first trading day of a given year is at the bottom of the plot. For ease in presentation, the density functions are only shown in the range $D = [-0.1, 0.1]$. See Tsay (2016) for further details. From the plots, it is clear that the behavior of cross-sectional stock returns evolves over time and the return distributions can be treated as functional time series. Alternatively, one can model the time evolution of the quantiles of cross-sectional stock returns using distributional time series. Lin and Chen (2017) used distributional time series to study the dynamics of cross-section stock returns. They modeled the cross-sectional distribution by a skewed t-distribution with four parameters $\boldsymbol{\theta}_t = (\mu_t, \sigma_t, \nu_t, \lambda_t)$, the location, scale, skewness, and degrees of freedom parameters, respectively, and established an SSM as in (6.16), where the $\boldsymbol{\theta}_t$ is serving as the dynamic driving process and is assumed to follow a VAR model.

6.2.8 Markov Regime Switching Models

The regime switching mechanism, similar to threshold crossing, provides a powerful tool for modeling nonlinear dynamics. It often serves as a first-step departure from the linearity. The Markov switching autoregressive models of Section 2.3 belong to a larger class of Markovian regime switching models. Model (2.16) can be viewed as an SSM in which S_t is the unobserved state variable, with the state equation specifying the transition probability structure $P(S_t = i \mid S_{t-1} = j)$ for $i, j = 1, 2$. The observation equation specifies the conditional distribution of x_t given the state S_t. When the state variable follows a Markov chain and takes value in a finite set, the corresponding SSM is also referred to as a hidden Markov model (HMM).

The Markov switching autoregressive model in Equation (2.16) can be extended so that the regime switching mechanism is enforced on a general SSM. Specifically, a Markov regime-switching SSM assumes the form

$$
\begin{aligned}
\text{State equation} \quad & x_t = h_{S_t}(x_{t-1}, \varepsilon_t), \\
& P(S_t = i \mid S_{t-1} = j) = p_{ij}, \\
\text{Observation equation} \quad & y_t = g_{S_t}(x_t, e_t),
\end{aligned}
\tag{6.17}
$$

where $S_t \in \{1, \ldots, k\}$ and the functions $h_i(\cdot)$ and $g_i(\cdot)$ belong to a finite family of k functions, where k is the number of regimes. In this case, the state variable is actually (x_t, S_t). We stress the regime indicator S_t for its important interpretation, but also for technical conveniences.

The model in Equation (6.17) is useful in many situations. For example, a Markov switching ARMA model can be written as a Markov regime switching SSM by rewriting the ARMA model in each regime in a state space form, as in Section 6.2.1. Tracking a maneuvering target problem in Section 6.2.4 can be viewed as a Markov regime switching SSM, in which the maneuvering status is treated as the regime. The time-varying coefficient linear models in Section 6.2.3, in which the coefficients may vary under a regime switching mechanism, can be extended.

The varying coefficient Fama–French model can be extended so that the coefficients follow a Markov switching autoregressive (MSA) model, since the coefficients may assume different dynamics when the market conditions change. We can write the model as

$$\begin{aligned}
\text{State equation} \quad & x_t = c_{s_t} + \boldsymbol{H}_{s_t} x_{t-1} + \varepsilon_t, \\
& P(s_t = i \mid s_{t-1} = j) = p_{ij}, \quad i,j = 1,2 \\
\text{Observation equation} \quad & y_t = \boldsymbol{F}_t x_t + e_t,
\end{aligned}$$

where $\boldsymbol{F}_t$ is a $1 \times k$ matrix consisting of the Fama–French factors at time t.

Note that model (6.17) is different from the HMM model in that the state variable is a hybrid of continuous and discrete variables. Inference for such a model is much more challenging than that of the standard HMM model, but the Markov switching SSM has a wider range of applications than the traditional HMM.

6.2.9 Stochastic Volatility Models

Stochastic volatility (SV) models are an alternative to modeling conditional heteroscedasticity, discussed in Section 5.4, and they are often referred to as an extension of the volatility models by allowing the variance (volatility) component of an asset return to vary according to a stochastic diffusion process. See, for instance, Hull and White (1987), Jacquier et al. (1994), Shephard and Pitt (1997), Aït-Sahalia and Jacod (2014), and Fouque et al. (2000, 2011). The models have been considered in both continuous time and discrete time.

In the ARCH and GARCH models of Section 5.4, the volatility changes deterministically, conditional on the past information. An alternative approach is to employ a stochastic process to govern the random propagation of volatility, resulting in a discrete-time stochastic volatility model. See Ghysels et al. (1996) and Shephard (1996).

We discuss a simple SV model here. Let y_t be the observed return of an asset at time t. Pitt and Shephard (1999) and Harvey et al. (1994) assumed that y_t follows a normal distribution with mean zero and variance σ_t^2. The time-varying volatility σ_t is an unobservable state variable, following an AR(1) process. To enforce the positivity of conditional variance, one can use the parametrization $\sigma_t = \exp(x_t)$, and form the following SSM:

$$\begin{aligned}
x_t &= \phi_0 + \phi x_{t-1} + \eta_t, \\
y_t &= \exp(x_t) v_t,
\end{aligned}$$

where $\eta_t \sim N(0, \sigma^2)$, $v_t \sim N(0, 1)$. This is a nonlinear SSM.

Another form is to replace the observation equation by $\ln(y_t^2) = \beta^* + x_t + v_t$ where $v_t = \ln(e_t^2)$, $e_t \sim N(0, 1)$. This is a linear but non-Gaussian SSM.

6.2.10 Non-Gaussian Time Series

Several of the non-Gaussian time series models of Chapter 5 can be casted into a SSM by carefully defining a proper state variable. For example, consider the Poisson count data $y_t \sim Poi(\lambda_t)$. We can define $x_t = \ln(\lambda_t)$ as a state variable and employ an ARMA model for x_t.

As another example, consider a proportional time series $y_t \in (0, 1)$. For instance, y_t may denote the market share of a product at time t. Here we can assume

$$\begin{aligned} x_t &\sim \text{ARMA}(p, q), \\ y_t &\sim Beta(\tau\mu_t, \tau(1 - \mu_t)), \end{aligned}$$

where $x_t = \ln(\mu_t/(1 - \mu_t))$ is the logit transformation of the conditional mean μ_t. Note that this model is different from a direct transformation $z_t = \ln(y_t/(1 - y_t))$ and fitting an ARMA model to z_t.

The model can be extended to the compositional time series, where $y_t = (s_{t1}, \dots, s_{td})$ is the observed compositions such as the monthly household expenditure among different categories, import/export proportions of different product categories, or word count (in percentage) in a document. Since $s_{ti} > 0$ and $\sum_{i=1}^d s_{ti} = 1$, we may assume

$$\begin{aligned} x_t &\sim \text{ARMA}(p, q), \\ y_t &\sim \text{Dirichelet}(\tau, p_{t1}, \dots, p_{td}), \end{aligned}$$

where $x_t = (\eta_{t1}, \dots, \eta_{t(d-1)})$, $\tau = \sum_{i=1}^d p_{ti}$, and $\eta_{ti} = \ln(p_{ti}/p_{td})$. For analysis of compositional time series, see Zheng and Chen (2017).

6.2.11 Mixed Frequency Models

Economic time series are often reported at different frequencies. For example, real GDP growth is reported quarterly, while inflation index is reported monthly. When these time series are modeled jointly, or one series is used as a covariate of the other, the analysis is called a mixed frequency data analysis. It is easy to convert high-frequency series to the lower-frequency ones, but such a converting procedure does not utilize fully the available information. In recent years, there has been substantial research interest in mixed-frequency data analysis (MIDAS). See, for instance, Ghysels et al. (2007), Gagliardini et al. (2017), and Ghysels (2016), among others.

Here we demonstrate, via a simple example, that the SSM approach can be used to analyze mixed-frequency data. Consider a mixed-frequency VAR(1) model with two components. The setting can easily be extended to a VAR(p) model with multiple components. Suppose y_t is an observed monthly series and z_t is an unobserved monthly series for which only the quarterly total $s_{3s} = z_{3s} + z_{3s-1} + z_{3s-2}$ is available. Suppose (y_t, z_t) jointly follows a VAR(1) model,

$$\begin{pmatrix} z_t \\ y_t \end{pmatrix} = \begin{bmatrix} \phi_{11} & \phi_{12} \\ \phi_{21} & \phi_{22} \end{bmatrix} \begin{pmatrix} z_{t-1} \\ y_{t-1} \end{pmatrix} + \begin{pmatrix} w_t \\ v_t \end{pmatrix}.$$

We can rewrite the model as:

$$\begin{aligned} z_t &= \phi_{11} z_{t-1} + \phi_{12} y_{t-1} + w_t && \text{where } w_t \sim N(0, \sigma^2), \\ z_{t-1} &= z_{t-1}, \\ z_{t-2} &= z_{t-2}, \\ y_t &= \phi_{21} z_{t-1} + \phi_{22} y_{t-1} + v_t, && \text{where } v_t \sim N\left(0, \sigma_v^2\right), \\ s_t &= z_t + z_{t-1} + z_{t-2}, && \text{if } t = 3s. \end{aligned}$$

Let $x_t = (y_t, z_t, z_{t-1}, z_{t-2})$ be the underlying state variable, then the model can be rewritten in an SSM. We analyze a mixed-frequency series in Example 6.3 under a slightly different setting.

6.2.12 Other Applications

SSMs, especially nonlinear and non-Gaussian SSMs, have been successfully designed for many other important applications, including speech recognition (Rabiner, 1989), sensor collaborations (Guo and Wang, 2004), movement intent inference (Ahmad et al., 2016), brain source localization (Chen et al., 2015), combinatorial optimizations (Wong and Liang, 1997), genetics (Irwing et al., 1994), DNA, RNA and protein sequence analysis (Churchill, 1989; Krough et al., 1994; Liu et al., 1999; Liang et al., 2002; Zhang et al., 2003, 2009), probabilistic expert systems (Kong et al., 1994; Spiegelhalter and Lauritzen, 1990; Berzuini et al., 1997), target recognition (Srivastava et al., 2001), freeway traffic vision (for vehicle control) (Huang et al., 1994), dynamic Bayesian networks (Koller and Lerner, 2001; Murphy and Russell, 2001; Naesseth et al., 2014; Bouchard-Côté et al., 2012), infectious disease dynamics (Smith et al., 2017), copula models (Creal and Tsay, 2015), on-line control of industrial production (Marrs, 2001), diffusion bridges (Lin et al., 2010), audio signal enhancing (Fong et al., 2002), data network analysis (Coates and Nowak, 2002), neural networks (Andrieu et al., 1999; de Freitas et al., 2000), and financial applications (Lopes and Tsay, 2011), among many others.

6.3 LINEAR GAUSSIAN STATE SPACE MODELS

When the functions are linear and the noise distributions are Gaussian, the Markovian SSM in Equations (6.3) and (6.4) takes the form

$$x_t = \boldsymbol{H}_t x_{t-1} + \boldsymbol{B}_t b_t + \boldsymbol{W}_t w_t, \tag{6.18}$$

$$y_t = \boldsymbol{G}_t x_t + \boldsymbol{C}_t c_t + \boldsymbol{V}_t v_t, \tag{6.19}$$

where $\boldsymbol{H}_t, \boldsymbol{G}_t, \boldsymbol{B}_t, \boldsymbol{C}_t, \boldsymbol{W}_t$ and $\boldsymbol{V}_t$ are matrices, c_t and b_t are known input series, and $w_t \sim N(0, \boldsymbol{I})$ and $v_t \sim N(0, \boldsymbol{I})$, with $\boldsymbol{I}$ being the identity matrix with proper dimension. The model in Equations (6.18)–(6.19) is called a *dynamic linear model* in West and Harrison (1989) and Harvey (1989). In the literature the error terms $\boldsymbol{W}_t w_t$ and $\boldsymbol{V}_t v_t$ are often replaced simply by ε_t and e_t with non-identity covariance matrices. We use $\boldsymbol{W}_t w_t$ and $\boldsymbol{V}_t v_t$ because they have the advantage of easy parameterization if there exist certain unknown parameters or imposed structures in the covariance matrices.

In the dynamic linear model, if the input series b_t and c_t are set to vectors of ones, and $\boldsymbol{B}_t$ and $\boldsymbol{C}_t$ are constant diagonal matrices, the terms $\boldsymbol{B}_t b_t$ and $\boldsymbol{C}_t c_t$ serve as the constant terms of the model. These constants can be useful in building the state dynamics.

6.3.1 Filtering and the Kalman Filter

Under models (6.18) and (6.19), if the initial distribution of x_0 is also Gaussian, $x_0 \sim N(\mu_0, \boldsymbol{\Sigma}_0)$, it can easily be seen that

$$p(x_t \mid y_1, \ldots y_t) \sim N(\mu_t, \boldsymbol{\Sigma}_t)$$

for all t. Hence, the filtering problem discussed in Section 6.1 only requires obtaining the mean and covariance matrix. The Kalman filter (Kalman, 1960) is the most efficient way to recursively obtain and update $(\mu_t, \boldsymbol{\Sigma}_t)$ at time t, given $(\mu_{t-1}, \boldsymbol{\Sigma}_{t-1})$.

Two simple properties of multivariate normal distributions are useful in understanding the Kalman filter.

Lemma 6.1 *If $X \sim N(\mu_x, \boldsymbol{\Sigma}_x)$ and $Y = c + \boldsymbol{G}X + \boldsymbol{V}v$, where $v \sim N(0, \boldsymbol{I})$ and is independent with X, then the joint distribution of (X, Y) is*

$$\begin{pmatrix} X \\ Y \end{pmatrix} \sim N\left(\begin{pmatrix} \mu_x \\ \mu_y \end{pmatrix}, \begin{bmatrix} \boldsymbol{\Sigma}_{xx} & \boldsymbol{\Sigma}_{xy} \\ \boldsymbol{\Sigma}_{yx} & \boldsymbol{\Sigma}_{yy} \end{bmatrix} \right),$$

where

$$\boldsymbol{\Sigma}_{xx} = \boldsymbol{\Sigma}_x,$$

$$\mu_y = E[Y] = E[c + \boldsymbol{G}X + \boldsymbol{V}v] = c + \boldsymbol{G}E[X] = c + \boldsymbol{G}\mu_x,$$

$$\boldsymbol{\Sigma}_{xy} = \boldsymbol{\Sigma}_x \boldsymbol{G}', \tag{6.20}$$

$$\boldsymbol{\Sigma}_{yx} = \boldsymbol{G}\boldsymbol{\Sigma}_x,$$

$$\boldsymbol{\Sigma}_{yy} = \boldsymbol{G}\boldsymbol{\Sigma}_x \boldsymbol{G}' + \boldsymbol{V}\boldsymbol{V}'. \tag{6.21}$$

Equation (6.20) holds because

$$\boldsymbol{\Sigma}_{xy} = E[(X-\mu_x)(Y-\mu_y)'] = E[(X-\mu_x)((X-\mu_x)'\boldsymbol{G}' + v'\boldsymbol{V}')] = \boldsymbol{\Sigma}_x\boldsymbol{G}',$$

and Equation (6.21) follows from

$$\begin{aligned}\boldsymbol{\Sigma}_{yy} &= E[(Y-\mu_y)(Y-\mu_y)'] = E[\boldsymbol{G}(X-\mu_x)(X-\mu_x)'\boldsymbol{G}' + \boldsymbol{V}vv'\boldsymbol{V}'] \\ &= \boldsymbol{G}\boldsymbol{\Sigma}_x\boldsymbol{G}' + \boldsymbol{V}\boldsymbol{V}'.\end{aligned}$$

Lemma 6.2 *If*

$$\begin{pmatrix} X \\ Y \end{pmatrix} \sim N\left(\begin{pmatrix} \mu_x \\ \mu_y \end{pmatrix}, \begin{bmatrix} \boldsymbol{\Sigma}_{xx} & \boldsymbol{\Sigma}_{xy} \\ \boldsymbol{\Sigma}_{yx} & \boldsymbol{\Sigma}_{yy} \end{bmatrix}\right),$$

and we assume that $\boldsymbol{\Sigma}_{yy}^{-1}$ *exists, then*

$$\begin{aligned} E(X \mid Y=y) &= \mu_x - \boldsymbol{\Sigma}_{xy}\boldsymbol{\Sigma}_{yy}^{-1}(y-\mu_y), \\ Var(X \mid Y=y) &= \boldsymbol{\Sigma}_{xx} - \boldsymbol{\Sigma}_{xy}\boldsymbol{\Sigma}_{yy}^{-1}\boldsymbol{\Sigma}_{yx}. \end{aligned}$$

The Kalman filter uses these two properties to obtain the distribution of $p(x_t \mid \boldsymbol{y}_t)$ recursively, where $\boldsymbol{y}_t = (y_1, \ldots, y_t)$. Suppose at time $t-1$ we have obtained μ_{t-1} and $\boldsymbol{\Sigma}_{t-1}$, that is,

$$p(x_{t-1} \mid \boldsymbol{y}_{t-1}) \sim N(\mu_{t-1}, \boldsymbol{\Sigma}_{t-1}).$$

Before we observe y_t, we can use the state equation to predict x_t. Since

$$x_t = \boldsymbol{H}_t x_{t-1} + \boldsymbol{B}_t b_t + \boldsymbol{W}_t w_t,$$

and both $p(x_{t-1} \mid \boldsymbol{y}_{t-1})$ and w_t are normal, $p(x_t \mid \boldsymbol{y}_{t-1})$ is normal. Denoting

$$p(x_t \mid \boldsymbol{y}_{t-1}) \sim N(\mu_{t|t-1}, \boldsymbol{\Sigma}_{t|t-1}), \tag{6.22}$$

we have

$$\begin{aligned} \mu_{t|t-1} &= E(\boldsymbol{H}_t x_{t-1} + \boldsymbol{B}_t b_t + \boldsymbol{W}_t w_t \mid \boldsymbol{y}_{t-1}) \\ &= \boldsymbol{H}_t E(x_{t-1} \mid \boldsymbol{y}_{t-1}) + \boldsymbol{B}_t b_t \\ &= \boldsymbol{H}_t \mu_{t-1} + \boldsymbol{B}_t b_t, \end{aligned} \tag{6.23}$$

$$\begin{aligned} \boldsymbol{\Sigma}_{t|t-1} &= \text{Var}(\boldsymbol{H}_t x_{t-1} + \boldsymbol{B}_t b_t + \boldsymbol{W}_t w_t \mid \boldsymbol{y}_{t-1}) \\ &= \text{Var}(\boldsymbol{H}_t x_{t-1} \mid \boldsymbol{y}_{t-1}) + \boldsymbol{W}_t \boldsymbol{W}_t' \\ &= \boldsymbol{H}_t \boldsymbol{\Sigma}_{t-1} \boldsymbol{H}_t' + \boldsymbol{W}_t \boldsymbol{W}_t'. \end{aligned} \tag{6.24}$$

Now consider the new observation y_t. Using (6.23), (6.24) and the observation equation $y_t = \boldsymbol{G}_t x_t + \boldsymbol{C}_t c_t + \boldsymbol{V}_t v_t$, Lemma 6.1 provides the joint distribution $p(x_t, y_t \mid y_1, \ldots, y_{t-1})$. Then using Lemma 6.2, we get $p(x_t \mid y_1, \ldots, y_{t-1}, y_t)$. Putting everything together, we have

Algorithm 6.1: Kalman Filter

$$\mu_{t|t-1} = \boldsymbol{H}_t \mu_{t-1} + \boldsymbol{B}_t b_t, \tag{6.25}$$
$$\boldsymbol{\Sigma}_{t|t-1} = \boldsymbol{H}_t \boldsymbol{\Sigma}_{t-1} \boldsymbol{H}_t' + \boldsymbol{W}_t \boldsymbol{W}_t', \tag{6.26}$$
$$\mu_t = \mu_{t|t-1} + \boldsymbol{K}_t (y_t - \boldsymbol{C}_t c_t - \boldsymbol{G}_t \mu_{t|t-1}), \tag{6.27}$$
$$\boldsymbol{\Sigma}_t = \boldsymbol{\Sigma}_{t|t-1} - \boldsymbol{K}_t \boldsymbol{G}_t \boldsymbol{\Sigma}_{t|t-1}, \tag{6.28}$$

where

$$\boldsymbol{K}_t = \boldsymbol{\Sigma}_{t|t-1} \boldsymbol{G}_t' [\boldsymbol{G}_t \boldsymbol{\Sigma}_{t|t-1} \boldsymbol{G}_t' + \boldsymbol{V}_t' \boldsymbol{V}_t]^{-1}. \tag{6.29}$$

The matrix $\boldsymbol{K}_t$ is often called the Kalman gain matrix.

Another way of understanding the Kalman filter is via the Bayes theorem. Before the arrival of y_t, the predictive distribution $p(x_t \mid y_1, \dots, y_{t-1})$ in Equation (6.22) can be thought as the prior distribution of x_t. The new observation y_t provides additional information about x_t, or correction to the prediction. Based on Bayes theorem

$$\begin{aligned} p(x_t \mid y_1, \dots, y_{t-1}, y_t) &\propto p(y_t \mid x_t, y_1, \dots, y_{t-1}) p(x_t \mid y_1, \dots, y_{t-1}) \\ &= p(y_t \mid x_t) p(x_t \mid y_1, \dots, y_{t-1}). \end{aligned}$$

The Kalman filter can be obtained by a detailed calculation of the above equation.

Intuitively, Equation (6.25) provides a one-step ahead prediction of x_t before the observation y_t arrives and Equation (6.26) gives the prediction variance. Once the observation y_t arrives, the prediction is updated by the correction term $\boldsymbol{K}_t(y_t - \boldsymbol{C}_t c_t - \boldsymbol{G}_t \mu_{t|t-1})$. Here $\boldsymbol{C}_t c_t + \boldsymbol{G}_t \mu_{t|t-1}$ is the prediction of y_t using the predicted state $\mu_{t|t-1}$. Hence the correction is proportional to $(y_t - \boldsymbol{C}_t c_t - \boldsymbol{G}_t \mu_{t|t-1})$, the difference between the predicted value of y_t and the observed y_t. Equation (6.28) shows that, with the new information y_t, the variance of x_t is reduced from its prediction variance $\boldsymbol{\Sigma}_{t|t-1}$ by the amount of $\boldsymbol{K}_t \boldsymbol{G}_t \boldsymbol{\Sigma}_{t|t-1}$.

6.3.2 Evaluating the likelihood function

When there are unknown parameters in the linear Gaussian system (6.18) and (6.19), one can evaluate the likelihood function of a given set of parameters $\boldsymbol{\theta}$. Specifically, as discussed in Section 6.1, the likelihood function has the form

$$L(\boldsymbol{\theta}) = p(y_1, \dots, y_T \mid \boldsymbol{\theta}) = \prod_{t=1}^{T} p(y_t \mid \boldsymbol{y}_{t-1}, \boldsymbol{\theta}).$$

The Kalman filter can be used to obtain $p(y_t \mid \boldsymbol{y}_{t-1}, \boldsymbol{\theta})$, given $\boldsymbol{\theta}$. Since $p(y_t \mid \boldsymbol{y}_{t-1}, \boldsymbol{\theta})$ is a normal distribution, we only need to obtain its mean and variance:

$$
\begin{aligned}
E(y_t \mid \boldsymbol{y}_{t-1}) &= E(E(y_t \mid x_t, \boldsymbol{y}_{t-1}) \mid \boldsymbol{y}_{t-1}) \\
&= E(\boldsymbol{G}_t x_t + \boldsymbol{C}_c c_t \mid \boldsymbol{y}_{t-1}) \\
&= \boldsymbol{G}_t \mu_{t|t-1} + \boldsymbol{C}_t c_t
\end{aligned}
$$

and

$$
\begin{aligned}
\mathrm{Var}(y_t \mid \boldsymbol{y}_{t-1}) &= \mathrm{Var}(E(y_t \mid x_t, \boldsymbol{y}_{t-1}) \mid \boldsymbol{y}_{t-1}) + E(\mathrm{Var}(y_t \mid x_t, \boldsymbol{y}_{t-1}) \mid \boldsymbol{y}_{t-1}) \\
&= \mathrm{Var}(\boldsymbol{G}_t x_t + \boldsymbol{C}_t c_t \mid \boldsymbol{y}_{t-1}) + E(\boldsymbol{V}_t \boldsymbol{V}_t' \mid \boldsymbol{y}_{t-1}) \\
&= \boldsymbol{G}_t \boldsymbol{\Sigma}_{t|t-1} \boldsymbol{G}_t' + \boldsymbol{V}_t \boldsymbol{V}_t'. \qquad (6.30)
\end{aligned}
$$

The terms $\mu_{t|t-1}$ and $\boldsymbol{\Sigma}_{t|t-1}$ can be obtained by the Kalman filter for a given $\boldsymbol{\theta}$.

In fact, the process

$$\varepsilon_t = y_t - \boldsymbol{G}_t \mu_{t|t-1} - \boldsymbol{C}_t c_t \qquad (6.31)$$

is often called the *innovation* process. It is the prediction error of y_t given $\boldsymbol{y}_{t-1}$. The above calculation shows that ε_t follows a Gaussian distribution with mean zero, and variance in Equation (6.30). In addition, $\{\varepsilon_t\}$ is an independent series by noting that $\varepsilon_t = y_t - \boldsymbol{G}_t \mu_{t|t-1} - \boldsymbol{C}_t c_t = \boldsymbol{G}_t(x_t - \mu_{t|t-1}) + \boldsymbol{V}_t v_t$, hence for $s < t$,

$$E(\varepsilon_s \varepsilon_t) = E[E(\varepsilon_s \varepsilon_t \mid \boldsymbol{y}_{t-1})] = E[\varepsilon_s E(\varepsilon_t \mid \boldsymbol{y}_{t-1})] = 0.$$

Since $p(\boldsymbol{y}_T \mid \boldsymbol{\theta}) = p(\varepsilon_1(\boldsymbol{\theta}), \ldots, \varepsilon_T(\boldsymbol{\theta})) = \prod_{t=1}^T p(\varepsilon_t(\boldsymbol{\theta}))$ (here we emphasize that $\varepsilon_t(\boldsymbol{\theta})$ is obtained through the Kalman filter, based on $\boldsymbol{\theta}$.) The log-likelihood function is then

$$l(\boldsymbol{\theta}) = -\frac{1}{2} \sum_{t=1}^{T} \log |\boldsymbol{\Sigma}_{(e)t}(\boldsymbol{\theta})| - \frac{1}{2} \sum_{t=1}^{T} \varepsilon_t(\boldsymbol{\theta})' \boldsymbol{\Sigma}_{(e)t}^{-1}(\boldsymbol{\theta}) \varepsilon_t(\boldsymbol{\theta}),$$

where $\boldsymbol{\Sigma}_{(e)t}(\boldsymbol{\theta})$ is the same as that in (6.30).

Hence it is relatively easy and fast to evaluate the log likelihood function for any given parameter $\boldsymbol{\theta}$. The maximum likelihood estimator of the parameter can then be obtained using optimization algorithms such as the Newton–Raphson algorithm. Caines (1988) and Hannan and Deistler (1988) (see also Douc et al., 2014) established the consistency and asymptotic normality for the maximum likelihood estimates (MLE) of a stable linear system, in which all matrices are assumed to be constant, the spectral norm of the state transition matrix $\boldsymbol{H}$ is less than 1 (so that the underlying state process is stable), the expanded observation coefficient matrix $[\boldsymbol{G}', \boldsymbol{G}'\boldsymbol{H}', \boldsymbol{G}'\boldsymbol{H}^{2'}, \ldots, \boldsymbol{G}'\boldsymbol{H}^{p'}]$ has full rank p where p is the dimension of the state x_t (so that the observation y_t has sufficient information about x_t), and some other conditions. The estimated Hessian matrix at the MLE can be used as an estimate of the asymptotic variance matrix of the MLE. Shumway and Stoffer (1982) used the EM algorithm in parameter estimation.

6.3.3 Smoothing

For a fixed dimensional problem in which the entire sequence $\boldsymbol{y}_T = (y_1, \ldots, y_T)$ is observed, it is of interest to make inference of x_t given the whole data. This is the smoothing problem discussed in Section 6.1. In other words, we are interested in finding $p(x_1, \ldots, x_T \mid \boldsymbol{y}_T)$ or its marginal distributions $p(x_t \mid \boldsymbol{y}_T)$ for any t. Again, as the system is linear and Gaussian, $p(x_t \mid \boldsymbol{y}_T)$ follows a normal distribution, denoted by $N(\mu_{t|T}, \boldsymbol{\Sigma}_{t|T})$. The following *Kalman smoother* recursively obtains $\mu_{t|T}$ and $\boldsymbol{\Sigma}_{t|T}$ in a forward and backward two-pass algorithm. Note that the main difference between the filtering distribution $p(x_t \mid \boldsymbol{y}_t)$ and the smoothing distribution $p(x_t \mid \boldsymbol{y}_T)$ is the additional information provided by $y_{t+1}, \ldots, y_T$. By the Markovian property, given x_{t+1}, the extra information $y_{t+1}, \ldots, y_T$ and x_t are conditionally independent. Hence the process can be *disentangled* by inserting x_{t+1}. If we obtain $p(x_{t+1} \mid \boldsymbol{y}_T)$, which provides a summary of all information in $y_{t+1}, \ldots, y_T$, then one can obtain $p(x_t \mid \boldsymbol{y}_t)$ by a correction of $p(x_t \mid \boldsymbol{y}_t)$ using $p(x_{t+1} \mid \boldsymbol{y}_T)$. Such a two-pass algorithm can also be used for nonlinear systems and is discussed in Chapter 8.

Specifically, the first pass is to run the Kalman filter forward from $t = 1$ to T and obtain $\mu_{t|t-1}$, $\mu_t = \mu_{t|t}$, $\boldsymbol{\Sigma}_{t|t-1}$, and $\boldsymbol{\Sigma}_t = \boldsymbol{\Sigma}_{t|t}$ for $t = 1, \ldots, T$. Then the Kalman smoother runs backward as follows.

Algorithm 6.2: Kalman smoother

For $t = T-1, T-2, \ldots, 1$,

$$\mu_{t|T} = \mu_t + \boldsymbol{J}_t(\mu_{t+1|T} - \mu_{t+1|t}), \tag{6.32}$$

$$\boldsymbol{\Sigma}_{t|T} = \boldsymbol{\Sigma}_t + \boldsymbol{J}_t(\boldsymbol{\Sigma}_{t+1|T} - \boldsymbol{\Sigma}_{t+1|t})\boldsymbol{J}_t', \tag{6.33}$$

where

$$\boldsymbol{J}_t = \boldsymbol{\Sigma}_t \boldsymbol{H}_{t+1}[\boldsymbol{\Sigma}_{t+1|t}]^{-1}. \tag{6.34}$$

The Kalman smoother can be derived by inserting x_{t+1} in a disentanglement operation:

$$\begin{aligned}
E(x_t \mid \boldsymbol{y}_T) &= E(E(x_t \mid x_{t+1}, \boldsymbol{y}_T) \mid \boldsymbol{y}_T) \\
&= E(E(x_t \mid x_{t+1}, \boldsymbol{y}_t) \mid \boldsymbol{y}_T), \\
\mathrm{Var}(x_t \mid \boldsymbol{y}_T) &= E[\mathrm{Var}(x_t \mid x_{t+1}, \boldsymbol{y}_T) \mid \boldsymbol{y}_T] + \mathrm{Var}[E(x_t \mid x_{t+1}, \boldsymbol{y}_T) \mid \boldsymbol{y}_T] \\
&= E[\mathrm{Var}(x_t \mid x_{t+1}, \boldsymbol{y}_t) \mid \boldsymbol{y}_T] + \mathrm{Var}[E(x_t \mid x_{t+1}, \boldsymbol{y}_t) \mid \boldsymbol{y}_T].
\end{aligned}$$

To obtain $E(x_t \mid x_{t+1}, \boldsymbol{y}_t)$, we use

$$\begin{bmatrix} x_t \mid \boldsymbol{y}_t \\ x_{t+1} \mid \boldsymbol{y}_t \end{bmatrix} \sim N\left(\begin{bmatrix} \mu_{t|t} \\ \mu_{t+1|t} \end{bmatrix}, \begin{bmatrix} \boldsymbol{\Sigma}_{t|t} & \boldsymbol{\Sigma}_{t|t}\boldsymbol{H}'_{t+1} \\ \boldsymbol{H}_{t+1}\boldsymbol{\Sigma}_{t|t} & \boldsymbol{\Sigma}_{t+1|t} \end{bmatrix} \right).$$

Using Lemma 6.2, we have

$$E(x_t \mid x_{t+1}, \boldsymbol{y}_t) = \mu_{t|t} + \boldsymbol{J}_t(x_{t+1} - \mu_{t+1|t}),$$
$$\mathrm{Var}(x_t \mid x_{t+1}, \boldsymbol{y}_t) = \boldsymbol{\Sigma}_{t|t} - \boldsymbol{\Sigma}_{t|t}\boldsymbol{H}'_{t+1}[\boldsymbol{\Sigma}_{t+1|t}]^{-1}\boldsymbol{H}_{t+1}\boldsymbol{\Sigma}_t = \boldsymbol{\Sigma}_{t|t} - \boldsymbol{J}_t\boldsymbol{\Sigma}_{t+1|t}\boldsymbol{J}'_t,$$

where $\boldsymbol{J}_t = \boldsymbol{\Sigma}_t\boldsymbol{H}_{t+1}[\boldsymbol{\Sigma}_{t+1|t}]^{-1}$. Hence

$$\begin{aligned} E[E(x_t \mid x_{t+1}, \boldsymbol{y}_t) \mid \boldsymbol{y}_T] &= E[\mu_{t|t} + \boldsymbol{J}_t(x_{t+1} - \mu_{t+1|t}) \mid \boldsymbol{y}_T] \\ &= \mu_{t|t} + \boldsymbol{J}_t(E(x_{t+1} \mid \boldsymbol{y}_T) - \mu_{t+1|t}) \\ &= \mu_{t|t} + \boldsymbol{J}_t(\mu_{t+1|T} - \mu_{t+1|t}) \end{aligned}$$

and

$$\begin{aligned} \mathrm{Var}[E(x_t \mid x_{t+1}, \boldsymbol{y}_t) \mid \boldsymbol{y}_T] &= \mathrm{Var}[\mu_{t|t} + \boldsymbol{J}_t(x_{t+1} - \mu_{t+1|t}) \mid \boldsymbol{y}_T] \\ &= \boldsymbol{J}_t[V(x_{t+1} \mid \boldsymbol{y}_T)]\boldsymbol{J}'_t \\ &= \boldsymbol{J}_t\boldsymbol{\Sigma}_{t+1|T}\boldsymbol{J}'_t, \end{aligned}$$

and

$$\begin{aligned} E[\mathrm{Var}(x_t \mid x_{t+1}, \boldsymbol{y}_t) \mid \boldsymbol{y}_T] &= E[\boldsymbol{\Sigma}_{t|t} - \boldsymbol{J}\boldsymbol{\Sigma}_{t+1|t}\boldsymbol{J}'_t \mid \boldsymbol{y}_T] \\ &= \boldsymbol{\Sigma}_{t|t} - \boldsymbol{J}_t\boldsymbol{\Sigma}_{t+1|t}\boldsymbol{J}'_t. \end{aligned}$$

Putting everything together, we have the smoother in Equations (6.32) and (6.33).

The smoother can also be derived directly by working on the joint distribution:

$$\begin{aligned} p(x_t, x_{t+1} \mid \boldsymbol{y}_T) &= p(x_t \mid x_{t+1}, \boldsymbol{y}_T)p(x_{t+1} \mid \boldsymbol{y}_T) \\ &= p(x_t \mid x_{t+1}, \boldsymbol{y}_t)p(x_{t+1} \mid \boldsymbol{y}_T) \\ &= \frac{p(x_{t+1} \mid x_t, \boldsymbol{y}_t)p(x_t \mid \boldsymbol{y}_t)}{p(x_{t+1} \mid \boldsymbol{y}_t)} p(x_{t+1} \mid \boldsymbol{y}_T) \\ &= \frac{p(x_{t+1} \mid x_t)p(x_t \mid \boldsymbol{y}_t)}{p(x_{t+1} \mid \boldsymbol{y}_t)} p(x_{t+1} \mid \boldsymbol{y}_T), \end{aligned}$$

where $p(x_{t+1} \mid x_t)$ is given by the model, $p(x_t \mid \boldsymbol{y}_t)$ and $p(x_{t+1} \mid \boldsymbol{y}_t)$ can be obtained by the forward Kalman filter, and $p(x_{t+1} \mid \boldsymbol{y}_T)$ is the result of backward smoothing at $t + 1$. Through a detailed calculation one can find the mean vector and covariance matrix of the joint normal distribution of $p(x_t, x_{t+1} \mid \boldsymbol{y}_T)$. Then using Lemma 6.2, the smoother Equations (6.32) and (6.33) can be obtained.

Intuitively, Equation (6.32) shows how the filtering mean μ_t is corrected when the future observations $y_{t+1}, \ldots, y_T$ are observed. Its correction size is proportional to the difference between the predicted mean of x_{t+1} based on $\boldsymbol{y}_t$ and the smoothed mean of x_{t+1} based on the whole sequence y_T. This difference is caused by the extra information. Equation (6.33) shows how the extra information reduces the estimation variance.

De Jong (1989) and Durbin and Koopman (2001) developed a more computationally friendly version of the Kalman smoother, though it is less intuitive. See also Tsay (2010, Chapter 11).

Algorithm 6.3: Kalman smoother (II)

Set $r_T = 0$ and $N_T = 0$. For $t = T-1, t-2, \ldots, 1$,

$$\begin{aligned} \mu_{t|T} &= \mu_t + \boldsymbol{\Sigma}_{t|t-1} r_t, \\ r_{t-1} &= \boldsymbol{H}_t \boldsymbol{F}_t^{-1} \varepsilon_t + (\boldsymbol{G}_t - \boldsymbol{G}_t \boldsymbol{\Sigma}_{t|t-1} \boldsymbol{H}_t \boldsymbol{F}_t^{-1}) r_t, \\ \boldsymbol{\Sigma}_{t|T} &= \boldsymbol{\Sigma}_{t|t-1} - \boldsymbol{\Sigma}_{t|t-1} N_{t-1} \boldsymbol{\Sigma}_{t|t-1}, \\ \boldsymbol{N}_{t-1} &= \boldsymbol{G}_t \boldsymbol{F}_t^{-1} \boldsymbol{G}_t' + \boldsymbol{L}_t \boldsymbol{N}_t \boldsymbol{L}_t', \end{aligned}$$

where $\varepsilon_t = y_t - \boldsymbol{G}_t \mu_{t|t-1} - \boldsymbol{C}_t c_t$ is the prediction error as in Equation (6.31), $\boldsymbol{L}_t = \boldsymbol{H}_t - \boldsymbol{K}_t \boldsymbol{G}_t$, and $\boldsymbol{F}_t = [\boldsymbol{G}_t \boldsymbol{\Sigma}_{t|t-1} \boldsymbol{G}_t' + \boldsymbol{V}_t \boldsymbol{V}_t']$.

The inverse of $\boldsymbol{F}_t$ has been obtained during the forward filtering step in Equation (6.29), hence the backward iterations do not involve any matrix inversion, though it requires the storage of $\boldsymbol{F}_t^{-1}$ for all t. This version of the Kalman smoother is difficult to interpret, but it avoids the inversion of $\boldsymbol{\Sigma}_{t|t-1}$ in Equation (6.34).

Delayed estimation: Delayed estimation is also called fixed lag smoothing. For a fixed delay $d > 0$, one is interested in $p(x_{t-d} \mid \boldsymbol{y}_t)$. This is a smoothing problem with $\boldsymbol{y}_t$ as the entire sequence. Moving from t to $t+1$ to get $p(x_{t-d+1} \mid \boldsymbol{y}_{t+1})$, one just needs to run a one-step Kalman filter from $(\mu_{t|t}, \boldsymbol{\Sigma}_{t|t})$ to get $(\mu_{t+1|t+1}, \boldsymbol{\Sigma}_{t+1|t+1})$, then run backward smoothing to get $p(x_{t-d+1} \mid \boldsymbol{y}_{t+1})$. One needs to store, at time t, $(\mu_{s|s}, \mu_{s+1|s}, \boldsymbol{\Sigma}_{s|s}, \boldsymbol{\Sigma}_{s+1|s})$ for $s = t-d, \ldots, t$.

6.3.4 Prediction and Missing Data

The prediction problem is to obtain $p(x_{t+d} \mid y_1, \ldots, y_t)$. This is easy as there is no new information available from $t+1$ to $t+d$. Hence we can recursively obtain

$$\begin{aligned} \mu_{t+d|t} &= E(x_{t+d} \mid \boldsymbol{y}_t) \\ &= E[E(x_{t+d} \mid x_{t+d-1}, \boldsymbol{y}_t) \mid \boldsymbol{y}_t] \\ &= E(\boldsymbol{H}_{t+d} x_{t+d-1} + \boldsymbol{B}_{t+d} b_{t+d} \mid \boldsymbol{y}_t) \\ &= \boldsymbol{H}_{t+d} \mu_{t+d-1|t} + \boldsymbol{B}_{t+d} b_{t+d}, \end{aligned}$$

and

$$
\begin{aligned}
\boldsymbol{\Sigma}_{t+d|t} &= \mathrm{Var}(x_{t+d} \mid \boldsymbol{y}_t) \\
&= \mathrm{Var}[E(x_{t+d} \mid x_{t+d-1}, \boldsymbol{y}_t) \mid \boldsymbol{y}_t] + E[\mathrm{Var}(x_{t+d} \mid x_{t+d-1}, \boldsymbol{y}_t) \mid \boldsymbol{y}_t] \\
&= \mathrm{Var}(\boldsymbol{H}_{t+d} x_{t+d-1} \mid \boldsymbol{y}_t) + E(\boldsymbol{W}'_{t+d} \boldsymbol{W}_{t+d} \mid \boldsymbol{y}_t) \\
&= \boldsymbol{H}'_{t+d} \boldsymbol{\Sigma}_{t+d-1|t} \boldsymbol{H}_{t+d} + \boldsymbol{W}'_{t+d} \boldsymbol{W}_{t+d},
\end{aligned}
$$

assuming $b_{t+1}, \ldots, b_{t+d}$ are available at time t.

Dealing with missing data is also relatively easy via the Kalman filter. If y_t is missing, then $\mu_{t|t} = \mu_{t|t-1}$ and $\boldsymbol{\Sigma}_{t|t} = \boldsymbol{\Sigma}_{t|t-1}$ in the Kalman filter operation, with no updating step. When only a component of y_t is missing, the matrices $\boldsymbol{G}_t$ and $\boldsymbol{V}_t$ need to be reconfigured to fully utilize the remaining components.

6.3.5 Sequential Processing

When the observational noise matrix $\boldsymbol{V}_t$ is diagonal, the noises of the components of the observation are independent. In this case it is often more computationally efficient to sequentially process each component of the observation, while holding the state unchanged, before moving to the next time period. This is particularly useful when the dimension of y_t changes over time. Sequential processing makes $\boldsymbol{G}_t$ one dimensional, hence avoiding the matrix inversion in calculating the Kalman gain matrix $\boldsymbol{K}_t$ in Equation (6.29). To achieve this, one needs to stretch the time and adjust the coefficient matrices. Specifically, for all inserted extra times, $\boldsymbol{H}_t$ should be set to the identity matrix and $\boldsymbol{W}_t = 0$. Hence at those extra times, there is no evolution of x_t, though its conditional distribution changes due to the processing of the components of the observation. Each $\boldsymbol{G}_t$ now is a $1 \times d$ matrix and $\boldsymbol{V}_t$ is a scalar.

Even when $\boldsymbol{V}_t$ is not diagonal, one can compute a Cholesky decomposition of the covariance matrix $\boldsymbol{V}_t\boldsymbol{V}'_t$ and make a proper transformation of y_t to make the observational noise independent.

6.3.6 Examples and R Demonstrations

In this section we present four examples to illustrate the analysis of linear and Gaussian systems.

Example 6.1: Three-dimensional tracking of a target with random acceleration in a clean environment In Section 6.2.4 we introduced the tracking problem. Here we demonstrate a detailed implementation of tracking a target (e.g., an airplane, a missile) moving in a three-dimensional space with no interference and with random acceleration. The state variable $x_t = (d_{1t}, d_{2t}, d_{3t}, s_{1t}, s_{2t}, s_{3t})' \equiv (x_{1t}, \ldots, x_{6t})'$ consists of the three-dimensional true location and the speed of the target at time t. The observation $y_t = (y_{1t}, y_{2t}, y_{3t})$ is the observed three-dimensional location contaminated with noise at time t. For simplicity, assume the time unit between two observations is $\Delta T = 1$. We also assume that the random acceleration and the observational noise are all three-dimensional and independent. The SSM for the problem can then be written as

$$\begin{aligned} x_t &= \boldsymbol{H} x_{t-1} + \boldsymbol{W} w_t, \\ y_t &= \boldsymbol{G} x_t + \boldsymbol{V} v_t, \end{aligned}$$

where

$$\boldsymbol{H} = \begin{bmatrix} 1 & 0 & 0 & 1 & 0 & 0 \\ 0 & 1 & 0 & 0 & 1 & 0 \\ 0 & 0 & 1 & 0 & 0 & 1 \\ 0 & 0 & 0 & 1 & 0 & 0 \\ 0 & 0 & 0 & 0 & 1 & 0 \\ 0 & 0 & 0 & 0 & 0 & 1 \end{bmatrix}, \quad \boldsymbol{W} = \begin{bmatrix} 0.5q_1 & 0 & 0 \\ 0 & 0.5q_2 & 0 \\ 0 & 0 & 0.5q_3 \\ q_1 & 0 & 0 \\ 0 & q_2 & 0 \\ 0 & 0 & q_3 \end{bmatrix},$$

and

$$\boldsymbol{G} = \begin{bmatrix} 1 & 0 & 0 & 0 & 0 & 0 \\ 0 & 1 & 0 & 0 & 0 & 0 \\ 0 & 0 & 1 & 0 & 0 & 0 \end{bmatrix}, \quad V = \begin{bmatrix} r_1 & 0 & 0 \\ 0 & r_2 & 0 \\ 0 & 0 & r_3 \end{bmatrix}.$$

There are six unknown parameters related to the random acceleration and observational noise. Figures 6.10 and 6.11 show the tracking and smoothing errors of the first element of the location (i.e., x_{1t}) and the speed (x_{4t}) based on 100 simulations of a system with $q_1 = q_2 = 1$, $q_3 = 0.02$, $r_1 = r_2 = r_3 = 1$, and a total time duration $T = 200$. The filter and smoothing are carried out using the true parameters. For ease in reading, the figures only show the errors of the first 20 time steps. From the plots, it is easy to see that, as expected, smoothing is much more accurate than filtering because the former uses information in all observations $y_{1:200}$. The figures also show that the errors are larger for smaller t because of the filter initiation. We used a diffusion initialization $x_0 \sim N(\mu_0, \boldsymbol{\Sigma}_0)$ where $\mu_0 = (0, 0, 0, 0, 0, 0)'$ and $\boldsymbol{\Sigma}_0 = 10000\boldsymbol{I}$. Figure 6.12 shows the histograms of MLE of q_1 and r_1 over 100 simulations. In the estimation we do not restrict the parameters to be positive, though one can take their absolute values as the estimates.

The following R commands are used for filtering, smoothing, and MLE estimation using the R package `dlm`. The function `buildTrackLinear(parm)` is for building the SSM for filtering, smoothing, and MLE estimation. Examples of how to obtain the filtered and smoothed states and MLE are shown.

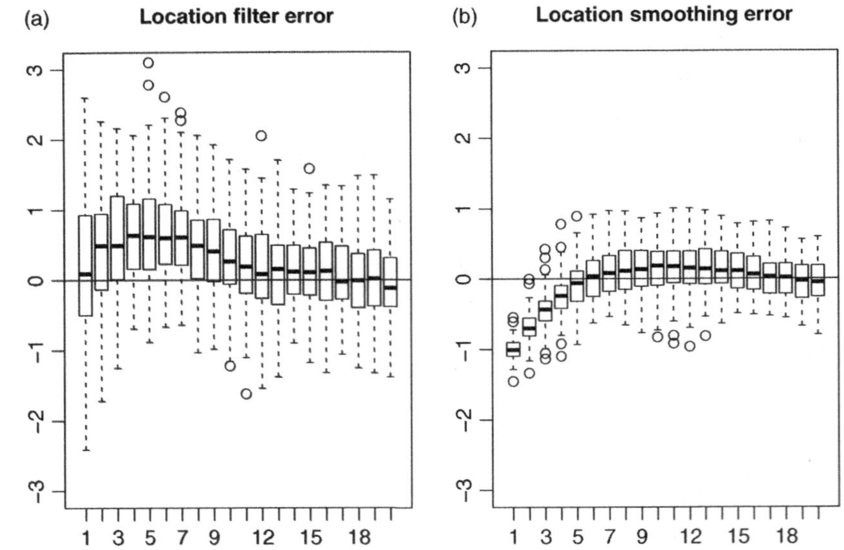

Figure 6.10 (a) Filtering and (b) smoothing errors of state x_{1t} (location) of Example 6.1 based on 100 simulations.

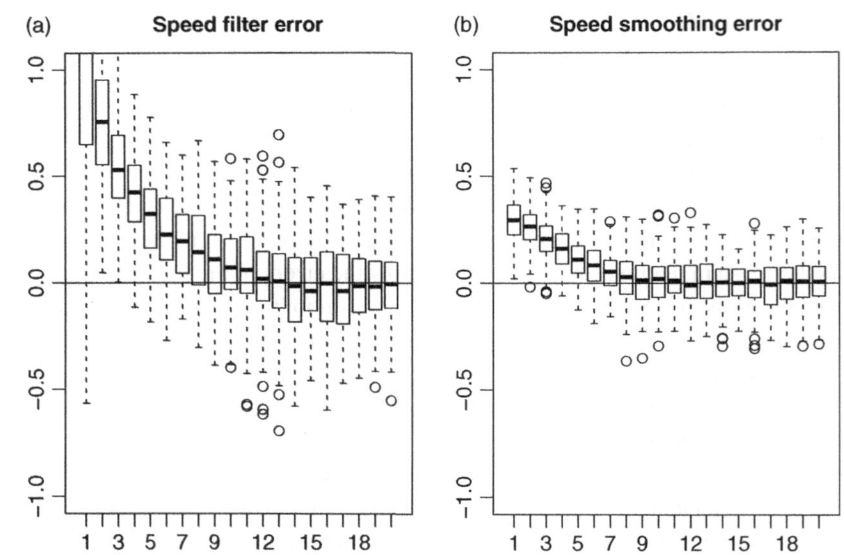

Figure 6.11 (a) Filtering and (b) smoothing errors of state x_{4t} (speed) of Example 6.1 based on 100 simulations.

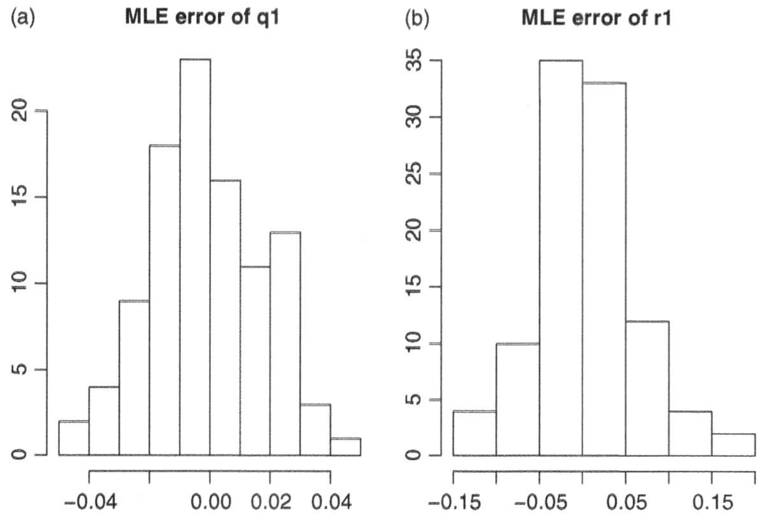

Figure 6.12 Histograms of MLE estimation errors of (a) q_1 and (b) r_1 for Example 6.1 based on 100 simulations.

R demonstration: Application of the Kalman filter.

```
#### Build function to MLE estimation
buildTrackLinear=function(parm){
    q <- parm[1:3]
    r <- parm[4:6]
    H <- c(1,0,0,1,0,0,
           0,1,0,0,1,0,
           0,0,1,0,0,1,
           0,0,0,1,0,0,
           0,0,0,0,1,0,
           0,0,0,0,0,1)
    H <- matrix(H,ncol=6,nrow=6,byrow=T)
    G <- c(1,0,0,0,0,0,
           0,1,0,0,0,0,
           0,0,1,0,0,0)
    G <- matrix(G,ncol=6,nrow=3,byrow=T)
    W <- c(0.5*q[1], 0,        0,
           0,        0.5*q[2], 0,
           0,        0,        0.5*q[3],
           q[1],     0,        0,
           0,        q[2],     0,
           0,        0,        q[3])
    W <- matrix(W,ncol=3,nrow=6,byrow=T)
```

```
    V <- diag(r)
    return(list(
            m0=rep(0,6),
            C0=1000000*diag(6),
            FF=G,
            GG=H,
            V=V%*%t(V),
            W=W%*%t(W)
        ))
}
#### end function ##
#true parameters
q <- c(0.1,0.1,0.02)
r <- c(1,1,1)
### Run Kalman filter with DLM package
library(dlm)
myModel <- buildTrackLinear(c(q,r))    #true model
outFilter <- dlmFilter(y,myModel)
outSmooth <- dlmSmooth(y,myModel)
###Estimation
par.init <- c(1,1,1,10,10,10)
outMLE <- dlmMLE(y,par.init,buildTrackLinear)
```

Example 6.2: Time-varying coefficient Fama–French factor models In Section 6.2.3 we briefly discussed the SSM setting that allows the coefficients of a regression model to vary over time. Specifically, we mentioned the time-varying coefficient Fama–French factor model. Here we present an empirical analysis of such a model. Fama and French (1993, 1996) introduced a size factor (SMB), an equity to book ratio factor (HML), and a momentum factor (MoM), in addition to the excess market return factor (Market), to model portfolio returns as an extension of the capital asset pricing model (CAPM) of Sharpe (1964) and Fama (1970).

Figure 6.13 shows the four monthly Fama–French factors from January 1946 to December 2010, obtained from http://mba.tuck.dartmouth.edu/pages/faculty/ken.french/Data Library. In the following analysis the time series of interest is the return series of the portfolios consisting of all stocks with capital sizes in the 20–40% range of all stocks and equity-to-book ratios also in the 20–40% range of all stocks, with equal weights, obtained from the same website. Figure 6.14 shows the return of the portfolio. A quick regression analysis shows the following results:

```
> out0 <- lm(y~xx)
> summary(out0)
```

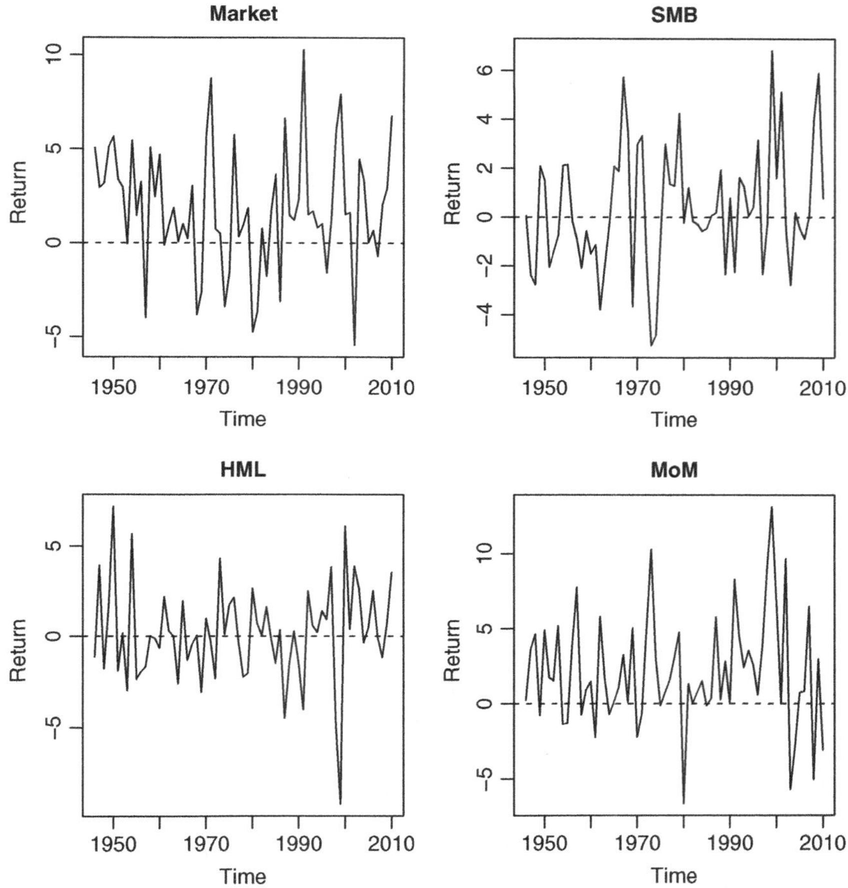

Figure 6.13 Time plots of monthly Fama–French four factors from 1946.1 to 2010.12.

```
Coefficients:
             Estimate Std. Error t value Pr(>|t|)
(Intercept)   0.03369    0.05432   0.620 0.535367
xxMkt.RF      0.99580    0.01268  78.544  < 2e-16 ***
xxSMB         0.86220    0.01902  45.330  < 2e-16 ***
xxHML         0.08927    0.01984   4.499 7.89e-06 ***
xxMoM        -0.04479    0.01348  -3.324 0.000929 ***
---
Residual standard error: 1.441 on 775 degrees of freedom
Multiple R-squared:  0.9358,    Adjusted R-squared:  0.9355
F-statistic:  2825 on 4 and 775 DF,  p-value: < 2.2e-16
```

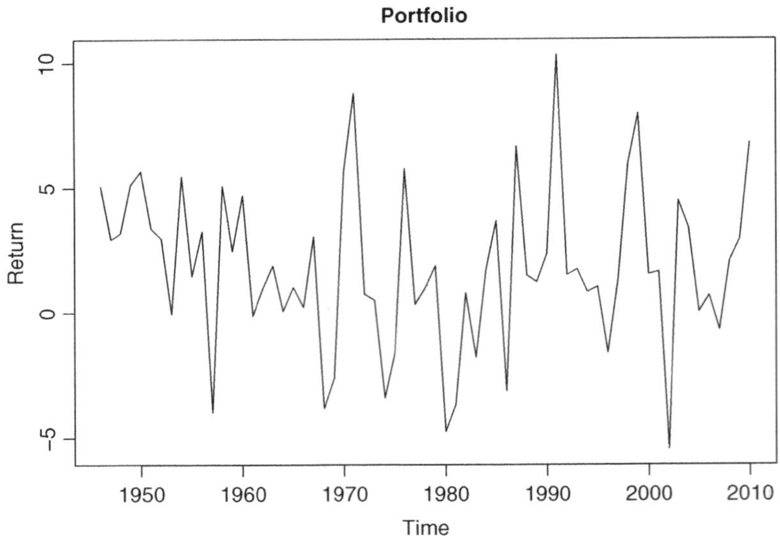

Figure 6.14 Return series of the portfolio of interest.

This shows that all four factors are highly significant. However, if one uses data in different subperiods, the estimation results are markedly different. Figure 6.15 shows the estimated parameters and their 95% confidence intervals using a moving 3-year window. The estimated coefficients and their 95% confidence intervals using the full data are also shown (straight lines). Although the estimated coefficients using the entire data stayed mostly within the 3-year rolling 95% confidence intervals, there are noticeable exceptions, especially for HML (around 1970, late 1980, and early 2000), and occasionally for SMB and MoM. In the following we use a varying-beta model to capture the changes in parameters. Specifically, we assume that

$$y_t = \beta_{0t} + \beta_{1t}\eta_{1t} + \beta_{2t}\eta_{2t} + \beta_{3t}\eta_{3t} + \beta_{4t}\eta_{5t} + e_t$$

where η_{it}, $i = 1, \ldots, 4$ are the four Fama–French factors. We further assume a simple random walk model for the coefficients $\beta_{it} = \beta_{i(t-1)} + \varepsilon_{it}$ for $i = 0, \ldots, 4$. The corresponding SSM becomes

$$\begin{aligned} \boldsymbol{x}_t &= \boldsymbol{H}\boldsymbol{x}_{t-1} + \boldsymbol{W}w_t, \\ y_t &= \boldsymbol{G}_t\boldsymbol{x}_t + \boldsymbol{V}v_t, \end{aligned}$$

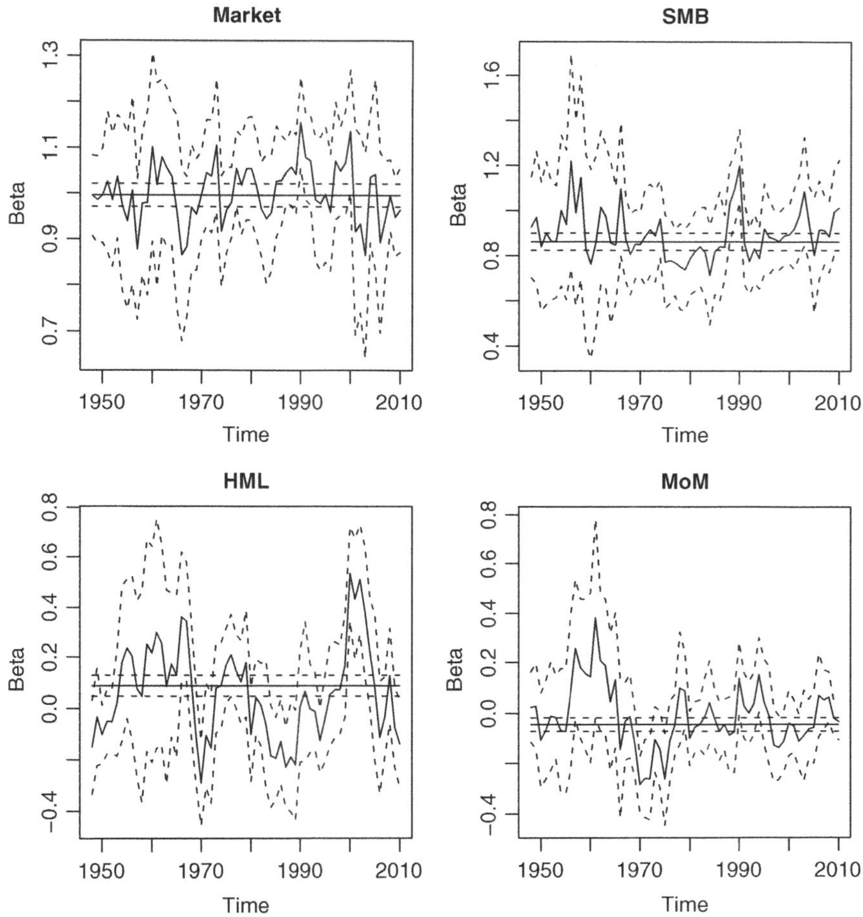

Figure 6.15 Estimated coefficients using a 3-year moving window. The dashed lines are the estimated coefficients and their corresponding 95% confidence intervals using all data.

where $\boldsymbol{x}_t = (\beta_{0t}, \dots, \beta_{4t})'$, $\boldsymbol{H} = I_{5\times5}$, $\boldsymbol{W} = diag(w_1, \dots, w_5)$, and the time-varying $\boldsymbol{G}_t$ is

$$\boldsymbol{G}_t = \begin{bmatrix} 1, & \eta_{1t}, & \eta_{2t}, & \eta_{3t}, & \eta_{4t} \end{bmatrix},$$

and $\boldsymbol{V} = v$.

The MLEs of the parameters $(w_1, \dots, w_5, v)$ are $(0, 0, 0, 0.044, -0.001, 1.25)$, with likelihood value -640.5. Figure 6.16 shows the smoothed states (the β values) against the estimated values using the 3-year moving window. The state for β_{0t} is omitted. Under this model, β_{0t}, β_{1t} and β_{2t} are almost constant, while the

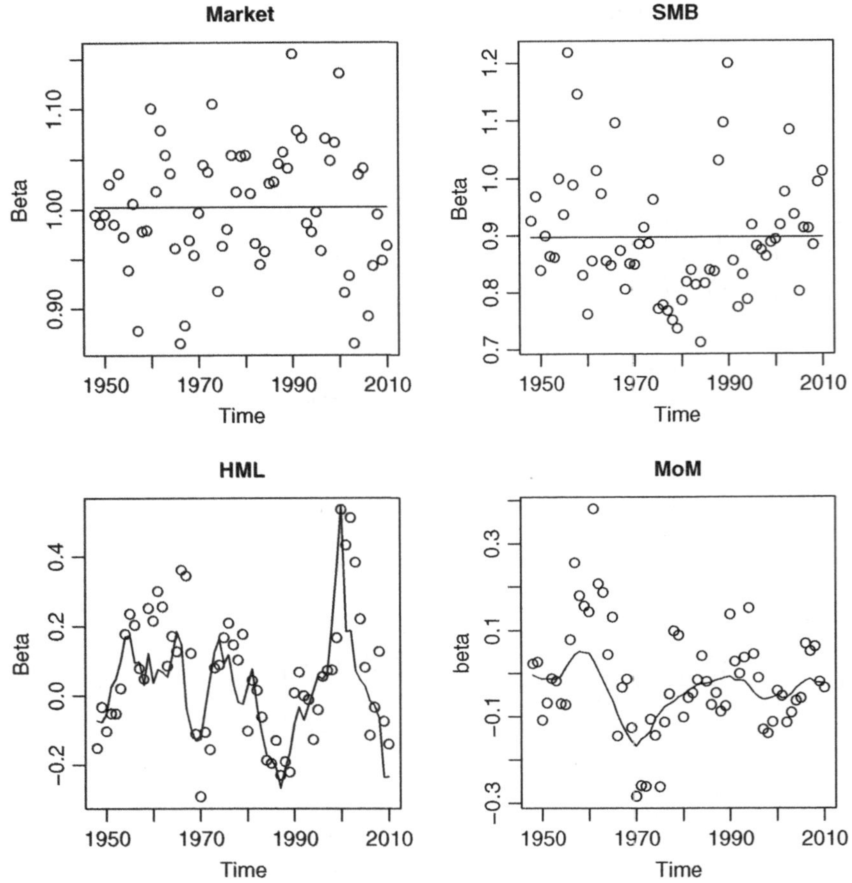

Figure 6.16 Smoothed states using a diffused state initial values (line) vs estimated coefficients with 3-year moving windows (dots).

coefficients corresponding to HML and MoM change markedly. The estimated $\hat{\nu} = 1.25$ is slightly smaller than the estimated standard deviation of residuals of the linear regression model. Note that the state variables are assumed to follow a random walk model, which is sensitive to the specification of the initial state μ_0 and $\mathbf{\Sigma}_0$. The MLE and filter/smoothing shown above were obtained using a diffused setting for the initial states. If instead we use a specific state initial value $\mu_0 = (1, 2, -0.2, 0.8, -0.3)'$ and $\mathbf{\Sigma}_0 = diag(0.1, 0.01, 0.1, 10, 10)$, one obtains a different set of MLEs $(\hat{w}_1, \ldots, \hat{w}_5, \hat{\nu}) = (0.0147, 0.025, 0.0275, 0.0397, 0.011, 1.230)$ with likelihood value of -670. The results of these particular initial values are not as good as those starting with the diffused setting. However, the resulting smoothed

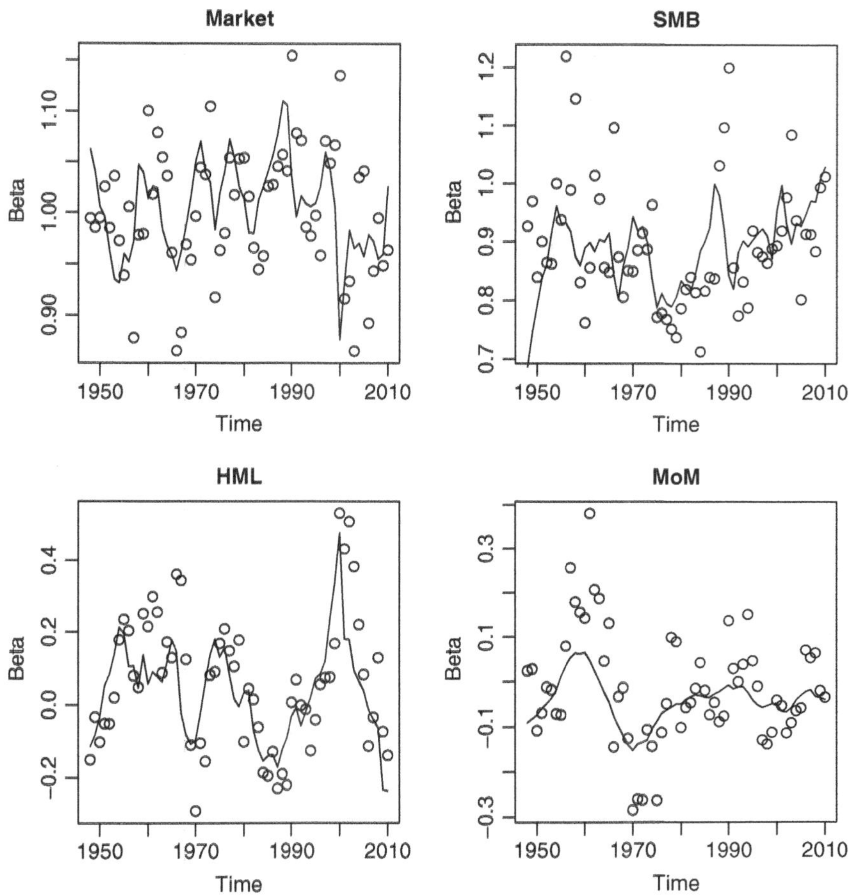

Figure 6.17 Smoothed states using a specific state initial value (lines) vs estimated coefficients with 3-year moving windows (dots).

state variable is more interesting, with all five β values changing over time (see Figure 6.17).

R demonstration: Time-varying coefficient model.

```
library(dlm)
#### Build function to MLE estimation:
buildVaryingBeta1=function(parm){
   phiW <- parm[1:5]
   phiV <- parm[6]
   H <- diag(c(1,1,1,1,1))
   W0 <- diag(phiW)
```

```
   W0 <- W0%*%t(W0)
   W <- matrix(W0,ncol=5,nrow=5,byrow=T)
   G <- c(1,1,1,1,1)
   G <- matrix(G,ncol=5,nrow=1,byrow=T)
   JGG <- c(0,1:4)
        #JGG indicates which element is time varying
   JGG <- matrix(JGG,ncol=5,nrow=1,byrow=T)
   V <- matrix(phiV[1]**2,ncol=1,nrow=1)
   TXX <- xx[,1:4]
        #TXX provides the time varying inputs
   return(list(m0=mm0,C0=CC0,FF=G,GG=H,V=V,W=W,JFF=JGG,X=TXX))
}
### end function  ###
xxx <- read.csv(file="FF_example.csv")
y <- xxx[,2]
xx <- xxx[,3:6]

## FF factor plot
year0 <- 1946:2010
y <- xxx[,2]
xx <- xxx[,3:6]
name <- c('cnst','market','SMB','HML','MoM')
par(mfrow=c(2,2),mai=0.7*c(1,1,1,1))
for(j in 1:4){
plot(year0,xx[(year0-1945)*12,j],type='l',xlab='time',ylab='return',
main=name[j-1])
abline(h=0,lty=2)
}
## fig_varying_beta0.pdf
## FF factor plot
## y-plot
par(mfrow=c(1,1),mai=1*c(1,1,1,1))
plot(year0,x[(year0-1945)*12,1],type='l',xlab='time',ylab='return',
main='portfolio')

# overall FF model
out0 <- lm(y~xx)
summary(out0)

#rolling window estimation
ntotal <- length(y)
nyear <- ntotal/12
beta <- matrix(ncol=5,nrow=(nyear-2))
for(i in 3:nyear){
  t0 <- (i-3)*12+(1:36)
  y0 <- y[t0]    ##excess return of portfolio 22
  xx0 <- x[t0,]
  out <- lm(y0~xx0)
  beta[i-2,] <- out$coef
}
```

```
year <- 1948:2010
par(mfrow=c(2,2),mai=0.7*c(1,1,1,1))
for(j in 2:5){
ylim <- range(cbind(ucl[,j],lcl[,j]))
plot(year,beta[,j],type='l',xlab='time',ylab='beta',
     main=name[j-1],ylim=ylim)
lines(year,year*0+summary(out0)$coef[j,1])
lines(year,year*0+summary(out0)$coef[j,1]+summary(out0)$coef[j,2]*2,lty=2)
lines(year,year*0+summary(out0)$coef[j,1]-summary(out0)$coef[j,2]*2,lty=2)
lines(year,ucl[,j],lty=2)
lines(year,lcl[,j],lty=2)
}
## fig_varying_beta1.pdf
## 3-year regression

##++++++ use diffused initial value
mm0 <- c(1,1,1,1,1)
CC0 <- diag(c(1,1,1,1,1))*1000
ww <- c(1,1,1,1,1)
vv <- 1
init.par <- c(ww,vv)
MLE2 <- dlmMLE(y,init.par,buildVaryingBeta1,debug=F)
MLE2
MyModel2 <- buildVaryingBeta1(MLE2$par)
ss2 <- dlmSmooth(y,MyModel2)

#--- smoothed state plot
year <- 1948:2010
par(mfrow=c(2,2),mai=0.7*c(1,1,1,1))
for (j in 2:5){
plot(year,beta[,j],type='p',xlab='time',ylab='beta',main=name[j-1])
lines(year,ss2$s[(3:65)*12+1,j])
}

##++++++ use an arbitrary state initial value
mm0 <- c(1,2,-0.2,0.8,-0.3)
CC0 <- diag(c(0.1,0.01,0.1,10,10))
ww <- c(30,30,30,10,10)
vv <- 1
init.par <- c(ww,vv)
MLE <- dlmMLE(y,init.par,buildVaryingBeta1,debug=F)
MyModel <- buildVaryingBeta1(MLE$par)
ss <- dlmSmooth(y,MyModel)
#
year <- 1948:2010
par(mfrow=c(2,2),mai=0.7*c(1,1,1,1))
for (j in 2:5){
plot(year,beta[,j],type='p',xlab='time',ylab='beta',main=name[j-1])
lines(year,ss$s[(3:65)*12+1,j])
}
```

Example 6.3: Modeling macroeconomic series with a mixed-frequency VAR model and measurement error In Section 6.3 we discussed using the SSM to model mixed-frequency time series. Typically the analysis employs the observed series directly, assuming they are observed without noises. Here we use SSM to model the low- and high-frequency series jointly. In addition, we assume that the data are contaminated by noises. Three monthly series and one quarterly series are used. The monthly series are the consumer price index (CPI), the industrial production index (IPI), and the unemployment rate. The quarterly series is the US real gross domestic product (GDP). The unemployment rate data are obtained from https://data.bls.gov/pdq/SurveyOutputServlet (Series ID: LNS14000000). The industrial production index is obtained from FRED at https://fred.stlouisfed.org (code:INDPRO). The CPI data are from http://www.bls.gov/cpi/tables.htm (CPI All Urban), and the real GDP data from http://www.bea.gov/national/.

We use the 12-month difference of monthly series y_{1t}, y_{2t}, y_{3t} and the four-quarter difference of log GDP z_{3t}. The processes are also mean-removed and multiplied by 100 (except the unemployment rate series, which is multiplied by 10). The observed series are plotted (with dots) in Figure 6.18.

We assume the corresponding true underlying monthly series are s_{it}, $i = 1, \ldots, 4$ and they follow a vector AR model

$$s_t = \boldsymbol{\Phi} s_{t-1} + \boldsymbol{W}_* \varepsilon_t,$$

where ε_t is a vector white noise distributed as $N(\boldsymbol{0}, \boldsymbol{I})$. The coefficient matrix $\boldsymbol{\Phi}$ is 4×4, and the $\boldsymbol{W}_*$ is an upper triangle matrix. We expand them into a state variable $x_t = (s_{1t}, s_{2t}, s_{3t}, s_{4t}, s_{4(t-1)}, s_{4(t-2)})'$. Hence the state equation becomes

$$x_t = \boldsymbol{H} x_{t-1} + \boldsymbol{W} w_t,$$

in which

$$\boldsymbol{H} = \begin{bmatrix} \phi_{11} & \phi_{12} & \phi_{13} & \phi_{14} & 0 & 0 \\ \phi_{21} & \phi_{22} & \phi_{23} & \phi_{24} & 0 & 0 \\ \phi_{31} & \phi_{32} & \phi_{33} & \phi_{34} & 0 & 0 \\ \phi_{41} & \phi_{42} & \phi_{43} & \phi_{44} & 0 & 0 \\ 0 & 0 & 0 & 1 & 0 & 0 \\ 0 & 0 & 0 & 0 & 1 & 0 \end{bmatrix}, \quad \boldsymbol{W} = \begin{bmatrix} w_{11} & w_{12} & w_{13} & w_{14} \\ 0 & w_{22} & w_{23} & w_{24} \\ 0 & 0 & w_{33} & w_{34} \\ 0 & 0 & 0 & w_{44} \\ 0 & 0 & 0 & 0 \\ 0 & 0 & 0 & 0 \end{bmatrix}. \quad (6.35)$$

Assume $y_{4t} = z_t$ if $t = 3s$ (when the quarterly GDP were observed) and $y_{4t} = 0$ if t is not a multiple of 3 (when the GDP was missing). The observation equation is

$$y_{it} = s_{it} + e_{it}, \quad \text{for} \quad i = 1, 2, 3$$

and for $t = 3s$,

$$z_t = s_{4t} + s_{4(t-1)} + s_{4(t-2)} + e_{4t},$$

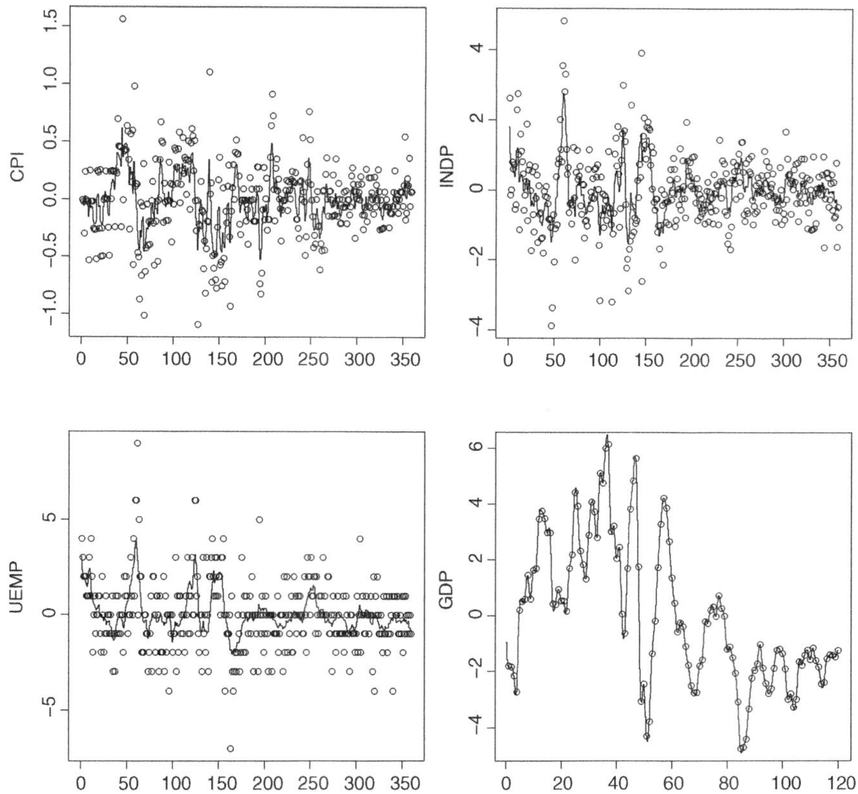

Figure 6.18 Plots of the mixed-frequency macroeconomic series (dots) and smoothed states (lines).

where e_{it} are independent noises following $N(0, \eta_i^2)$. Let $y_t = (y_{1t}, y_{2t}, y_{3t})$ when t is not a multiple of 3, and $y_t = (y_{1t}, y_{2t}, y_{3t}, z_t)$ when t is a multiple of 3. Then the observation equation can be written as

$$y_t = \boldsymbol{G}_t x_t + \boldsymbol{V}_t e_t,$$

where when $t = 3s$,

$$\boldsymbol{G}_t = \begin{bmatrix} 1 & 0 & 0 & 0 & 0 & 0 \\ 0 & 1 & 0 & 0 & 0 & 0 \\ 0 & 0 & 1 & 0 & 0 & 0 \\ 0 & 0 & 0 & 1 & 1 & 1 \end{bmatrix}, \quad \text{and} \quad \boldsymbol{V}_t = \text{diag}(\eta_1, \eta_2, \eta_3, \eta_4),$$

	Φ				W_*				V
CPI	0.70	0	0	0.02	0.17	−0.14	0.019	0.022	0.238
INDP	0	0.80	0	−0.00	0	0.42	0.436	−0.032	0.774
UEMP	0	0	0.91	0.07	0	0	0.155	−0.249	1.547
GDP	−0.16	−0.24	0.10	0.93	0	0	0	0.001	0.081

Table 6.1 MLEs of the mixed-frequency data example.

and when t is not a multiple of 3,

$$\boldsymbol{G}_t = \begin{bmatrix} 1 & 0 & 0 & 0 & 0 & 0 \\ 0 & 1 & 0 & 0 & 0 & 0 \\ 0 & 0 & 1 & 0 & 0 & 0 \end{bmatrix}, \quad \text{and} \quad \boldsymbol{V}_t = \text{diag}(\eta_1, \eta_2, \eta_3).$$

Because the dimension of the observations changes over time, it is easier to use the sequential processing approach of Section 6.3.5. To simplify the model and because we are more interested in the low-frequency component GDP, we assume that $\boldsymbol{\Phi}$ satisfies the following simplifying structure

$$\boldsymbol{\Phi} = \begin{bmatrix} \phi_{11} & 0 & 0 & \phi_{14} \\ 0 & \phi_{22} & 0 & \phi_{24} \\ 0 & 0 & \phi_{33} & \phi_{34} \\ \phi_{41} & \phi_{42} & \phi_{43} & \phi_{44} \end{bmatrix}.$$

Table 6.1 shows the MLEs of the system with log-likelihood value −360.

In Figure 6.18, the lines show the smoothed states using the MLEs. For the GDP series, the line shown is the sum of the three consecutive monthly smoothed states corresponding to the observed GDP series.

The following R codes are used to estimate the model using the maximum likelihood method via the `dlm` package. The function `buildMixedFreqSeq4(parm)` uses the parameter values stored in `parm` to build the state space model by specifying the time-varying system matrices. The program is constructed to sequentially process one observation at a time, hence one quarter is stretched into 10 steps, three monthly observations from each of the three monthly series and one observation for the quarterly GDP series. At steps 1, 4 and 7 the $\boldsymbol{H}_t$ matrix is that in (6.35). At all other steps, the state does not evolve, hence $\boldsymbol{H}_t$ is the 6×6 identity matrix and the corresponding $\boldsymbol{W}_t = \boldsymbol{0}$. The matrix $\boldsymbol{G}_t$ and $\boldsymbol{V}_t$ change according to which series the observation comes from.

R demonstration: Maximum likelihood estimation of an SSM.

```
### Build function to MLE estimation: sequential processing
buildMixedFreqSeq4=function(parm){
   nyear <- 30
   phiH <- parm[1:10]
   phiW <- parm[11:20]
   phiV <- parm[21:24]
   H <- c(phiH[1],0,0,phiH[2],0,0,
         0,phiH[3],0,phiH[4],0,0,
         0,0,phiH[5],phiH[6],0,0,
         phiH[7:10],0,0,
         0,0,0,1,0,0,
         0,0,0,0,1,0)
   H <- matrix(H,ncol=6,nrow=6,byrow=T)
   JHH <- c(1,0,0,2,0,0,
            0,3,0,4,0,0,
            0,0,5,6,0,0,
            7:10,0,0,
            0,0,0,11,12,0,
            0,0,0,0,13,14)
   JHH <- matrix(JHH,ncol=6,nrow=6,byrow=T)
   W0 <- matrix(c(phiW[1:4],
                 0,phiW[5:7],
                 0,0,phiW[8:9],
                 0,0,0,phiW[10]),ncol=4,nrow=4,byrow=T)
   W0 <- W0%*%t(W0)
   W1 <- rep(0,16)
   W <- c(c(W0[1,]),0,0,
          c(W0[2,]),0,0,
          c(W0[3,]),0,0,
          c(W0[4,]),0,0,
          rep(0,6),
          rep(0,6))
   W <- matrix(W,ncol=6,nrow=6,byrow=T)
   JWW <- c(15:18,0,0,
            19:22,0,0,
            23:26,0,0,
            27:30,0,0,
            rep(0,6),
            rep(0,6))
   JWW <- matrix(JWW,ncol=6,nrow=6,byrow=T)
   G <- c(1,0,0,0,0,0)
   G <- matrix(G,ncol=6,nrow=1,byrow=T)
   JGG <- c(31:36)
```

```
   JGG <- matrix(JGG,ncol=6,nrow=1,byrow=T)
   V <- matrix(phiV[1]**2,ncol=1,nrow=1)
   JVV <- matrix(37,ncol=1,nrow=1)
   xx1 <- c(phiH[1:10],1,0,1,0,
           c(W0),
           1,rep(0,5), phiV[1]**2)
   xx2 <- c(1,0,1,0,1,0,0,0,0,1,0,1,0,1,
           c(W1),
           0,1,rep(0,4), phiV[2]**2)
   xx3 <- c(1,0,1,0,1,0,0,0,0,1,0,1,0,1,
           c(W1),
           0,0,1,rep(0,3), phiV[3]**2)
   xx4 <- c(1,0,1,0,1,0,0,0,0,1,0,1,0,1,
           c(W1),
           0,0,0,1,1,1,phiV[4]**2)
   xx <- c(rep(c(xx1,xx2,xx3),3),xx4)
   TXX <- matrix(rep(xx,nyear),ncol=37,nrow=nyear*40,byrow=T)
   return(list(
             m0=rep(0,6),
             C0=1000000*diag(6),
             FF=G,
             GG=H,
             V=V,
             W=W,
             JFF=JGG,
             JGG=JHH,
             JWW=JWW,
             JVV=JVV,
             X=TXX
         ))
}
### end function ###
# Analysis
y_start=22 ## 1948+22=1970
nyear <- 30
#
temp <- c(t(cbind(CPIuse_m,INDPuse_m,UEMPuse_m)))
yseq <- 1:(nyear*40)
for(i in 1:(nyear*4)){
  yseq[(i-1)*10+(1:9)] <- temp[(i-1)*9+(1:9)]
  yseq[(i-1)*10+10] <- GDPuse_m[i]
}
#
phi <- c(0.7,-0.3,0.7,-0.3,0.7,-0.3,0.1,0.1,0.1,0.8)
ww <- c(0.3,0,0,0,0.5,0,0,0.5,0,0.3)
vv <- c(1,1,1,1)
```

```
parm <- c(phi,ww,vv)
MLE <- dlmMLE(yseq,parm,buildMixedFreqSeq4)
#
MyModel <- buildMixedFreqSeq4(MLE$par)
ss <- dlmSmooth(yseq,MyModel)  ## ss provides smoothed states
```

Example 6.4: Analysis of dynamic factor model Finally, we demonstrate the use of SSMs in the estimation of a DFM for a panel of macroeconomic time series. Figure 6.19 shows the GDP growth series of Germany (DEU), France (FRA), the United States (USA), Great Britain (GBR), and Canada (CAN) from the second quarter of 1990 to the fourth quarter of 2016 for a total of 107 observations. The series is demeaned and in percentage.

Table 6.2 shows the results of a PCA analysis of the panel of series, using its covariance matrix. It reveals two (possibly three) important factors. Figures 6.21 to 6.23 show the ACF, PACF, and cross-correlation plots of the first two empirical

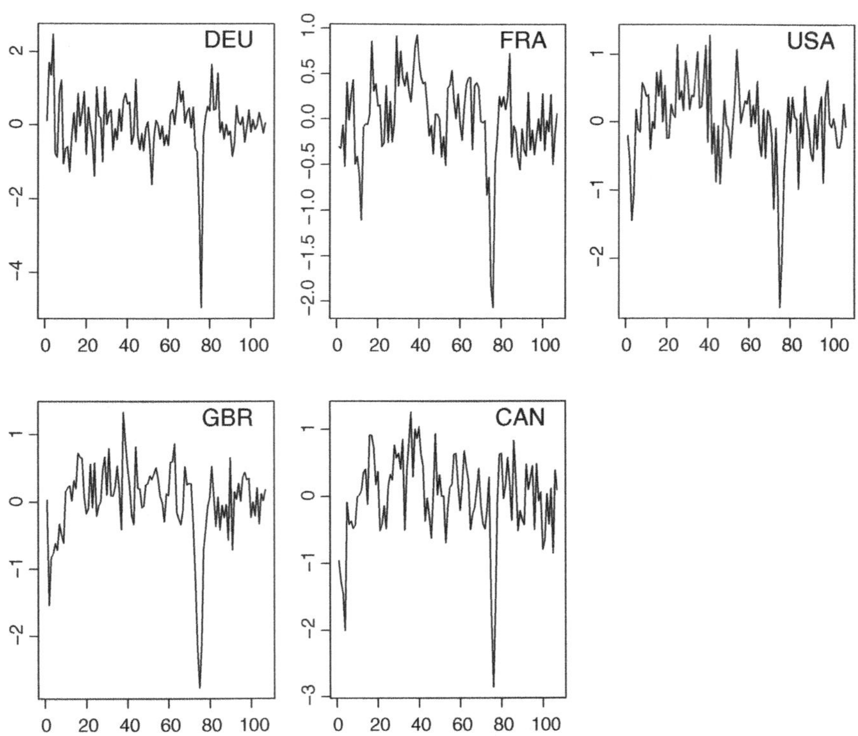

Figure 6.19 Growth rates of gross domestic products of five countries (Germany, France, the USA, the UK, and Canada) (demeaned and in percentages).

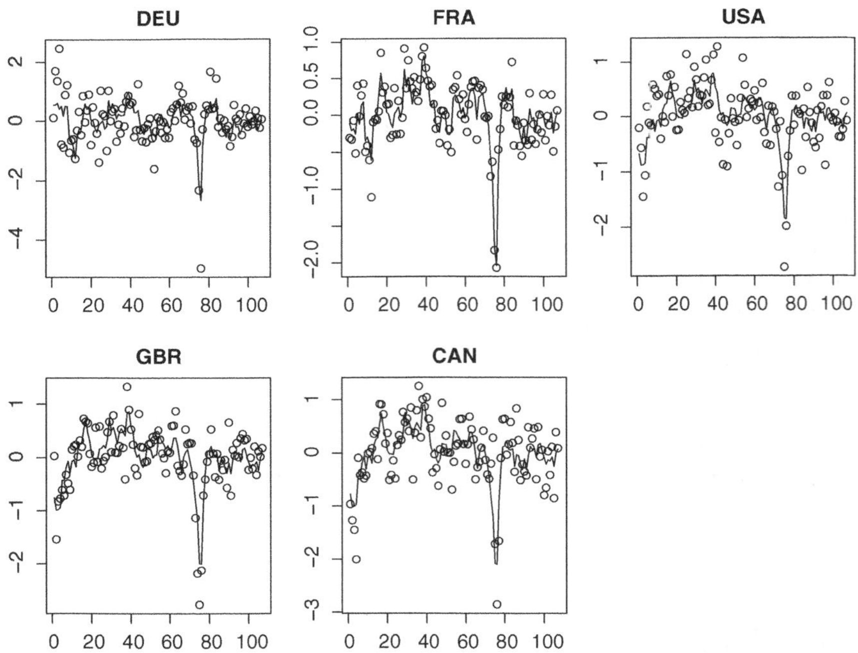

Figure 6.20 GDP growth rates of five countries (Germany, France, the USA, the UK, and Canada) (dots) and SSM fitted lines.

principle components. From the plots it seems that a VAR(1) model would be appropriate for the two estimated factors. Table 6.3 shows the estimated results of a VAR(1) model $\eta_t = c + \Phi\eta_{t-1} + e_t$ fitted to the estimated factors.

Next we set up a corresponding SSM based on the results of the PCA exercise, with two dynamic factors following a VAR(1) model. Following the discussion in Section 6.2.6, let y_t be the five-dimensional observed time series, and

	Component				
	1	2	3	4	5
Standard deviation	1.085	0.735	0.406	0.392	0.285
Proportion of variance	0.556	0.255	0.078	0.073	0.038
Cumulative proportion	0.556	0.811	0.889	0.962	1.000

Table 6.2 PCA analysis of the GDP growth rates of five countries.

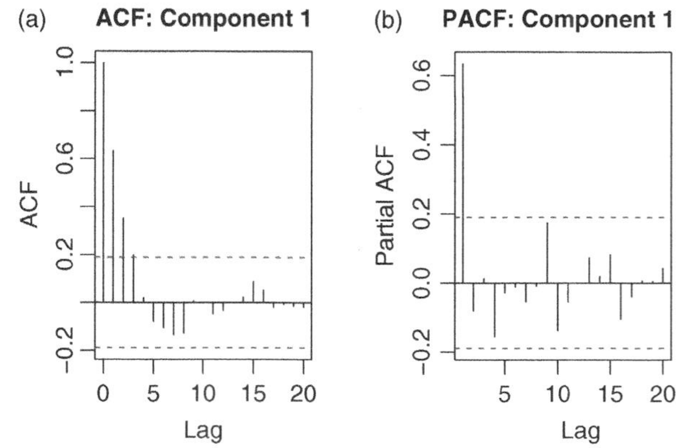

Figure 6.21 (a) ACF and (b) PACF plots of the first principle component of the GDP growth rates.

$x_t = (\eta_{1t}, \eta_{2t})'$. We have

$$x_t = \boldsymbol{H}x_{t-1} + \boldsymbol{W}w_t,$$
$$y_t = \boldsymbol{G}x_t + \boldsymbol{V}v_t,$$

where $\boldsymbol{G} = (g_{ij})$ is a 5×2 loading matrix, $\boldsymbol{V}$ is a 5×5 diagonal matrix of standard deviations of the measurement errors, $\boldsymbol{H} = (h_{ij})$ is a 2×2 VAR(1) coefficient matrix, $\boldsymbol{W}$ is a 2×2 square root matrix of the VAR(1) innovation process, and v_t

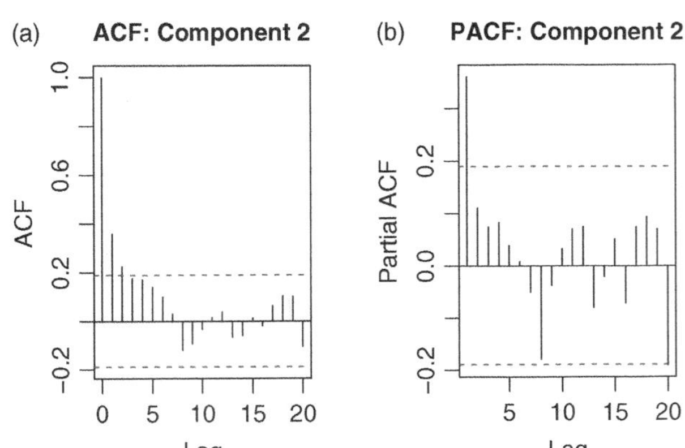

Figure 6.22 (a) ACF and (b) PACF plots of the second principle component of the GDP growth rates.

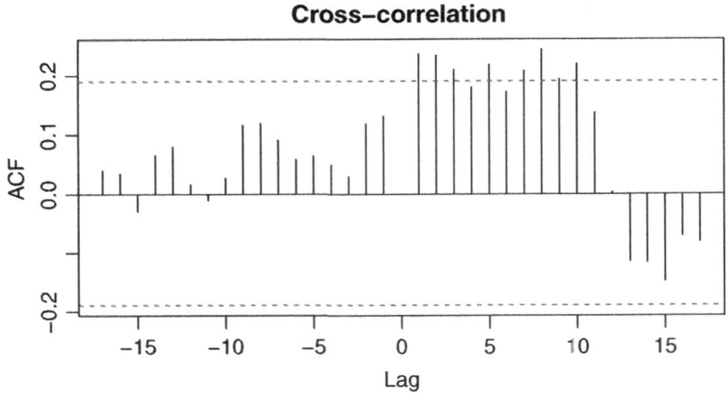

Figure 6.23 Cross-correlation plot of the first two principle components of the GDP growth rates.

and w_t are independent Gaussian white noises of corresponding dimensions. Since for any invertible matrix $\mathbf{\Gamma}$, the model can be written as

$$y_t = (\boldsymbol{G}\mathbf{\Gamma}^{-1})(\mathbf{\Gamma}x_t) + \boldsymbol{V}v_t \text{ and } \mathbf{\Gamma}x_t = (\mathbf{\Gamma}\boldsymbol{H}\mathbf{\Gamma}^{-1})(\mathbf{\Gamma}x_{t-1}) + \mathbf{\Gamma}\boldsymbol{W}w_t,$$

therefore the model is not uniquely defined. Hence we impose the constraints that $g_{11} = g_{12} = 1$ and $h_{21} = 0$. The MLEs of the model are shown in Table 6.4.

The solid lines in Figure 6.20 show the fitted series using the smoothed states and the estimated loading matrix. The total sum of squares of residuals over the total sum of squares of the original demeaned series is 0.352. Figure 6.24 shows the smoothed states. Note that the first two principal components explain approximately 81% of the total variation, as shown in Table 6.2. This percentage is higher than that of the DFM. Such a result is not surprising because the DFM does not try to minimize the contemporary residual sum of squares. Instead, it tries to establish the dynamics of the system. To see the benefit of the DFM, Table 6.5 shows the

	Constant	Φ	
Component 1	−0.006	0.634	0.350
	(0.079)	(0.072)	(0.107)
Component 2	−0.006	0.089	0.360
	(0.067)	(0.061)	(0.090)

Table 6.3 Estimated VAR(1) model of the first two PCA components of the GDP growth rates of five countries. Standard errors are shown in parenthesis.

	G		V	H		WW'	
DEU	1*	1*	0.62	0.695	0.050	0.415	−0.142
FRA	0.75	1.20	0.23	0*	0.794	−0.142	0.069
USA	0.68	1.56	0.42				
GBR	0.73	1.73	0.35				
CAN	0.77	1.78	0.42				

Table 6.4 MLEs of the DFM for the five GDP growth rates. The entries marked * are fixed a priori.

comparisons of out-of-sample one-step to three-step ahead forecasts of the GDP growth rates of each country using the SSM and PCA-VAR, respectively. For SSM, ℓ-step prediction of the states (factors) is obtained by running the Kalman forecasting step using the MLE, then using the estimated loading matrix to predict the GDP growth rates. The PCA-VAR forecast is obtained by obtaining the first two principle components, fitting a VAR(1) model to the components, obtaining the ℓ-step prediction of the factors, then using the estimate loading matrix for the ℓ-step ahead prediction. We also fit individual AR(1) models for each series and

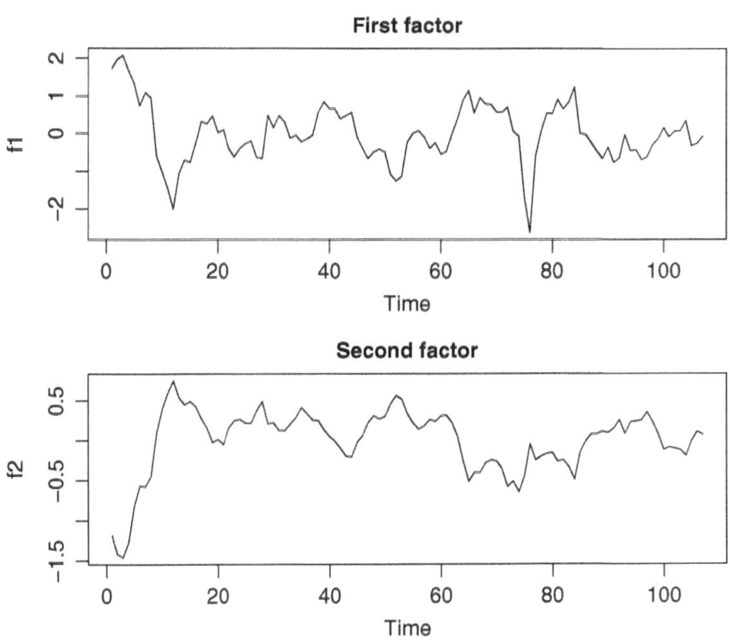

Figure 6.24 Time plots of the smoothed factors of the DFM for GDP growth rates.

ℓ		DEU	FRA	USA	GRB	CAN	Average
1	DFM	1.94	1.69	2.86	2.52	4.61	2.724
	PCA-VAR	1.78	1.61	3.40	2.79	5.14	2.943
	Univ-AR	2.30	2.20	3.33	4.14	5.49	3.489
2	DFM	2.24	1.87	2.52	2.30	4.22	2.628
	PCA-VAR	2.27	2.23	2.70	2.55	4.18	2.787
	Univ-AR	2.27	1.60	3.08	1.68	4.05	2.535
3	DFM	2.02	1.59	2.68	2.42	4.39	2.619
	PCA-VAR	2.34	1.66	2.55	2.49	4.46	2.699
	Univ-AR	2.17	1.72	2.63	2.66	4.04	2.646

Table 6.5 Out-of-sample prediction comparison between SSM, PCA-VAR, and scalar AR models. Mean squared errors are used and ℓ denotes step.

obtain ℓ-step out-of-sample predictions. For each of $s = 88 - \ell, \dots, 107 - \ell$, the respective models are estimated using data from $t = 1$ to s only, and ℓ-step prediction is obtained and compared with the true observation at time $s + \ell$. Table 6.5 shows the mean squared errors of out-of-sample ℓ-step predictions. It can be seen that the DFM outperforms the other two competing models for USA, GRB, and CAN, but its performance was marginally worse for DEU and FRA.

Figure 6.25 shows the one-step ahead prediction of US GDP growth rate for the three different methods employed.

The following R codes are used to estimate the models using the maximum likelihood method and to obtain the out-of-sample predictions. The function `buildDFM3(parm)` uses the parameter values stored in `parm` to build the state space model used in the analysis. The functions `outsample.prediction.DFM`, `outsample.prediction.PCA`, and `outsample.prediction.uAR` are used to obtain the out-of-sample predictions for the three methods employed in the analysis. We also include some essential commands, but do not display output of the commands.

R demonstration: DFM and model comparison.

```
####### buildDFM3
buildDFM3=function(parm){
   hh <- c(parm[1],parm[2],0,parm[3])
   gg <- c(1,1,parm[4:11])
   qq <- c(parm[12:13],0,parm[14])
   rr <- parm[15:19]
   H <- matrix(hh,ncol=2,nrow=2,byrow=T)
```

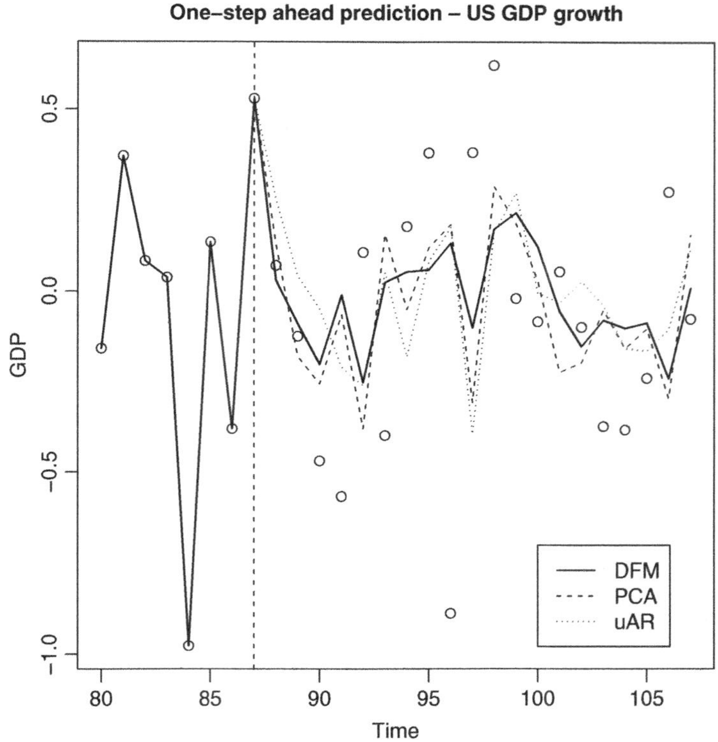

Figure 6.25 Out-of-sample one-step ahead prediction of US GDP growth rates.

```
   G <- matrix(gg,ncol=2,nrow=5,byrow=T)
   W <- matrix(qq,ncol=2,nrow=2,byrow=T)
   V <- diag(rr)
   return(list(
         m0=rep(0,2),
         C0=5*diag(2),
         FF=G,
         GG=H,
         V=V%*%t(V),
         W=W%*%t(W)
      ))
}
### end function ###
### DFM outsample prediction
outsample.prediction.DFM=function(x,np,n.Start,n.End,par.init){
   xpred <- x
```

```
  SS <- rep(0,5)
  for(i in (n.Start-np):(n.End-np)){
    MLE <- dlmMLE(x[1:i,],par.init,buildDFM3)
    myMode <- buildDFM3(MLE$par)
    filter <- dlmFilter(x[1:i,],myMode)
    pp <- dlmForecast(filter,nAhead=np)
    xpred[i+np,] <- pp$f[np,]
    SS <- SS+(x[i+np,]-xpred[i+np,])**2
  }
  return(list(xpred=xpred,SS=SS,meanSS=mean(SS)))
}
### PCA outsample prediction
outsample.prediction.PCA=function(x,np,n.Start,n.End){
  xpred <- x
  SS <- rep(0,5)
  for(i in (n.Start-np):(n.End-np)){
    PCA <- princomp(x[1:i,], cor=F)
    ff <- PCA$score[,1:2]
    out <- VAR(ff)
    pp <- predict(out,n.ahead=np)
    pp0 <- c(pp$fcst$Comp.1[np,1],pp$fcst$Comp.2[np,1])
    xpred[i+np,] <- PCA$loading[,1:2]%*%pp0
    SS <- SS+(x[i+np,]-xpred[i+np,])**2
  }
  return(list(xpred=xpred,SS=SS,meanSS=mean(SS)))
}
### univariate AR prediction
outsample.prediction.uAR=function(x,np,n.Start,n.End){
  xpred <- x
  SS <- rep(0,5)
  for(i in (n.Start-np):(n.End-np)){
    for(j in 1:5){
      out <- arima(x[1:i,j],c(1,0,0))
      xpred[i+np,j] <- predict(out,n.ahead=np)$pred[np]
    }
    SS <- SS+(x[i+np,]-xpred[i+np,])**2
  }
  return(list(xpred=xpred,SS=SS,meanSS=mean(SS)))
}
###  Analysis ###
xx.nm <- matrix(scan("GDP_panel_nm.dat"),ncol=5)
#------- PCA analysis --------------#
PCAout <- princomp(xx.nm, cor=F)
ff <- PCAout$scores[,1:2]
library(vars)
out <- VAR(ff)
```

```
#-------- DFM analysis ---------------#
hh3 <- c(0.7,0.2,0.8)
gg3 <- rep(c(0.7,0.7), 4)
qq3 <- c(1,-1,1)
rr3 <- rep(0.3,5)
parm3 <- c(hh3,gg3,qq3,rr3)  # initial parameter values
outMLE3 <- dlmMLE(xx.nm,parm3,buildDFM3)
myModel3 <- buildDFM3(outMLE3$par)
outSmooth3 <- dlmSmooth(xx.nm,myModel3)
fit3 <- xx.nm
for(i in 1:5){
   fit3[,i] <- outSmooth3$s[-1,]%*%myModel3$F[i,]
}
print(sum(xx.nm-fit3)**2)/sum(xx.nm**2))
#---------- Prediction ---------------#
for(np in 1:3){
   pred.d <- outsample.prediction.DFM(xx.nm,np,88,107,outMLE3$par)
   pred.p <- outsample.prediction.PCA(xx.nm,np,88,107)
   pred.ar <- outsample.prediction.uAR(xx.nm,np,88,107)
}
```

6.4 EXERCISES

6.1 Consider the linear zero-mean ARMA(p,p) model

$$y_t = \sum_{i=1}^{p} \phi_i y_{t-i} + \epsilon_t + \sum_{i=1}^{p} \theta_i \epsilon_{t-i},$$

where ϵ_t is a sequence of iid random variates following $N(0, \sigma^2)$. Derive the Shumway and Stoffer state space representation for y_t given in Section 6.2.1.

6.2 Again, consider the linear zero-mean ARMA(p,p) model of question 1. Let $y_{t|h} = E(y_t \mid y_1, y_2, \ldots, y_h)$ be the conditional expectation of y_t given $F_h = \sigma\{y_1, \ldots, y_h\}$, where $h \le t$. Let $x_t = (y_{t|t}, y_{t+1|t}, \cdots, y_{t+p-1|t})'$ be a p-dimensional state vector. Derive an SSM for y_t using the state variable x_t. This state space representation of ARMA models was proposed by Akaike (1975).

6.3 The file `m-dec5FF-6117.txt` contains the monthly returns of the CRSP Decile 5 portfolio and the Fama–French factors from January 1961 to June 2017. The data are in percentages. The Decile 5 returns are from CRSP and have been adjusted by removing the risk-free interest rates. Fama–French factors are from Professor K. French's website.

(a) Fit a linear regression model relating the Decile 5 excess returns to the Fama–French factors. Write down the fitted model and perform model checking.

(b) Fit a time-varying coefficient model relating the Decile 5 excess returns to the Fama–French three factors as shown in Example 6.2.

(c) Show the time plots of smoothed states using a diffused initial value.

6.4 The file `m-stateunrate.txt` contains the monthly unemployment rates of the states of California, Illinois, Michigan, New York, Pennsylvania, and Texas from January 1976 to August 2017 for 500 observations. The data are seasonally adjusted and available from FRED. For simplicity, focus on the first differenced series and denote the differenced series by $\boldsymbol{y}_t$.

(a) Perform a principal component analysis on $\boldsymbol{y}_t$ using a sample covariance matrix. Obtain the scree plot. What percentage of total variations is explained by the first two principal components?

(b) Let $\boldsymbol{z}_t$ be the first two principal component series. Obtain a suitable VAR model to $\boldsymbol{z}_t$.

(c) Put the fitted VAR model for $\boldsymbol{z}_t$ into a SSM.

(d) Fit a dynamic factor model to $\boldsymbol{y}_t$ using two common factors similar to that of Example 6.4.

(e) Obtain the time plots of smoothed factors based on the fitted SSM.

6.5 Seasonal adjustment via SSM. The file `CAURN.csv` contains the monthly unemployment rates of California from January 1976 to August 2017. The rates are not seasonally adjusted. Let y_t be the unemployment rate and consider the model

$$y_t = T_t + S_t + I_t,$$

where T_t is the trend component following the local trend model, S_t denotes the seasonal component following the model $\sum_{i=1}^{12} S_{t-i+1} = \epsilon_t$, where ϵ_t is a Gaussian white noise series with mean zero and variance σ_ϵ^2, and I_t is the irregular component following an AR(5) model.

(a) Write down an SSM for y_t. What is the dimension of the state vector?

(b) Fit the specified SSM model and write down the parameter estimates.

(c) Obtain a smoothed seasonal component S_t. Denote the series by $\widehat{S}_t$.

(d) Let $z_t = y_t - \widehat{S}_t$ be the seasonally adjusted series of California monthly unemployment rate. Let ρ_{12} be the lag-12 autocorrelation of z_t. Test $H_0 : \rho_{12} = 0$ versus $H_a : \rho_{12} \neq 0$ and draw a conclusion using the 5% type-I error.

REFERENCES

Adrian, T. and Franzoni, F. (2009). Learning about beta: Time-varying factor loadings, expected returns, and the conditional CAPM. *Journal of Empirical Finance* **16**: 537–556.

Ahmad, B. I., Murphy, J. K., Langdon, P. M., Godsill, S. J., Hardy, R., and Skrypchuk, L. (2016). Intent inference for hand pointing gesture-based interactions in vehicles. *IEEE Transactions on Cybernetics* **46**: 878–889.

Aït-Sahalia, Y. and Jacod, J. (2014). *High-Frequency Financial Econometrics*. Princeton University Press, Princeton, NJ.

Akaike, H. (1975). Markovian representation of stochastic processes by canonical variables. *Siam Journal on Control* **13**: 162–173.

Andrieu, C., De Freitas, N., and Doucet, A. (1999). Sequential MCMC for Bayesian model selection. *IEEE Signal Processing Workshop on Higher-Order Statistics*, pp 130–134.

Aoki, M. (2013). *State Space Modeling of Time Series*. Universitext, Springer-Verlag, Berlin, Heidelberg.

Arulampalam, M. S., Maskell, S., Gordon, N., and Clapp, T. (2002). A tutorial on particle filters for online nonlinear/non-Gaussian Bayesian tracking. *IEEE Transactions on Signal Processing* **50**: 174–188.

Arulampalam, M. S., Ristic, B., Gordon, N., and Mansell, T. (2004). Bearings-only tracking of maneuvering targets using particle filters. *EURASIP Journal of Applied Signal Processing* **2004**: 2351–2365.

Avitzour, D. (1995). Stochastic simulation Bayesian approach to multitarget tracking. *IEE Proceedings on Radar, Sonar and Navigation* **142**: 41–44.

Bai, J. and Ng, S. (2002). Determining the number of factors in approximate factor models. *Econometrica* **70**: 191–221.

Bai, J., and Ng, S. (2008). Large dimensional factor snalysis. *Foundations and Trends in Econometrics* **3**(2): 89–163

Bar-Shalom, Y. and Fortmann, T. (1988). *Tracking and Data Association*. Academic Press, Boston.

Berzuini, C. and Best, N.G., Gilks, W.R., and Larizza, C. (1997). Dynamic conditional independence models and Markov chain Monte Carlo methods. *Journal of the American Statistical Association* **92**: 1403–1412.

Blake, A., Bascle, B., Isard, M., and MacCormick, J. (1998). Statistical models of visual shape and motion. *Philosophical Transactions of the Royal Society A* **356**: 1283–1302.

Blake, A., Isard, M., and MacMormick, J. (2001). Statistic models of visual shape and motion. In Doucet, A. and de Freitas, J. F. G. and Gordon, N. J. (eds), *Sequential Monte Carlo in Practice*. Springer-Verlag, New York.

Bouchard-Côté, A., Sankararaman, S., and Jordan, M. I. (2012). Phylogenetic inference via sequential Monte Carlo. *Systematic Biology* **61**: 579–593.

Box, G. E. P. and Jenkins, G. M. (1976). *Time Series Analysis, Forecasting and Control*, revised edition. Holden Day, San Francisco.

Bruno, M. G. and Moura, J. M. (2000). Multiframe Bayesian tracking of cluttered targets with random motion. *Proceedings of the 2000 International Conference on Image Processing* **3**: 90–93.

Bruno, M. G., Araújo, R. V., and Pavlov, A. G. (2007). Sequential Monte Carlo methods for joint detection and tracking of multiaspect targets in infrared radar images. *EURASIP Journal of Advances in Signal Processing* **2008**: 217–373.

Caines, P. E. (1988). *Linear Stochastic Systems*. Wiley, New York.

Campbell, J. Y., Lettau, M., Malkiel, B. G. , and Xu, Y. (2001). Have individual stocks become more volatile? An empirical exploration of idiosyncratic risk. *Journal of Finance* **56**: 1–43.

Chamberlain, G. (1983). Funds, Factors, and Diversication in Arbitrage Pricing Models. *Econometrica* **51**(5): 1305–1323.

Chamberlain, G. and Rothschild, M. (1983). Arbitrage, factor structure, and mean-variance analysis on large asset markets. *Econometrica* **51**(5): 1281–1304. 304.

Chen, R., Wang, X., and Liu, J.S. (2000). Adaptive joint detection and decoding in flat-fading channels via mixture Kalman filtering. *IEEE Transactions on Information Theory* **46**: 2079–2094.

Chen, X., Särkkä, S., and Godsill, S. (2015). A Bayesian particle filtering method for brain source localisation. *Digital Signal Processing* **47**: 192–204.

Chitode, J. S. (2009). *Digital Communications*. Technical Publications.

Churchill, G. A. (1989). Stochastic models for heterogeneous DNA sequences. *Bulletin of Mathematical Biology* **51**: 79–94.

Coates, M. J. and Nowak, R. (2002). Sequential Monte Carlo inference of internal delays in nonstationary data networks. *IEEE Transactions on Signal Processing* **50**: 366–376.

Cox, J. C., Ingersoll, J. E., and Ross, S. A. (1985). A theory of the term structure of interest rates. *Econometrica* **53**: 385–407.

Creal, D. D. and Tsay, R. S. (2015). High dimensional dynamic stochastic copula models. *Journal of Econometrics* **189**: 335–345.

de Freitas, J. F. G., Niranjan, M., Gee, A. H., and Doucet, A. (2000). Sequential Monte Carlo methods to train neural network models. *Neural Computation* **12**: 955–993.

De Jong, P. (1989). Smoothing and interpolation with the state-space model. *Journal of the American Statistical Association* **84**: 1085–1088.

Dellaert, F., Burgard, W., Fox, D., and Thrun, S. (1999). Using the condensation algorithm for robust, vision-based mobile robot localization. *IEEE International Conference on Computer Vision and Pattern Recognition*. IEEE, Fort Collins, CO.

Diebold, F. X. and Li, C. (2006). Forecasting the term structure of goverment bond yields. *Journal of Econometrics* **130**: 337–364.

Djuric, P. (2001). Applications of Monte Carlo particle filters in signal processing. *Proceedings of the International Statistical Institute*. Seoul, South Korea.

Djuric, P. M., Kotecha, J. H., Zhang, J., Hunag, Y., Ghirmai, T., Bugallo, M. F., and Miguez, J. (2003). Particle filtering. *IEEE Signal Processing Magazine* **20**: 19–38.

Douc, R., Moulines, E., and Stoffer, D. (2014). *Nonlinear Time Series: Theory, Methods and Applications with R Examples*. Taylor & Francis.

Doz, C., Giannone, D., and Reichlin, L. (2007). *A Two-Step Estimator for Large Approximate Dynamic Factor Models Based on Kalman Filtering*. CEPR Discussion Papers 6043, CEPR.

Duffie, D. and Kan, R. (1998). A yield factor model of interest rates. *Mathematical Finance* **6**: 379–406.

Durbin, J. and Koopman, S. J. (2001). *Time Series Analysis by State Space Methods*. Oxford University Press, Oxford.

Faff, R. W., Hillier, D., and Hillier, J. (2000). Time varying beta risk: An analysis of alternative modeling techniques. *Journal of Business Finance & Accounting* **27**: 523–554.

Fama, E. F. (1970). Efficient capital markets: A review of theory and empirical work. *Journal of Finance* **25**: 383–417.

Fama, E. F. and French, K. R. (1993). The cross-section of expected stock returns. *Journal of Finance* **47**: 427–465.

Fama, E. F. and French, K. R. (1996). Multifactor explanations of asset pricing anomalies. *Journal of Finance* **51**: 55–84.

Fong, W., Godsill, S.J., Doucet, A., and West, M. (2002). Monte Carlo smoothing with application to speech enhancement. *IEEE Transactions on Signal Processing* **50**: 438–449.

Forni, M., Giannone, D., Lippi, M., and Reichlin, L. (2009). Opening the Black Box: Structural factor models with large cross sections. *Econometric Theory* **25**(05): 1319–1347.

Forni, M., Hallin, M., Lippi, M., and Reichlin, L. (2000). The generalized dynamic-factor model: Identification and estimation. *The Review of Economics and Statistics* **82**(4): 540–554.

Fouque, J. P., Papanicolaou, G., and Sircar, K. R. (2000). *Derivatives in Financial Markets with Stochastic Volatility*. Cambridge University Press, Cambridge.

Fouque, J. P., Papanicolaou, G., Sircar, R., and Sølna, K. (2011). *Multiscale Stochastic Volatility for Equity, Interest Rate, and Credit Derivatives*. Cambridge University Press, Cambridge.

Fox, D., Burgard, W., Dellaert, F., and Thrun, S. (1999). Monte Carlo localization: Efficient position estimation for mobile robots. *Proceedings of National Conference on Artificial Intelligence*. AAAI, Orlando, FL.

Fox, D., Thrun, S., Burgard, W., and Dellaert, F. (2001). Particle filters for mobile robot localization. In Doucet, A. and de Freitas, J. F. G. and Gordon, N. J. (eds), *Sequential Monte Carlo in Practice*. Springer-Verlag, New York.

Gagliardini, P., Ghysels, E., and Rubin, M. (2017). Indirects inference estimation of mixed frequency stochastic volatility state space models using MIDAS regressions and ARCH models. *Journal of Financial Econometrics*, nbw013.

Ghysels, E. (2016). Macroeconomics and the reality of mixed frequency data. *Journal of Econometrics* **193**: 294–314.

Ghysels, E., Harvey, A. C., and Renault, E. (1996). 5 Stochastic volatility. *Handbook of Statistics* **14**: 119–191.

Ghysels, E., Sinko, A., and Valkanov, R. (2007). MIDAS regressions: Further results and new directions. *Econometric Reviews* **16**: 53–90.

González-Rivera, G. (1997). The pricing of time-varying beta. *Empirical Economics* **22**: 345–363.

Grelaud, A., Mitra, P., Xie, M., and Chen, R. (2018). State space approach for nuclear surveillance with a mobile sensor network. *Applied Stochastic Models in Business and Industry*. **34**: 3–19.

Guo, D. and Wang, X. (2004). Dynamic sensor collaboration via sequential Monte Carlo. *IEEE Journal on Selected Areas in Communications* **22**: 1037–1047.

Gustaffson, F., Gunnarsson, F., Bergman, N., Forssell, U., Jansson, J., Karlsson, R., and Nordlund, P. J. (2002). Particle filters for positioning, navigation, and tracking. *IEEE Transaction on Signal Processing* **50**: 425–437.

Haggan, V. and Ozaki, T. (1981). Modeling nonlinear vibrations using an amplitude-dependent autoregressive time series model. *Biometrika* **68**: 189–196.

Hallin, M. and Liška, R. (2007). Determining the number of factors in the general dynamic factor model. *Journal of the American Statistical Association* **102**: 603–617.

Hamilton, J. D. and Lin, G. (1996). Stock market volatility and the business cycle. *Journal of Applied Econometrics* **11**: 573–593.

Hannan, E. J. and Deistler, M. (1988). *The Statistical Theory of Linear Systems*. Wiley, Hoboken, NJ.

Harvey, A. (1989). *Forecasting, Structure Time Series Models and the Kalman Filter*. Cambridge University Press, Cambridge.

Harvey, A. C. (1993). *Time Series Models*. Harvester Wheatsheaf, New York.

Harvey, A., Ruiz, E., and Shephard, N. (1994). Multivariate stochastic variance models. *The Review of Economic Studies* **61**: 247–264.

Haug, A. J. (2012). *Bayesian Estimation and Tracking: A Practical Guide*. Wiley, Hoboken, NJ.

Huang, T., Koller, D., Malik, J., Ogasawara, G., Rao, B., Russell, S., and Weber, J. (1994). Automatic symbolic traffic scene analysis using belief networks. *Proceedings of the Twelfth National Conference on Artificial Intelligence (AAAI)*, pp. 966–972.

Heath, D., Jarrow, R., and Morton, A. (1992). Bond pricing and the term structure of interest rates: a new methodology for contingent claims valuation. *Econometrica* **60**: 77–105.

Hue, C., Le Cadre, J-P., and Perez, P. (2002). Sequential Monte Carlo methods for multiple target tracking and data fusion. *IEEE Transactions on Signal Processing* **50**: 309–325.

Hull, J. and White, A. (1987). The pricing of options on assets with stochastic volatilities. *Journal of Finance* **42**: 281–300.

Hull, J. and White, A. (1990). Pricing interest-rate-derivative securities. *Review of Financial Studies* **3**: 573–592.

Irwin, M. E., Cressie, N., and Johannesson, G. (2002). Spatial-temporal nonlinear filtering based on hierarchical statistical models. *Test* **11**: 249–302.

Irwing, M., Cox, N., and Kong, A. (1994). Sequential imputation for multilocus linkage analysis. *Proceedings of the National Academy of Science, USA* **91**: 11684–11688.

Isard, M. and Blake, A. (1996). Contour tracking by stochastic propagation of conditional density. *European Conference on Computer Vision*, pp 343–356. Springer, Berlin, Heidelberg.

Itohara, T., Otsuka, T., Mizumoto, T., Ogata, T., and Okuno, H.G. (2011). Particle-filter based audio-visual beat-tracking for music robot ensemble with human guitarist. *IEEE/RSJ International Conference on Intelligent Robots and Systems*. pp 118–124.

Jacquier, E., Polson, N. G., and Rossi, P. E. (1994). Bayesian analysis of stochastic volatility models (with discussion). *Journal of Business and Economic Statistics* **12**: 371–417.

Kalman, R. E. (1960). A new approach to linear filtering and prediction problems. *Journal of Basic Engineering* **82**: 35–45.

Koller, D. and Lerner, U. (2001). Sampling in factored dynamic systems. In *Sequential Monte Carlo in Practice*, A. Doucet, J. F. G. de Freitas, and N. J. Gordon (eds), Springer-Verlag, New York.

Kong, A., Liu, J., and Wong, W. (1994). Sequential imputations and Bayesian missing data problems. *Journal of the American Statistical Association* **89**: 278–288.

Krough, A., Brown, M., Mian, I.S., Sjolander, K., and Haussler, D. (1994). Hidden Markov models in computational biology: Applications to protein modeling. *Journal of Molecular Biology* **235**: 1501–1531.

Kronhamn, T. R. (1998). Bearings-only target motion analysis based on a multihypothesis Kalman filter and adaptive ownship motion control. *IEE Proceedings – Radar, Sonar and Navigation* **145**: 247–252.

Kung, S. Y. (1988). *VLSI Array Processors*. Prentice Hall, Englewood Cliffs, NJ.

Liang, J., Chen, R., and Zhang, J. (2002). On statistical geometry of packing defects of lattice chain polymer. *Journal of Chemical Physics* **117**: 3511–3521.

Lillo, F. and Mantegna, R. N. (2000a). Statistical properties of statistical ensemble of stock returns. *International Journal of Theoretical and Applied Finance* **3**: 405–408.

Lillo, F. and Mantegna, R. N. (2000b). Symmetry alteration of ensemble return distribution in crash and rally days of financial markets. *European Physical Journal B – Condensed Matter and Complex Systems* **15**: 603–606.

Lin, M. and Chen, R. (2017). *Functional and distributional time series driven by dynamic processes*. Technical Report, Department of Statistics, Rutgers University.

Lin, M., Chen, R., and Mykland, P. (2010). On generating Monte Carlo samples of continuous diffusion bridges. *Journal of the American Statistical Association* **105**: 820–838.

Liu, T., Bahl, P., and Chlamtac, I. (1998). Mobility modeling, location tracking, and trajectory prediction in wireless ATM networks. *IEEE Journal on Selected Areas in Communications* **16**: 922–936.

Liu, J. S., Neuwald, A. F., and Lawrence, C. E. (1999). Markovian structures in biological sequence alignments. *Journal of the American Statistical Association* **94**: 1–15.

Lopes, H. and Tsay, R. S. (2011). Particle filters and Bayesian inference in financial econometrics. *Journal of Forecasting* **30**: 168–209.

Marrs, A. D. (2001). In-Situ ellipsometry solutions using sequential Monte Carlo. In *Sequential Monte Carlo in Practice*, A. Doucet, J. F. G. de Freitas, and N. J. Gordon (eds), Springer-Verlag, New York.

Murphy, K. and Russell, S. (2001). Rao–Blackwellised particle filtering for dynamic Bayesian networks. In *Sequential Monte Carlo in Practice*, A. Doucet, J. F. G. de Freitas, and N. J. Gordon (eds), Springer-Verlag, New York.

Naesseth, C. A., Lindsten F., and Schön, T. B. (2014). Sequential Monte Carlo for graphical models. *Advances in Neural Information Processing Systems* **27**: 1862–1870.

Nelson, C. and Seigel, A. (1987). Parsimonious modeling of yield curve. *Journal of Business* **60**: 473–489.

Nickel, K., Gehrig, T., Ekenel, H. K., McDonough, J., and Stiefelhagen, R. (2007). An audio-visual particle filter for speaker tracking on the CLEAR'06 evaluation dataset. In *Multimodal Technologies for Perception of Humans: First International Evaluation Workshop on Classification of Events, Activities and Relationships, CLEAR 2006, Southampton, UK, April 6–7, 2006, Revised Selected Papers*, R. Stiefelhagen and J. Garofolo (eds), pp. 69–80. Springer, Berlin.

Orton, M. and Fitzgerald, W. (2002). A Bayesian approach to tracking multiple targets using sensor arrays and particle filters. *IEEE Transactions on Signal Processing* **50**: 216–223.

Paris, S. and Jauffret, C. (2004). Frequency line tracking using HMM-based schemes [passive sonar]. *IEEE Transactions on Aerospace and Electronic Systems* **39**: 439–449.

Peach, N. (1995). Bearings-only tracking using a set of range-parameterized extended Kalman filters. *IEE Proceedings – Control Theory and Applications* **142**: 73–80.

Pitt, M. K. and Shephard, N. (1999). Filtering via simulation: auxiliary particle filters. *Journal of the American Statistical Association* **94**: 590–599.

Proakis, J. G. (2001). *Digital Communications.* Electrical Engineering Series. McGraw Hill, New York.

Ristic, B., Arulampalam, S., and Gordon, N. (2004). *Beyond the Kalman filter: Particle filters for tracking applications.* Artech House Publishers.

Salmond, D. and Gordon, N. (2001). Particles and mixtures for tracking and guidance. In *Sequential Monte Carlo in Practice*, pp 517–532. Springer, New York.

Sargent, T. J. and Sims, C. A. (1977). Business cycle modeling without pretending to have too much a priori economic theory. In *New Methods in Business Research*, C. A. Sims (ed.).

Schwert, G. W. (1989). Why does stock market volatility change over time? *Journal of Finance* **44**: 1115–1153.

Sharpe, W. F. (1964). Capital asset prices: a theory of market equilibrium under conditions of risk. *Journal of Finance* **19**: 425–442.

Shephard, N. (1996). Statistical aspects of ARCH and stochastic volatility. *Monographs on Statistics and Applied Probability* **65**: 1–68.

Shephard, N. and Pitt, M. K. (1997). Likelihood analysis of non-Gaussian measurement time series. *Biometrika* **84**: 653–667.

Shumway, R. and Stoffer, D. (1982). An approach to time series smoothing and forecasting using the EM algorithm. *Journal of Time Series Analysis* **3**: 253–264.

Shumway, R. and Stoffer, D. (1991). Dynamic linear models with switching. *Journal of the American Statistical Association* **86**: 763–769.

Shumway, R. and Stoffer, D. (2006). *Time Series Analysis and Its Applications: With R Examples*. Springer.

Smith, R. A., Ionides, E. L., and King, A. A. (2017). Infectious disease dynamics inferred from genetic data via sequential Monte Carlo. *Molecular Biology and Evolution* **34**: 2065–2084.

Spiegelhalter, D. J. and Lauritzen, S. L. (1990). Sequential updating of conditional probabilities on directed graphical structures. *Network* **20**: 579-605.

Srivastava, A., Lanterman, A. D., Grenander, U., Loizeaux, M., and Miller, M. I. (2001). Monte Carlo techniques for automated target recognition. *Sequential Monte Carlo in Practice*, pp 533–552. Springer, New York.

Stock, J. H. and Watson, M. W. (2006). Forecasting with Many Predictors. In *Handbook of Economic Forecasting*, G. Elliott, C. Granger, and A. Timmermann (eds), pp. 515–554. North Holland.

Stock, J. H. and Watson, M. W. (2012). Dynamic Factor Models. In *The Oxford Handbook of Economic Forecasting*, M. P. Clements and D. F. Hendry (eds), Oxford University Press.

Tong, H. (1990). *Nonlinear Time Series Analysis: A Dynamical System Approach*. Oxford University Press, London.

Torma, P. and Szepesvari, C. (2003). *Sequential importance sampling for visual tracking reconsidered*. International Workshop on Artificial Intelligence and Statistics (AISTATS), pp. 271–278, January 2003.

Tsay, R. S. (2005). *Analysis of Financial Time Series*, 2nd edition. Wiley, Hoboken, NJ.

Tsay, R. S. (2010). *Analysis of Financial Time Series*, 3rd edition. Wiley, Hoboken, NJ.

Tsay, R. S. (2016). Some methods for analyzing big dependent data. *Journal of Business & Economic Statistics* **34**: 673–688.

Wang, X., Chen, R., and Liu, J. S. (2002). Monte Carlo Bayesian signal processing for wireless communication. *Journal of VLSI Signal Processing* **30**: 89–105.

West, M. and Harrison, J. (1989). *Bayesian Forecasting and Dynamic Models*. Springer-Verlag, New York.

Wong, W. H. and Liang, F. (1997). Dynamic weighting in Monte Carlo and optimization. *Proceedings of the National Academy of Sciences* **94**: 14220–14224.

Zhang, J., Chen, R., Tang, C., and Liang, J. (2003). Origin of scaling behavior of protein packing density: A sequential Monte Carlo study of compact long chain polymer. *Journal of Chemical Physics* **118**: 6102–6109.

Zhang, J., Dundas, J., Lin, M., Chen, R., Wang, W., and Liang, J. (2009). Prediction of geometrically feasible three-dimensional structures of pseudo-knotted RNA through free energy estimation. *RNA* **15**: 2248–2263.

Zheng, T. and Chen, R. (2017). Dirichlet ARMA models for compositional time series. *Journal of Multivariate Analysis* **158**: 31–46.

CHAPTER 7

NONLINEAR STATE SPACE MODELS

This chapter is a continuation of Chapter 6, focusing on analysis and applications of nonlinear state space models (NLSSMs), including hidden Markov models (HMMs). The goal of the chapter is to demonstrate further that state space models (SSMs) and their nonlinear extensions are widely applicable. Our discussions start with using linear and Gaussian SSMs to approximate NLSSMs and introducing various extensions of the Kalman filter. Some simulated examples are used in the demonstration for ease in understanding. We then turn to hidden Markov models and discuss the associated filtering and smoothing methods. Two real examples are used to demonstrate the applications of HMM.

7.1 LINEAR AND GAUSSIAN APPROXIMATIONS

In many applications, the underlying system is often nonlinear and non-Gaussian, resulting in an NLSSM. For such a model, although $p(x_1, \ldots, x_t \mid \boldsymbol{y}_t)$ can explicitly

Nonlinear Time Series Analysis, First Edition. Ruey S. Tsay and Rong Chen.

Companion website: www.wiley.com/go/tsay/nonlineartimeseries

be written, as shown in Equation (6.5), it is often impossible to obtain any meaningful inference in an analytical form, such as computing the expectation of the state x_t given $\boldsymbol{y}_t = (y_1, \ldots, y_t)$,

$$E(x_t \mid \boldsymbol{y}_t) = \int \cdots \int x_t p(x_1, \ldots, x_t \mid \boldsymbol{y}_t) dx_1 \ldots dx_t,$$

for which a high dimensional integration is needed. Traditional approaches to studying systems like this one are to apply linear approximations.

7.1.1 Kalman Filter for Linear Non-Gaussian Systems

The Kalman filter can be used as the best linear unbiased estimator (BLUE) for linear non-Gaussian systems. When the system is linear but the noise is not Gaussian, $\mu_{t|t}$ obtained through running the Kalman filter is the BLUE of x_t (at time t without future observations) and $\mu_{t|T}$ obtained through running the Kalman smoother is also the BLUE, given the entire sequence of data $y_1, \ldots, y_T$.

Assume that we have the system in Equations (6.18) and (6.19), but the noises v_t and w_t are not Gaussian. We still assume $E(v_t) = 0$, $\text{Var}(v_t) = \boldsymbol{I}$, $E(w_t) = 0$, $\text{Var}(w_t) = \boldsymbol{I}$ and start with the initial values μ_0 and $\boldsymbol{\Sigma}_0$. Duncan and Horn (1972) showed that the Kalman filter is the BLUE of x_t given $\boldsymbol{y}_t$. Their argument is similar to that of the Gauss–Markov theorem. Durbin and Koopman (2001, Section 4.3) provided an alternative argument. For two random variables $(\boldsymbol{x}, \boldsymbol{y})$, the BLUE of $\boldsymbol{x}$ given $\boldsymbol{y}$ can be shown to be in the form of $\boldsymbol{\mu}_x + \boldsymbol{\Sigma}_{xy}\boldsymbol{\Sigma}_{yy}^{-1}(\boldsymbol{y} - \boldsymbol{\mu}_y)$. This is exactly in the same form as the BLUE under the joint normal distribution. Since the Kalman filter and the Kalman smoother are recursive ways of finding the solution of such an estimator, they also provide the BLUE for non-Gaussian systems. Note that the BLUE property of the Kalman filter and the Kalman smoother only holds for linear systems (with non-Gaussian innovations and noises), since only in linear systems are the mean and variance structure of the joint distribution the same as in the Gaussian case, and the conditional mean and variance formula under the Gaussian case remain BLUE for non-Gaussian linear systems.

7.1.2 Extended Kalman Filters for Nonlinear Systems

Gelb (1974) used the Taylor expansion to approximate nonlinear but Gaussian systems. Specifically, consider the Markovian SSM in Equations (6.3) and (6.4). We assume $E(\varepsilon_t) = \boldsymbol{0}$, $E(e_t) = \boldsymbol{0}$, $\text{Var}(\varepsilon_t) = \boldsymbol{R}_{1t}$, and $\text{Var}(e_t) = \boldsymbol{R}_{2t}$. Let $\varepsilon_t = \boldsymbol{R}_{1t}^{1/2} w_t$ and $e_t = \boldsymbol{R}_{2t}^{1/2} v_t$ and let the functions $h_t(\cdot)$ and $g_t(\cdot)$ absorb the covariance matrices $\boldsymbol{R}_{1t}$ and $\boldsymbol{R}_{2t}$. Then we have

$$x_t = h_t(x_{t-1}, w_t), \tag{7.1}$$

$$y_t = g_t(x_t, v_t), \tag{7.2}$$

where $w_t \sim N(0, \boldsymbol{I})$ and $v_t \sim N(0, \boldsymbol{I})$ of proper dimensions.

Suppose at time $t-1$ we have obtained $\mu_{t-1} = E(x_{t-1} \mid \boldsymbol{y}_{t-1})$ and $\boldsymbol{\Sigma}_{t-1} = \mathrm{Var}(x_{t-1} \mid \boldsymbol{y}_{t-1})$ (or their approximations). Then at time t, using the first-order Taylor expansion of $h_t(\cdot,\cdot)$ around $(\mu_{t-1}, 0)$, we have

$$h_t(x_{t-1}, w_t) \approx h_t(\mu_{t-1}, 0) + \boldsymbol{H}_{t|t-1}(x_{t-1} - \mu_{t-1}) + \boldsymbol{W}_{t|t-1} w_t, \tag{7.3}$$

where

$$\boldsymbol{H}_{t|t-1} = \left. \frac{\partial h_t(x_{t-1}, w_t)}{\partial x_{t-1}} \right|_{(x_{t-1}, w_t) = (\mu_{t-1}, 0)}$$

and

$$\boldsymbol{W}_{t|t-1} = \left. \frac{\partial h_t(x_{t-1}, w_t)}{\partial w_t} \right|_{(x_{t-1}, w_t) = (\mu_{t-1}, 0)}.$$

Let $\mu_{t|t-1} = h_t(\mu_{t-1}, 0)$ and we use the Taylor expansion of $g_t(\cdot,\cdot)$ around $(\mu_{t|t-1}, 0)$ to get

$$g_t(x_t, v_t) \approx g_t(\mu_{t|t-1}, 0) + \boldsymbol{G}_{t|t-1}(x_t - \mu_{t|t-1}) + \boldsymbol{V}_{t|t-1} v_t, \tag{7.4}$$

where

$$\boldsymbol{G}_{t|t-1} = \left. \frac{\partial g_t(x_t, v_t)}{\partial x_t} \right|_{(x_t, v_t) = (\mu_{t|t-1}, 0)}$$

and

$$\boldsymbol{V}_{t|t-1} = \left. \frac{\partial g_t(x_t, v_t)}{\partial v_t} \right|_{(x_t, v_t) = (\mu_{t|t-1}, 0)}.$$

Applying the Kalman filter to the approximated system, we have the following (one-step) extended Kalman filter algorithm.

Algorithm 7.1: Extended Kalman filter (EKF)

$$\mu_{t|t-1} = h_t(\mu_{t-1}, 0), \tag{7.5}$$

$$\boldsymbol{\Sigma}_{t|t-1} = \boldsymbol{H}_{t|t-1} \boldsymbol{\Sigma}_{t-1} [\boldsymbol{H}_{t|t-1}]' + \boldsymbol{W}_{t|t-1} \boldsymbol{W}'_{t|t-1}, \tag{7.6}$$

$$\mu_t = \mu_{t|t-1} + \boldsymbol{K}_t (y_t - g_t(\mu_{t|t-1}, 0)), \tag{7.7}$$

$$\boldsymbol{\Sigma}_t = \boldsymbol{\Sigma}_{t|t-1} - \boldsymbol{K}_t \boldsymbol{G}_{t|t-1} \boldsymbol{\Sigma}_{t|t-1}, \tag{7.8}$$

where

$$\boldsymbol{K}_t = \boldsymbol{\Sigma}_{t|t-1} \boldsymbol{G}'_{t|t-1} [\boldsymbol{G}_{t|t-1} \boldsymbol{\Sigma}_{t|t-1} \boldsymbol{G}'_{t|t-1} + \boldsymbol{V}_{t|t-1} \boldsymbol{V}'_{t|t-1}]^{-1}. \tag{7.9}$$

The extended Kalman filter only uses the first-order Taylor expansions. The approximation accuracy can be poor in some cases. For example, (7.5) uses the approximation in the form of

$$\begin{aligned} E(x_t \mid y_{t-1}) &= E(h(x_{t-1}, w_t) \mid y_{t-1}) \\ &\approx h(E(x_{t-1} \mid y_{t-1}), E(w_t \mid y_{t-1})) \\ &= h_t(\mu_{t-1}, 0). \end{aligned}$$

For certain nonlinear functions $h_t(\cdot)$, such an approximation should be avoided. One can extend the EKF to include the second-order terms in the Taylor expansion to obtain a better approximation. See Tanizaki (1996) and Roth and Gustafsson (2011) for further details.

The extended Kalman smoother is the same as the standard Kalman smoother, with the approximation (7.3). Note the Kalman smoother in Equations (6.32) and (6.33) only involves $\boldsymbol{H}_{t+1}$.

7.1.3 Gaussian Sum Filters

Gaussian sum filters (GSFs) belong to the class of *density-based* filters in which the procedure concentrates on the propagation of the underlying distribution of the states of a nonlinear and non-Gaussian system. In GSFs, the underlying conditional distribution of x_t (given y_t or y_T) is approximated by a mixture normal distribution, and the system is locally linearized at the center of each component of the mixture distribution in the same way as that in the EKF. Hence GSFs can be viewed as a mixture of EKFs, providing improvements via using a mixture Gaussian distribution to approximate the underlying conditional distribution, instead of a single Gaussian distribution. This extension has a close relationship to the mixture Kalman filter algorithm, which is discussed in Section 8.6.2

Specifically, in the filtering case, assume $p(x_{t-1} \mid y_{t-1})$ is approximated by $\sum_{j=1}^{m} w_{j,t-1} N(\mu_{j,t-1|t-1}, \Sigma_{j,t-1|t-1})$, where the mixture weights $w_{j,t-1}$ are non-negative and $\sum_{j=1}^{m} w_{j,t-1} = 1$. At time t, each component j is updated using the EKF to $N(\mu_{j,t|t}, \boldsymbol{\Sigma}_{j,t|t})$, with linear localization done around the jth component ($\mu_{j,t-1}$ and $\mu_{j,t|t-1}$), ignoring all other components. The weights need to be updated as well. In the prediction step, each component j produces a prediction of y, $\mu^{y}_{j,t|t-1} = g_t(\mu_{j,t|t-1}, 0)$, with a covariance matrix

$$\boldsymbol{\Sigma}^{y}_{j,t} = \boldsymbol{G}_{j,t|t-1}\boldsymbol{\Sigma}_{j,t|t-1}\boldsymbol{G}'_{j,t|t-1} + \boldsymbol{V}_{j,t|t-1}\boldsymbol{V}'_{j,t|t-1}.$$

Then the weight $w_{j,t}$ is updated using

$$w_{j,t} = \frac{w_{j,t-1}\phi(y; \mu^y_{j,t}, \boldsymbol{\Sigma}^y_{j,t|t-1})}{\sum_{i=1}^m w_{i,t-1}\phi(y; \mu^y_{i,t}, \boldsymbol{\Sigma}^y_{i,t|t-1})}.$$

This is similar to Bayesian updating of the posterior proportion of mixture distributions. More detailed discussions can be found in Anderson and Moore (1979) and Sorenson and Alspach (1971).

7.1.4 The Unscented Kalman Filter

The GSF uses a mixture Gaussian distribution to approximate the underlying conditional distributions and has to rely on local linearization for the propagation of each Gaussian component of the mixture Gaussian distribution. An alternative approach is to use a discrete distribution to approximate the conditional distributions, with matching first and second moments. The discrete distribution can be propagated through the nonlinear system without any local linearization. With a carefully constructed approximation, a Kalman-filter type of recursive updating scheme can be used to obtain the conditional means and conditional variances of the filtering and smoothing distributions. The unscented Kalman filter (UKF) of Julier and Uhlmann (1997, 2004) is such an algorithm.

Julier and Uhlmann (1997) noted that the first and second moments of $Y = f(X)$, a nonlinear transformation of the random variable X, can be approximated by

$$\mu_y = \sum_{i=-m}^{m} w_i f(x_i) \text{ and } \boldsymbol{\Sigma}_y = \sum_{i=-m}^{m} w_i (f(x_i) - \mu_y)'(f(x_i) - \mu_y),$$

where x_i, $i = -m, \ldots, m$ is a set of points in the domain of X and w_i is a set of positive weights such that $\sum_i w_i = 1$, $\sum_i w_i x_i = \mu_x$, and $\sum_i w_i (x_i - \mu_x)'(x_i - \mu_x) = \boldsymbol{\Sigma}_x$, with given μ_x and $\boldsymbol{\Sigma}_x$. In addition, the covariance between X and $Y = f(X)$ can be approximated by

$$\boldsymbol{\Sigma}_{xy} = \sum_{i=-m}^{m} w_i (f(x_i) - \mu_y)'(x_i - \mu_x). \tag{7.10}$$

The accuracy of such an approximation depends on the choice of (x_i, w_i). Julier and Uhlmann (1997) suggested using

$$x_0 = \mu_x;\ x_{\pm i} = \mu_x \pm \sqrt{m+k}\sqrt{\boldsymbol{\Sigma}_{x,i}};$$
$$w_0 = \frac{k}{m+k};\ w_{\pm i} = \frac{1}{2(m+k)}, \tag{7.11}$$

where m is the dimension of x and $\sqrt{\Sigma_{x,i}}$ is the ith column of the matrix square root of Σ_x. The authors showed that, with this construction, the approximation of Σ_y is accurate to the second order. The extra parameter k can be used to match some higher-order moments, if needed. Note that this approximation is deterministic and is different from using a (weighted) random sample of X to approximate the distribution of X, which we will discuss in Chapter 8.

Consider the SSM model with additive noises:

$$x_t = h_t(x_{t-1}) + \boldsymbol{W}_t w_t, \tag{7.12}$$

$$y_t = g_t(x_t) + \boldsymbol{V}_t v_t, \tag{7.13}$$

where $\text{Var}(w_t) = \boldsymbol{I}$ and $\text{Var}(v_t) = \boldsymbol{I}$. Armed with the discrete approximation in Equation (7.11), an unscented Kalman filter engages a Kalman-filter type of recursion. Specifically, suppose at time $t-1$ we have obtained μ_{t-1} and Σ_{t-1}, with a corresponding discrete approximation $(x_{i,t-1}, w_{i,t-1})$, $i = -m, \ldots, m$. Then the Kalman prediction step is carried out as

$$\mu_{t|t-1} = \sum_{i=-m}^{m} w_{i,t-1} h(x_{i,t-1}),$$

$$\Sigma_{t|t-1} = \sum_{i=-m}^{m} w_{i,t-1}(h(x_{i,t-1}) - \mu_{t|t-1})'(h(x_{i,t-1}) - \mu_{t|t-1}) + \boldsymbol{W}_t'\boldsymbol{W}_t,$$

based on the designed approximation. With given $\mu_{t|t-1}$ and $\Sigma_{t|t-1}$, we obtain the new set of $(x_{i,t|t-1}, w_{i,t|t-1})$, $i = -m, \ldots, m$, for the distribution $p(x_t \mid \boldsymbol{y}_{t-1})$, using the approximation in Equation (7.11). Then the updating step can be carried out by noting that the approximation in Equation (7.10) implies

$$\Sigma_{xy,t|t-1} = \text{Cov}(x_t, y_t \mid \boldsymbol{y}_{t-1}) = \sum_{i=-m}^{m} w_{i,t|t-1}(g(x_{i,t|t-1}) - \mu^{(y)}_{t|t-1})'(x_{i,t|t-1} - \mu_{t|t-1}),$$

where $\mu^{(y)}_{t|t-1} = \sum_{i=-m}^{m} w_{i,t|t-1} g(x_{i,t|t-1})$, and

$$\begin{aligned}\Sigma_{yy,t|t-1} &= \text{Var}(y_t \mid \boldsymbol{y}_{t-1}) \\ &= \sum_{i=-m}^{m} w_{i,t|t-1}(g(x_{i,t|t-1}) - \mu^{(y)}_{t|t-1})'(g(x_{i,t|t-1}) - \mu^{(y)}_{t|t-1}) + \boldsymbol{V}_t\boldsymbol{V}_t'.\end{aligned}$$

Then the BLUEs of the first and second moments of $p(x_t \mid \boldsymbol{y}_t)$ are

$$\mu_{t|t} = \mu_{t|t-1} + \Sigma_{xy,t|t-1}\Sigma^{-1}_{yy,t|t-1}(y_t - \mu^{(y)}_{t|t-1}),$$

$$\Sigma_{t|t} = \Sigma_{t|t-1} - \Sigma_{xy,t|t-1}\Sigma^{-1}_{yy,t|t-1}\Sigma'_{xy,t|t-1}.$$

The unscented Kalman smoother can be developed along the same lines as the Kalman smoother.

7.1.5 Ensemble Kalman Filters

In many applications of SSMs, such as those in geophysics and climate modeling, the dimension d of the underlying state variable x_t can be extremely large, sometimes in the range of millions. In such cases, updating and storing the $d \times d$ filtering and smoothing covariance matrices in the Kalman filter becomes an infeasible task. The ensemble Kalman filter (EnKF) of Evensen (1994, 2003, 2007) is a class of approximation methods in which the filtering (and smoothing) distribution is represented and stored by a set of samples (called ensembles) from the corresponding distributions. The number of samples m is assumed to be much smaller than d, hence achieving dimension reduction. In EnKF, instead of updating the underlying filtering and smoothing distributions, the ensembles are updated, mimicking the updating procedure of the Kalman filter and UKF, and any inference is carried out using the ensembles.

Assume that the SSM is linear and Gaussian,

$$x_t = \boldsymbol{H}_t x_{t-1} + \boldsymbol{W}_t w_t,$$

$$y_t = \boldsymbol{G}_t x_t + \boldsymbol{V}_t v_t.$$

Assume also that at time $t-1$ we have stored an ensemble $x_{t-1}^{(j)}, j = 1, \dots, m$ such that $x_{t-1}^{(j)} \sim N(\mu_{t-1}, \boldsymbol{\Sigma}_{t-1}) = p(x_{t-1} \mid \boldsymbol{y}_{t-1})$. At time t, we have the prediction step

$$x_{t|t-1}^{(j)} = \boldsymbol{H}_t x_{t-1}^{(j)} + \boldsymbol{W}_t w_t^{(j)},$$

where $w_t^{(j)}$ is simulated from $N(0, \boldsymbol{I})$. Then $x_{t|t-1}^{(j)} \sim N(\mu_{t|t-1}, \boldsymbol{\Sigma}_{t|t-1})$. The corresponding (would-be) realization of y_t is $\hat{y}_t^{(j)} = \boldsymbol{G}_t x_{t|t-1}^{(j)} + \boldsymbol{V}_t v_t^{(j)}$, where $v_t^{(j)}$ is simulated from $N(0, \boldsymbol{I})$. Next the update step can be achieved by

$$x_t^{(j)} = x_{t|t-1}^{(j)} + \boldsymbol{K}_t (y_t - \hat{y}_t^{(j)})$$

where $\boldsymbol{K}_t$ is the Kalman gain matrix in Equation (6.29). It can easily be shown that $x_t^{(j)} \sim N(\boldsymbol{\mu}_t, \boldsymbol{\Sigma}_t)$.

However, the above updating step requires knowing the Kalman gain matrix $\boldsymbol{K}_t$, which requires knowing the high dimensional $\boldsymbol{\Sigma}_{t|t-1}$, something we try to avoid in the first place. In practice, an estimate of $\boldsymbol{\Sigma}_{t|t-1}$ based on the sample covariance matrix of the predicted $x_{t|t-1}^{(j)}$ is used. (In fact, a more accurate estimate is the sample covariance matrix of $\boldsymbol{H} x_{t-1}^{(j)}$ plus $\boldsymbol{W}_t \boldsymbol{W}_t'$.) Since m is much smaller than d, the rank of the estimator is much smaller than that of $\boldsymbol{\Sigma}_{t|t-1}$, inducing some bias. Often the covariance inflation (Anderson, 2007) and localization (Furrer et al., 2006) techniques are used to improve the accuracy and stabilize the algorithm.

The updating step can be approximated with a deterministic approach as well. See Tippett et al. (2003), Anderson (2001), and Bishop et al. (2001). Smoothing and parameter estimation using EnKF have been studied by Stroud et al. (2010). EnKF can be used in conjunction with the EKF for nonlinear and non-Gaussian systems.

7.1.6 Examples and R implementations

In this section we consider some examples and the associated analyses using R packages.

Example 7.1: Bearings-only tracking using the EKF In Section 6.2.4 we briefly introduced the problem of tracking a target with passive sonars, in which the observations are the bearing of the target related to the sensors, but no distance information can be obtained. The system is nonlinear. Here we present an implementation of filtering and smoothing using EKF.

Suppose two sensors are placed as $\eta_1 = (0, 0)$ and $\eta_2 = (20, 0)$. A target moves with a random acceleration with variance 0.03^2 in both directions and $\Delta T = 1$. Figure 7.1 shows a simulated trajectory for $t = 1, \ldots, 200$, which contains a

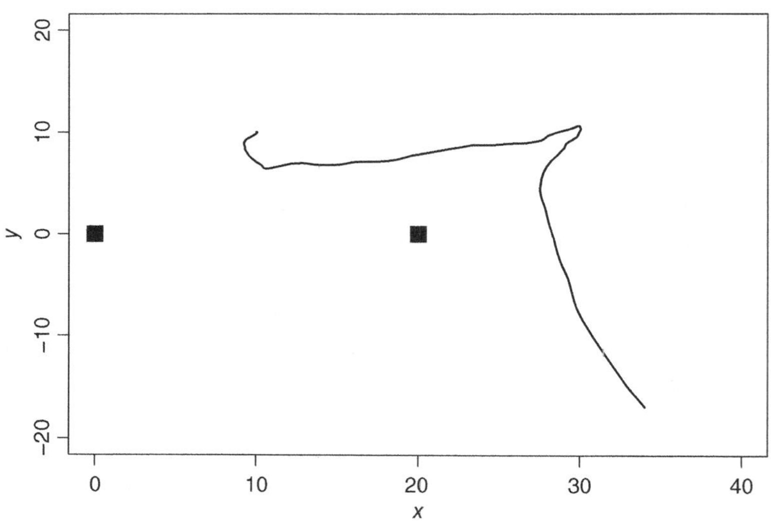

Figure 7.1 Trajectory of a simulated target. Two sensors are marked by solid squares.

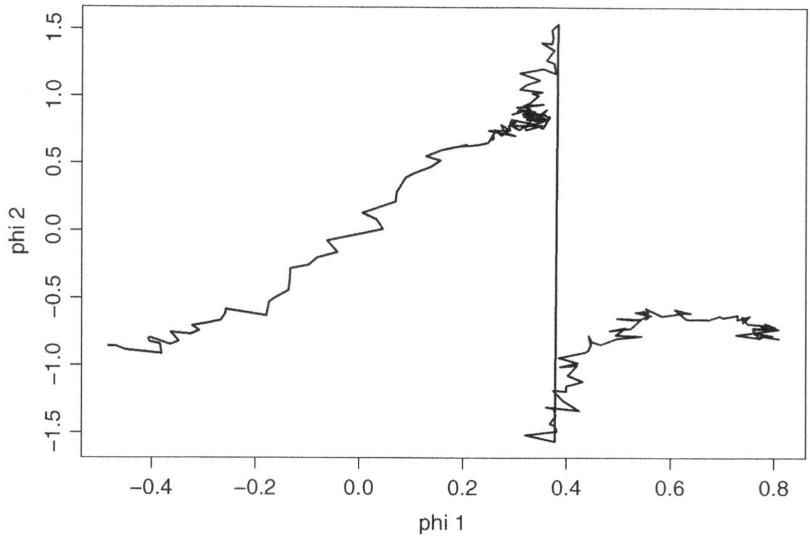

Figure 7.2 Observed bearings of a simulated target by two sensors.

sharp turn in the middle of the observational period. Assume that the sensors have measurement errors e_{1t} and e_{2t} that follow $N(0, 0.02^2)$. Figure 7.2 shows the observed bearings (y_{1t}, y_{2t}). The bearings are restricted to $[-\pi/2, \pi/2]$ by modulation. The jump of the path in the middle is due to the modulation. Figure 7.3 shows the true simulated trajectory, and the filtered and the smoothed trajectories from $t = 100$ to 200, the more difficult part of the tracking task due to the turns. Tracking is actually quite accurate during the first sharp turn, but becomes less accurate during the second turn. Smoothing was able to correct some of the filtering errors. Note that the largest errors are in the x direction, when the target is close to the x axis. This is because both sensors are placed on the x axis, hence the angles are small when the target is close to the x axis. During this period, the signal-to-noise ratio is small, resulting in large tracking errors.

Figures 7.4 and 7.5 show the tracking errors of locations in the x and y directions, respectively. Figures 7.6 and 7.7 show the tracking errors of speeds in the x and y directions, respectively. It is seen that the error in the x direction is larger than that of the y direction, again due to the placement of the sensors.

The following R codes are used in the example. Note that in this particular instance, the state equation is linear and Gaussian. Hence, to use EKF, approximation is only needed for the observation equation, in the form of Equation (7.4).

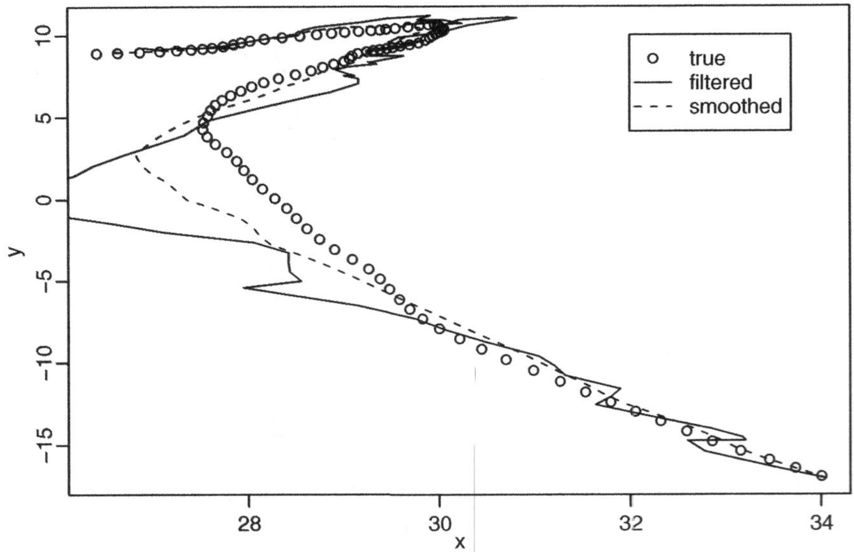

Figure 7.3 The filtered and smoothed trajectories of a simulated target, for $t = 100, \ldots, 200$.

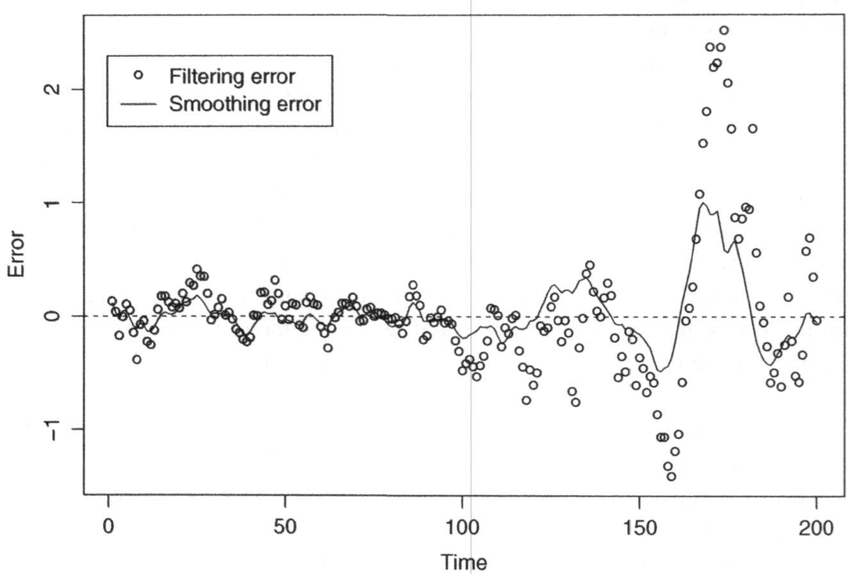

Figure 7.4 Tracking error in x direction locations.

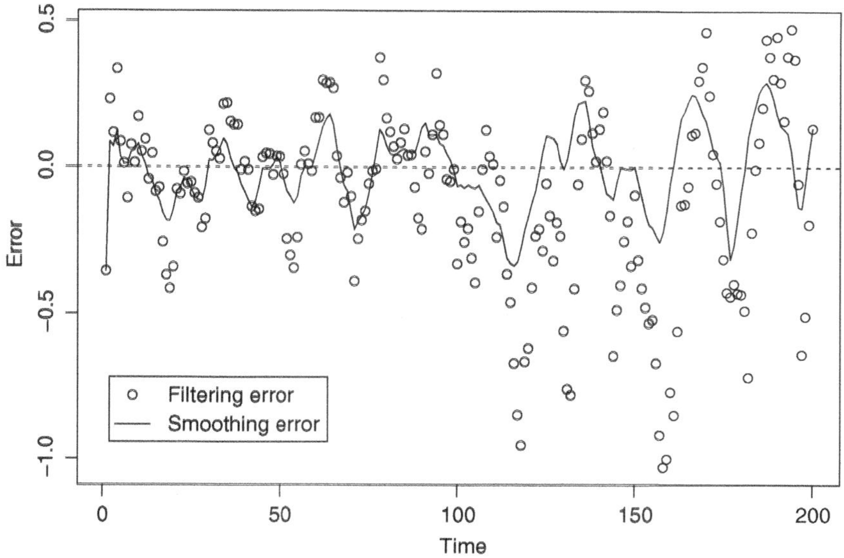

Figure 7.5 Tracking error in y direction locations.

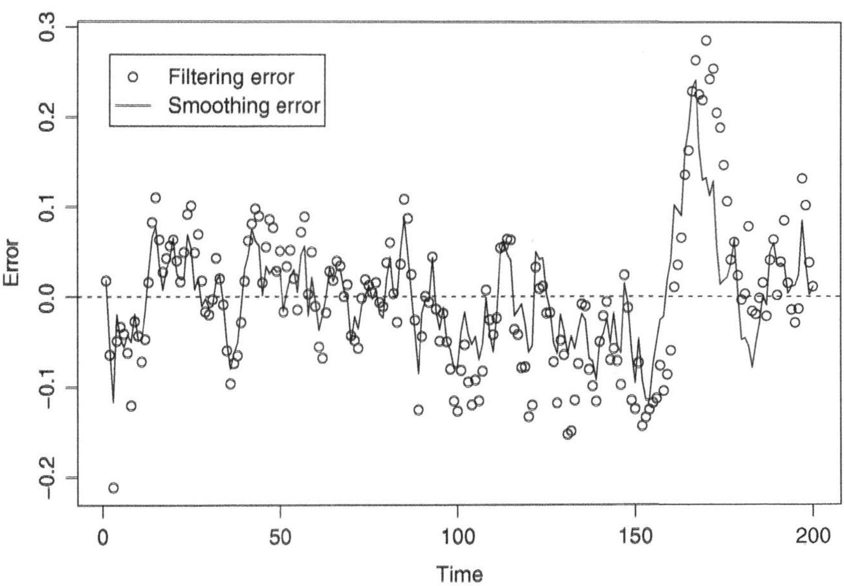

Figure 7.6 Tracking error in x direction speeds.

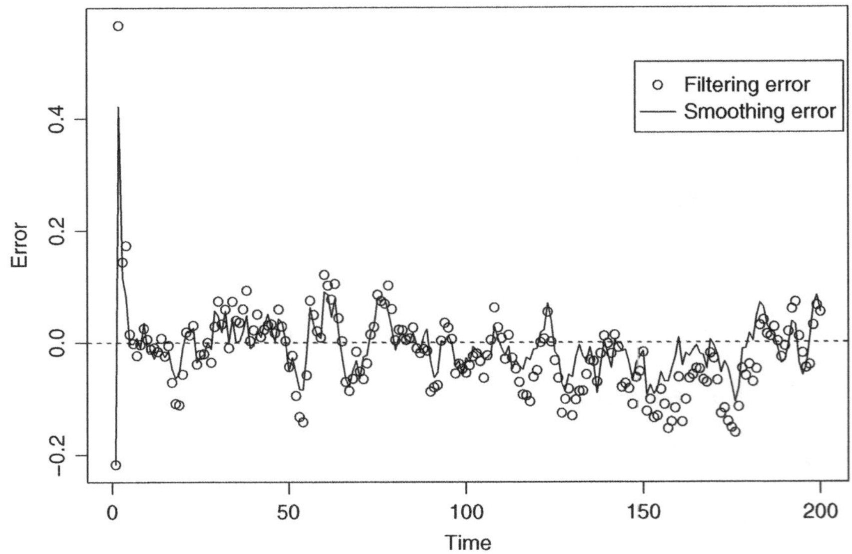

Figure 7.7 Tracking error in y direction speeds.

Specifically, with two sensors located at $\eta_i = (\eta_{i1}, \eta_{i2})$ $(i = 1, 2)$, and with given $\mu_{t|t-1} = (\mu_{1,t|t-1}, \mu_{2,t|t-1}, \mu_{3,t|t-1}, \mu_{4,t|t-1})$, we have

$$\boldsymbol{G}_{t|t-1} = \begin{bmatrix} -\dfrac{\tau_{1,t|t-1}}{(1+\tau^2_{1,t|t-1})(\mu_{1,t|t-1}-\eta_{11})} & \dfrac{1}{(1+\tau^2_{1,t|t-1})(\mu_{1,t|t-1}-\eta_{11})} & 0 & 0 \\ -\dfrac{\tau_{2,t|t-1}}{(1+\tau^2_{2,t|t-1})(\mu_{1,t|t-1}-\eta_{21})} & \dfrac{1}{(1+\tau^2_{2,t|t-1})(\mu_{1,t|t-1}-\eta_{21})} & 0 & 0 \end{bmatrix},$$

where

$$\tau_{i,t|t-1} = \frac{\mu_{2,t|t-1} - \eta_{i2}}{\mu_{1,t|t-1} - \eta_{i1}}, \qquad i = 1, 2.$$

The function `simPassiveSonar` generates a sample trajectory of the target and the corresponding observations, with sensor locations at (0, 0) and (20, 0). The functions `KF1pred`, `KF1update` and `KF1smoother` perform one-step ahead prediction in Equations (6.25) and (6.26), one-step updating in Equations (6.27) and (6.28), and one-step smoothing in Equations (6.32) and (6.33), of the Kalman filter and Kalman smoothing, respectively. The functions `EKF1pred` and `EKF1update` perform one-step ahead prediction and updating of EKF in Equations (7.5) to (7.6), respectively. In our setting, since the state equation is linear and Gaussian, only `EKF1update` is used. In addition, since the state transition matrix H_t does not change and there is no approximation, the extended Kalman smoother is the same as the Kalman smoother.

R demonstration: Extended Kalman filter.

```
## 2-D passive sonar  tracking example
###
# function to simulate the data
###
simPassiveSonar=function(nn=200,q,r,start,s2,seed){
  set.seed(seed)
###  s2 : second sonar location at (s2,0)
  H <- c(1,0,1,0,
         0,1,0,1,
         0,0,1,0,
         0,0,0,1)
  H <- matrix(H,ncol=4,nrow=4,byrow=T)
  G <- c(1,0,0,0,
         0,1,0,0)
  G <- matrix(G,ncol=4,nrow=2,byrow=T)
  W <- c(0.5*q[1], 0,
         0, 0.5*q[2],
         q[1],0,
         0,q[2])
  W <- matrix(W,ncol=2,nrow=4,byrow=T)
  V <- diag(r)
  x <- matrix(nrow=4,ncol=nn)
  y <- matrix(nrow=2,ncol=nn)
  for(ii in 1:nn){
    if(ii == 1) x[,ii] <- start
    if(ii > 1) x[,ii] <- H%*%x[,ii-1]+W%*%rnorm(2)
    y[1,ii] <- atan(x[2,ii]/x[1,ii])
    y[2,ii] <- atan(x[2,ii]/(x[1,ii]-s2))
    y[,ii] <- y[,ii]+V%*%rnorm(2)
    y[,ii] <- (y[,ii]+0.5*pi)%%pi-0.5*pi
  }
  return(list(xx=x,yy=y,G=G,H=H,W=W,V=V))
}
####
### KF-single step ahead prediction
####
KF1pred=function(mu,SS,HH,bb,WW){
   mu <- HH%*%mu+bb
   SS <- HH%*%SS%*%t(HH)+WW%*%t(WW)
   return(list(mu=mu,SS=SS))
}
```

```
####
### KF-single step update, no likelihood
####
KF1update=function(mu,SS,yy,GG,cc,VV){
   KK <- SS%*%t(GG)%*%solve(GG%*%SS%*%t(GG)+VV%*%t(VV))
   mu <- mu+KK%*%(yy-cc-GG%*%mu)
   SS <- SS-KK%*%GG%*%SS
   return(list(mu=mu,SS=SS))
}
####
### KF-single step smoother
####
KF1smoother=function(mu,SS,mu1,SS1,muT,SST,HH,VV){
   JJ <- SS%*%HH%*%solve(SS1)
   mu <- mu+JJ%*%(muT-mu1)
   SS <- SS+JJ%*%(SST-SS1)%*%t(JJ)
   return(list(mu=mu,SS=SS))
}
####
### EKF-single step ahead prediction, with input mu_pred
####
EKF1pred=function(mu,SS,HH,bb,WW){
   SS <- HH%*%SS%*%t*t(HH)+WW%*%t(WW)
   return(list(mu=mu,SS=SS))
}
####
### EKF-single step update, with input yy_pred, no likel:hood
####
EKF1update=function(mu,SS,yy,yy_pred,GG,VV){
   KK <- SS%*%t(GG)%*%solve(GG%*%SS%*%t(GG)+VV%*%t(VV))
   res <- (yy-yy_pred+0.5*pi)%%pi-0.5*pi
   mu <- mu+KK%*%res
   SS <- SS-KK%*%GG%*%SS
   return(list(mu=mu,SS=SS))
}
##### KF for passive sonar
##-- parameters
s2 <- 20  #second sonar location at (s2,0)
q <- c(0.03,0.03)
r <- c(0.02,0.02)
nn <- 200
##-- simulation
start <- c(10,10,0.01,0.01)
simu_out <- simPassiveSonar(nn,q,r,start,s2,seed=20)
yy <- simu_out$yy
```

```
##-- parameter for filtering and smoothing
HH <- c(1,0,1,0,
      0,1,0,1,
      0,0,1,0,
      0,0,0,1)
HH <- matrix(HH,ncol=4,nrow=4,byrow=T)
WW <- c(0.5*q[1], 0,
      0, 0.5*q[2],
      q[1],0,
      0,q[2])
WW <- matrix(WW,ncol=2,nrow=4,byrow=T)
GG <- matrix(rep(0,8),ncol=4,nrow=2)
VV <- diag(r)
##-- setup storage
muall <- matrix(nrow=4,ncol=nn)
mu1all <- matrix(nrow=4,ncol=nn)
muTall <- matrix(nrow=4,ncol=nn)
SSall <- array(dim=c(4,4,nn))
SS1all <- array(dim=c(4,4,nn))
SSTall <- array(dim=c(4,4,nn))
##initial value
mu0 <- c(10,10,0.1,0.1)
SS0 <- diag(c(1,1,1,1))*10000
bb <- 0*mu0
yy_pred <- yy[,1]*0
##start
for(ii in 1:nn){
  if(ii==1) out_pred <- KF1pred(mu0,SS0,HH,bb,WW)
  if(ii > 1) out_pred <- KF1pred(muall[,ii-1],SSall[,,ii-1],HH,bb,WW)
  mu1all[,ii] <- out_pred$mu
  SS1all[,,ii] <- out_pred$SS
  temp1 <- mu1all[2,ii]/mu1all[1,ii]
  temp2 <- mu1all[2,ii]/(mu1all[1,ii]-s2)
  yy_pred[1] <- atan(temp1)
  yy_pred[2] <- atan(temp2)
  GG[1,1] <- -mu1all[2,ii]/(1+temp1**2)/mu1all[1,ii]**2
  GG[1,2] <- 1/(1+temp1**2)/mu1all[1,ii]
  GG[2,1] <- -mu1all[2,ii]/(1+temp2**2)/(mu1all[1,ii]-s2)**2
  GG[2,2] <- 1/(1+temp2**2)/(mu1all[1,ii]-s2)
  out_update <- EKF1update(mu1all[,ii],SS1all[,,ii],yy[,ii],yy_pred,GG,VV)
  muall[,ii] <- out_update$mu
  SSall[,,ii] <- out_update$SS
}
### smoothing
muTall[,nn] <- muall[,nn]
```

```
SSTall[„nn] <- SSall[„nn]
for(ii in (nn-1):1){
  out_smooth <- KF1smoother(muall[,ii],SSall[„ii],
                            mu1all[,ii+1],SS1all[„ii+1],
                            muTall[,ii+1],SSTall[„ii+1],HH,VV)
  muTall[,ii] <- out_smooth$mu
  SSTall[„ii] <- out_smooth$SS
}
### Plotting
plot(simu_out$xx[1,],simu_out$xx[2,],type='l',lwd=2,
     xlim=c(0,40),ylim=c(-20,20),xlab="x",ylab="y")
points(c(0,s2),c(0,0),pch=15,cex=2)
title("target trajectory")
#Figure 7.1

plot(simu_out$yy[1,],simu_out$y[2,],type='l',lwd=2,
     xlab="phi 1",ylab="phi 2")
title("observed bearings")
#Figure 7.2

tt <- 100:200
plot(simu_out$xx[1,tt],simu_out$xx[2,tt],xlab="x",ylab="y",
     ylim=c(-18,13),xlim=c(26,34))
title("target trajectory, filtering and smoothing")
lines(muall[1,tt],muall[2,tt])
lines(muTall[1,tt],muTall[2,tt],lty=2)
legend(32,10,legend=c("true","filtered","smoothed"),
      pch=c(1,NA,NA),lty=c(0,1,2))
#Figure 7.3

plot(simu_out$xx[1,]-muall[1,],ylab='error',xlab='time')
lines(simu_out$xx[1,]-muTall[1,])
abline(h=0,lty=2)
title("x-director tracking error")
legend(2,2.3,legend=c("filtering error","smoothing error"),
       pch=c(1,NA),lty=c(0,1))
#Figure 7.4

plot(simu_out$xx[2,]-muall[2,],ylab='error',xlab='time')
lines(simu_out$xx[2,]-muTall[2,])
abline(h=0,lty=2)
title("y-director tracking error")
legend(2,-0.7,legend=c("filtering error","smoothing error"),
      pch=c(1,NA),lty=c(0,1))
#Figure 7.5
```

```
plot(simu_out$xx[3,]-muall[3,],ylab='error',xlab='time')
lines(simu_out$xx[3,]-muTall[3,])
abline(h=0,lty=2)
title("x-director speed error")
legend(2,0.27,legend=c("filtering error","smoothing error"),
      pch=c(1,NA),lty=c(0,1))
#Figure 7.6

plot(simu_out$xx[4,]-muall[4,],ylab='error',xlab='time')
lines(simu_out$xx[4,]-muTall[4,])
abline(h=0,lty=2)
title("y-director speed error")
legend(150,0.5,legend=c("filtering error","smoothing error"),
      pch=c(1,NA),lty=c(0,1))
#Figure 7.7
```

7.2 HIDDEN MARKOV MODELS

The HMM is a special case of the Markovian SSM in Equations (6.3) and (6.4), when the state x_t is a Markov chain taking values in a discrete set $\{\eta_1, \ldots, \eta_K\}$. The state equation is specified by the conditional transition probability matrix $\boldsymbol{P}$. Specifically, an HMM takes the form

$$x_t \sim \text{Markov chain}(\boldsymbol{P}), \tag{7.14}$$

$$y_t = g_t(x_t, e_t) \quad \text{or} \quad y_t \sim f_t(\cdot \mid x_t), \tag{7.15}$$

where the (i,j)th element of the transition probability matrix $\boldsymbol{P}$ is $p[i,j] = p(x_t = \eta_j \mid x_{t-1} = \eta_i)$.

HMM has a wide range of applications in various scientific fields. In time series analysis, the Markov switching autoregressive model is one of the most commonly used HMMs. A detailed treatment of such a model with $K = 2$ has been given in Chapter 2, in parallel with the threshold AR model.

7.2.1 Filtering

Filtering of the state of an HMM gives the marginal probability $p_{t|t}[i] = p(x_t = \eta_i \mid \boldsymbol{y}_t)$. It can be carried out similarly to the approach of the Kalman filter, except it is done in discrete space under the transition matrix $\boldsymbol{P}$. Specifically, at time $t-1$

suppose we have obtained $p_{t-1|t-1}[i] = p(x_{t-1} = \eta_i \mid \boldsymbol{y}_{t-1})$ for $i = 1, \ldots, K$. Then the predictive probability is

$$\begin{aligned} p_{t|t-1}[j] &= p(x_t = \eta_j \mid \boldsymbol{y}_{t-1}) \\ &= \sum_{i=1}^{K} p(x_t = \eta_j \mid x_{t-1} = \eta_i) p(x_{t-1} = \eta_i \mid \boldsymbol{y}_{t-1}) \\ &= \sum_{i=1}^{K} p[i,j] p_{t-1|t-1}[i]. \end{aligned}$$

To incorporate the new information y_t, we note that

$$\begin{aligned} p_{t|t}[i] &= p(x_t = \eta_i \mid \boldsymbol{y}_{t-1}, y_t) \\ &\propto p(y_t \mid x_t = \eta_i, \boldsymbol{y}_{t-1}) p(x_t = \eta_i \mid \boldsymbol{y}_{t-1}) \\ &= f_t(y_t \mid x_t = \eta_i) p_{t|t-1}[i]. \end{aligned}$$

Hence, we have the update equation

$$p_{t|t}[i] = \frac{f_t(y_t \mid x_t = \eta_i) p_{t|t-1}[i]}{\sum_{j=1}^{K} f_t(y_t \mid x_t = \eta_j) p_{t|t-1}[j]}, \qquad i = 1, \ldots, K.$$

In summary, suppose at time $t-1$ we have obtained $p_{t-1|t-1}[i] = p(x_{t-1} = \eta_i \mid \boldsymbol{y}_{t-1})$. Then the one-step HMM filter at time t is as follows.

Algorithm 7.2: HMM filtering

$$p_{t|t-1}[j] = \sum_{i=1}^{K} p[i,j] p_{t-1|t-1}[i], \quad j = 1, \ldots, K$$

$$p_{t|t}[i] = \frac{f_t(y_t \mid x_t = \eta_i) p_{t|t-1}[i]}{\sum_{j=1}^{K} f_t(y_t \mid x_t = \eta_j) p_{t|t-1}[j]}, \qquad i = 1, \ldots, K.$$

7.2.2 Smoothing

Smoothing of an HMM gives the marginal distribution of the state given the entire observed series, i.e. $p_{t|T}[i] = p(x_t = \eta_i \mid \boldsymbol{y}_T)$. It can be done by using a procedure similar to the forward and backward two-pass disentanglement of the Kalman smoother. Specifically, we first carry out the forward filtering from $t = 1$ to T and obtain $p_{t|t}[i]$ for $i = 1, \ldots, K$, $t = 1, \ldots, T$. We then use the disentanglement operation by inserting x_{t+1}. Specifically, we notice that

$$\begin{aligned}
p_{t|T}[i] &= p(x_t = \eta_i \mid \boldsymbol{y}_T) \\
&= \sum_{j=1}^{K} p(x_t = \eta_i \mid x_{t+1} = \eta_j, \boldsymbol{y}_T) p(x_{t+1} = \eta_j \mid \boldsymbol{y}_T) \\
&= \sum_{j=1}^{K} p(x_t = \eta_i \mid x_{t+1} = \eta_j, \boldsymbol{y}_t) p(x_{t+1} = \eta_j \mid \boldsymbol{y}_T) \\
&= \sum_{j=1}^{K} p_{t+1|T}[j] \gamma_t[i,j],
\end{aligned}$$

where

$$\begin{aligned}
\gamma_t[i,j] &= p(x_t = \eta_i \mid x_{t+1} = \eta_j, \boldsymbol{y}_t) \\
&= \frac{p(x_t = \eta_i, x_{t+1} = \eta_j \mid \boldsymbol{y}_t)}{p(x_{t+1} = \eta_j \mid \boldsymbol{y}_t)} \\
&= \frac{p(x_{t+1} = \eta_j \mid x_t = \eta_i, \boldsymbol{y}_t) p(x_t = \eta_i \mid \boldsymbol{y}_t)}{p(x_{t+1} = \eta_j \mid \boldsymbol{y}_t)} \\
&= \frac{p[i,j] p_{t|t}[i]}{\sum_{\ell=1}^{K} p[\ell,j] p_{t|t}[\ell]}.
\end{aligned} \tag{7.16}$$

In summary, we have the following algorithm.

Algorithm 7.3: HMM smoothing

For $t = T-1, \ldots, 1$,

$$p_{t|T}[i] = \sum_{j=1}^{K} p_{t+1|T}[j] \gamma_t[i,j]$$

where

$$\gamma_t[i,j] = \frac{p[i,j] p_{t|t}[i]}{\sum_{\ell=1}^{K} p[\ell,j] p_{t|t}[\ell]}.$$

Another approach is to use $p(y_{t+1}, \dots, y_T \mid x_t = \eta_i)$. Specifically, we have

$$\begin{aligned} p_{t|T}[i] &= p(x_t = \eta_i \mid \boldsymbol{y}_T) \\ &\propto p(y_{t+1}, \dots, y_T \mid x_t = \eta_i) p(x_t = \eta_i \mid \boldsymbol{y}_t) \\ &\propto p(y_{t+1}, \dots, y_T \mid x_t = \eta_i) p_{t|t}[i]. \end{aligned}$$

Let

$$\beta_t[i] = p(y_{t+1}, \dots, y_T \mid x_t = \eta_i).$$

Starting with $\beta_T[i] = 1$, for $i = 1, \dots, K$, and by inserting x_{t+1}, we have, for $t = T-1, T-2, \dots, 1$,

$$\begin{aligned} \beta_t[i] &= p(y_{t+1}, \dots, y_T \mid x_t = \eta_i) \\ &= \sum_{j=1}^{K} p(y_{t+1}, \dots, y_T \mid x_{t+1} = \eta_j, x_t = \eta_i) p(x_{t+1} = \eta_j \mid x_t = \eta_i) \\ &= \sum_{i=1}^{K} p(y_{t+1}, \dots, y_T \mid x_{t+1} = \eta_j) p(x_{t+1} = \eta_j \mid x_t = \eta_i) \\ &= \sum_{i=1}^{K} p(y_{t+2}, \dots, y_T \mid y_{t+1}, x_{t+1} = \eta_j) p(y_{t+1} \mid x_{t+1} = \eta_j) p[i,j] \\ &= \sum_{i=1}^{K} \beta_{t+1}[j] p[i,j] f_t(y_{t+1} \mid x_{t+1} = \eta_j). \end{aligned}$$

Hence we have the following algorithm.

Algorithm 7.4: HMM smoothing (II)

Starting with $\beta_T[i] = 1$, for $i = 1, \dots, K$, let

$$\beta_t[i] = \sum_{i=1}^{K} \beta_{t+1}[j] p[i,j] f_t(y_{t+1} \mid x_{t+1} = \eta_j)$$

and

$$p_{t|T}[i] = \frac{\beta_t[i] p_{t|t}[i]}{\sum_{k=1}^{K} \beta_t[k] p_{t|t}[k]}.$$

For parameter estimation via the EM algorithm, it is often necessary to obtain the joint smoothing probability $p(x_t, x_{t-1} \mid \boldsymbol{y}_T)$. This can be done by the following:

$$\begin{aligned} p(x_{t+1} = \eta_j, x_t = \eta_i \mid \boldsymbol{y}_T) &= p(x_t = \eta_i \mid x_{t+1} = \eta_j, \boldsymbol{y}_T) p(x_{t+1} = \eta_j \mid \boldsymbol{y}_T) \\ &= p(x_t = \eta_i \mid x_{t+1} = \eta_j, \boldsymbol{y}_t) p(x_{t+1} = \eta_j \mid \boldsymbol{y}_T) \\ &= \gamma_t[i,j] p_{t+1|T}[j], \end{aligned}$$

where $\gamma_t[i,j]$ is given in (7.16).

7.2.3 The Most Likely State Path: the Viterbi Algorithm

Suppose one wishes to obtain the most likely state path $\boldsymbol{s}_T^* = (x_1^*, \ldots, x_T^*)$ that maximizes $p(\boldsymbol{x}_T \mid \boldsymbol{y}_T)$. This can be done efficiently with the Viterbi algorithm. See Viterbi (1967) and Forney (1973). Although the maximization is over K^T possible paths (where K is the number of possible values x_t may take), the computing cost of the Viterbi algorithm is linear in T. It utilizes the fact that x_t only takes a finite number of possible values and the Markovian property, which implies that

$$p(\boldsymbol{x}_T \mid \boldsymbol{y}_T) \propto p(\boldsymbol{x}_T, \boldsymbol{y}_T) = \prod_{t=1}^{T} p(y_t \mid x_t) p(x_t \mid x_{t-1}).$$

Specifically, let

$$\begin{aligned} \alpha_t[i] &= \max_{\boldsymbol{x}_{t-1}} p(\boldsymbol{x}_{t-1}, x_t = \eta_i, \boldsymbol{y}_t), \\ \beta_t[i,j] &= \max_{\boldsymbol{x}_{t-1}} p(\boldsymbol{x}_{t-1}, x_t = \eta_i, x_{t+1} = \eta_j, y_{t+1}, \boldsymbol{y}_t). \end{aligned}$$

We have

$$\begin{aligned} \beta_t[i,j] &= \max_{\boldsymbol{x}_{t-1}} p(\boldsymbol{x}_{t-1}, x_t = \eta_i, x_{t+1} = \eta_j, y_{t+1}, \boldsymbol{y}_t) \\ &= \max_{\boldsymbol{x}_{t-1}} p(y_{t+1} \mid x_{t+1} = \eta_j) p(x_{t+1} = \eta_j \mid x_t = \eta_i) p(\boldsymbol{x}_{t-1}, x_t = \eta_i, \boldsymbol{y}_t) \\ &= p(y_{t+1} \mid x_{t+1} = \eta_j) p(x_{t+1} = \eta_j \mid x_t = \eta_i) \max_{\boldsymbol{x}_{t-1}} p(\boldsymbol{x}_{t-1}, x_t = \eta_i, \boldsymbol{y}_t) \\ &= f_{t+1}(y_{t+1} \mid x_{t+1} = \eta_j) p[i,j] \alpha_t[i]. \end{aligned}$$

Hence,

$$\begin{aligned} \alpha_{t+1}[i] &= \max_{\boldsymbol{x}_t} p(\boldsymbol{x}_t, x_{t+1} = \eta_i, \boldsymbol{y}_{t+1}) \\ &= \max_{x_t} \max_{\boldsymbol{x}_{t-1}} p(\boldsymbol{x}_{t-1}, x_t, x_{t+1} = \eta_i, \boldsymbol{y}_{t+1}) \\ &= \max_{\ell} \beta_t[\ell, i]. \end{aligned} \tag{7.17}$$

Let $\alpha_1[i] = f_1(y_1 \mid x_1 = \eta_i)p[i]$ and $\boldsymbol{x}_1^*[i] = \eta_i$, for $i = 1, \ldots, K$, and iterating from $t = 1, \ldots, T$, we obtain $\boldsymbol{x}_t^*[i] = (\boldsymbol{x}_{t-1}^*[\ell_i^*], \eta_i)$ where ℓ_i^* is the index ℓ that maximizes $\beta_t[\ell, i]$ in Equation (7.17). Once we reach T, let i^* be the index that maximized $\alpha_T[i]$, then the most likely path is $\boldsymbol{x}_T^*[i^*]$.

In summary, we have the following algorithm.

Algorithm 7.5: Viterbi algorithm

- Let $\alpha_1[i] = f_1(y_1 \mid x_1 = \eta_i)p[i]$ and $\boldsymbol{x}_1^*[i] = \eta_i$,
- For $t = 1, \ldots, T$, let

$$\begin{aligned} \beta_t[i,j] &= f_{t+1}(y_{t+1} \mid x_{t+1} = \eta_j)p[i,j]\alpha_t[i]. \\ \alpha_{t+1}[i] &= \max_{\ell} \beta_t[\ell, i]. \\ \ell_t^*[i] &= \arg\max_{\ell} \beta_t[\ell, i] \\ \boldsymbol{x}_t^*[i] &= (\boldsymbol{x}_{t-1}^*[\ell_t^*[i]], \eta_i). \end{aligned}$$

  for $i = 1, \ldots, K, j = 1, \ldots, K$.
- Let i^* be the index that maximized $\alpha_T[i]$, then the most likely path is $\boldsymbol{x}_T^*[i^*]$.

This algorithm belongs to a larger class of algorithms called dynamic programming (Bellman, 2003) and is useful in many applications.

7.2.4 Parameter Estimation: the Baum–Welch Algorithm

To estimate the unknown parameters of the HMM model in Equations (7.14) and (7.15), including the transition probability $\boldsymbol{P}$ and unknown parameters in $f_t(\cdot)$, the EM algorithm of Dempster et al. (1977) can be used. This procedure is also referred to as the Baum–Welch algorithm. See Baum and Petrie (1966), Baum et al. (1970), and Baggenstoss (2001). For convergence of the EM algorithm, see Wu (1983).

Specifically, the full data log likelihood function is

$$\ln p(\boldsymbol{x}_T, \boldsymbol{y}_T, \theta) = \sum_{t=1}^{T} \ln p(x_t \mid x_{t-1}, \theta) + \sum_{t=1}^{T} \ln p(y_t \mid x_t, \theta).$$

The EM algorithm requires us to obtain the θ that maximizes

$$\begin{aligned} &E[\ln p(\boldsymbol{x}_T, \boldsymbol{y}_T, \theta \mid \boldsymbol{y}_T, \theta^*)] \\ &= E\left[\sum_{t=1}^{T} \ln p(x_t \mid x_{t-1}, \theta) \mid \boldsymbol{y}_T, \theta^*\right] + E\left[\sum_{t=1}^{T} \ln p(y_t \mid x_t, \theta) \mid \boldsymbol{y}_T, \theta^*\right] \end{aligned} \quad (7.18)$$

where θ^* is the parameter value of the current iteration.

The evaluation of $E[\ln\, p(x_t \mid x_{t-1}, \boldsymbol{y}_T, \theta^*)]$ involves the joint smoothing probabilities $p(x_t, x_{t-1} \mid \boldsymbol{y}_T)$ in Section 7.2.2, that is,

$$p(x_t \mid x_{t-1}, \boldsymbol{y}_T, \theta^*) = \frac{p(x_t, x_{t-1} \mid \boldsymbol{y}_T, \theta^*)}{p(x_{t-1} \mid \boldsymbol{y}_T, \theta^*)} = \frac{\gamma_t[i,j] p_{t+1|T}[j]}{p_{t|T}[i]},$$

where $\gamma_t[i,j]$ and $p_{t|T}[i]$ are given in Section 7.2.2, obtained under θ^*. Since the second summation in Equation (7.18) does not involve the transition matrix, one can maximize the first term to obtain an updated estimate of $p[i,j]$ using

$$\hat{p}[i,j] = \frac{\sum_{t=1}^{T} p(x_t = \eta_i, x_{t+1} = \eta_j \mid \boldsymbol{y}_T, \theta^*)}{\sum_{t=1}^{T} \sum_{\ell=1}^{K} p(x_t = \eta_\ell, x_{t+1} = \eta_j \mid \boldsymbol{y}_T, \theta^*)}.$$

To evaluate the second summation, we notice that

$$\begin{aligned} E\left[\sum_{t=1}^{T} \ln p(y_t \mid x_t, \theta \mid \boldsymbol{y}_T, \theta^*)\right] &= \sum_{t=1}^{T} E\left[\sum_{i=1}^{K} I(x_t = \eta_i) \ln p(y_t \mid x_t = \eta_i, \theta \mid \boldsymbol{y}_T, \theta^*)\right] \\ &= \sum_{t=1}^{T} \sum_{i=1}^{K} E[I(x_t = \eta_i) \mid \boldsymbol{y}_T, \theta^*] \ln p(y_t \mid x_t = \eta_i, \theta) \\ &= \sum_{t=1}^{T} \sum_{i=1}^{K} p_{t|T}[i] \ln f_t(y_t \mid x_t = \eta_i, \theta). \end{aligned}$$

This term only involves the parameters in the observation Equation (7.15) and can be viewed as a weighted sum (or finite mixture) of the likelihood of the observations. For complex observations, the Newton–Raphson method can be used to update the parameters. For simpler cases, the maximum can be obtained analytically. For example, if y_t is also discrete, taking values in $\{\xi_1, \dots, \xi_M\}$ and the unknown parameters are the assignment probability $p^*[i,j] = p(y_t = \xi_j \mid x_t = \eta_i)$, then they can be estimated by

$$\hat{p}^*[i,j] = \frac{\sum_{t=1}^{T} \sum_{i=1}^{K} p_{t|n}[i] I(y_t = \xi_j)}{\sum_{t=1}^{T} \sum_{i=1}^{K} p_{t|n}[i]},$$

for $i = 1, \dots, K$ and $j = 1, \dots, M$.

For a hidden Markov mixture Gaussian model with $f_t(y_t \mid x_t = \eta_i) \sim N(\mu_i, \boldsymbol{\Sigma}_i)$, we have

$$\hat{\mu}_i = \frac{\sum_{t=1}^{T} p_{t|n}[i] y_t}{\sum_{t=1}^{T} p_{t|n}[i]} \quad \text{and} \quad \hat{\boldsymbol{\Sigma}}_i = \frac{\sum_{t=1}^{T} p_{t|n}[i](y_t - \hat{\mu}_i)(y_t - \hat{\mu}_i)'}{\sum_{t=1}^{T} p_{t|n}[i]}.$$

A hidden Markov regression model with Gaussian errors can be written as

$$y_t = z_t' \beta_i + \sigma_i e_t, \quad \text{if} \quad x_t = i,$$

where x_t is the regime indicator, z_t is a set of observed (p-dimensional) co-variates, and (β_i, σ_i^2) are the corresponding coefficients and noise variance in the ith regime. When z_t includes the lag variables of y_t, this model becomes the Markov switching autoregressive model we discussed in detail in Section 2.3.

In this case, the observation equation is $f_t(y_t \mid x_t = i) \sim N(z_t\beta_i, \sigma_i^2)$. We have

$$\hat{\beta}_i = \left(\sum_{t=1}^{T} p_{t|n}[i] z_t' z_t\right)^{-1} \left(\sum_{t=1}^{T} p_{t|n}[i] z_t' y_t\right) \quad \text{and} \quad \hat{\sigma}_i^2 = \frac{\sum_{t=1}^{T} p_{t|n}[i](y_t - z_t\hat{\beta}_i)^2}{\sum_{t=1}^{T} p_{t|n}[i]}.$$

7.2.5 HMM Examples and R Implementation

In this section we show two detailed examples of applying HMMs.

Example 7.2: Multi-regime stock return mean and volatility The stock market often shows periods of growth (bull market), neutral, and contraction (bear market). It is commonly expected that the mean and volatility of stock returns are different in different periods. A hidden Markov mixture Gaussian model can be used to model such phenomenon.

Figure 7.8(a) shows the monthly log return of S&P 500 index from January 1987 to December 2016. The associated histogram is shown in Figure 7.8(b), indicating clearly certain non-normality in the log returns. A Markovian three-state mixture Gaussian model, $x_t \sim N(\mu_{I_t}, \sigma_{I_t}^2)$ with $I_t \in \{1, 2, 3\}$ is fitted to the data using the Baum–Welch algorithm. The estimated transition probability matrix and the estimated parameters of the normal distribution in each regime are shown in Tables 7.1 and 7.2.

It can be seen from the estimated parameters that state 1 corresponds to a state with a negative average return and high volatility. State 2 has a medium positive

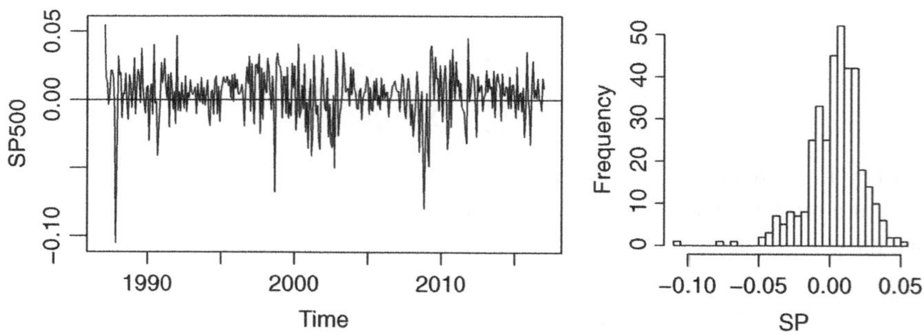

Figure 7.8 Log returns of monthly S&P 500 index and their histogram. The period is from January 1987 to December 2016.

	State 1	State 2	State 3
μ	−0.0216	0.005	0.0068
σ	0.0303	0.0099	0.0185

Table 7.1 Parameter estimates of fitting a three-state HMM to the monthly log returns of the S&P 500 Index from 1987.1 to 2016.12.

state	1	2	3
1	0.638	0	0.362
2	0	0.972	0.028
3	0.0663	0.0264	0.907

Table 7.2 Estimated state transition probability matrix of fitting a three-state HMM to the monthly log returns of the S&P 500 Index from 1987.1 to 2016.12.

mean return and low volatility, and state 3 has the largest positive mean return with medium volatility. Furthermore, state 1 represents a relatively rare state since its probability to transit out to other states is large. States 2 and 3 are relatively stable, with state 2 being the most persistent one. Most interestingly, state 1 only transits to state 3 (from negative mean return with high volatility to positive mean return with medium volatility). Such a mean-reverting phenomenon has been well documented in the finance literature. It is also interesting to see that there exists a certain probability for a bull market to be followed by a bear market.

Figure 7.9 shows the smoothed probabilities of $P(I_t = k)$ for $k = 1, 2, 3$. It is reassuring to see that state 1 occurs only occasionally whereas the other two states are relatively stable. Figure 7.10 shows the estimated most likely states using the Viterbi algorithm. The corresponding smoothed probabilities of each state on the most likely path are shown in Figure 7.11. Some of these states have marginal smoothed probability less than 0.5.

Figure 7.12 compares the partial autocorrelation function (PACF) of the squared residuals of the fitted three-state HMM model (a) and that of the squared S&P 500 return series (b). Figure 7.13 compares the normal Q-Q plots of the residual series (a) and the original series (b). It seems that the fitted HMM model is effective in handling the autocorrelations in the squared return series and its residuals follow more closely a Gaussian distribution.

A two-state HMM and a four-state HMM were also fitted to the log return series. For the two-state HMM model, the estimation results show that in

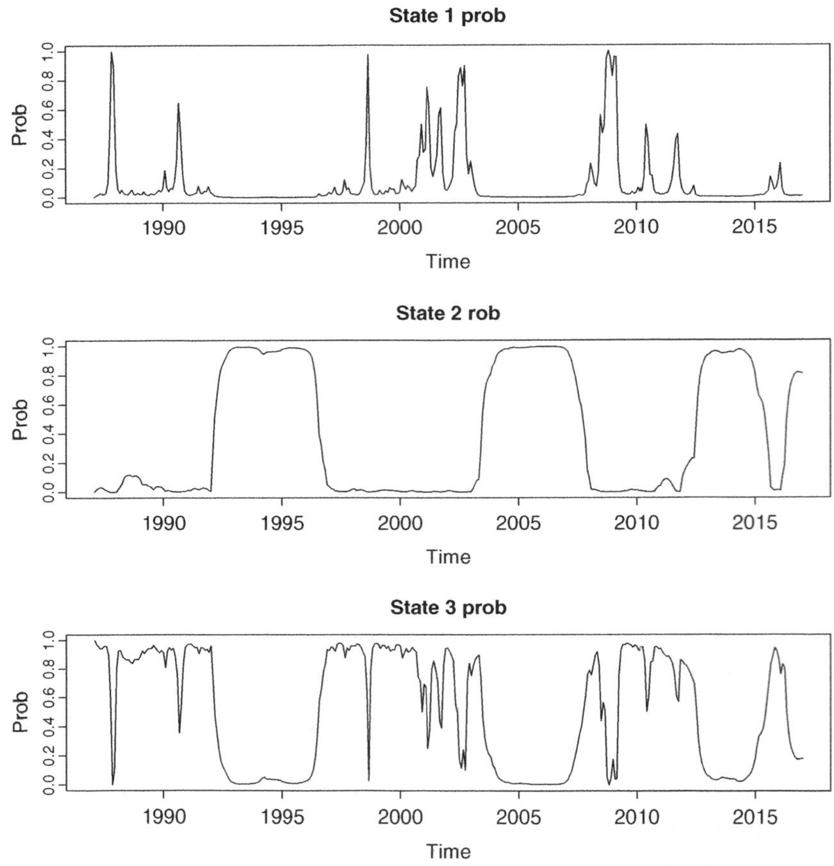

Figure 7.9 Smoothed probabilities of states 1, 2, and 3 for the monthly log returns of the S&P 500 index.

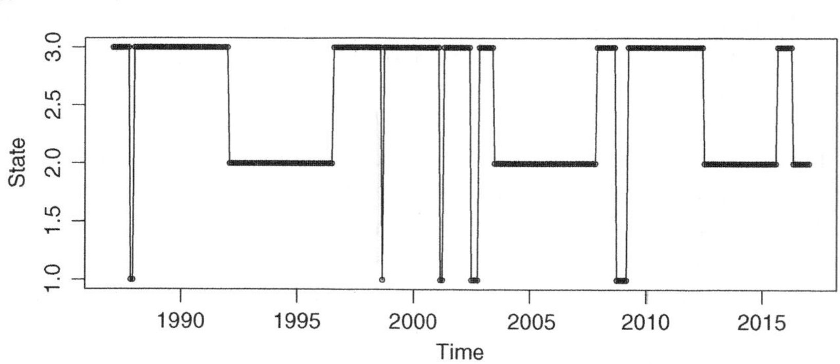

Figure 7.10 The most likely state path for the monthly log returns of the S&P 500 index.

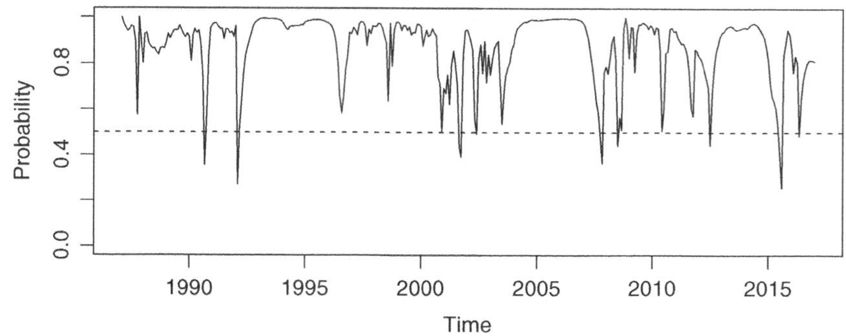

Figure 7.11 Smoothed probabilities of states on the most likely path for the monthly log returns of the S&P 500 index.

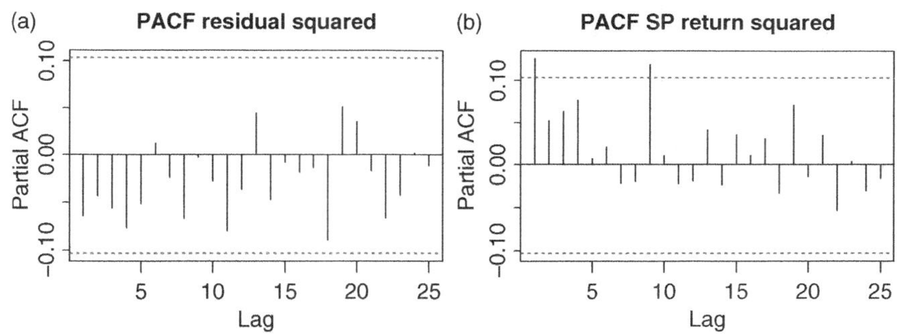

Figure 7.12 Partial autocorrelations of (a) the squared residuals and (b) the original series of the monthly log returns of the S&P 500 index.

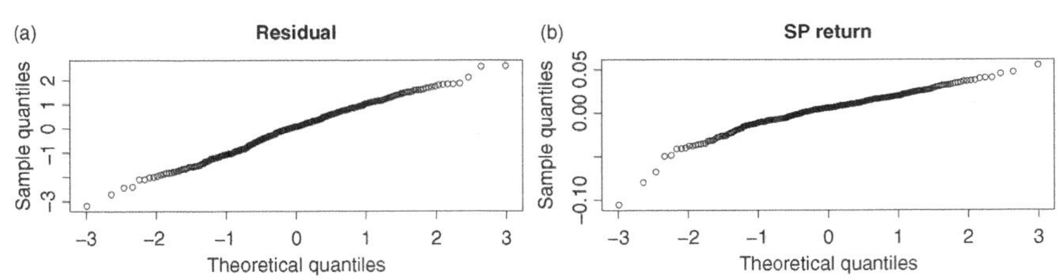

Figure 7.13 Normal Q-Q plot of (a) the residual series and (b) the original series for the monthly log returns of the S&P 500 index.

	State 1	State 2	State 3	State 4
μ	−0.0170	0.0056	0.0273	0.0372
σ	0.0252	0.0124	0.0060	0.0081

Table 7.3 Parameter estimates of fitting a four-state HMM to the monthly log returns of the S&P 500 Index from 1987.1 to 2016.12.

State	1	2	3	4
1	0.730	0	0.091	0.180
2	0.055	0.945	0	0
3	0.156	0.312	0.533	0
4	0	0.967	0.011	0.021

Table 7.4 Estimated transition probabilities of fitting a four-state HMM to the monthly log returns of the S&P 500 Index from 1987.1 to 2016.12.

state 1 the returns follow $N(-0.0034, 0.028^2)$, i.e. a negative mean return and high volatility state. In state 2 the distribution is $N(0.0065, 0.0124^2)$, hence it has positive mean return with relatively lower volatility. Both states are relatively stable, with the probability of staying in the same state being 0.86 and 0.94, respectively. Tables 7.3 and 7.4 show the MLEs of a four-state HMM model. The results are difficult to interpret. In particular, state 4 is extremely transitive. The log likelihood values of the two-, three- and four-state HMM models are 949, 961, and 963, respectively. Taking into account the number of parameters used, it seems that the three-state model is preferred.

R demonstration: Analysis of the monthly log returns of the S&P 500 index via HMM with the `HiddenMarkov` package.

```
> ## SP500 HMM Mixture Gaussian Example
> data <- read.csv(file="SP500.csv",header=T)
> sp <- data[,3]
> tt <- as.Date(data[,1],format="%d-%b-%Y")
> plot(tt,sp,type='l',xlab='Time',ylab='SP500')
> abline(h=0)
> hist(sp,nclass=40,main="")
> ## HMM modeling
> require("HiddenMarkov")
> #-- three-regime --
> ## ------ setting initial values
```

```
> Pi <- diag(rep(0.8,3))+matrix(rep(1,9)/3*0.2,ncol=3,nrow=3)
> delta <- rep(1,3)/3
> dist <- 'norm'      # mixture normal
> pm <- list(mean=c(-0.3,0.05,0.2),sd=c(0.1,0.1,0.5))
> temp <- dthmm(x,Pi,delta,dist,pm)   #setting model
> out3 <- BaumWelch(temp,control=bwcontrol(prt=FALSE))
> #Baum Welch estimation
> summary(out3)
$delta
[1]  1.222375e-79 6.275890e-243  1.000000e+00
$Pi
             [,1]          [,2]        [,3]
[1,] 6.379404e-01 2.597030e-22 0.36205961
[2,] 7.020327e-08 9.717455e-01 0.02825443
[3,] 6.626644e-02 2.645275e-02 0.90728081
$nonstat
[1] TRUE
$distn
[1] "norm"
$pm
$pm$mean
[1] -0.021637295  0.005144475  0.006810866
$pm$sd
[1] 0.030305953 0.009929002 0.018494987

> # smoothed State  probability
> par(mfrow=c(3,1))
> plot(tt,out3$u[,1],type='l',xlab='time',ylim=c(0,1),
      ylab='prob',main='State 1 Prob')
> plot(tt,out3$u[,2],type='l',xlab='time',ylim=c(0,1),
      ylab='prob',main='State 2 Prob')
> plot(tt,out3$u[,3],type='l',xlab='time',ylim=c(0,1),
      ylab='prob',main='State 3 Prob')
> par(mfrow=c(1,1))

> logLik(out3)    # log likelihood value
[1] 961.0367
> v3 <- Viterbi(out3)   # most likely path: Viterbi
> plot(tt,v3,cex=0.5,xlab='time',ylab='state'); lines(tt,v3)
> #plot most likely path
> pp <- 1:360; for(i in 1:360)pp[i] <- out3$u[i,v3[i]]
> # obtain smoothed probability of the Viterbi states
> plot(tt,pp,type='l',ylim=c(0,1),xlab='time',ylab='prob')
> abline(h=0.5,lty=2)
> res3 <- residuals(out3)  # get residuals
> pacf(res3**2,main='PACF residual squared')    # pacf residual squared
```

```
> pacf(sp**2, main='PACF SP return squared')     # pacf sp return squared
> qqnorm(res3, main='residual')     # qq plot residual
> qqnorm(sp,main='SP return')       # qq plot sp return
> #-- two-regime --
> Pi <- diag(c(0.9,0.9))+matrix(rep(1,4)/20,ncol=2,nrow=2)
> delta <- rep(1,2)/2
> dist <- 'norm'    # mixture normal
> pm <- list(mean=c(-0.1,0.1),sd=c(0.1,0.1))
> temp <- dthmm(sp,Pi,delta,dist,pm,nonstat=FALSE)  #setting model
> out2 <- BaumWelch(temp,control=bwcontrol(prt=FALSE))
> #Baum Welch estimation
> summary(out2)
$delta
[1] 0.289552 0.710448
$Pi
           [,1]      [,2]
[1,] 0.86500964 0.1349904
[2,] 0.05501701 0.9449830
$nonstat
[1] FALSE
$distn
[1] "norm"
$pm
$pm$mean
[1] -0.003436932  0.006498526
$pm$sd
[1] 0.02793554 0.01243074

> logLik(out2)
[1] 949.0639

> #-- four-regime --
> Pi <- diag(rep(0.8,4))+matrix(rep(1,16)/4*0.2,ncol=4,nrow=4)
> delta <- rep(1,4)/4
> dist <- 'norm'
> pm <- list(mean=c(-0.1,0,0.1,0.3),sd=c(0.1,0.1,0.1,0.1))
> temp <- dthmm(x,Pi,delta,dist,pm,nonstat=FALSE)
> out4 <- BaumWelch(temp,control=bwcontrol(prt=FALSE))
> summary(out4)
$delta
[1] 0.17424141 0.75917705 0.03457448 0.03200705
$Pi
             [,1]         [,2]         [,3]         [,4]
[1,] 7.295873e-01 6.368364e-12 9.077792e-02 1.796348e-01
[2,] 5.497348e-02 9.450265e-01 8.427622e-12 1.581627e-16
[3,] 1.556781e-01 3.115264e-01 5.327956e-01 5.263778e-19
```

```
[4,] 4.188804e-09 9.674036e-01 1.050020e-02 2.209623e-02
$nonstat
[1] FALSE
$distn
[1] "norm"
$pm
$pm$mean
[1] -0.017055821  0.005591701  0.027254630  0.037185901
$pm$sd
[1] 0.025168944 0.012357184 0.005992786 0.008122566
> logLik(out4)
[1] 963.0156
```

Example 7.3: Multi-regime Fama–French multi-factor models Example 6.2 illustrates the analysis of time-varying coefficient regression of the Fama–French multi-factor model. See Fama and French (1993, 1996). The coefficients there change constantly at each time point. Here we adopt an alternative approach assuming that the coefficients follow a Markov regime-switching mechanism. In other words, the coefficients change stochastically according to a transition probability matrix.

Tables 7.5 and 7.6 show the estimation results of a two-state HMM using the same data as in Example 6.2. Figure 7.14 shows the smoothed probabilities $p_{t|T}[1]$, the probabilities that the state is in State 1, given the entire series. Note that $p_{t|T}[2] = 1 - p_{t|T}[1]$. From the plot, it is clear that the system stays in State 1 significantly more often than in State 2. The coefficients associated with

	Constant	Market	SMB	HML	MoM	σ
State 1	0.225	1.015	0.963	0.441	−0.081	2.171
State 2	0.012	1.009	0.880	−0.039	−0.032	1.110

Table 7.5 Parameter estimates of a two-state HMM of the Fama–French multi-factor model. The sample period is from January 1946 to December 2010.

State	1	2
1	0.854	0.146
2	0.024	0.976

Table 7.6 Estimated state transition probabilities of a two-state HMM of the Fama–French multi-factor model. The sample period is from January 1946 to December 2010.

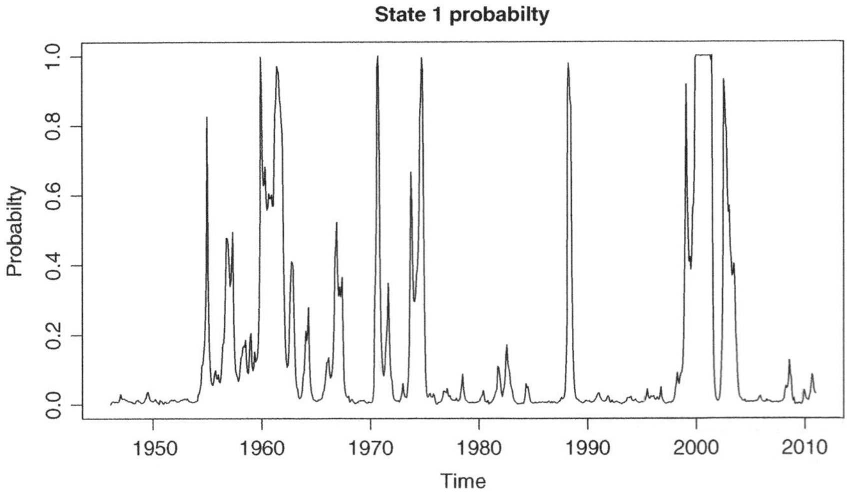

Figure 7.14 Smoothed probability of states 1 and 2 of a fitted two-state HMM of the Fama–French multi-factor model.

the intercept and the market are almost the same for the two regimes. On the other hand, the coefficients associated with HML and MoM change significantly between the States. Figure 7.15 shows the most likely path obtained via the Viterbi algorithm using the estimated parameters. To compare with the time-varying

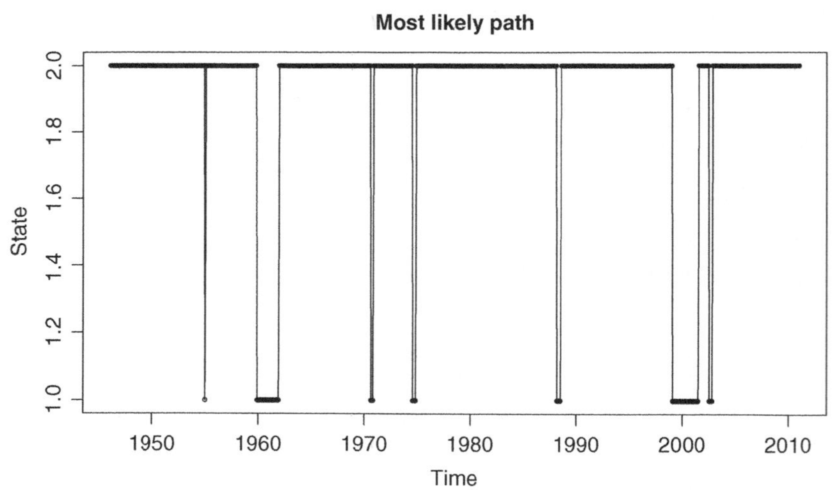

Figure 7.15 The path of most likely states based on a two-state HMM of the Fama–French multi-factor model.

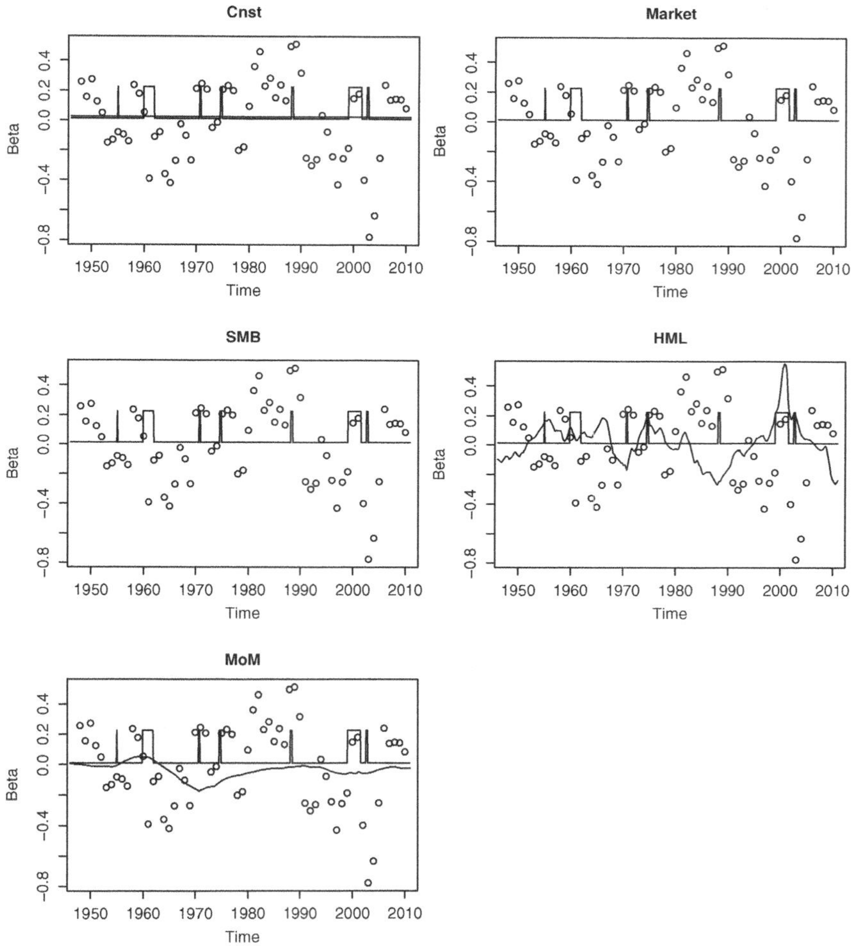

Figure 7.16 Comparison between a two-state HMM and the time-varying coefficient model of the Fama–French multi-factor model. Dots are the estimated coefficients with a 3-year moving window. The thin line is the smoothed state estimate under the time-varying coefficient model and the thick line is the estimated coefficients of the most likely path of regime switchings.

coefficient analysis in Example 6.2, in Figure 7.16 we show the estimated coefficients using a three-year moving window, the smoothed states under the varying coefficient model, and the estimated coefficients under the most likely regime-switching path of the HMM model. It seems that for the coefficients of the HML, both the time-varying coefficient model and the HMM model try to capture the same time-varying feature of the data.

	Constant	Market	SMB	HML	MoM	σ
State 1	0.988	0.823	1.535	0.390	0.040	2.094
State 2	−0.015	1.017	0.863	−0.076	−0.026	1.100
State 3	−0.006	1.006	0.827	0.214	−0.079	1.525

Table 7.7 Parameter estimates of a three-state HMM of the Fama–French multi-factor model. The sample period is from January 1946 to December 2010.

State	1	2	3
1	0.816	0.184	0
2	0.006	0.977	0.016
3	0.006	0.025	0.969

Table 7.8 Estimated state transition probabilities of a three-state HMM of the Fama–French multi-factor model. The sample period is from January 1946 to December 2010.

Tables 7.7 and 7.8 show the estimation results of fitting a three-state HMM. Figure 7.17 shows the smoothed probabilities of states. Comparing with the two-state HMM model, it is seen that the three-state model essentially splits the major state of the two-state model into two subregimes, with almost the same coefficients in both subregimes. The likelihood value of the three-state model is $-1,281$, whereas that of the two-state model is $-1,301$. The improvement in the third state seems to be marginal given that the three-state model employs 10 additional parameters over the two-state model.

R demonstration: HMM modeling via the `HiddenMarkov` package.

```
## FF HMM Mixture Regression Example
xxx <- read.csv(file="FF_example.csv",header=T)
year0 <- 1946:2010
y <- xxx[,2]
xx <- xxx[,3:6]
xx <- as.matrix(xx)
ntotal <- length(y)

## HMM modeling
require("HiddenMarkov")
#-- two-regime --
Pi <- diag(rep(0.95,2))+matrix(rep(1,4)/2*0.05,ncol=2,nrow=2)
delta <- rep(1,2)/2
```

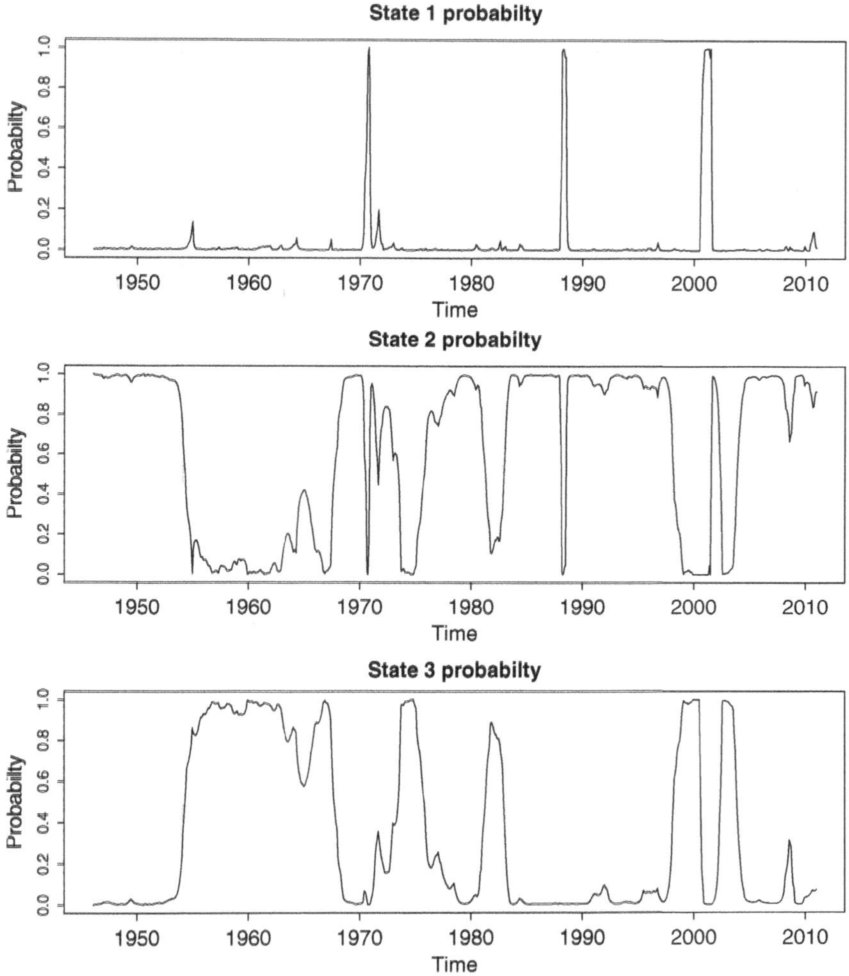

Figure 7.17 Smoothed probabilities of states 1, 2, and 3 in a three-state HMM of the Fama–French multi-factor model.

```
sd <- c(1,1)*10
beta1 <- c(-1,-1,-1,-1,-1)
beta2 <- c(1,1,1,1,1)
beta <- cbind(beta1,beta2)

glmformula <- formula(~xx)
glmfamily <- gaussian(link="identity")
ones <- rep(1,ntotal)
```

```
data <- data.frame(xx=cbind(ones,xx))
Xdesign <- model.matrix(glmformula,data)
MyModel <- mmglm1(y,Pi,delta,glmfamily,beta,Xdesign,sigma=sd)
out <- BaumWelch(MyModel, bwcontrol(posdiff=FALSE, maxiter=200))
summary(out)

tt <- (1:780)/12+1946
par(mfrow=c(1,1),mai=0.9*c(1,1,1,1))
plot(tt,out$u[,1],type='l',xlab='time',ylim=c(0,1),ylab='prob')
title('State 1 Prob')  # smoothed State 1 probability
## Figure 7.14

par(mfrow=c(1,1),mai=c(1,1,1,1))
logLik(out)     # log likelihood value
v2 <- Viterbi(out)   # most likely path: Viterbi
plot(tt,v2,cex=0.5,xlab='time',ylab='state'); lines(tt,v2)
title("Most Likely Path")
#plot most likely path  #Figure 7.15

### compare with time-varying_beta example
ntotal <- length(y)
nyear <- ntotal/12
beta.v <- matrix(ncol=5,nrow=(nyear-2))
for(i in 3:nyear){
  t0 <- (i-3)*12+(1:36)
  y0 <- y[t0]    ##excess return of portfolio 22
  xx0 <- as.matrix(xx[t0,])
  out.v <- lm(y0~xx0)
  beta.v[i-2,] <- out.v$coef
}
year <- 1948:2010
par(mfrow=c(3,2),mai=0.7*c(1,1,1,1))
for(jj in 1:5){
plot(year,beta.v[,1],type='p',xlab='time',ylab='beta',main=name[jj])
pp <- 1:780; for(i in 1:780)pp[i] <- out$beta[1,v2[i]]
lines(tt,pp,lwd=2);
lines(tt,ss2$s[-1,jj])
}
## ss2 from time-varying beta example
## Figure 7.16

### --- three regimes
## ------ setting initial values
Pi <- diag(rep(0.8,3))+matrix(rep(1,9)/3*0.2,ncol=3,nrow=3)
delta <- rep(1,3)/3
```

```
sd <- c(1,1,1)*10
beta1 <- c(-1,-1,-1,-1,-1)
beta2 <- c(1,-1,-1,1,1)
beta3 <- c(1,1,1,1,1)
beta <- cbind(beta1,beta2,beta3)

glmformula <- formula(~xx)
glmfamily <- gaussian(link="identity")
Xdesign <- model.matrix(glmformula,data)
MyModel <- mmglm1(y,Pi,delta,glmfamily,beta,Xdesign,sigma=sd)
out3 <- BaumWelch(MyModel, bwcontrol(posdiff=FALSE, maxiter=200))
summary(out)

par(mfrow=c(3,1),mai=0.5*c(1,1,1,1))
plot(tt,out3$u[,1],type='l',xlab='time',ylim=c(0,1),ylab='prob',
      main='State 1 Prob')
# smoothed State 1 probability
plot(tt,out3$u[,2],type='l',xlab='time',ylim=c(0,1),ylab='prob',
       main='State 2 Prob')
# smoothed State 2 probability
plot(tt,out3$u[,3],type='l',xlab='time',ylim=c(0,1),ylab='prob',
       main='State 3 Prob')
# smoothed State 3 probability
## Figure 7.17
```

7.3 EXERCISES

7.1 Let seed = 11. Repeat the simulation study of Example 7.1 with 200 observations and a second sonar at (25,0), i.e. s2 = 25 in R, using the same initial value and parameters.

7.2 Consider the weekly crude oil price from January 3, 1986 to October 6, 2017. The data are available from FRED and are also in the file `w-coilwtico.csv`. Let y_t be the first differenced series of the weekly crude oil price. Part of the series was analyzed in Example 1.2 of Chapter 1. Apply a two-state HMM to the series y_t.

   (a) Obtain parameter estimates, including estimates of the transition probability matrix.

   (b) Show the time plot of smoothed probabilities of state 1.

   (c) Obtain the most likely state path.

   (d) Compare with the analysis of Example 1.2.

7.3 Consider again the first differenced series y_t of the weekly crude oil prices from January 3, 1986 to October 6, 2017. Suppose that y_t follows the exponential autoregressive model

$$y_t = \exp(b)y_{t-1} + \exp(c + dy_{t-1}^2)y_{t-8} + \epsilon_t,$$

where ϵ_t is a sequence of iid $N(0, \sigma_e^2)$.

(a) Put the exponential autoregressive model into a nonlinear Gaussian state space model.

(b) Fit the model using the extended Kalman filter with the first-order Taylor expansion. Write down the parameter estimates.

(c) Check the model using the residuals $\hat{\epsilon}_t = y_t - \exp(\hat{b})y_{t-1} - \exp(\hat{c} + \hat{d}y_{t-1}^2)y_{t-8}$. Are there serial correlations in the residuals? You may use Ljung–Box statistics with 10 lags.

7.4 Consider the daily gold fixing prices at 10:30am in the London Bullion Market (in US dollars) from October 19, 2012 to October 18, 2017. The data are available from FRED and are also in the file `d-goldamGBD228nlbm.csv`. There are some missing values in the FRED data and we have deleted those missing values. Let y_t be the log returns of the daily gold price, i.e. the first-differenced of the log prices. Fit a two-state HMM model to y_t. You may use y_{t-1} and y_{t-6} as input variables. Write down the fitted model, including estimates of the transition probabilities. Obtain the time-plot of the smoothed state probabilities as well.

7.5 Consider the data set `m-dec5FF-6117.txt` used in question 3 of the exercise in Chapter 6. Fit a two-state HMM Fama–French multi-factor model to the excess returns of Decile 5 portfolio. Perform similar analysis to that in Example 7.3.

REFERENCES

Anderson, J. L. (2001). An Ensemble Adjustment Kalman Filter for Data Assimilation. *Monthly Weather Review* **129**: 2884–2903.

Anderson, J. L. (2007). An adaptive covariance inflation error correction algorithm for ensemble filters. *Tellus A* **59**: 210–224.

Anderson, B. D. O. and Moore, J. B. (1979). *Optimal filtering*. Prentice-Hall, Englewood Cliffs, NJ.

Baggenstoss, P. M. (2001). A modified Baum–Welch algorithm for hidden Markov models with multiple observation spaces. *IEEE Transactions on Speech and Audio Processing* **9**: 411–416.

Baum, L. E. and Petrie, T. (1966). Statistical inference for probabilistic functions of finite state Markov chains. *Annals of Mathematical Statistics* **37**: 1554–1563.

Baum, L. E., Petrie, T., Soules, G., and Weiss, N. (1970). A maximization technique occurring in the statistical analysis of probabilistic functions of Markov chains. *Annals of Mathematical Statistics* **41**: 164–171.

Bellman, R. E. (2003). *Dynamic Programming*. Dover Publications, Inc., New York.

Bishop, C.H., Etherton, B.J., and Majumdar, S.J. (2001). Adaptive sampling with the ensemble transform Kalman filter. Part I: Theoretical aspects. *Monthly Weather Review* **129**: 420–436.

Dempster, A. P., Laird, N. M., and Rubin, D. B. (1977). Maximum-likelihood from incomplete data via the EM algorithm. *Journal of the Royal Statistical Society Series B* **39**: 1–38.

Duncan, D. B. and Horn, S. D. (1972). Linear dynamic recursive estimation from the viewpoint of regression analysis. *Journal of the American Statistical Association* **67**: 815–821.

Durbin, J. and Koopman, S. J. (2001). *Time Series Analysis by State Space Methods*. Oxford University Press, Oxford.

Evensen, G. (1994). Sequential data assimilation with a nonlinear quasi-geostrophic model using Monte Carlo methods to forecast error statistics. *Journal of Geophysical Research* **99**: 10143–10162.

Evensen, G. (2003). The ensemble Kalman filter: theoretical formulation and practical implementation. *Ocean Dynamics* **53**: 343–367.

Evensen, G. (2007). *Data Assimilation: the Ensemble Kalman Filter*. Springer, New York.

Fama, E. F. and French, K. R. (1993). The cross-section of expected stock returns. *Journal of Finance* **47**: 427–465.

Fama, E. F. and French, K. R. (1996). Multifactor explanations of asset pricing anomalies. *Journal of Finance* **51**: 55–84.

Forney, G. D. (1973). The Viterbi algorithm. *Proceedings of the IEEE* **61**: 268–278.

Furrer, R., Genton, M. G., and Nychka, D. (2006). Covariance tapering for interpolation of large spatial datasets. *Journal of Computational and Graphical Statistics* **15**: 502–523.

Gelb, A. (1974). *Applied Optimal Estimation*. MIT Press, Cambridge, MA.

Julier, S. J. and Uhlmann, J. K. (1997). New extension of the Kalman filter to nonlinear systems. *Proceedings of the SPIE* **3068**: 182–193.

Julier, S. J. and Uhlmann, J. K. (2004). Unscented filtering and nonlinear estimation. *Proceedings of the IEEE* **92**: 401–422.

Roth, M. and Gustafsson, F. (2011). An efficient implementation of the second order extended Kalman filter. In *Proceedings of the 14th International Conference on Information Fusion*, pp. 1–6. IEEE.

Sorenson, H. W. and Alspach, D. L. (1971). Recursive Bayesian estimation using Gaussian sums. *Automatica* **7**: 465–479.

Stroud, J.R., Stein, M.L., Lesht, B.M., Schwab, D.J., and Beletsky, D. (2010). An ensemble Kalman filter and smoother for satellite data assimilation. *Journal of the American Statistical Association* **105**: 978–990.

Tanizaki, H. (1996). *Nonlinear Filters: Estimation and Applications*. Springer, New York.

Tippett, M. K., Anderson, J. L., Bishop, C. H., Hamill, T. M., and Whitaker, J. S. (2003). Ensemble Square Root Filters. *Monthly Weather Review* **131**: 1485–1490.

Viterbi, A. J. (1967). Error bounds for convolutional codes and an asymptotically optimum decoding algorithm. *IEEE Transactions on Information Theory* **13**: 260–269.

Wu, C. F. (1983). On the convergence properties of the EM algorithm. *Annals of Statistics* **11**: 95–103.

CHAPTER 8

SEQUENTIAL MONTE CARLO

In Chapter 7 we discussed several approaches for dealing with nonlinear and non-Gaussian state space models (SSMs). Except for the hidden Markov model (HMM), all the approaches discussed use approximations as there are no analytic solutions for filtering or smoothing in most nonlinear and non-Gaussian models. In some applications, these approximation approaches may not be sufficient or be able to provide the required accuracy. Some alternatives must be sought. With the advances in computational capability, Monte Carlo methods have become a popular and useful tool in solving complex problems. In this chapter we discuss an effective Monte Carlo approach designed specifically for dealing with SSMs. It is referred to as the *sequential Monte Carlo* (SMC) method. This method tries to utilize fully the sequential and dynamic nature of SSMs, yet enjoys the flexibility of the powerful Monte Carlo approaches. The SMC approach is parallel to that of the Kalman filter in which the conditional distribution of the state x_t given the entire history of the observations $\boldsymbol{y}_t$ is obtained and updated based on the results obtained for x_{t-1} given $\boldsymbol{y}_{t-1}$ at the previous time step. Because there are no parametric and

Nonlinear Time Series Analysis, First Edition. Ruey S. Tsay and Rong Chen.

Companion website: www.wiley.com/go/tsay/nonlineartimeseries

analytical forms available for the conditional distributions, SMC uses discrete random samples to represent those conditional distributions. This approach belongs to the class of *density-based filtering* approaches. The Gaussian sum filters and the unscented Kalman filters introduced in Chapter 7 also belong to this class. The difference is that SMC uses random samples instead of a deterministic mixture distribution approximation or a fixed-point approximation.

A Monte Carlo method tries to generate a set of samples $x^{(1)}, \ldots, x^{(m)}$ from a target distribution $\pi(x)$. Statistical inferences, such as estimating $E_\pi(h(x))$, can then be obtained through the Monte Carlo samples such as the sample average estimator

$$E_\pi[h(x)] \approx \frac{\sum_{i=1}^{m} h(x^{(i)})}{m}.$$

SMC achieves estimation and optimization tasks by recursively generating Monte Carlo samples of the state variables of a system. The basic principle behind SMC dates back to the "growth Monte Carlo" method known in molecular physics in the 1950s; see Hammersley and Morton (1954) and Rosenbluth and Rosenbluth (1955). With modern computing power it has generated significant interest in recent years.

In this chapter we start with a brief introduction to general Monte Carlo methods, then establish a general framework of SMC, followed by some detailed discussions of its implementation. Several specific implementations are discussed for a wide range of nonlinear non-Gaussian SSMs.

8.1 A BRIEF OVERVIEW OF MONTE CARLO METHODS

Monte Carlo methods are essentially a class of numerical methods to evaluate (or estimate) some integrals. They rely on random sampling, which is the key principle behind all statistical analyses, and use statistical methods to assess the estimation accuracy. Specifically, we focus on two types of integrals:

(I) Normalizing constant:

$$Z = \int \pi^*(x)dx, \tag{8.1}$$

where $\pi^*(x)$ is a non-negative integrable function.

(II) Expectation:

$$\mu_h = \int h(x)\pi(x)dx, \tag{8.2}$$

where $\pi(x)$ is a probability density and $h(x)$ is an π-integrable real function.

Many statistical inference problems can be reformulated as an integration problem with a properly chosen function $h(x)$ and a target distribution $\pi(x)$ or a non-negative integrable function $\pi^*(x)$. The integral Z is essentially the normalizing constant of a probability density $\pi(x) = \pi^*(x)/Z$. It has a broad range of applications, including likelihood calculation, entropy calculation (Gelman and Meng, 1998), and communication systems (Smith et al., 1997).

The integral μ_h is the expectation of $h(x)$ under the distribution $\pi(x)$. By choosing a proper $h(x)$, various moments and the probability of a certain (maybe rare) event can be estimated. For example, if $h(x) = x$, μ_h is the mean of the distribution $\pi(x)$, higher-order moments can be obtained using $h(x) = x^k$. To obtain the probability $P(X \in \Omega)$, if $X \sim \pi(x)$, one can use $h(x) = I_\Omega(x)$, the indicator function. A histogram of $\pi(x)$ can be obtained by changing Ω. The expected shortfall often used in financial risk management can be obtained using $h(x) = xI(x \geq c)$. If $h(x) = K((x - x_0)/h)$, a kernel function around a fixed point x_0, then μ_h is essentially a (unnormalized) kernel estimation of $\pi(x_0)$.

In a typical statistical inference problem, one is interested in estimating and assessing the estimation accuracy of some unknown parameter θ of a distribution $p(\cdot, \theta)$ based on a set of random samples $X_1, \ldots, X_n$ drawn from the distribution $p(\cdot, \theta)$. Often the random samples are physically observed (e.g., survey sampling, physical measurements, etc.). Monte Carlo methods start with a known distribution $\pi(\cdot, \theta)$ with a known parameter θ and try to estimate μ_h or Z. In fact μ_h depends on θ and can be viewed as a re-parameterized parameter to be estimated. Mimicking the traditional estimation approach, we would need a set of random samples from the given distribution $\pi(\cdot, \theta)$. Instead of trying to get physical samples, the Monte Carlo method tries to generate pseudo random samples (random numbers) using computers. Using these samples, the integrals can be estimated and the estimation accuracy can be assessed.

For example, to estimate μ_h, if we can generate an iid random sample $(x_1, \ldots, x_m)$ from $\pi(x)$, then

$$\hat{\mu}_h = \frac{1}{m} \sum_{i=1}^{m} h(x_i)$$

is an unbiased estimator of μ_h with variance V_h/m, where $V_h = \int h^2(x)\pi(x)dx - \mu_h^2$ is the variance of $h(X)$. Note that V_h can also be estimated using

$$\widehat{V}_h = \frac{1}{m} \sum_{i=1}^{m} h^2(x_i) - \hat{\mu}_h^2.$$

The key to Monte Carlo methods is to find ways to efficiently generate the pseudo numbers from a distribution $\pi(x)$, and if it is difficult to generate such random samples directly from $\pi(x)$, then find an alternative way to achieve the same objectives. Often the random samples themselves are of interest. For example, if one is able to generate samples $\{x_1^{(j)}, x_2^{(j)}, \ldots, x_d^{(j)}\}$, $(j = 1, \ldots, m)$, from a joint distribution $\pi(x_1, \ldots, x_d)$, then the marginal samples $\{x_1^{(1)}, \ldots, x_1^{(m)}\}$ can be used to study the marginal distribution $\pi(x_1)$, such as a kernel density estimation of $\pi(x_1)$.

8.1.1 General Methods of Generating Random Samples

In this section we consider some methods for generating random samples.

1. Computer-generated pseudo random numbers: Monte Carlo methods depend on generating random numbers. Physical devices such as coins, dice, colored balls in an urn, or lottery machines are not practical for large experiments. Except for some high-end physical random number generating devices such as those in Fischer and Drutarovský (2003) and Majzoobi et al. (2011), computer-generated pseudo random numbers are often used in Monte Carlo methods. The resulting random numbers are often from a deterministic sequence, but behave like a random sequence following a uniform distribution on [0, 1]. Robert and Casella (2004) provide a good introduction. Pseudo random number generators are widely available on almost all platforms and softwares.

2. Transformation: If a complicated random variable Y is a function of a simpler random variable X, then a set of random samples of Y can be obtained through a set of random samples from X. Specifically, if $Y = f(X)$ and $x \sim X$, then $y = f(x) \sim Y$.

For example, random samples of a normal distribution can be obtained via those from a uniform distribution. Specifically, if X_1 and X_2 are independent Unif(0, 1) random variables, and

$$Y_1 = \sqrt{-2\ln(X_1)}\cos(2\pi X_2), \quad Y_2 = \sqrt{-2\ln(X_1)}\sin(2\pi X_2),$$

then Y_1 and Y_2 are two independent $N(0, 1)$ random variables. This transformation is often referred to as the Box–Muller algorithm.

To generate samples from $N(\mu, \sigma^2)$, one can use the transformation $Y = \mu + \sigma X$, where $X \sim N(0, 1)$. If $Z_1, \ldots, Z_k$ are independent $N(0, 1)$ random variables, then $Y = \sum_{i=1}^k Z_i^2$ follows a χ_k^2 distribution. Hence a χ_k^2 random sample can be obtained from k standard Normal random samples. When k is small, this method of generating χ_k^2 samples is efficient, but it becomes inefficient when k is large.

Many commonly used distributions can be obtained using transformations based on the standard exponential random variable. Let X_i be iid standard exponential random samples, which are easy to generate using the inverse cumulative distribution function (CDF) method discussed below. Then χ_{2k}^2 distribution can be obtained using $Y = 2\sum_{i=1}^k X_i$, gamma(α, β) distribution with shape parameter α

and scale parameter β can be obtained using $Y = \beta \sum_{i=1}^{\alpha} X_i$, and beta$(a, b)$ distribution can be obtained using $Y = A/(A + B)$ with $A = \sum_{i=1}^{a} X_i$ and $B = \sum_{i=a+1}^{a+b} X_i$). For a χ^2 distribution with $2k + 1$ degrees of freedom, one can use $Y = X_1 + X_2^2$, where $X_1 \sim \chi^2_{2k}$ and $X_2 \sim N(0, 1)$.

To generate a random sample from a Student-t distribution with k degrees of freedom, we notice the transformation $T = \frac{Z}{\sqrt{X}}$ where $Z \sim N(0, 1)$ and $X \sim \chi^2_k$, and Z and X independent. Random samples of an F distribution can be obtained by generating two independent χ^2 random samples with proper degrees of freedom.

3. Inverse CDF: The method of inverse CDF is a general transformation method to generate random samples. The following lemma provides a direct connection between any random variable and the uniform distribution, the simplest and most commonly available distribution.

Lemma 8.1 *If a random variable X has a CDF $F(\cdot)$, then $Y = F(X) \sim Unif(0, 1)$. Furthermore, $\tilde{F}^{-1}(U) \sim X$ where $U \sim Unif(0, 1)$ and*

$$\tilde{F}^{-1}(u) = \arg\min_x \{F(x) \geq u\}, \quad 0 \leq u \leq 1$$

is the generalized inverse of $F(x)$.

To see why this is true, let $Y = F(X)$. Then for any $0 \leq u \leq 1$,

$$P(Y \leq u) = P(F(X) \leq u) = P(X \leq \tilde{F}^{-1}(u)) = F(\tilde{F}^{-1}(u)) = u.$$

Hence $Y = F(X) \sim Unif(0, 1)$. In addition, let $X^* = \tilde{F}^{-1}(U)$, then

$$P(X^* < x) = P(\tilde{F}^{-1}(U) < x) = P(U < F(x)) = F(x).$$

If the CDF $F(x)$ has an analytical inverse, the method is very effective. For example, if X has a probability density function (pdf) $p(x) = 2x$, $(0 < x < 1)$, its CDF is $F(x) = x^2$. Then X can be generated using the transformation $X = \sqrt{U}$, $U \sim Unif(0, 1)$. The most important distribution in this class is the standard exponential distribution with pdf $p(x) = e^{-x}$ for $x > 0$. Its CDF is $F(x) = 1 - e^{-x}$, hence $F^{-1}(u) = -\ln(1 - u)$. Then a sample of the standard exponential distribution exp(1) can be obtained by generating U from $Unif[0, 1]$ and $X = -\ln(1 - U)$ or, equivalently, $X = -\ln(U)$. For general exponential distributions, we note that $Y = X/\lambda \sim \exp(\lambda)$ with pdf $p(y) = \lambda \exp(-\lambda y)$, where $X \sim \exp(1)$. This provides a simple approach to generate random samples from χ^2_{2k}, gamma, beta, and other distributions.

4. Rejection method: The rejection method is an effective approach to generate random samples from complicated distributions. It is based on the following two facts.

Lemma 8.2 *Suppose $p(x)$ is a probability density function with a domain Ω. If (X, Y) is uniformly distributed in $\Omega^* = \{(x, y), x \in \Omega, 0 \leq y \leq p(x)\}$, then the marginal distribution of X is $p(x)$.*

This can easily be seen from the fact that $\int_0^{p(x)} 1dy = p(x)$ for all $x \in \Omega$.

Lemma 8.3 *If a random variable is uniformly distributed in a region, then it is also uniformly distributed in any of its subregions.*

Note that, by definition, if a random variable is uniformly distributed in the region, then it has the same probability to be in any area of the same size (volume) in that region. It is then obvious that the same is true in a subregion.

The rejection method utilizes a combination of these two lemmas. For a distribution $p(x)$, Lemma 8.2 says that if we can generate (X, Y) uniformly in the region $\Omega^* = \{(x, y), x \in \Omega, 0 \leq y \leq p(x)\}$, then X follows $p(x)$. However, the region Ω^* can be irregular, and it might be difficult to generate uniform random samples in it. Lemma 8.3 can be used to solve the problem by generating uniform random samples in an enlarged but more regular region Ω^{**} that covers Ω^* as its subregion.

Figure 8.1 depicts the process of the rejection method. Figure 8.1a shows 100 random samples uniformly distributed in the rectangular region Ω^{**}. The solid line is the pdf $p(x)$ from which we wish to generate samples. Figure 8.1b shows the retained samples in Ω^*, the shaded area under the pdf $p(x)$. By Lemma 8.3, these samples are uniformly distributed in Ω^*. The samples outside the shaded area Ω^* are rejected. Figure 8.1c shows the marginal samples of x values projected from the samples shown in the middle panel.

Formally, suppose $\pi(x)$ is the target distribution from which we wish to generate samples. Suppose there is a distribution with pdf $g(x)$ that is easy to generate sample from, and we can find a constant c such that $\pi(x) \leq cg(x)$ for all $x \in \Omega$.

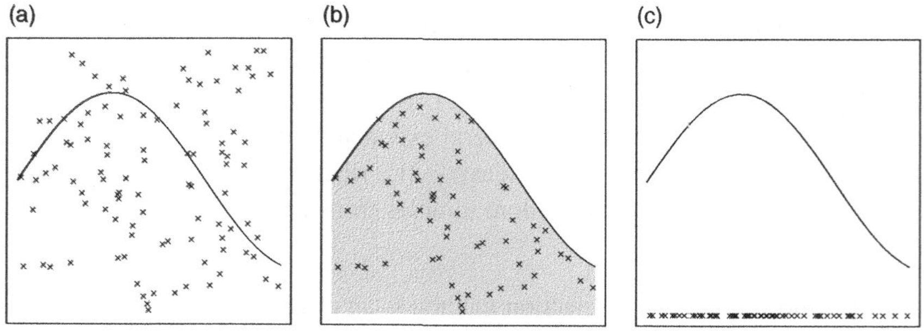

Figure 8.1 The method of rejection sampling.

The density $g(x)$ is then called an envelop density of $\pi(x)$. The rejection method can be summarized as follows.

Algorithm 8.1: Rejection Sampling

- Generate m samples $x_1, \ldots, x_m$ according to the distribution $g(x)$.
- Generate m samples $u_1, \ldots, u_m$ following Unif[0, 1].
- Retain the samples x_i that satisfy $u_i < \pi(x_i)/cg(x_i)$.

The above approach is equivalent to generating uniform random samples in the region $\Omega^{**} = \{(x, y), x \in \Omega, 0 \le y \le cg(x)\}$, then rejecting those outside $\Omega^* = \{(x, y), x \in \Omega, 0 \le y \le \pi(x)\}$.

The efficiency of the rejection sampling depends on the proportion of the rejected samples, as rejections waste computation. Hence the sampling region Ω^{**} needs to be as small as possible, or the ratio of the areas $|\Omega^*|/|\Omega^{**}|$ needs to be as large as possible. For the given $\pi(x)$ and $g(x)$, the smaller c, the higher the efficiency.

Uniform distribution on a rectangle is easy to generate samples from. In this case $g(x)$ is a uniform distribution. It is often used when $\pi(x)$ has a finite support. In many other situations one needs to carefully choose a $g(x)$ that balances the efficiency and the ease in generating samples from it.

We demonstrate the rejection method with some examples. We generate a truncated distribution $\pi(x) = c\pi^*(x)I(x > a)$, where $\pi^*(x)$ is a probability density function and $c = 1/\int_a^\infty \pi^*(x)dx$ is the normalizing constant of $\pi^*(x)I(x > a)$. The obvious envelop distribution is $g(x) = \pi^*(x)$ and we have $\pi(x) \le c\pi^*(x)$ for all x. Since $\pi(x)/c\pi^*(x) = 1$ for all $x > a$ and $\pi(x)/c\pi^*(x) = 0$ for all $x < a$, the rejection method accepts all samples x_j generated from $\pi^*(x)$ satisfying $x_j > a$ and rejects those smaller than a. This is simple if it is easy to generate random samples from $\pi^*(x)$. However, when a is very large (or small) so c is very large, this approach is not efficient. For example, for a truncated normal distribution, if $a = \mu + 3\sigma$, one only accepts about 0.13% of the samples. For large a, a better trial distribution in this case is the translated exponential distribution. Consider the truncated standard normal distribution

$$\pi(x) = \frac{1}{(1 - \Phi(a))\sqrt{2\pi}} \exp(-x^2/2)I(x > a),$$

where $1 - \Phi(a) = P(X > a)$ is the normalizing constant. We can use $g(x) = \lambda e^{-\lambda(x-a)}I(x > a)$, a shifted exponential distribution. One can easily generate random samples from this distribution by simply adding the constant a to an exponential $\exp(\lambda)$ random variate (using the inverse CDF method). Note that

$$\frac{\exp\{-x^2/2\}}{\exp\{-\lambda(x-a)\}} I(x>a) = \exp\left\{-\frac{1}{2}(x^2-2\lambda x+\lambda^2)+\frac{1}{2}\lambda^2-\lambda a\right\} I(x>a)$$

$$\leq \exp\left\{\frac{1}{2}\lambda^2-\lambda a\right\} \quad \text{for all} \quad x>a.$$

Choosing $\lambda = a + \frac{1}{2}\sqrt{a^2+4}$, we minimize the envelop constant

$$c = \frac{1}{\lambda(1-\Phi(a))\sqrt{2\pi}} \exp\{\frac{1}{2}\lambda^2 - \lambda a\},$$

and $\pi(x) \leq cg(x)$ for all x. For $a = 3$ and $a = 4$, the acceptance rates are 6% and 1.9%, respectively, compared to 0.13% and 3.16×10^{-5}, respectively, when using the standard normal trial distribution.

5. Sequential sampling: Sequential sampling is one of the most commonly used approaches for generating Monte Carlo samples of a high dimensional distribution. It sequentially generates low dimensional components one at a time until the sample is completed. It is based on the chain rule

$$p(X_1, X_2, \ldots, X_n) = p(X_1)p(X_2 \mid X_1) \cdots p(X_n \mid X_1, \ldots, X_{n-1}).$$

Specifically, if $(X, Y) \sim p(x, y)$, sequential sampling starts with sampling $X = x$ from the marginal distribution $p(x) = \int p(x, y)dy$, then sampling $Y = y$ from the conditional distribution $p(y \mid X = x) = p(x, y)/p(x)$. Then (x, y) is a sample from the joint distribution of $p(x, y)$.

For example, suppose

$$(X, Y) \sim N\left(\begin{bmatrix}\mu_1\\ \mu_2\end{bmatrix}, \begin{bmatrix}\sigma_1^2 & \rho\sigma_1\sigma_2\\ \rho\sigma_1\sigma_2 & \sigma_2^2\end{bmatrix}\right).$$

To generate a sample of (X, Y), one first generates $X = x$ from $N(\mu_1, \sigma_1^2)$, then generates $Y = y$ from $p(Y \mid X = x) \sim N(\mu_2 - \rho\frac{\sigma_2}{\sigma_1}(x - \mu_1), (1-\rho^2)\sigma_2^2)$.

To generate an AR(1) process, $X_t = \phi X_{t-1} + e_t$ where $e_t \sim N(0, \sigma^2)$, one starts with simulating $X_0 = x_0$ from $N(\mu_0, \sigma_0^2)$ (often the stationary distribution with $\mu_0 = 0$ and $\sigma_0^2 = \sigma^2/(1-\phi^2)$), then iteratively generates $X_t = x_t$ from $N(\phi x_{t-1}, \sigma^2)$. The sample $(x_1, x_2, \ldots, x_n)$ forms a realization of the time series.

For sequential sampling, one may choose the order of the variables or components to generate random samples from. The criterion is to make the sampling of the marginal and conditional distributions easier.

6. Augmentation: In many cases it is not easy to generate random samples directly from a distribution. However, by augmenting an additional component and using sequential sampling, one can simplify the problem. Specifically, if $p_X(x)$ is not easy to generate sample from, one can increase the dimension by augmenting an extra component Y so that the joint distribution $p(X, Y)$ is easy to generate sample from through the sequential sampling of $p_Y(y)$ and $p(X \mid Y = y)$. This is based on the fact that if $(X, Y) \sim p(x, y)$, then $X \sim p_X(x)$. Hence if we have a sample $(x, y) \sim p(x, y)$, then the marginal sample x (dropping y) follows $p_X(x)$. This method is particularly useful when one can easily generate random samples from $p_Y(y)$ and $p(X \mid Y = y)$.

Generating samples from a mixture distribution is often done by augmenting an indicator representing which of the mixture components the sample comes from. For example, if $Y \sim (1 - p)N(\mu_0, \sigma_0^2) + pN(\mu_1, \sigma_1^2)$, one samples $I = i$ from Bernoulli(p), where p is the mixture proportion, then generates a sample Y from $N(\mu_i, \sigma_i^2)$. It can easily be verified that the marginal distribution of Y follows the mixture distribution.

The augmentation method also works naturally for continuous compound mixture distribution $\pi(y) = \int \omega(x)\pi_2(y, x)dx$ where $\omega(x) \geq 0$ and $\int \omega(x)dx = 1$ (basically a density function) and $\pi_2(y, x)$ is a density function for every x (basically a conditional density function). For example, the non-central χ^2 distribution with m degrees of freedom with a non-central parameter λ is a Poisson mixture of central χ^2 distributions

$$X_m \sim \sum_{k=0}^{\infty} P(K = k)\chi^2_{2k+m},$$

where $K \sim Poisson(\lambda)$. Hence a random sample of X_m can be obtained by generating a Poisson random sample $K = k$ first, then a χ^2_{2k+m} random sample. In fact there is another formulation to make the problem easier. It can be shown that

$$X_m = (Z + \lambda)^2 + X^*_{m-1},$$

where Z follows a standard normal distribution and X^*_{m-1} follows the central χ^2 distribution with $m - 1$ degrees of freedom. This representation provides an easier way to generate X_m. The method can be extended to non-central gamma distributions.

A remark is in order. The rejection method utilizes the idea of augmentation. Since $p(x)$ is difficult to generate samples from directly, we augment an additional component Y, so (X, Y) follows the uniform distribution in the region of $\Omega^* = \{(x, y), x \in \Omega, 0 \leq y \leq p(x)\}$. Lemma 8.2 makes sure that the marginal distribution of X follows $p(x)$. The samples of the joint uniform distribution in Ω^* are drawn by generating a set of samples uniformly distributed in a larger region Ω^{**}.

8.1.2 Variance Reduction Methods

Another important issue in using Monte Carlo methods is variance reduction. The goal of variance reduction is to reduce the variation in Monte Carlo samples while maintain the consistency in estimation. In this section, we consider some approaches to variance reduction.

1. Control variates: Suppose we have iid samples $X_1, \ldots, X_m$ from $\pi(x)$ and use $\hat{\mu} = m^{-1} \sum_{i=1}^{m} h(X_i)$ to estimate $\mu_h = \int h(x)\pi(x)dx$. The variance of the estimator is

$$m\text{Var}(\hat{\mu}) = \text{Var}[h(X)].$$

An alternative estimator of μ_h is

$$\hat{\mu}_C = \frac{1}{m} \sum_{i=1}^{m} [h(X_i) - c(\eta(X_i) - \mu^*)],$$

where $\eta(\cdot)$ is a function with a known expectation $\mu^* = \int \eta(x)\pi(x)dx$ and c is a constant to be chosen. Clearly, $E(\hat{\mu}_C) = \mu_h$ and, hence, $\hat{\mu}_C$ is a valid estimator of μ_h. Its variance is

$$m\text{Var}(\hat{\mu}_C) = \text{Var}[h(X)] + c^2\text{Var}[\eta(X)] - 2c\text{Cov}[h(X), \eta(X)].$$

For a properly chosen $\eta(x)$ and a constant c, the variance of $\hat{\mu}_C$ can be reduced. Specifically, for a given $\eta(\cdot)$, the optimal choice of c is

$$c = \text{Cov}[h(X), \eta(X)]/\text{Var}[\eta(X)].$$

The variance then becomes $(1 - \rho^2)\text{Var}[h(X)]$, where ρ is the correlation coefficient between $h(X)$ and $\eta(X)$. For $\rho \neq 0$, $\text{Var}(\hat{\mu}_C)$ is smaller than $\text{Var}(\hat{\mu})$.

The basic requirement of using this approach is the ability to find a function $\eta(x)$ with a known expectation $\mu^* = E_\pi[\eta(X)] = \int \eta(x)\pi(x)dx$. One would choose a $\eta(\cdot)$ to make the correlation coefficient between $h(X)$ and $\eta(X)$ large (in absolute value). In practice, the optimal c can be estimated after sampling, using the estimated sample correlation between $h(X)$ and $\eta(X)$.

One commonly used $\eta(\cdot)$ is

$$\eta(x) = \frac{h(x)g(x)}{\pi(x)},$$

where $g(x)$ is a known and *easy* density so that $\int h(x)g(x)dx = E_g(h(X)) = \mu^*$ can be obtained analytically. In this case,

$$\int \eta(x)\pi(x)dx = \int h(x)g(x)dx = \mu^*.$$

Ideally, $g(x)$ should be *close* to $\pi(x)$ so the correlation between $\eta(x)$ and $h(x)$ is large. For example, $g(x)$ can be a normal approximation of $\pi(x)$, with its mode close to that of $\pi(x)$, and with (approximately) matching curvature around the mode.

If the first two moments of $\pi(x)$ are available, another effective $\eta(x)$ to use is

$$\eta(x) = h(\tilde{\mu}) + h'(\tilde{\mu})(x - \tilde{\mu}) + \frac{1}{2}h''(\tilde{\mu})(x - \tilde{\mu})^2,$$

which uses the first two terms of Taylor expansion of $h(x)$ around $\tilde{\mu}$, where $\tilde{\mu}$ is a rough estimate of μ.

More generally, if one can generate control variates $Y_1, \ldots, Y_n$ such that $(h(X_i), Y_i)$ are correlated (e.g., $Y_i = \eta(X_i)$ in the above discussion), then we can achieve variance reduction. Specifically,

1. If $E(Y_i) = b$ is known, we can use

$$\hat{\mu}_C = \frac{1}{m}\sum_{i=1}^{m}[h(X_i) - c(Y_i - b)].$$

2. If $E(Y_i) = a\mu$ with a known constant a (μ is to be estimated), we can use

$$\hat{\mu}_C = \left(\frac{1}{1-ac}\right)\frac{1}{m}\sum_{i=1}^{m}(h(X_i) - cY_i).$$

3. If $E(Y_i) = a\mu + b$ with known constants a and b, we can use

$$\hat{\mu}_C = \left(\frac{1}{1-ac}\right)\frac{1}{m}\sum_{i=1}^{m}[h(X_i) - cY_i - b].$$

Further information can be found in Rubinstein and Kroese (2008).

2. Antithetic variates: To reduce Monte Carlo variation in some cases we can use pairs of negatively correlated random samples.

Consider again the estimation of $\mu_h = \int h(x)\pi(x)dx$. We can generate iid pairs (X_1, Y_1), ..., (X_n, Y_n) such that $(h(X_i), h(Y_i))$ are negatively correlated and X_i and Y_i both follow $\pi(x)$. Then an *antithetic variate estimator* of μ_h takes the form

$$\hat{\mu}_C = \frac{1}{2n}\sum_{i=1}^{n}[h(X_i) + h(Y_i)].$$

Its variance (comparing to $2n$ iid samples) is

$$\frac{1}{2n}\left[\mathrm{Var}(h(X_1)) + \mathrm{Cov}(h(X_1), h(Y_1))\right] < \frac{1}{2n}\mathrm{Var}(h(X_1)).$$

In many cases, antithetic variate estimator requires less computation and has smaller variance. For example, if $U_i \sim Unif[0, 1]$, then $1 - U_i$ is its antithetic variate. Or, more generally, under the CDF inverse method, if $X = F^{-1}(U)$, then $Y = F^{-1}(1-U)$ is its antithetic variate. A more commonly used situation is when $\pi(x)$ is symmetric around a constant c, then X_i and $c - X_i$ are antithetic and can be used.

Of course, how useful the antithetic pair (X, Y) is depends on $h(\cdot)$, since the efficiency improvement depends on the correlation between $h(X)$ and $h(Y)$. For example, if $\pi(x)$ is symmetric around 0, then $(X, -X)$ is extremely effective if $h(x) = x$ (or any odd function of x), but has no impact at all if $h(x) = |x|$ (or any even function of x).

3. Stratified sampling: For a mixture distribution such as

$$\pi(x) = p\pi_1(x) + (1-p)\pi_2(x), \tag{8.3}$$

it is often more efficient to use stratified sampling, as often used in the survey sampling literature. Specifically, to obtain m samples from the mixture distribution in Equation (8.3), one can generate $m_1 = mp$ samples $x_1, \ldots, x_{m_1}$ from $\pi_1(x)$ and generate $m_2 = m(1-p)$ samples $y_1, \ldots, y_{m_2}$ from $\pi_2(x)$. Then $\mu = \int h(x)\pi(x)dx$ can be estimated by

$$\hat{\mu} = \frac{1}{m}\left[\sum_{i=1}^{m_1} h(x_i) + \sum_{i=1}^{m_2} h(y_i)\right] = p\hat{\mu}_1 + (1-p)\hat{\mu}_2,$$

where $\hat{\mu}_1 = (1/m_1)\sum_{i=1}^{m_1} h(x_i)$ and $\hat{\mu}_2 = (1/m_2)\sum_{i=1}^{m_2} h(y_i)$. Clearly, $E(\hat{\mu}_1) = \int h(x)\pi_1(x)dx$ and $E(\hat{\mu}_2) = \int h(x)\pi_2(x)dx$, hence $E(\hat{\mu}) = \int h(x)\pi(x)dx$. The variance of the estimator is

$$m\text{Var}(\hat{\mu}) = pV_1 + (1-p)V_2,$$

where $V_1 = \int h^2(x)\pi_1(x)dx - \mu_1^2$ and $V_2 = \int h^2(x)\pi_2(x)dx - \mu_2^2$. On the other hand, one can generate $x_1, \ldots, x_m$ directly from $\pi(x)$, via the augmentation method, by generating an indicator $I_i = 1$ with probability p and $I_i = 2$ with probability $1-p$, then generating x_i from $\pi_{I_i}(x)$. The variance of the sample mean of such a random sample is $\text{Var}(X_1)/m$, where

$$\begin{aligned}\text{Var}(X_1) &= E[\text{Var}(X_1 \mid I_1)] + \text{Var}[E(X_1 \mid I_1)] \\ &= pV_1 + (1-p)V_2 + (\mu_1 - \mu_2)^2 p(1-p).\end{aligned}$$

Hence the augmented estimator has a larger variance due to the variation in Bernoulli sampling of the indicator I_i. In addition, the stratified sampling saved the computation by avoiding generating the indicators.

Sometimes if a non-mixture distribution can be reformulated into a mixture distribution (e.g., partition the space into non-overlapping regions with known mixture proportions), stratified sampling may have some advantages.

8.1.3 Importance Sampling

The importance sampling (IS) of Marshall (1956) is one of the most commonly used methods in applying Monte Carlo samples to perform numerical integration and to make statistical inference. It is often mentioned on an equal footing with the rejection method discussed in Section 8.1.1, but since it is mostly related to the SMC methods of this chapter, we provide a more detailed introduction here.

IS is designed for approximating analytically intractable integrals. Again, we focus on the two types of integrals Z and μ_h in Equations (8.1) and (8.2).

Suppose a set of Monte Carlo samples, $\{x^{(j)}, j = 1, \ldots, m\}$, has been generated from a trial distribution $g(\cdot)$, which is different from the target distribution $\pi(\cdot)$. Note that, if the support of $g(x)$ contains that of integral function $\pi^*(x)$ (or $\pi^*(x)$ is absolutely continuous with respect to $g(x)$), then

$$Z = \int \pi^*(x)dx = \int \left[\frac{\pi^*(x)}{g(x)}\right] g(x)dx.$$

Hence Z is essentially the expectation of the function $\pi^*(x)/g(x)$ under the density $g(x)$. With the set of Monte Carlo samples $\{x^{(j)}, j = 1, \ldots, m\}$ from $g(x)$, Z can be approximated by the sample average

$$\hat{Z} = \frac{1}{m} \sum_{j=1}^{m} \frac{\pi^*(x^{(j)})}{g(x^{(j)})}.$$

That is, the integral of $\pi^*(\cdot)$ can be estimated using the random samples generated from the distribution $g(\cdot)$. Note that in this case we need to be able to evaluate the functions $\pi^*(x)$ and $g(x)$ explicitly and be able to generate samples from $g(x)$ easily.

For approximating μ_h, we use

$$\mu_h = \int h(x)\pi(x)dx = \int \left[h(x)\frac{\pi(x)}{g(x)}\right] g(x)dx = E_g[h(X)w(X)], \tag{8.4}$$

where $w(x) = \pi(x)/g(x)$. Using samples generated from $g(x)$, we have

$$\hat{\mu}_h = \frac{1}{m} \sum_{j=1}^{m} w^{(j)} h(x^{(j)}),$$

where $w^{(j)} = \pi(x^{(j)})/g(x^{(j)})$. Here the expectation with respect to π is estimated using the samples generated from the distribution $g(x)$. However, the samples need to be properly weighted.

Note that even if one can generate samples from $\pi(x)$ easily, using a different sampling distribution $g(x)$ can be more efficient for some $h(x)$. For example, we can make $g(x)$ to concentrate more in the areas of importance, where both $h(x)$ and $\pi(x)$ are large. Note that there is no need to have samples in the area where $h(x)$

is zero. In areas where $h(x)$ is relatively flat, only a few samples are needed. On the other hand, in the area where $h(x)$ is large and changes significantly locally, one would need more samples to zoom in to capture the differences. Of course the value of $\pi(x)$ is important, since it is the baseline distribution. If one used fewer samples in an area than what $\pi(x)$ demands, one may need to adjust the *weight* of the samples so they can properly reflect the underlying distribution.

Mathematically, the estimation error, measured by the sample variance of the estimator $\hat{\mu}_h$, is in the form

$$m\mathrm{Var}_g(\hat{\mu}_h) = \int \left[\frac{h(x)^2\pi(x)^2}{g(x)^2}\right] g(x)dx - \mu_h^2,$$

which is minimized if $g(x) \propto |h(x)\pi(x)|$. Similarly,

$$m\mathrm{Var}_g(\hat{Z}) = Z^2 \int \left[\frac{\pi(x)^2}{g(x)^2} - 1\right] g(x)dx.$$

For estimating Z, to increase the estimation accuracy, one would like to have $g(x)$ as close to $\pi(x)$ as possible. However, any samples directly from $\pi(x)$ cannot be used because we do not know how to evaluate $g(x) = \pi(x) = \pi^*(x)/Z$ for Z being unknown.

In certain cases, such as the rare event probability calculation (e.g., when $h(x) = I(x > c)$ for large c), the improvement in IS can be of several orders of magnitude. See, for instance, Hesterberg (1995).

Often, calculating the exact value of $w(x) = \pi(x)/g(x)$ in Equation (8.4) is difficult, since it involves the evaluation of the normalizing constants for both $\pi(x)$ and $g(x)$. A simple solution is then to use weighted averages in estimation. If we can only evaluate the weights $w(x)$ up to a multiplicative constant $w_*(x) = cw(x)$, we have

$$E_g[w_*(x)] = E_g\left[\frac{c\pi(x)}{g(x)}\right] = c,$$

so $E_g(\sum_{j=1}^m w_*^{(j)}) = cm$. As a result, the weighted average

$$\frac{1}{\sum_{j=1}^m w_*^{(j)}} \sum_{j=1}^m w_*^{(j)} h(x^{(j)}) = \sum_{j=1}^m \left[\frac{w_*^{(j)}}{\sum_{j=1}^m w_*^{(j)}}\right] h(x^{(j)}) \approx E_\pi(h(X)), \qquad (8.5)$$

can be used for estimation. With this formulation, the weight $w(x)$ only needs to be evaluated up to a multiplicative constant, hence avoiding the evaluation of the normalization constants in the distributions of $\pi(x)$ and $g(x)$. In the following, unless specifically mentioned otherwise, we always use the weighted average estimator and for notational simplicity $w(x)$ is used to denote either normalized or unnormalized weight.

Another way of understanding IS is from a distribution approximation point of view. In IS, the distribution $\pi(x)$ is essentially being approximated by a discrete distribution taking values at the set $\{x^{(j)}, j = 1, \ldots, m\}$ with probabilities proportional to the $w^{(j)}$, or the CDF of $\pi(x)$ is approximated by

$$\hat{F}(x) = \hat{P}(X \leq x) = \frac{1}{m}\sum_{j=1}^{m} w^{(j)} I(x^{(j)} \leq x) \quad \text{or} \quad \hat{F}(x) = \frac{\sum_{j=1}^{m} w^{(j)} I(x^{(j)} \leq x)}{\sum_{j=1}^{m} w^{(j)}}.$$

Note that this approximation is independent of the function $h(x)$, though the efficiency in estimating μ_h depends on the choice of $g(x)$.

The following concept is useful.

Definition: A random sample $\{(x^{(j)}, w^{(j)}), j = 1, \ldots, m\}$ is said to be *properly weighted* with respect to the distribution $\pi(\cdot)$ if for all π-integrable function $h(\cdot)$, we have

$$\frac{1}{\sum_{j=1}^{m} w^{(j)}} \sum_{j=1}^{m} w^{(j)} h(x^{(j)}) \to E_{\pi}[h(x)],$$

and

$$\sup_{j=1,\ldots,m} w^{(j)} < \infty,$$

as $m \to \infty$.

Lemma 8.4 *If $\pi(x)$ is absolutely continuous with respect to $g(x)$, let $w(x) = \pi(x)/g(x)$. If $\sup w(x) < \infty$, then $\{(x^{(j)}, w^{(j)}), j = 1, \ldots, m\}$ is properly weighted with respect to $\pi(x)$, where $x^{(j)}$ are iid following $g(x)$.*

Lemma 8.5 *Suppose $\{(x^{(j)}, w^{(j)}), j = 1, \ldots, m\}$ is properly weighted with respect to $\pi_1(x)$. If $\pi_2(x)$ is absolutely continuous with respect to $\pi_1(x)$ and $u(x) = \pi_2(x)/\pi_1(x)$ with $\sup_x u(x) < \infty$, then $\{(x^{(j)}, \tilde{w}^{(j)}), j = 1, \ldots, m\}$ is properly weighted with respect to $\pi_2(x)$, where $\tilde{w}^{(j)} = w^{(j)} u(x^{(j)})$.*

Lemma 8.5 can be seen from the following:

$$\begin{aligned}
\frac{1}{m}\sum_{j=1}^{m} \tilde{w}^{(j)} h(x^{(j)}) &= \frac{1}{m}\sum_{j=1}^{m} w^{(j)} [u(x^{(j)}) h(x^{(j)})] \\
&\to \int u(x) h(x) \pi_1(x) dx \\
&= \int h(x) \pi_2(x) dx \\
&= E_{\pi_2}[h(x)].
\end{aligned}$$

Lemma 8.6 *If $\mathbf{x} = (x_1, \dots, x_d)$ and $\{(\mathbf{x}^{(j)}, w^{(j)}), j = 1, \dots, m\}$ is properly weighted with respect to $\pi(\mathbf{x})$, then the weighted marginal samples $\{(x_i^{(j)}, w^{(j)}), j = 1, \dots, m\}$ is properly weighted with respect to the marginal distribution $\pi_i(x_i)$ given by $\pi_i(x_i) = \int \pi(\mathbf{x})dx_1 \dots dx_{i-1}dx_{i+1} \dots dx_d$.*

The lemma can be seen from the following: Let $h^*(\mathbf{x}) = h(x_i)$.

$$\frac{1}{m}\sum_{j=1}^{m} w^{(j)}h(x_i^{(j)}) = \frac{1}{m}\sum_{j=1}^{m} w^{(j)}h^*(\mathbf{x}^{(j)}) \to \int h^*(\mathbf{x})\pi(\mathbf{x})d\mathbf{x} = \int h(x_i)\pi(\mathbf{x})d\mathbf{x}$$

$$= \int h(x_i)\left[\int \pi(x_1, \dots, x_{i-1}, x_i, x_{i+1} \dots x_d)dx_1 \dots dx_{i-1}dx_{i+1} \dots dx_d\right] dx_i$$

$$= \int h(x_i)\pi_i(x_i)dx_i$$

$$= E_{\pi_i}[h(x_i)].$$

The efficiency of a Monte Carlo integration is often measured by the estimation variation. Kong et al. (1994) proposed a general purpose measure of efficiency that does not depend on the function $h(\cdot)$. It is termed the *effective sample size* (ESS):

$$\text{ESS} = \frac{m}{1 + cv^2(w)}, \tag{8.6}$$

where

$$cv(w) = \frac{\sum_{j=1}^{m}(w^{(j)} - \overline{w})^2}{m\overline{w}^2}, \quad \overline{w} = \frac{1}{m}\sum_{j=1}^{m} w^{(j)}.$$

Heuristically, it says that the weighted samples are equivalent to a set of samples of size ESS directly generated from π.

In the following, we discuss some implementation issues of IS.

1. **Selecting the proposal distribution:** There are four key considerations in selecting a proposal distribution $g(\cdot)$ using IS. A poorly chosen one not only may result in loss of efficiency, but, in some cases, may fail to work at all. The four main considerations are as follows.

(1) The support: A key condition for IS to work properly is that the support of $g(\cdot)$ should cover that of $\pi(x)$. However, in some cases this condition may hold theoretically, yet it fails empirically with a finite sample size. For example, if $g(x) \sim N(\mu, \sigma^2)$, then its *empirical support* (i.e., empirical range) of a sample of size m is approximately $[\mu - \sigma\sqrt{2\ln m}, \mu + \sigma\sqrt{2\ln m}]$, since the maximum or minimum of m iid samples from a standard normal distribution is in the order

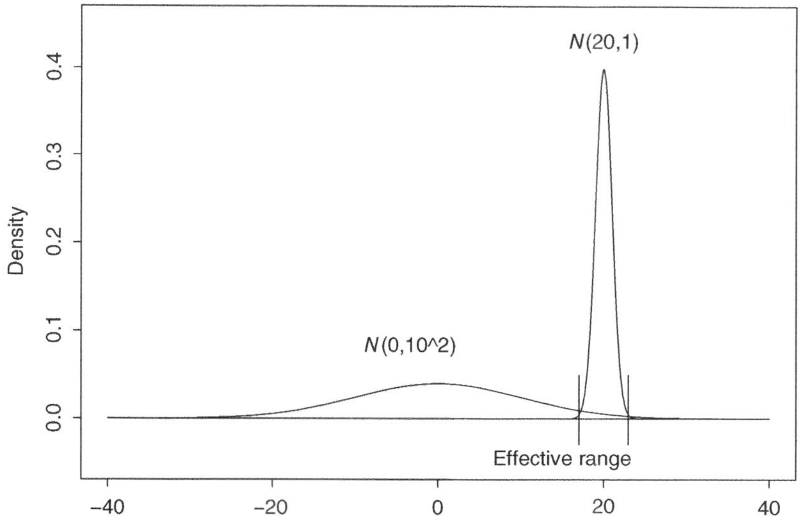

Figure 8.2 Illustration of empirical range: $g(\cdot) \sim N(20, 1)$ and $\pi(\cdot) \sim N(0, 10^2)$.

of $\sqrt{2 \ln m}$. IS requires the empirical support to have an effective coverage of the target distribution $\pi(x)$. Figure 8.2 shows a situation where the empirical support is too small. The other side of the same question is that one needs to have sufficient sample size in order to have the empirical support covering the support of the target distribution.

(2) Covering important areas: It is important to find a $g(x)$ that covers all the important parts of $h(x)\pi(x)$. For example, when the function $h(\cdot)\pi(x)$ is multimodal (often when $\pi(x)$ itself is multimodal or $h(\cdot)$ is a periodic function), a unimodal $g(x)$ will not be efficient. See, for instance, Owen and Zhou (1999) and Ford and Gregory (2007).

(3) Tail dominance: If the target distribution $\pi(x)$ has an infinite support, then $g(x)$ cannot have a tail thinner than that of $\pi(x)$, in order to fulfill the requirement that $\sup_j w^{(j)} < \infty$.

For example, if $\pi(x) \sim N(0, \sigma_\pi^2)$ and $g(x) \sim N(0, \sigma_g^2)$, and we use $\hat{\mu} = \frac{1}{m} \sum_{j=1}^m x_j w(x_j)$ to estimate $\mu = E_\pi(X)$, the variance of the estimator is

$$m\text{Var}_g(\hat{\mu}) = \int x^2 \frac{\pi^2(x)}{g^2(x)} g(x)dx = \frac{\sqrt{\sigma_g}}{\sqrt{2\pi\sigma_\pi}} \int x^2 \exp\left\{ -\frac{x^2}{2} \left(\frac{2\sigma_g^2 - \sigma_\pi^2}{\sigma_\pi^2 \sigma_g^2} \right) \right\} dx.$$

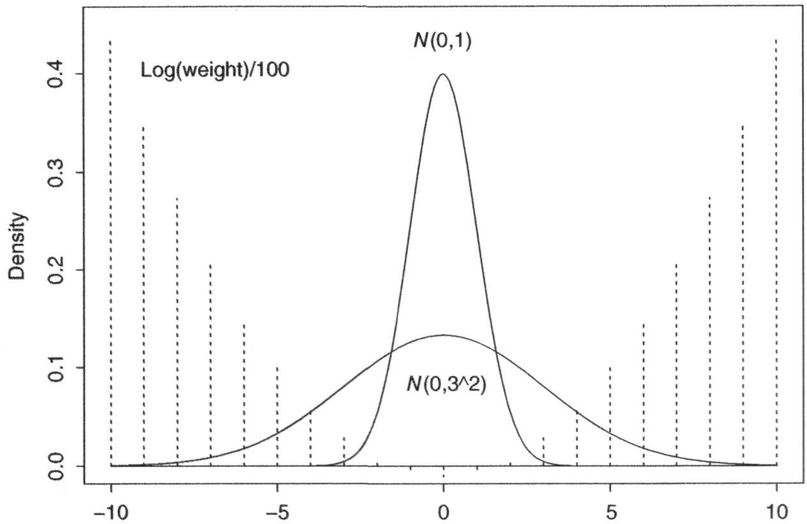

Figure 8.3 Importance weight distribution for $\pi(x) \sim N(0, 3^2)$ and $g(x) \sim N(0, 1)$.

If $2\sigma_g^2 < \sigma_\pi^2$, the estimator has infinite variance. Figure 8.3 shows the weight distribution for $\pi(x) \sim N(0, 3^2)$ and $g(x) \sim N(0, 1)$.

If $\pi(x) \sim t_\alpha$ and $g(x) \sim t_\beta$, then it is required to have $\beta < 2(\alpha - 1)$. Figure 8.4 shows the weight distribution for $\pi(x) \sim t_5$ and $g(x) \sim t_9$.

(4) Efficiency: The requirement that the proposal distribution needs to dominate the tails of the target distribution can easily be achieved by using some heavy-tail distributions as the trial distribution. However, this choice may be less efficient since there may be too many samples in the tails and too few samples in the middle and more important part of the distribution, that is, one cannot be too conservative. If $g(x)$ has a very large empirical support and a tail that is too heavy, one may lose efficiency significantly. Table 8.1 shows the effective sample size for various target and trial distributions.

2. Using a mixture of multiple proposal distributions: Multiple proposal distributions are often used to address all the aforementioned concerns. The main idea of using multiple proposal distributions is to use simple, but different, distributions to cover different regions of importance in order to gain efficiency. Typically, one uses light-tail proposals (e.g., normal) for the *main* part of the target distribution to gain efficiency while safeguarding the tails with a heavy tail proposal. For example, Oh and Berger (1993) and Owen and Zhou (1999) used a family of

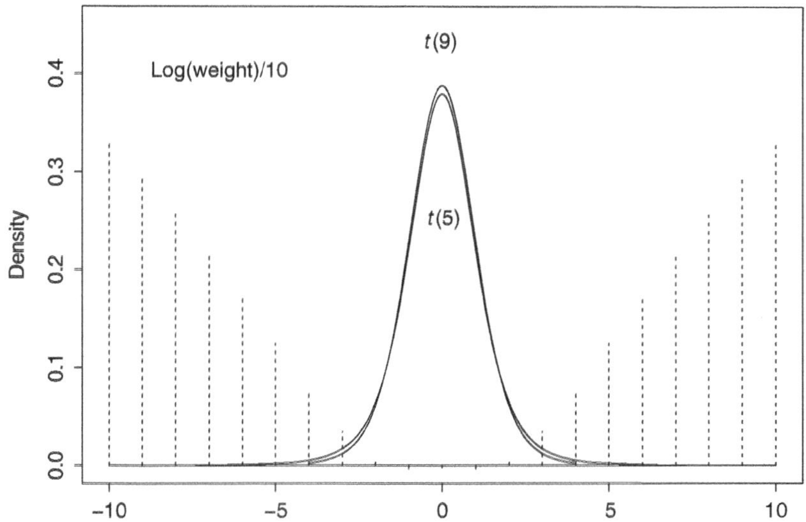

Figure 8.4 Importance weight distribution for $\pi(x) \sim t_5$ and $g(x) \sim t_9$.

t-distributions and beta distributions to cover each mode of $h(x)\pi(x)$ individually. West (1993) and Givens and Raftery (1996) used a mixture of normal or Student-t distributions as the proposal distribution, where the mixture proportion is obtained from a kernel estimate of $h(x)\pi(x)$ with a pre-sampling exercise.

Even for a unimodal target distribution, one can construct a mixture of two proposals with one mimicking the center of the target and the other dominating the tails. This method was used in Giordani and Kohn (2010), although in a different scenario. Such a construction is useful to safeguard the tails without losing too much efficiency in the important support area of the target. For example, if

Target $\pi(x)$	Proposal $g(x)$	ESS($\times m$)
$N(20, 1)$	$N(0, 10^2)$	3.3%
$N(0, 1)$	$N(0, 10^2)$	23%
$N(0, 1)$	$N(0, 100^2)$	2.36%
$N(0, 1)$	$50\sqrt{2}t_4$	1.59%
$N_{10}(\mathbf{0}, \mathbf{I})$	$N_{10}(\mathbf{0}, 10^2\mathbf{I})$	4×10^{-7}

Table 8.1 Effective sample size under various target and trial distributions.

one includes the target distribution itself as one of the proposal distributions, then one obtains an upper bound for the estimation variance. See Hesterberg (1995). For distributions with a bounded domain, including a uniform distribution in the domain can safeguard the variance (Owen and Zhou, 2000). Other approaches can be found in Liang et al. (2007) and Ford and Gregory (2007).

Given multiple potentially useful proposals, a straightforward combination method is to use their mixture as the proposal distribution. Specifically, suppose $g_1(x), \ldots, g_p(x)$ are p probability densities serving as proposals. Given a mixture proportion vector $\boldsymbol{\alpha} = (\alpha_1, \ldots, \alpha_p)$ satisfying $\sum_{k=1}^{p} \alpha_k = 1$ $(0 \le \alpha_k < 1)$, we can use the mixture distribution $g_{\alpha}(x) = \sum_{k=1}^{p} \alpha_k g_k(x)$ as the proposal distribution and estimate μ and Z by

$$\hat{\mu}_{MIS} = \frac{1}{m}\sum_{i=1}^{m}\frac{h(x_i)\pi(x_i)}{g_{\alpha}(x_i)}, \quad \hat{Z}_{MIS} = \frac{1}{m}\sum_{i=1}^{m}\frac{\pi(x_i)}{g_{\alpha}(x_i)},$$

where $\{x_1, \ldots, x_m\}$ are generated from $g_{\alpha}(x)$.

Hesterberg (1988) proposed that, instead of generating samples directly from the mixture distribution, one can use stratified samples $\{x_{k1}, \ldots, x_{km}\}$, discussed in Section 8.1.2, with $m_k = m\alpha_k$ samples from the kth proposal $g_k(\cdot)$. The estimators then become

$$\hat{\mu}_{SIS} = \frac{1}{m}\sum_{k=1}^{p}\sum_{i=1}^{m_k}\frac{h(x_{ki})\pi(x_{ki})}{g_k(x_{ki})}, \quad \hat{Z}_{SIS} = \frac{1}{m}\sum_{k=1}^{p}\sum_{i=1}^{m_k}\frac{\pi(x_{ki})}{g_k(x_{ki})}.$$

Veach and Guibas (1995) and Raghavan and Cox (1998) proposed an extension of this approach.

Owen and Zhou (2000) pointed out that it is possible that using a mixture proposal distribution may lose some efficiency due to mixing good (efficient) proposal distributions with poor (inefficient) but necessary ones (for safeguarding). It is possible that a mixture can lose efficiency by several orders of magnitude if the original proposal is nearly perfect. The impact can be reduced by combining IS and control variates discussed in Section 8.1.2. Specifically, Owen and Zhou (2000) proposed using $\mathbf{g}(x) = (g_2(x) - g_1(x), \ldots, g_p(x) - g_1(x))'$ as the control variates to modify the estimator $\hat{Z}_{SIS}$ to

$$\hat{Z}_{Reg} = \frac{1}{m}\sum_{k=1}^{p}\sum_{i=1}^{m_k}\frac{\pi(x_{ki}) - \hat{\boldsymbol{\beta}}'_{\alpha}\mathbf{g}(x_{ki})}{g_{\alpha}(x_{ki})}, \tag{8.7}$$

where

$$\hat{\boldsymbol{\beta}}_{\alpha} = \widetilde{Var}\left[\frac{\mathbf{g}(X)}{g_{\alpha}(X)}\right]^{-1}\widetilde{Cov}'\left[\frac{\pi(X)}{g_{\alpha}(X)}, \frac{\mathbf{g}(X)}{g_{\alpha}(X)}\right],$$

and $\widetilde{Var}$ and $\widetilde{Cov}$ denote the pooled-sample variance and covariance estimators. It can be easily seen that $\boldsymbol{g}(x)$ serves as a set of valid control variates. It can also be shown that, using this set of control variates, the asymptotic variance of $\widehat{Z}_{Reg}$ is zero when $\pi(x)$ is indeed a linear combination of the proposals. More importantly, it can be shown that $\widehat{Z}_{Reg}$ has a smaller asymptotic variance than every IS estimator constructed solely by g_k with m_k samples, $k = 1, \ldots, p$. That is, $\widehat{Z}_{Reg}$ is always at least as good as the best one among the individual proposals.

An important issue of using mixture distributions is to determine the mixture weights $\boldsymbol{\alpha}$. Emond et al. (2001) showed that proper mixture proportions can increase the efficiency by an order of magnitude. Fan et al. (2006) and Raghavan and Cox (1998) used some heuristic rules derived from experience or interpretation of the proposals.

It is natural to use a pilot sample to select the mixture proportions before the actual sampling and estimation. Oh and Berger (1993) and Hesterberg (1995) studied the numerical properties of such an idea. Li et al. (2013) proposed a more sophisticated two-stage approach. In the first stage, pilot samples are obtained using a mixture proposal with predetermined proportions. Based on the pilot samples, the optimal mixture proportions are then estimated by minimizing the estimated asymptotic variance of the final estimator, with control variates. In the second stage, the sample is drawn from the mixture proposal with the estimated proportions. Final estimators are based on all observations, including those from the pilot stage. They show that the two-stage estimators are consistent and asymptotically normal with minimum asymptotic variance over all mixture proportions. This two-stage procedure can also serve as a selection tool among many potential proposal candidates.

3. **Weight equivalence:** Another special feature of IS is that a sample can be accompanied by different sets of weights while still being properly weighted with respect to the same target distribution. For example, if $\{(x_i, y_i), i = 1, \ldots, m\}$ iid from $g(x, y)$ and $w_i = \pi(x_i, y_i)/g(x_i, y_i)$, then $\{((x_i, y_i), w_i), i = 1, \ldots, m\}$ is properly weighted with respect to $\pi(x, y)$. Hence the weighted sample $\{(y_i, w_i), i = 1, \ldots, m\}$ is properly weighted with respect to $\pi_Y(y)$, the marginal distribution derived from $\pi(x, y)$, according to Lemma 8.6. On the other hand, by discarding x_i, the marginal sample y_i is iid from $g_Y(y)$, where $g_Y(y)$ is the marginal distribution derived from $g(x, y)$. Let $w_i^* = \pi_Y(y)/g_Y(y)$. Then $\{(y_i, w_i^*), i = 1, \ldots, m\}$ is also properly weighted with respect to $\pi(y)$. Note that the samples $\{y_i, i = 1, \ldots, m\}$ are the same, but the weights are different, even though the target distribution is the same. One may wonder which set of weights is better. We

notice that

$$\begin{aligned}
\mathrm{Var}_g(w_i) &= \mathrm{Var}_g\left[\frac{\pi(x_i,y_i)}{g(x_i,y_i)}\right] \\
&= \mathrm{Var}_{g_Y}\left[E_g\left\{\frac{\pi(x_i,y_i)}{g(x_i,y_i)} \mid y_i\right\}\right] + E_{g_Y}\left[\mathrm{Var}_g\left\{\frac{\pi(x_i,y_i)}{g(x_i,y_i)} \mid y_i\right\}\right] \\
&= \mathrm{Var}_{g_Y}\left[\frac{\pi(y_i)}{g(y_i)}\right] + E_{g_Y}\left[\mathrm{Var}_g\left\{\frac{\pi(x_i,y_i)}{g(x_i,y_i)} \mid y_i\right\}\right] \\
&= \mathrm{Var}_{g_Y}(w_i^*) + E_{g_Y}\left[\mathrm{Var}_g\left\{\frac{\pi(x_i,y_i)}{g(x_i,y_i)} \mid y_i\right\}\right].
\end{aligned}$$

Hence w_i^* is better. It has a smaller variance since it removes the Monte Carlo variation in x_i when calculating the weights. However, one would need to be able to evaluate the marginal distributions $g_Y(y)$ and $\pi_Y(y)$. This is often called the *marginal weighting*.

A particularly useful application of the marginal weighting approach is in generating samples from a mixture distribution. Suppose $\pi(x) = (1-p)\pi_1(x) + p\pi_2(x)$ and we use trial distribution $g(x) = (1-p)g_1(x) + pg_2(x)$. To generate samples from $g(x)$, it is often easier to use the sequential sampling approach by generating an indicator $I_i \sim$ *Bernulli*(p) first, then generating $x_i \sim g_{I_i}(\cdot)$. Hence $(x_i, I_i) \sim g_{I_i}(x_i)p(I_i)$. Its corresponding target distribution is then $\pi_{I_i}(x_i)p(I_i)$, hence $w_i = \pi_{I_i}(x_i)/g_{I_i}(x_i)$. On the other hand, since $x_i \sim g(\cdot)$, we have

$$w_i^* = \frac{\pi(x_i)}{g(x_i)} = \frac{(1-p)\pi_1(x_i) + p\pi_2(x_i)}{(1-p)g_1(x_i) + pg_2(x_i)}.$$

We have shown that w_i^* is a better set of weights. The basic principle is that one should carry out as much analytical calculation as possible to improve the efficiency. We use this feature later in Section 8.6.

4. Resampling in IS: As discussed before, a useful interpretation of IS is that it essentially approximates the distribution $\pi(x)$ by a discrete distribution taking values in the set $\{x^{(j)}, j = 1, \ldots, m\}$ with probabilities proportional to $w^{(j)}$. It leads to the following resampling approach to generate (unweighted) samples that follow the distribution $\pi(x)$. Since the CDF of $\pi(x)$ is approximated by

$$\hat{F}(x) = \hat{P}(X \leq x) = \frac{1}{m}\sum_{j=1}^{m} w^{(j)} I(x^{(j)} \leq x),$$

generating samples from $\hat{F}(x)$ can be done by sampling $X = x$ from the set $\{x^{(j)}, j = 1, \ldots, m\}$ with probabilities proportional to $w^{(j)}$. Then X approximately follows $\pi(x)$. Particularly, such a sample satisfies

$$E_{\hat{F}}(h(X)) = E_{\pi}(h(X)),$$

since $\{(x^{(j)}, w^{(j)}), j = 1, \ldots, m\}$ is properly weighted with respect to π.

The following lemma extends the results to any set of resampling probabilities $0 < \alpha(x^{(j)}) < 1$, called the *priority* score.

Lemma 8.7 *Suppose* $\{(x^{(j)}, w^{(j)}), j = 1, \ldots, m\}$ *are independent and properly weighted with respect to* $\pi(x)$. *Let* $\{I_j, j = 1, \ldots, k\}$ *be a set of samples generated from the set* $\{1, \ldots, m\}$ *with probability* $\alpha^{(j)}$, *where* $0 < \alpha^{(j)} < 1$ *and* $\sum_j \alpha^{(j)} = 1$. *Then the set* $\{(x^{(I_j)}, \tilde{w}^{(I_j)}), j = 1, \ldots, k\}$ *is properly weighted with respect to* $\pi(x)$, *where* $\tilde{w}^{(j)} = w^{(j)}/\alpha^{(j)}$.

To see why this is true, let $S = \{x^{(1)}, \ldots, x^{(m)}\}$. For any π-integrable function $h(\cdot)$, we have

$$\begin{aligned}
E_{I,g}\left[\frac{1}{k}\sum_{j=1}^{k} \tilde{w}^{(I_j)} h(x^{(I_j)})\right] &= E_g\left[E_{I_1}(\tilde{w}^{(I_1)} h(x^{(I_1)}) \mid S)\right] \\
&= E_g\left[\sum_{j=1}^{m} \alpha^{(j)} \tilde{w}^{(j)} h(x^{(j)})\right] \\
&= E_g\left[\sum_{j=1}^{m} w^{(j)} h(x^{(j)})\right] \\
&= cE_{\pi}(h(x)),
\end{aligned}$$

where

$$c = E_{I,g}\left[\frac{1}{k}\sum_{j=1}^{k} \tilde{w}^{(I_j)}\right] = E_g\left[\sum_{j=1}^{m} w^{(j)}\right].$$

For a more rigorous proof, see Douc et al. (2014, Chapter 10).

If one generated $x^{(1)}, \ldots, x^{(k)}$ using this method, with $\alpha^{(j)} = w^{(j)}$, the set of samples can be treated as iid samples from $\pi(\cdot)$ for inferences, if $k \ll m$. The resampled weight in this case is constant 1. However, if k is not small (or in many cases, $k = m$), even though they are independent given the original weighted samples $\{(x^{(j)}, w^{(j)}), j = 1, \ldots, m\}$, the samples are marginally dependent, since there will be duplicates, the same $x^{(j)}$ being sampled multiple times. The samples with large weights $w^{(j)}$ tend to be duplicated and the samples with small weights may not be sampled. Note that inference should be done using the original weighted samples, instead of the resampled (unweighted) samples $x_1, \ldots, x_m$, because

$$m\mathrm{Var}\left[\frac{1}{m}\sum_{j=1}^{m} \tilde{w}^{(I_j)} h(x^{(I_j)})\right] \geq m\mathrm{Var}\left[\frac{1}{m}\sum_{j=1}^{m} w^{(j)} h(x^{(j)})\right].$$

Hence for making inference, one should not use the resampled samples. However, resampling is a critical step in the SMC method, which we discuss later.

5. Comparison with rejection sampling: Rejection sampling and IS are similar in that both use a trial distribution $g(\cdot)$ and both require that $\pi(x)/g(x) < c$, though for IS the condition can be relaxed if one focuses on a specific $h(x)$. In addition, rejection sampling does not use the samples that are rejected. IS keeps these samples, but uses importance weights as adjustments. The rejection sampling produces a set of iid samples, which can be useful in certain situations and makes theoretical investigation easier when it is embedded as a component in a procedure. In a way, ESS for IS is the acceptance rate for the rejection sampling.

8.1.4 Markov Chain Monte Carlo

Markov chain Monte Carlo (MCMC) methods are a class of powerful methods for generating samples (not necessarily iid) from a complex and possibly high dimensional target distribution $\pi(x)$. The key idea is to construct a Markov chain such that its equilibrium (stationary) distribution is $\pi(x)$. Then the Monte Carlo realizations of such a Markov chain can be used as samples of the target distribution $\pi(x)$.

In the appendix of Chapter 2 we discussed some important concepts and properties of a Markov chain taking values in a finite set. Here we expand the theory to the general spaces.

A Markov chain (MC) is a stochastic process $X_1, \ldots, X_t, \ldots$ with the property that the conditional density

$$p(X_t \mid X_{t-1}, X_{t-2}, \ldots) = p(X_t \mid X_{t-1}).$$

Hence a homogeneous Markov chain is completely specified by the one-step transition probability density $p(\cdot \mid x) = p(X_{t+1} \mid X_t = x)$ for all t. Clearly, the k-step transition density $p^{(k)}(\cdot \mid x) = p(X_{t+k} \mid X_t = x)$ can be obtained through the Chapman–Kolmogorov equation

$$p^{(k)}(y \mid x) = \int p^{(i)}(y \mid z) p^{(j)}(z \mid x) dz,$$

where $i + j = k$.

MCMC methods are based on the following key result.

Lemma 8.8 *For an ϕ-irreducible and aperiodic Markov chain with transition density $p(x_t \mid x_{t-1})$, if it is invariant respect to $\pi(x)$, then the Markov chain is ergodic, with $\pi(x)$ as its equilibrium distribution.*

Some brief descriptions of the terms used in Lemma 8.8 are in order. They are parallel to the discussion on the finite-space Markov chain in the appendix of Chapter 2, hence the discussion here is brief. A Markov chain is said to be *irreducible* if it has a connected support. Conceptually, it requires that, for any region A with non-zero ϕ-measure in the support of the chain Ω, there is a positive probability that, starting with $X_0 = x$ for any $x \in \Omega$, the chain will reach A in finite steps. Alternatively, there is a $t < \infty$ such that $P(X_t \in A \mid X_0 = x) > 0$. A Markov chain is said to be *aperiodic* if it is not periodic, that is, the largest common divider of $\{t : P(X_t \in A \mid X_0 = x) > 0\}$ is 1, for any A and starting location $x \in \Omega$. A Markov chain is *invariant* if there is a density $\pi(x)$ such that $\pi(y) = \int p(y \mid x)\pi(x)dx$. This means that, if at time t, $x_t \sim \pi(x)$, then at time $t+1$, x_{t+1} still follow $\pi(x)$, under the transition density $p(y \mid x)$. The invariance condition is satisfied if one can show a perfect balance condition

$$p(y \mid x)\pi(x) = p(x \mid y)\pi(y). \tag{8.8}$$

This is because, under the balance condition, we have

$$\int p(y \mid x)\pi(x)dx = \int p(x \mid y)\pi(y)dx = \pi(y).$$

Ergodicity is one of the most useful properties of a stochastic process (if the chain is ergodic). One key result we use here is the *ergodic theorem*, which says that if a Markov chain is ergodic, then the law of large number holds. That is $\bar{x} \to E_\pi(x)$, where $\pi(\cdot)$ is the equilibrium distribution. It also says that, starting with $x_0 \sim g(x)$, the distribution of X_t converges to $\pi(x)$. For more detailed discussions, see Robert and Casella (2004) and Liu (2001), among others.

Using the above result, MCMC methods construct a transition rule $p(y \mid x)$ so the resulting Markov chain is ergodic, with the target distribution $\pi(\cdot)$ as its equilibrium distribution. Given an appropriate $p(y \mid x)$ that is easy to sample y from x, we start with $X_0 = x_0 \in \Omega$, and iteratively generate samples $x_t \sim p(y \mid X_{t-1} = x_{t-1})$, $t = 1, \dots$. The resulting samples $x_{K+1}, \dots, x_{K+m}$ approximately follow $\pi(x)$, after removing samples in the *burning period* ($t = 1, \dots, K$) to reduce the impact of the initial value. Then $\mu_h = E_\pi(x)$ can be estimated by

$$\hat{\mu}_h = \frac{1}{m} \sum_{j=K+1}^{K+m} h(x_j).$$

There are two widely used constructions of the transition density $p(y \mid x)$.

1. Metropolis–Hastings algorithm. Consider a target distribution $\pi(x)$ that can be evaluated up to a normalizing constant. Let $T(x, y)$ be a *proposal move rule* that proposes a (random) move from x to y. In practice one should choose a $T(x, y)$ that is easy to generate the move.

Algorithm 8.2: Metropolis–Hastings Algorithm

- Start with any $X_0 = x_0$.
- Suppose at time t, $X_t = x_t$. At time $t + 1$,
  - Draw $y \sim T(x_t, y)$ (i.e., propose a move for the next step).
  - Compute the metropolis ratio (or 'goodness' ratio)
  $$r = \frac{\pi(y)}{\pi(x_t)} \frac{T(y, x_t)}{T(x_t, y)}.$$
  - Acceptance/rejection decision: let
  $$X_{t+1} = \begin{cases} y & \text{with prob} \quad p = \min\{1, r\} \\ x_t & \text{with prob} \quad 1 - p. \end{cases}$$

With such a movement, the Markov chain is invariant with respect to $\pi(x)$. Note that the actual transition probability from x to y is

$$p(y \mid x) = T(x, y) \min\left\{1, \frac{\pi(y)}{\pi(x)} \frac{T(y, x)}{T(x, y)}\right\}$$

and the transition probability from y to x is

$$p(x \mid y) = T(y, x) \min\left\{1, \frac{\pi(x)}{\pi(y)} \frac{T(x, y)}{T(y, x)}\right\}.$$

To check the balance condition in (8.8), for $x \neq y$, we have

$$\begin{aligned} \pi(x)p(y \mid x) &= \pi(x)T(x, y) \min\left\{1, \frac{\pi(y)}{\pi(x)} \frac{T(y, x)}{T(x, y)}\right\} \\ &= \min\{\pi(x)T(x, y), \pi(y)T(y, x)\} \\ &= \pi(y)T(y, x) \min\left\{1, \frac{\pi(x)}{\pi(y)} \frac{T(x, y)}{T(y, x)}\right\} \\ &= \pi(y)p(x \mid y). \end{aligned}$$

Hence the balance condition is satisfied. Along with irreducible and aperiodic conditions (by the design of $T(x, y)$), the Markov chain is ergodic.

In the special case of symmetric move $T(x, y) = T(y, x)$, we have

$$r = \frac{\pi(y)}{\pi(x)} \quad \text{and} \quad p = \min\{1, r\}.$$

Hence if $\pi(y) \geq \pi(x_t)$, we move to y with probability 1 (so $x_{t+1} = y$). If $\pi(y) < \pi(x_t)$, we move to y with probability r. The smaller $\pi(y)$ is, the lower the chance of moving. Unlike deterministic accent (decent) algorithms, we always move uphill with probability one, but also allow some probability to move downhill.

A different acceptance rule (Baker's acceptance rule) can also be used:

$$\rho(x, y) = \frac{r}{1+r} = \frac{\pi(y)T(y, x)}{\pi(x)T(x, y) + \pi(y)T(y, x)}.$$

Some commonly used proposal move rules include the *random-walk metropolis* which uses proposal $y = x_t + \varepsilon_t$, where $\varepsilon_t \sim N(0, \Sigma_0)$ in continuous cases, or $\varepsilon_t \sim$ *Bernoulli*$\{-1, 1\}$ in discrete cases, with $r = \min\{1, \pi(y)/\pi(x_t)\}$. The *metropolized independent sampler* uses proposal $y \sim g(y)$ (a proposal distribution) with

$$r = \min\left\{1, \frac{w(y)}{w(x)}\right\} \quad \text{where} \quad w(x) = \frac{\pi(x)}{g(x)}.$$

This is similar to the importance weight in IS.

2. Gibbs sampler: Suppose the target distribution is multidimensional $\pi(x) = \pi(x_1, \ldots, x_p)$. Sequential sampling requires the ability to generate samples from a marginal distribution $\pi(x_1)$ and the partial conditional distributions $p(x_i \mid x_1, \ldots, x_{i-1})$ of the joint distribution $\pi(x_1, \ldots, x_p)$, which are not always available. Gibbs sample is a MCMC implementation in the cases where marginal distribution and partial conditional distributions are not available, but sampling from the full conditional distributions $\pi(x_i \mid x_1, \ldots, x_{i-1}, x_{i+1}, \ldots, x_n)$ (denoted as $\pi(x_i \mid x_{-i})$) are easy for each i.

Algorithm 8.3: Gibbs Sampler

- Start with an initial value $x^{(1)} = (x_1^{(1)}, \ldots, x_p^{(1)})$.
- At iteration $j + 1$,
  - sample $x_1^{(j+1)} \sim \pi(x_1 \mid x_2^{(j)}, \ldots, x_p^{(j)})$
  - sample $x_2^{(j+1)} \sim \pi(x_2 \mid x_1^{(j+1)}, x_3^{(j)}, \ldots, x_p^{(j)})$
  - $\vdots$
  - sample $x_p^{(j+1)} \sim \pi(x_p \mid x_1^{(j+1)}, x_2^{(j+1)}, \ldots, x_{p-1}^{(j+1)})$.

Under some mild conditions, such a Markov chain $\{(x_1^{(j)}, \ldots, x_p^{(j)}), j = 1, 2, \ldots\}$ is ergodic with $\pi(x_1, \ldots, x_p)$ as its invariant distribution.

Note that, if $(x_1^{(j)}, \ldots, x_p^{(j)}) \sim \pi(x_1, \ldots, x_p)$, then the marginal distribution (without x_1) follows $(x_2^{(j)}, \ldots, x_p^{(j)}) \sim \pi(x_2, \ldots, x_p)$. Hence

$$\begin{aligned}(x_1^{(j+1)}, x_2^{(j)}, \ldots, x_p^{(j)}) &\sim \pi(x_1^{(j+1)} \mid x_2^{(j)}, \ldots, x_p^{(j)})\pi(x_2^{(j)}, \ldots, x_p^{(j)}) \\ &= \pi(x_1^{(j+1)}, x_2^{(j)}, \ldots, x_p^{(j)}).\end{aligned}$$

Similarly, if $(x_1^{(j+1)}, x_2^{(j)}, \ldots, x_p^{(j)}) \sim \pi(x_1^{(j+1)}, x_2^{(j)}, \ldots, x_p^{(j)})$, then

$$(x_1^{(j+1)}, x_3^{(j)}, \ldots, x_p^{(j)}) \sim \pi(x_1^{(j+1)}, x_3^{(j)}, \ldots, x_p^{(j)}).$$

Hence

$$\begin{aligned}&(x_1^{(j+1)}, x_2^{(j+1)}, x_3^{(j)}, \ldots, x_p^{(j)}) \\ &\sim \pi(x_2^{(j+1)} \mid x_1^{(j+1)}, x_3^{(j)}, \ldots, x_p^{(j)})\pi(x_1^{(j+1)}, x_3^{(j)}, \ldots, x_p^{(j)}) \\ &= \pi(x_1^{(j+1)}, x_2^{(j+1)}, x_3^{(j)}, \ldots, x_p^{(j)}).\end{aligned}$$

Repeating this process, we have that if $(x_1^{(j)}, \ldots, x_p^{(j)}) \sim \pi(x_1, \ldots, x_p)$, then $(x_1^{(j+1)}, \ldots, x_p^{(j+1)}) \sim \pi(x_1, \ldots, x_p)$. Hence the Markov chain is invariant under such a transition rule.

There are many more advanced implementations of MCMC methods. Details can be found in Robert and Casella (2004), and Liu (2001).

8.2 THE SMC FRAMEWORK

Consider the SSM:

$$\begin{aligned}&\text{State equation:} \quad && x_t = s_t(x_{t-1}, \varepsilon_t) \quad \text{or} \quad x_t \sim q_t(\cdot \mid x_{t-1}),\\ &\text{Observation equation:} \quad && y_t = h_t(x_t, e_t) \quad \text{or} \quad y_t \sim f_t(\cdot \mid x_t).\end{aligned}$$

When the system is nonlinear and non-Gaussian, filtering and smoothing can be difficult as there are no analytic solutions for the predicting and the updating steps. In Chapter 7 we discussed several approaches to deal with nonlinear and non-Gaussian systems based on approximations. SMC follows the prediction and updating steps of the Kalman filter and maintains the recursive nature of the algorithm for fast on-line processing, achieving real-time filtering. It performs the prediction and updating steps based on Monte Carlo approximations. The approach focuses on the conditional distributions in the filtering and smoothing steps. As no analytical forms are available, they are represented by Monte Carlo samples. In the literature, SMC methods or some of its specific implementations are also referred to as particle filtering and particle smoothing, e.g. Douc et al. (2014). The Monte Carlo samples are often called *particles* in this context.

Suppose $\{x_{t-1}^{(1)}, \dots, x_{t-1}^{(m)}\}$ is a set of random samples following $p(x_{t-1} \mid \boldsymbol{y}_{t-1})$ where $\boldsymbol{y}_t = (y_1, \dots, y_t)$. At time t, with a new observation y_t, we would like to obtain a set of samples $\{x_t^{(1)}, \dots, x_t^{(m)}\}$ following the distribution $p(x_t \mid \boldsymbol{y}_t)$. There are two things that have changed here. The first change is that the samples are now about the state x_t, instead of x_{t-1}. The second is that the information set changes from $\boldsymbol{y}_{t-1}$ to $\boldsymbol{y}_t$, with an extra observation y_t. A simple approach is to deal with the two changes separately. The first step is to move into the space of x_t from x_{t-1}, without changing the information set $\boldsymbol{y}_{t-1}$, similar to the prediction step of the Kalman filter. To generate x_t from $p(x_t \mid \boldsymbol{y}_{t-1})$, we note that

$$\begin{aligned} p(x_t, x_{t-1} \mid \boldsymbol{y}_{t-1}) &= p(x_t \mid x_{t-1}, \boldsymbol{y}_{t-1}) p(x_{t-1} \mid \boldsymbol{y}_{t-1}) \\ &= p(x_t \mid x_{t-1}) p(x_{t-1} \mid \boldsymbol{y}_{t-1}). \end{aligned}$$

Hence we can use the sequential sampling approach by first generating samples from $p(x_{t-1} \mid \boldsymbol{y}_{t-1})$ (which has been done at time $t-1$), then drawing x_t from the conditional distribution $p(x_t \mid x_{t-1})$, which is the state equation. Let

$$x_t^{(j)} \sim q_t(x_t \mid x_{t-1}^{(j)}).$$

Then $\{(x_t^{(1)}, x_{t-1}^{(1)}), \dots, (x_t^{(m)}, x_{t-1}^{(m)})\}$ is a set of random samples following the distribution $p(x_t, x_{t-1} \mid \boldsymbol{y}_{t-1})$. The marginal samples $\{x_t^{(1)}, \dots, x_t^{(m)}\}$ then follow $p(x_t \mid \boldsymbol{y}_{t-1})$. However, this set of samples is not exactly what we wanted, as we wish to have a set of samples from $p(x_t \mid \boldsymbol{y}_{t-1}, y_t)$, that is, the sample we have is from a wrong distribution. Not all is lost, however, as proper adjustments can be made. The adjustment can be carried out using the IS idea. The samples $\{x_t^{(1)}, \dots, x_t^{(m)}\}$ can be viewed as having been generated from a trial distribution $p(x_t \mid \boldsymbol{y}_{t-1})$ and our target distribution is $p(x_t \mid \boldsymbol{y}_{t-1}, y_t)$. Hence the weight function is

$$\begin{aligned} w_t(x_t) &= \frac{p(x_t \mid \boldsymbol{y}_{t-1}, y_t)}{p(x_t \mid \boldsymbol{y}_{t-1})} \\ &\propto \frac{p(x_t, y_t \mid \boldsymbol{y}_{t-1})}{p(x_t \mid \boldsymbol{y}_{t-1})} \\ &= \frac{p(y_t \mid x_t, \boldsymbol{y}_{t-1}) p(x_t \mid \boldsymbol{y}_{t-1})}{p(x_t \mid \boldsymbol{y}_{t-1})} \\ &= p(y_t \mid x_t). \end{aligned}$$

$$\begin{array}{ccccc}
t-1 & & t-1+\frac{1}{2} & & t \\
\hline
p(x_{t-1} \mid \boldsymbol{y}_{t-1}) & & p(x_t \mid \boldsymbol{y}_{t-1}) & & p(x_t \mid \boldsymbol{y}_{t-1}, y_t) \\
\searrow & & \nearrow \quad \searrow & & \nearrow \\
& q_t(\cdot \mid x_{t-1}) & & w_t{=}f_t(y_t|x_t) & \\
\cdots\cdots & \cdots\cdots & \cdots\cdots & \cdots\cdots & \cdots\cdots \\
x_{t-1}^{(1)} & \longrightarrow & x_t^{(1)} & \longrightarrow & (x_t^{(1)}, w_t^{(1)}) \\
\vdots & & \vdots & & \vdots \\
x_{t-1}^{(m)} & \longrightarrow & x_t^{(m)} & \longrightarrow & (x_t^{(m)}, w_t^{(m)})
\end{array}$$

Figure 8.5 Illustration of a simple SMC step.

Note that $p(y_t \mid x_t) = f_t(y_t \mid x_t)$ is the likelihood from the observation equation. Hence $w_t^{(j)} = f(y_t \mid x_t^{(j)})$ provides an assessment on how likely one would observe y_t if the state x_t were indeed at the sample location $x_t^{(j)}$.

Under the IS framework, $\{(x_t^{(j)}, w_t(x_t^{(j)})), j = 1, \ldots, m\}$ is properly weighted with respected to $p(x_t \mid \boldsymbol{y}_t)$. The weighting step can be viewed as the Monte Carlo version of the updating step of the Kalman filter in Equation (6.26). Figure 8.5 depicts the above process.

To continue moving from t to $t + 1$, with a properly weighted sample of x_t, we generate $x_{t+1} \sim q_{t+1}(x_{t+1} \mid x_t)$ and update the weight to $w_{t+1} = w_t f_{t+1}(y_{t+1} \mid x_{t+1})$. To see why this is true, suppose $x_t^{(j)} \sim p(x_t \mid \boldsymbol{y}_{t-1})$ and $x_{t+1}^{(j)} \sim q_{t+1}(x_{t+1} \mid x_t^{(j)})$. We have $(x_t^{(j)}, x_{t+1}^{(j)}) \sim p(x_t, x_{t+1} \mid \boldsymbol{y}_{t-1})$. Under the IS framework and considering the joint distribution of (x_t, x_{t+1}), we have

$$w_{t+1}(x_t, x_{t+1}) = \frac{p(x_t, x_{t+1} \mid \boldsymbol{y}_{t-1}, y_t, y_{t+1})}{p(x_t, x_{t+1} \mid \boldsymbol{y}_{t-1})}$$

$$\propto \frac{p(x_t, x_{t+1}, y_t, y_{t+1} \mid \boldsymbol{y}_{t-1})}{p(x_t, x_{t+1} \mid \boldsymbol{y}_{t-1})} \tag{8.9}$$

$$= p(y_{t+1} \mid x_{t+1}) p(y_t \mid x_t) \tag{8.10}$$

$$= w_t(x_t) p(y_{t+1} \mid x_{t+1}). \tag{8.11}$$

From Equation (8.9) to Equation (8.10), we notice that the numerator of Equation (8.9) equals

$$p(y_{t+1} \mid x_t, x_{t+1}, y_t, \boldsymbol{y}_{t-1}) p(x_{t+1} \mid x_t, y_t, \boldsymbol{y}_{t-1}) p(y_t \mid x_t, \boldsymbol{y}_{t-1}) p(x_t \mid \boldsymbol{y}_{t-1})$$

and the denominator equals

$$p(x_{t+1} \mid x_t, \boldsymbol{y}_{t-1}) p(x_t \mid \boldsymbol{y}_{t-1}).$$

Since (x_t, x_{t+1}, w_{t+1}) is properly weighted with respect to $p(x_t, x_{t+1} \mid \boldsymbol{y}_{t+1})$, we also have (x_t, w_{t+1}) properly weighted with respect to $p(x_t \mid \boldsymbol{y}_{t+1})$. In fact, (x_1, w_{t+1}) is properly weighted with respect to $p(x_1 \mid \boldsymbol{y}_{t+1})$. The whole process can be carried out recursively starting with a set of initial samples x_0 and weight w_0.

Figures 8.6 to 8.8 illustrate the aforementioned procedure with a simple SSM

$$x_t = x_{t-1} + e_t, \quad y_t = x_t + \varepsilon_t,$$

where $e_t \sim N(0,1)$, $\varepsilon_t \sim N(0,1)$, and $x_0 \sim N(0,1)$. Seven samples of $x_0 \sim N(0,1)$ are generated. Their initial weights are set to 1, as they are iid from the prior distribution $p(x_0)$. They are shown as dots on the left ($t = 0$) line in Figure 8.6. Their corresponding vertical lines depict their weights. Moving from $t = 0$ to $t = 1$, $x_1^{(j)}$ is simulated from $x_1^{(j)} \sim p(x_1 \mid x_0 = x_0^{(j)})$, or equivalently $x_1^{(j)} = x_0^{(j)} + \varepsilon_1^{(j)}$, where $\varepsilon_1^{(j)} \sim N(0,1)$. They follow $p(x_1)$ without using the observation y_1 and their weights remain one. These samples are shown on the middle ($t = 1/2$) line in Figure 8.6. The corresponding vertical lines depict their weights. The observation y_1 is shown as the diamond dot on the right ($t = 1$) line in Figure 8.6. Based on this observation, we obtained the weights of the seven samples $w_t^{(j)} \propto \exp\{-0.5(y_1 - x_1^{(j)})^2\}$ according to the observation equation $y_t = x_t + e_t$. The samples and their weights

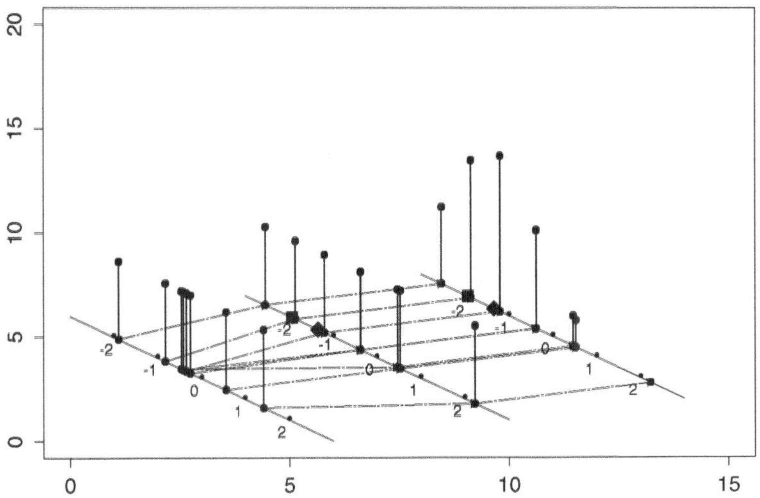

Figure 8.6 Simple illustration of updating scheme (I).

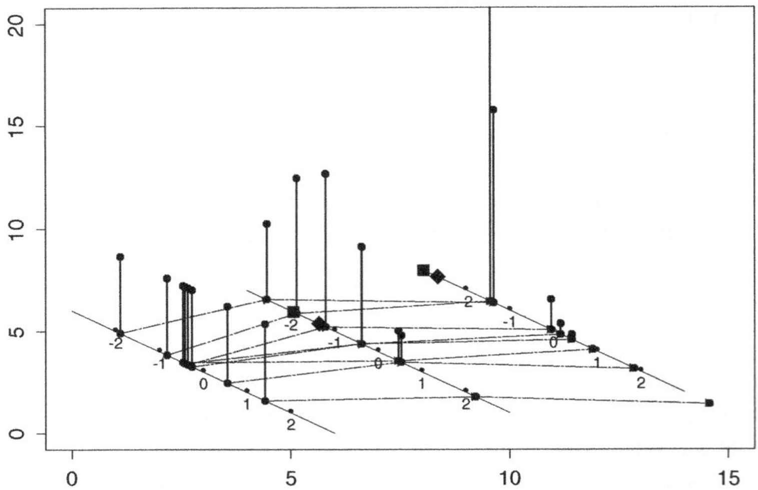

Figure 8.7 Simple illustration of updating scheme (II).

are shown on the $t = 1$ line. It is seen that the samples closer to the observation have larger weights.

Figure 8.7 shows how the samples move from $t = 1$ to $t = 2$. The middle ($t = 1$) line shows the weighted samples of x_1, the same as that in Figure 8.6. The dots on the right ($t = 2$) line are samples of $x_2^{(j)}$, obtained by $x_2^{(j)} = x_1^{(j)} + \varepsilon_2^{(j)}$, where $\varepsilon_2^{(j)} \sim N(0, 1)$. The weights are adjusted by $w_1^{(j)} f_2(y_2 \mid x_2^{(j)})$, where y_2 is marked by the diamond dot on the line. One feature stands out in the figure: the weights of the samples at $t = 2$ are now concentrated on only two or three of the samples, while all other samples have weights almost zero. For example, the sample on the far right is far away from the observed y_2, and it comes from the sample of x_1 that is also far away from the observed y_1, with a very small weight w_1. The accumulating effect makes its weight extremely small. Since inference (e.g., estimating $E(x_2 \mid y_1, y_2)$) is based on the weighted averages, samples with almost zero weights have negligible contributions to the inference. It is seen that among the seven samples, there are only three samples that are *important* enough to make contributions to the inference. In other words, the effective sample size defined in Equation (8.6) has been reduced to less than 30% of the original sample size in just two time steps ($ESS = 2.8$ in this case). That is, this set of seven samples is equivalent to only having 2.8 iid samples directly from $p(x_2 \mid y_1, y_2)$. One can imagine that moving from $t = 2$ to $t = 3$ we may lose even more effective samples. The reason for this is that the samples of x_1 are generated without using the information of y_1 and

y_2. The variation in the weights reflects the *distance* between the sampling distribution $p(x_1)$ and the target distribution $p(x_1 \mid y_1, y_2)$. As t increases, the distance between $p(x_1)$ and $p(x_1 \mid y_t)$ would increase. Since (x_1, w_2) is properly weighted with respect to $p(x_1 \mid y_1, y_2)$, such a discrepancy is reflected in the weight distribution. Hence the effective sample size will decrease, and often decreases quickly as t increases. Kong et al. (1994) showed that the effective sample size decreases (stochastically) as t increases. Hence, the estimation accuracy would decrease as t increases. In addition, continuing propagating the samples with small weights forward is a waste of computational resources as their weights are unlikely to get better. Discarding these samples reduces the total number of sample points, and regenerating new samples starting from $t = 0$ is computationally expensive. One solution is to use the resampling schemes introduced in Section 8.1.3 to duplicate the important samples and to eliminate the ones with low weights. Resampling is done randomly so the resampled samples remain properly weighted with respect to the target distribution.

Figure 8.8 shows the results of an improved sampling procedure in which the samples of x_1 are generated utilizing the information of y_1, and the samples of x_2 are generated utilizing the information of y_2. We see clearly the improvements. At $t = 2$, the effective sample size is 5.1. It shows that proper propagation (moving samples through time) is very important in terms of efficiency. Note that it is not necessary to use $q_t(x_t \mid x_{t-1})$ for moving from x_{t-1} to x_t in the predictive step. Since we are using IS adjustment in the updating step, any proposal distribution can be

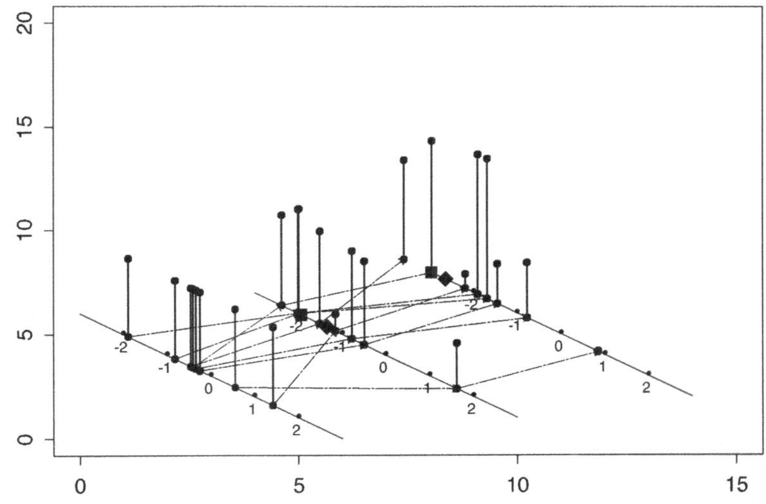

Figure 8.8 Simple illustration of updating scheme (III).

used here, though the weighting needs to be modified accordingly. Specifically, if we have $(x_{t-1}^{(j)}, w_{t-1}^{(j)})$ properly weighted with respect to $p(x_{t-1} \mid \boldsymbol{y}_{t-1})$ and if we use $x_t^{(j)} \sim g_t(x_t \mid x_{t-1}^{(j)}, y_t)$ for propagation, then $((x_t^{(j)}, x_{t-1}^{(j)}), w_{t-1}^{(j)})$ is properly weighted with respect to $g_t(x_t \mid x_{t-1}^{(j)}, y_t)p(x_{t-1} \mid \boldsymbol{y}_{t-1})$. Our target distribution is $p(x_t, x_{t-1} \mid \boldsymbol{y}_t)$. Hence by Lemma 8.5, the weight function is

$$w_t^{(j)} \propto w_{t-1}^{(j)} \frac{p(x_t^{(j)}, x_{t-1}^{(j)} \mid \boldsymbol{y}_t)}{p(x_{t-1}^{(j)} \mid \boldsymbol{y}_{t-1})g_t(x_t^{(j)} \mid x_{t-1}^{(j)}, y_t)} \propto w_{t-1}^{(j)} \frac{p(y_t \mid x_t^{(j)})p(x_t^{(j)} \mid x_{t-1}^{(j)})}{g_t(x_t^{(j)} \mid x_{t-1}^{(j)}, y_t)}.$$

Once the samples are generated, one also needs to use an efficient estimator for inference. Based on these observations, Liu and Chen (1998) formulated the following SMC framework.

Suppose $(\boldsymbol{x}_{t-1}^{(j)}, w_{t-1}^{(j)})$, $j = 1, \ldots, m$, are properly weighted with respect to $p(\boldsymbol{x}_{t-1} \mid \boldsymbol{y}_{t-1})$. When the system evolves from stage $t-1$ to stage t, three operations can be performed. Note that not all three operations have to be performed and the order of the operations can also vary.

Algorithm 8.4: Propagation Step

At time t, for $j = 1, \ldots, m$:

(A) Draw $x_t^{(j)}$ from a trial distribution $g_t(x_t \mid \boldsymbol{x}_{t-1}^{(j)}, y_t)$. Attach it to $\boldsymbol{x}_{t-1}^{(j)}$ to form $\boldsymbol{x}_t^{(j)} = (\boldsymbol{x}_{t-1}^{(j)}, x_t^{(j)})$.

(B) Compute the incremental weight

$$u_t^{(j)} \propto \frac{q_t(x_t^{(j)} \mid x_{t-1}^{(j)})f_t(y_t \mid x_t^{(j)})}{g_t(x_t^{(j)} \mid x_{t-1}^{(j)}, y_t)}$$

and the new weight

$$w_t^{(j)} = w_{t-1}^{(j)} u_t^{(j)}.$$

The resulting sample $\{(\boldsymbol{x}_t^{(j)}, w_t^{(j)}), j = 1, \ldots, m\}$ is properly weighted with respect to $p(\boldsymbol{x}_t \mid \boldsymbol{y}_t)$. The partial sample $\{(x_t^{(j)}, w_t^{(j)}), j = 1, \ldots, m\}$ is properly weighted with respect to $p(x_t \mid \boldsymbol{y}_t)$ for filtering problems. See Section 8.3 for a more detailed discussion. For more general dynamic systems, similar propagation step can be performed.

Algorithm 8.5: Resampling Step

(A) Obtain a new set of indices $\{I_1, \dots, I_m\}$, where $I_k \in \{1, \dots, m\}$ according to a set of priority scores $\alpha_t^{(j)}, j = 1, \dots, m$.

(B) Return $\{(x^{(I_j)}, \tilde{w}^{(I_j)}), j = 1, \dots, m\}$, where $\tilde{w}_t^{(j)} = w_t^{(j)} / \alpha_t^{(j)}$.

The resulting sample $\tilde{S}_t$ is properly weighted with respect to $p(x_t \mid \boldsymbol{y}_t)$, if Step (A) is unbiased. See Section 8.4 for further discussions.

Algorithm 8.6: Inference Step

(A) Estimation of $E_{\pi_t}(h(\boldsymbol{x}_t))$ for some integrable function $h(\cdot)$ using the generated weighted samples $(\boldsymbol{x}_t^{(j)}, w_t^{(j)}), j = 1, \dots, m$.

See Section 8.5 for a more detailed discussion.

Although in SSMs, particularly in filtering, interest is often placed on $p(x_t \mid \boldsymbol{y}_t)$, here we expand the scope to cover the entire history $p(\boldsymbol{x}_t \mid \boldsymbol{y}_t)$. If one is interested in $p(x_t \mid \boldsymbol{y}_t)$ only, one can simply keep the marginal samples $x_t^{(j)}$ instead of the entire history $\boldsymbol{x}_t^{(j)}$. Note that the weight depends on $\boldsymbol{x}_t^{(j)}$ since $x_t^{(j)}$ is generated using $\boldsymbol{x}_{t-1}^{(j)}$.

SMC recursively applies a proper combination of the three steps mentioned above. Again, not all three steps are required at each time step. There are many important issues involved in the design of an efficient implementation of SMC. Specifically, the selection of the propagation trial distribution $g_t(x_t \mid \boldsymbol{x}_{t-1}, y_t)$ is critical. The basic principle here is to effectively utilize as much information as possible, with reasonable computational complexity. Another issue is to choose the priority scores $\alpha_t^{(j)}$ in the resampling step. The scores are used not only to deal with samples with very small weights in the process, but also to achieve other objectives, including fulfilling constraints in the systems. The resampling schedule (how often and when to carry out a resampling operation) is also important in designing an efficient SMC. The third important issue is how to make inference as efficient as possible. Rao–Blackwellization (R-B) is one of the main tools that can be used in this step.

8.3 DESIGN ISSUE I: PROPAGATION

The propagation step in Algorithm 8.4 is the key to an efficient implementation of SMC. Its design is mainly concerned with the selection of the trial distribution $g_t(\cdot)$.

The following alternative interpretation of Algorithm 8.4 is useful. Consider the problem of generating samples from the target distribution $p(\boldsymbol{x}_t \mid \boldsymbol{y}_t)$. This is a high dimensional and difficult problem. To employ the IS approach, we need to find a trial distribution $g(\boldsymbol{x}_t)$. Since $\boldsymbol{x}_t$ is high dimensional, sequential sampling is a standard approach, and we can construct $g(\boldsymbol{x}_t)$ sequentially, building one component at a time, conditioning on the previous ones. Specifically, we use

$$g(\boldsymbol{x}_t) = g_1(x_1)g_2(x_2 \mid x_1)\ldots g_t(x_t \mid \boldsymbol{x}_{t-1}). \tag{8.12}$$

Each of the component conditional distribution $g_s(x_s \mid \boldsymbol{x}_{s-1})$ is of low dimension. The associated weight function is

$$w_t = \frac{p(\boldsymbol{x}_t \mid \boldsymbol{y}_t)}{g(\boldsymbol{x}_t)}.$$

The following expression is useful in obtaining some guidance for selecting $g_s(x_s \mid \boldsymbol{x}_{s-1})$ (ignoring $p(x_0)$).

$$\begin{aligned}
w_t &= \frac{p(\boldsymbol{x}_t \mid \boldsymbol{y}_t)}{\prod_{s=1}^t g_s(x_s \mid \boldsymbol{x}_{s-1})} \\
&= \prod_{s=1}^t \frac{p(\boldsymbol{x}_s \mid \boldsymbol{y}_s)}{p(\boldsymbol{x}_{s-1} \mid \boldsymbol{y}_{s-1})g_s(x_s \mid \boldsymbol{x}_{s-1})} \\
&\propto \prod_{s=1}^t \frac{p(\boldsymbol{x}_s, y_s \mid \boldsymbol{y}_{s-1})}{p(\boldsymbol{x}_{s-1} \mid \boldsymbol{y}_{s-1})g_s(x_s \mid \boldsymbol{x}_{s-1})} \\
&\propto \prod_{s=1}^t \frac{p(x_s \mid \boldsymbol{x}_{s-1}, \boldsymbol{y}_{s-1})p(y_s \mid \boldsymbol{x}_s)p(\boldsymbol{x}_{s-1} \mid \boldsymbol{y}_{s-1})}{p(\boldsymbol{x}_{s-1} \mid \boldsymbol{y}_{s-1})g_s(x_s \mid \boldsymbol{x}_{s-1})} \\
&\propto \prod_{s=1}^t \frac{p(x_s \mid x_{s-1})p(y_s \mid x_s)}{g_s(x_s \mid \boldsymbol{x}_{s-1})}.
\end{aligned}$$

Hence, letting

$$u_s = \frac{p(x_s \mid x_{s-1})p(y_s \mid x_s)}{g_s(x_s \mid \boldsymbol{x}_{s-1})} = \frac{q_s(x_s \mid x_{s-1})f_s(y_s \mid x_s)}{g_s(x_s \mid \boldsymbol{x}_{s-1})},$$

we have $w_s = w_{s-1}u_s, s = 1, \ldots, t$. If the trial distribution $g_s(x_s \mid \boldsymbol{x}_{s-1})$ only involves x_{s-1} and y_s, then the procedure can be made sequentially for $t = 1, 2, \ldots$.

Since the efficiency of the process depends on the weight distribution and the incremental weight is

$$u_t^{(j)} \propto \frac{q_t(x_t^{(j)} \mid x_{t-1}^{(j)})f_t(y_t \mid x_t^{(j)})}{g_t(x_t^{(j)} \mid x_{t-1}^{(j)}, y_t)}, \tag{8.13}$$

one would naturally want to choose $g_t(\cdot)$ so the ratio in Equation (8.13) is close to a constant.

8.3.1 Proposal Distributions

In Kong et al. (1994) and Liu and Chen (1995, 1998), the trial distribution

$$g_t(x_t \mid \boldsymbol{x}_{t-1}) = p(x_t \mid \boldsymbol{x}_{t-1}, \boldsymbol{y}_t) \propto q_t(x_t \mid x_{t-1})f_t(y_t \mid x_t) \tag{8.14}$$

was recommended for the SSM. This selection of g_t makes the ratio in Equation (8.13) a constant. Specifically, we have the following algorithm.

Algorithm 8.7: Full Information Propagation Step

At time t, for $j = 1, \ldots, m$:

(A) Draw $x_t^{(j)}$ from the *local posterior distribution*,

$$\begin{aligned} g_t(x_t \mid x_{t-1}^{(j)}, y_t) &= p(x_t \mid x_{t-1}^{(j)}, y_t) \\ &\propto f_t(y_t \mid x_t)q_t(x_t \mid x_{t-1}^{(j)}). \end{aligned} \tag{8.15}$$

(B) Compute the incremental weight

$$u_t^{(j)} = \int f_t(y_t \mid x_t)q_t(x_t \mid x_{t-1}^{(j)})dx_t$$

and the new weight

$$w_t^{(j)} = w_{t-1}^{(j)} u_t^{(j)}.$$

The incremental weight formula is based on Equation (8.13). Note that the denominator of that equation is a normalized version of the numerator, hence $u_t^{(j)}$ is the normalizing constant of $f_t(y_t \mid x_t)q_t(x_t \mid x_{t-1}^{(j)})$ as a density of x_t. It does not depend on the sampled $x_t^{(j)}$. This is a nice feature that we will make use of later in estimation.

The proposal distribution in Equation (8.14) utilizes information from both the state and observation equations, hence is termed a *full information propagation step*. However, in many cases this trial distribution requires complicated computations and can be difficult to generate samples from. In certain cases, the incremental weight can be difficult to evaluate as well.

Berzuini et al. (1997) proposed using local MCMC algorithms to generate samples from the full information trial distribution $q_t(x_t \mid x_{t-1})f_t(y_t \mid x_t)$ when direct sampling is difficult.

One can also use a suitable approximation of the full information trial distribution to achieve efficiency with reasonable computational complexity. For example, one can use

$$g_t(x_t \mid \boldsymbol{x}_{t-1}) \propto \hat{q}_t(x_t \mid x_{t-1})\hat{f}_t(y_t \mid x_t),$$

with incremental weight

$$u_t \propto \frac{q_t(x_t \mid x_{t-1})f_t(y_t \mid x_t)}{\hat{q}_t(x_t \mid x_{t-1})\hat{f}_t(y_t \mid x_t)} \int \hat{q}_t(x_t \mid x_{t-1})\hat{f}_t(y_t \mid x_t)dx_t,$$

where $\hat{q}_t$ and $\hat{f}_t$ are approximations of q_t and f_t, respectively. For ease of sampling and evaluating the integral $\int \hat{q}_t(x_t \mid x_{t-1})\hat{f}_t(y_t \mid x_t)dx_t$, normal or a mixture of normal distributions is often used for $\hat{q}_t$ and $\hat{f}_t$. However, since the weight depends on the sampled $x_t^{(j)}$, this is technically not a full information propagation.

Another algorithm that uses approximation to the full information proposal distribution is the *unscented particle filter* (van der Merwe et al., 2002). This is closely related to the unscented Kalman filter discussed in Section 7.1.4. The key idea is to use the unscented transformation (UT) to estimate the mean and variance of the full information proposal distribution $g_t(x_t \mid x_{t-1}, y_t)$ in Equation (8.14). Specifically, suppose the random noise terms $\boldsymbol{r}_t = (\varepsilon_t, e_t)$ at step t has mean $\mu_{\boldsymbol{r}}$ and covariance matrix $\Sigma_{\boldsymbol{r}}$. The UT chooses a set of weighted points $\{(\xi_\ell, \omega_\ell), \ell = 0, \pm 1, \ldots, \pm n\}$ as follows:

$$\xi_0 = \mu_{\boldsymbol{r}}; \qquad \omega_0 = \kappa/(2n+\kappa),$$

$$\xi_{\pm\ell} = \mu_{\boldsymbol{r}} \pm \sqrt{2n+\kappa} Q_\ell; \qquad \omega_{\pm\ell} = 1/(2n+\kappa),\ \ell = 1, \ldots, n$$

where n is the dimension of $\boldsymbol{r}_t$, κ is a constant, and Q_ℓ is the ℓth column of the square root matrix of $\Sigma_{\boldsymbol{r}}$. With a given $x_{t-1}^{(j)}$, the state and observation combination (x_t, y_t) is a nonlinear function of $\boldsymbol{r}_t = (\varepsilon_t, e_t)$ and the value of x_{t-1}. Let $(x_t, y_t) = \varphi^{(j)}(\boldsymbol{r}_t) = (s_t(x_{t-1}^{(j)}, \varepsilon_t), h_t(s_t(x_{t-1}^{(j)}, \varepsilon_t), e_t))$, corresponding to $x_{t-1}^{(j)}$. Then the mean and covariance of $p(x_t, y_t \mid x_{t-1}^{(j)})$ can be estimated by:

$$\hat{\mu}^{(j)} = \sum_{\ell=-n}^{n} \omega_\ell \varphi^{(j)}(\xi_l),$$

$$\hat{\Sigma}^{(j)} = \sum_{\ell=-n}^{n} \omega_\ell (\varphi^{(j)}(\xi_\ell) - \hat{\mu}^{(j)})(\varphi^{(j)}(\xi_\ell) - \hat{\mu}^{(j)})'.$$

We can then use $\hat{p}(x_t, y_t \mid x_{t-1}^{(j)}) \sim N(\mu^{(j)}, \Sigma^{(j)})$ as an approximation to $p(x_t, y_t \mid x_{t-1}^{(j)})$ and derive an approximation of the full information trial distribution $g_t(x_t \mid y_t, x_{t-1}^{(j)})$ from the approximate normal distribution.

The bootstrap filter of Gordon et al. (1993) and Kitagawa (1996) uses the state equation only, with $g_t(x_t \mid x_{t-1}, y_t) = q_t(x_t \mid x_{t-1})$ and weight $w_t = w_{t-1} f_t(y_t \mid x_t)$.

Algorithm 8.8: Propagation Step in a Bootstrap Filter

At time t, for $j = 1, \ldots, m$:

(A) Draw $x_t^{(j)}$ using the state equation

$$x_t^{(j)} \sim q_t(x_t \mid x_{t-1}^{(j)}).$$

This is the same as generating $\varepsilon_t^{(j)}$ first and obtaining $x_t^{(j)} = s_t(x_{t-1}^{(j)}, \varepsilon_t^{(j)})$.

(B) Compute the new weight $w_t^{(j)} = w_{t-1}^{(j)} f_t(y_t \mid x_t^{(j)})$.

This algorithm is usually easy to implement and fast to compute when $\varepsilon_t^{(j)}$ is easy to generate and $s_t(\cdot)$ and $f_t(\cdot)$ are easy to evaluate. However, it is often not efficient since the generation of x_t does not utilize the information in the current observation y_t, especially if y_t contains significant information on x_t (e.g., if y_t has a very small measurement error). The information in y_t is only used in calculating the weight. As a result, the trial distribution may be further away from the target distribution than the full information propagation, and the weight may degenerate faster.

At the other extreme, the *independent particle filters* (IPF) of Lin et al. (2005) use the observation equation only as the trial distribution, that is,

$$g_t(x_t \mid x_{t-1}, y_t) \propto f_t(y_t \mid x_t),$$

with incremental weight $u_t \propto q_t(x_t \mid x_{t-1})$. This choice is convenient when the state dynamics $q_t(\cdot)$ is weak and the information from the observation $f_t(\cdot)$ is strong. It is called the independent particle filter because the samples $x_t^{(j)}$ are independent of each other and independent of the past samples $x_{t-1}^{(i)}$ too. Hence there is no direct connection between the current sample $x_t^{(j)}$ with the past sample $x_{t-1}^{(j)}$, but the weight calculation $u_t \propto q_t(x_t^{(j)} \mid x_{t-1}^{(j)})$ requires a matching. Lin et al. (2005) proposed using a multiple matching scheme and showed its advantages.

Algorithm 8.9: Propagation Step in Independent Particle Filter

At time t, for $j = 1, \ldots, m$:

(A) Draw $x_t^{(j)}$ from the observation equation $g_t(x_t \mid y_t) \propto f_t(y_t \mid x_t)$.

(B) Select L different permutations of $\{1, \ldots, m\}$: $K_l = \{k_{l,1}, \ldots, k_{l,m}\}$, $l = 1, \ldots, L$. Compute the incremental weight

$$u_t^{(k_{l,j},j)} \propto q_t(x_t^{(j)} | x_{t-1}^{(k_{l,j})})$$

for each permutation. Compute the multiple matching weight

$$w_t^{(j)} = \frac{1}{L} \sum_{l=1}^{L} u_t^{(k_{l,j},j)} w_{t-1}^{(k_{l,j})}. \tag{8.16}$$

The permutation does not need to be random since the samples are random already. The number of multiple matches L controls the efficiency and computational cost of the procedure. A very large L achieves disengagement similar to that discussed in IS weight equivalence in Section 8.1.3. Specifically, if one uses full sample matching $L = m$ and

$$w_t^{(j)} = \frac{1}{m} \sum_{i=1}^{m} u_t^{(i,j)} w_{t-1}^{(i)},$$

then we have

$$\begin{aligned} w_t^{(j)} &\approx \int q_t(x_t^{(j)} \mid x_{t-1}) p(x_{t-1} \mid \boldsymbol{y}_{t-1}) dx_{t-1} \\ &= p(x_t^{(j)} \mid \boldsymbol{y}_{t-1}) \\ &= \frac{p(x_t^{(j)} \mid \boldsymbol{y}_{t-1}) f_t(y_t \mid x_t^{(j)})}{f_t(y_t \mid x_t^{(j)})} \\ &\propto \frac{p(x_t \mid \boldsymbol{y}_t)}{g_t(x_t \mid \boldsymbol{y}_t)}. \end{aligned} \tag{8.17}$$

Note that if one generates $x_t \sim g_t(x_t \mid \boldsymbol{y}_t)$ and is able to evaluate $p(x_t \mid \boldsymbol{y}_t)$ and $g_t(x_t \mid \boldsymbol{y}_t)$, then Equation (8.17) would be the weight and it does not involve $\boldsymbol{x}_{t-1}$ at all (disengagement). Hence the full matching scheme approximately achieves disengagement. This is a version of the marginal weight we discussed in Section 8.1.3, and it is better than if we use $L = 1$ and

$$w_t = w_{t-1}^{(j)} q_t(x_t^{(j)} \mid x_{t-1}^{(j)}) = \frac{p(\boldsymbol{x}_t \mid \boldsymbol{y}_t)}{\prod_{s=1}^{t} f_s(y_s \mid x_s^{(j)})}.$$

Of course, it is more computationally expensive to use $L = m$ matchings. Lin et al. (2005) showed that partial matching improves efficiency while maintaining low computational cost.

If it is difficult to generate samples x_t from $f_t(y_t \mid x_t)$ (as a distribution of x_t with y_t fixed), one can use a trial distribution $g_t(x_t \mid y_t)$ that is close to $f_t(y_t \mid x_t)$ but easy to generate samples from. Here $g_t(x_t \mid y_t)$ does not depend on x_{t-1}. The incremental weight becomes

$$u_t^{(k_{l,j},j)} \propto \frac{q_t(x_t^{(j)} | x_{t-1}^{(k_{l,j})}) f_t(y_t \mid x_t^{(j)})}{g_t(x_t^{(j)} \mid y_t)}.$$

Some special cases of IPFs can be found in Isard and Blake (2004) and Fox et al. (2001).

Mixed proposal distribution: In our discussion of IS in Section 8.1.3, a useful approach to ensure the quality of the proposal distribution is to use a mixture of multiple proposal distributions in finding the important areas of the target distribution and in defending the tails. Li et al. (2016) studied the use of such an approach in SMC.

8.3.2 Delay Strategy (Lookahead)

In Section 8.1.3, we discussed that it is in general good practice to include as much information as possible. The full information propagation step uses both the information from the state dynamics (from x_{t-1}) and the current information y_t to generate the samples of x_t. Since SSMs often process strong memory in which future observations contain substantial information on the current state, it is often beneficial if the inference on the current state x_t is delayed until future observations y_{t+1} to y_{t+d} become available. In practice, a slight delay is often tolerable. In this case, the target distribution at time t becomes

$$\begin{aligned}\pi_t^*(x_t) &= p(x_t \mid y_1, \ldots, y_t, y_{t+1}, \ldots, y_{t+d}) \\ &= \int p(x_t, x_{t+1}, \ldots, x_{t+d} \mid y_1, \ldots, y_t, y_{t+1}, \ldots, y_{t+d}) dx_{t+1} \ldots dx_{t+d}.\end{aligned}$$

This distribution is of course closer to the ultimate target distribution $p(x_t \mid y_1, \ldots, y_T)$ when all observations are available. Hence this is a special case of the smoothing problem to be discussed in Section 8.7.

There are several different approaches to achieve the borrowing of future information.

1. Delayed weighting method

As discussed in Section 8.1.3, if $(\boldsymbol{x}_{t+d}^{(k)}, w_{t+d}^{(k)})$ is properly weighted with respect to $p(\boldsymbol{x}_{t+d} \mid \boldsymbol{y}_{t+d})$, then the partial sample $(x_t^{(k)}, w_{t+d}^{(k)})$ is properly weighted with respect to the marginal distribution $p(x_t \mid \boldsymbol{y}_{t+d})$. Hence, inference on x_t can be made using $(x_t^{(k)}, w_{t+d}^{(k)})$ at time $t+d$ when $\boldsymbol{y}_{t+d}$ is available. Specifically, suppose $(\boldsymbol{x}_{t+d-1}^{(j)}, w_{t+d-1}^{(j)})$ is properly weighted with respect to $p(\boldsymbol{x}_{t+d-1} \mid \boldsymbol{y}_{t+d-1})$.

Algorithm 8.10: Propagation Step in Delayed Weighting

At time t, for $j = 1, \ldots, m$:

(A) Draw $x_{t+d}^{(j)}$ from a trial distribution $g_{t+d}(x_{t+d} \mid \boldsymbol{x}_{t+d-1}^{(j)}, \boldsymbol{y}_{t+d})$.

(B) Compute the incremental weight

$$u_{t+d}^{(j)} \propto \frac{q_{t+d}(x_{t+d}^{(j)} \mid \boldsymbol{x}_{t+d-1}^{(j)}) f_{t+d}(y_{t+d} \mid \boldsymbol{x}_{t+d}^{(j)})}{g_{t+d}(x_{t+d}^{(j)} \mid \boldsymbol{x}_{t+d-1}^{(j)}, \boldsymbol{y}_{t+d})}$$

and the new weight

$$w_{t+d}^{(j)} = w_{t+d-1}^{(j)} u_{t+d}^{(j)}.$$

(C) Return $(\boldsymbol{x}_{t+d}^{(j)}, w_{t+d}^{(j)})$ $(j = 1, \ldots, m)$ for further propagation. Return $(x_t^{(j)}, w_{t+d}^{(j)})$ $(j = 1, \ldots, m)$ for inference on x_t. Retain samples $\{x_{t+k}^{(j)}\}$ $(j = 1, \ldots, m;\ k = 0, \ldots, d)$.

The trial distribution $g_{t+d}(\cdot)$ can be any of those discussed in Section 8.3.1. Inference is discussed in Section 8.5. A typical estimator of $E(h(x_t) \mid \boldsymbol{y}_{t+d})$ would be $m^{-1} \sum_{j=1}^{m} w_{t+d}^{(j)} h(x_t^{(j)})$. This approach is based on the standard propagation step and only requires slight additional buffer space and no additional computation. However, the future information $\{y_{t+1}, \ldots, y_{t+d}\}$ is only utilized in the weight calculation, not in the generation of the samples of x_t, since these samples of x_t are generated at time t, with information $\boldsymbol{y}_t$. For more information, see Wang et al. (2002).

2. Exact delayed sampling method

To fully utilize the future information $\boldsymbol{y}_{t+d}$ in generating the samples of x_t, instead of only using it for weight calculation, the exact delayed sample method aims to generate samples from the target distribution $\pi_t(x_t) = p(x_t \mid \boldsymbol{y}_{t+d})$ at time t.

Specifically, suppose at time $t-1$, $(x_{t-1}^{(j)}, w_{t-1}^{(j)})$ is properly weighted with respect to $p(x_{t-1} \mid \boldsymbol{y}_{t+d-1})$. At time t, we use the following algorithm.

Algorithm 8.11: Propagation Step in Delayed Sampling

At time t, for $j = 1, \ldots, m$:

(A) Draw $x_t^{(j)}$ from $g_t(x_t \mid \boldsymbol{x}_{t-1}^{(j)}, \boldsymbol{y}_{t+d})$.

(B) Compute the incremental weight

$$u_t^{(j)} = \frac{p(y_t, y_{t+1}, \ldots, y_{t+d} \mid x_{t-1}^{(j)})}{p(y_t, y_{t+1}, \ldots, y_{t+d-1} \mid x_{t-1}^{(j)})} \frac{p(x_t^{(j)} \mid x_{t-1}^{(j)}, \boldsymbol{y}_{t+d})}{g_t(x_t^{(j)} \mid x_{t-1}^{(j)}, \boldsymbol{y}_{t+d})} \quad (8.18)$$

and the new weight

$$w_{t+1}^{(j)} = u_{t+1}^{(j)} w_t^{(j)}.$$

The incremental weight in Equation (8.18) is due to

$$\begin{aligned} u_t^{(j)} &= \frac{p(x_t^{(j)}, x_{t-1}^{(j)} \mid \boldsymbol{y}_{t+d})}{p(x_{t-1}^{(j)} \mid \boldsymbol{y}_{t+d-1}) g_t(x_t^{(j)} \mid x_{t-1}^{(j)}, \boldsymbol{y}_{t+d})} \\ &= \frac{p(x_{t-1}^{(j)} \mid \boldsymbol{y}_{t+d})}{p(x_{t-1}^{(j)} \mid \boldsymbol{y}_{t+d-1})} \frac{p(x_t^{(j)} \mid x_{t-1}^{(j)}, \boldsymbol{y}_{t+d})}{g_t(x_t^{(j)} \mid x_{t-1}^{(j)}, \boldsymbol{y}_{t+d})} \\ &\propto \frac{p(y_t, y_{t+1}, \ldots, y_{t+d} \mid x_{t-1}^{(j)}, \boldsymbol{y}_{t-1})}{p(y_t, \ldots, y_{t+d-1} \mid x_{t-1}^{(j)}, \boldsymbol{y}_{t-1})} \frac{p(x_t^{(j)} \mid x_{t-1}^{(j)}, \boldsymbol{y}_{t+d})}{g_t(x_t^{(j)} \mid x_{t-1}^{(j)}, \boldsymbol{y}_{t+d})}. \end{aligned}$$

Note that

$$\begin{aligned} &p(y_t, y_{t+1}, \ldots, y_{t+d} \mid x_{t-1}^{(j)}) \\ &\quad = \int p(y_t, y_{t+1}, \ldots, y_{t+d}, x_t, \ldots, x_{t+d} \mid x_{t-1}^{(j)}) dx_t \ldots dx_{t+d} \\ &\quad = \int q_t(x_t \mid x_{t-1}^{(j)}) f_t(y_t \mid x_t) \\ &\quad\quad \times \prod_{s=1}^{d} q_{t+s}(x_{t+s} \mid x_{t+s-1}) f_{t+s}(y_{t+s} \mid x_{t+s}) dx_t \ldots dx_{t+d}. \end{aligned}$$

The full information proposal using information $\boldsymbol{y}_{t+d}$ is

$$g_t(x_t \mid x_{t-1}, \boldsymbol{y}_{t+d}) = p(x_t \mid x_{t-1}, \boldsymbol{y}_{t+d-1}, y_{t+d})$$
$$\propto p(y_t, y_{t+1}, \ldots, y_{t+d} \mid x_t^{(j)}) q_t(x_t^{(j)} \mid x_{t-1}^{(j)}).$$

Its corresponding incremental weight is

$$u_t \propto \frac{p(y_t, y_{t+1}, \ldots, y_{t+d} \mid x_{t-1}^{(j)})}{p(y_t, y_{t+1}, \ldots, y_{t+d-1} \mid x_{t-1}^{(j)})}$$
$$= \frac{\int p(y_t, y_{t+1}, \ldots, y_{t+d}, x_t, \ldots, x_{t+d} \mid x_{t-1}^{(j)}) dx_t \ldots dx_{t+d}}{\int p(y_t, y_{t+1}, \ldots, y_{t+d-1}, x_t, \ldots, x_{t+d-1} \mid x_{t-1}^{(j)}) dx_t \ldots dx_{t+d-1}}.$$

For example, if x_t is a discrete random variable taking values in the set $\{a_1, \ldots, a_K\}$, then let

$$c_t^{(j)}(a_i) = g_t(x_t = a_i \mid x_{t-1}^{(j)}, \boldsymbol{y}_{t+d})$$
$$= p(x_t = a_i \mid x_{t-1}^{(j)}, \boldsymbol{y}_{t+d})$$
$$\propto p(y_t, y_{t+1}, \ldots, y_{t+d} \mid x_t = a_i) q_t(x_t = a_i \mid x_{t-1}^{(j)})$$
$$= q_t(x_t = a_i \mid x_{t-1}^{(j)}) f_t(y_t \mid x_t = a_i) \sum_{x_{t+1}, \ldots, x_{t+d}} \prod_{s=t+1}^{t+d} q_s(x_s \mid x_{s-1}) f_s(y_s \mid x_s).$$

The incremental weight is

$$u_t^{(j)} = \sum_{i=1}^{K} c_t^{(j)}(a_i).$$

Figure 8.9 illustrates the method with $x_t \in \mathcal{A} = \{0, 1\}$ and $d = 2$. The trial distribution is, for $i = 0, 1$,

$$g_t(x_t = i \mid x_{t-1}^{(j)}, y_t, y_{t+1}, y_{t+2})$$
$$= \sum_{x_{t+1}} \sum_{x_{t+2}} p(x_t = i, x_{t+1}, x_{t+2} \mid x_{t-1}^{(j)}, y_t, y_{y+1}, y_{t+2})$$
$$\propto q_t(x_t = i \mid x_{t-1}^{(j)}) f_t(y_t \mid x_t = i) \sum_{x_{t+1}} \sum_{x_{t+2}} q_{t+1}(x_{t+1} \mid x_t = i) \tau(x_{t+1}, x_{t+2}),$$

where

$$\tau(x_{t+1}, x_{t+2}) = f_{t+1}(y_{t+1} \mid x_{t+1}) q_{t+2}(x_{t+2} \mid x_{t+1}) f_{t+2}(y_{t+2} \mid x_{t+2}).$$

Although this approach fully utilizes the future information in generating the samples, the d-dimensional summation or integration is often computationally intensive, if it is not analytically available. For example, in the discrete state space case, each iteration requires the evaluation of a summation of K^d terms for the

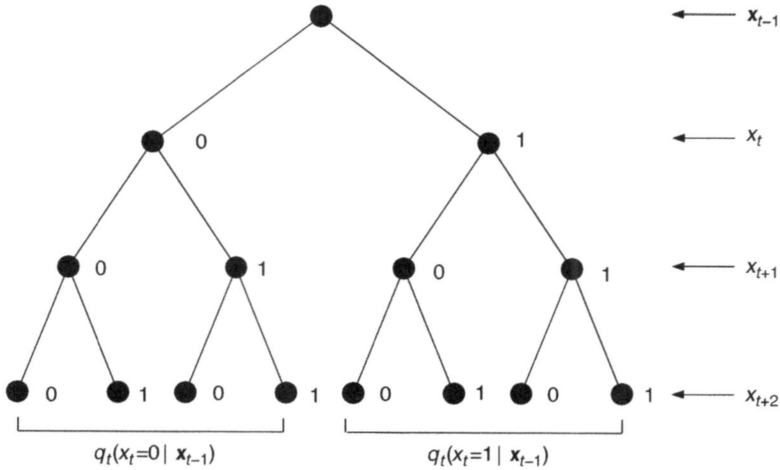

Figure 8.9 Illustration of the exact delayed sampling method.

exploration of the space of future states $(x_{t+1}, \ldots, x_{t+d})$. For more information, see Wang et al. (2002). Lin et al. (2013) showed that the delayed sampling method is more efficient than the delayed weighting method and, under mild conditions, the estimator is more accurate as the delay d increases.

3. Delayed pilot sampling method

To overcome the computational issue in the exact delayed sampling method, Wang et al. (2002) and Lin et al. (2013) proposed using pilot samples to partially explore the future space, while maintaining low computational cost. These are designed for discrete state space $x_t \in \{a_1, \ldots, a_K\}$. In this algorithm, pilot samples of future states are generated with the standard propagation step discussed in Section 8.3 and an approximation of the trial distribution $g_t(x_t = a_i \mid x_{t-1}^{(j)}, y_t, \ldots, y_{t+d})$ is obtained, along with an evaluation of the incremental weight. Specifically, suppose at time $t-1$ a set of properly weighted samples $(x_{t-1}^{(j)}, w_{t-1}^{(j)}), j = 1, \ldots, m$ with respect to $p(x_{t-1} \mid \mathbf{y}_{t-1})$ is available.

Algorithm 8.12: Propagation Step in Delayed Pilot Sampling

At time t, for $j = 1, \ldots, m$:

(A) For each $i \in \{a_1, \ldots, a_K\}$, send out a pilot, starting at $x_t = a_i$, to explore the space of future state $x_{t+1}, \ldots, x_{t+d}$ by using standard propagation steps. Specifically, set $x_t^{(i,j)} = a_i$ and $\gamma_t^{(i,j)} = 1$. For $s = 1, \ldots, d$:

(A.1) Generate $x_{t+s}^{(i,j)}$ from $g_{t+s}(x_{t+s} \mid x_{t+s-1}^{(i,j)}, y_{t+s})$.

(A.2) Calculate the incremental cumulative weight

$$\gamma_{t+s}^{(i,j)} = \gamma_{t+s-1}^{(i,j)} \frac{q_{t+s}(x_{t+s}^{(i,j)} \mid x_{t+s-1}^{(i,j)}) f_{t+s}(y_{t+s} \mid x_{t+s}^{(i,j)})}{g_{t+s}(x_{t+s}^{(i,j)} \mid x_{t+s-1}^{(i,j)}, y_{t+s})}.$$

(B) Draw $I_j \in \{1, \ldots, K\}$ with probability proportional to $\gamma_{t+d}^{(i,j)}$. Let $\boldsymbol{x}_t^{(j)} = (\boldsymbol{x}_{t-1}^{(j)}, x_t^{(I_j,j)})$.

(C) Return two weights:

$$w_t^{(j)} = w_{t-1}^{(j)} q_t(x_t^{(I_j,j)} \mid x_{t-1}^{(j)}) f_t(y_t \mid x_t^{(I_j,j)}) \frac{\sum_{i=1}^{K} \gamma_{t+d}^{(i,j)}}{\gamma_{t+s}^{(I_j,j)}},$$

$$w_t^{aux(j)} = w_{t-1}^{(j)} \sum_{k=1}^{K} \gamma_{t+d}^{(i,j)}.$$

The weight $w_t^{(j)}$ is the importance weight of $x_t^{(j)}$ with respect to $p(x_t \mid \boldsymbol{y}_t)$, that is, $(x_t^{(j)}, w_t^{(j)})$ is properly weighted with respect to $p(x_t \mid \boldsymbol{y}_t)$. It is used for propagation in the next step. The auxiliary weight $w_t^{aux(j)}$ is the importance weight of $x_t^{(j)}$ with respect to $p(x_t \mid \boldsymbol{y}_t, y_{t+1}, \ldots, y_{t+d})$, i.e. $(x_t^{(j)}, w_t^{aux(j)})$ is properly weighted with respect to $p(x_t \mid \boldsymbol{y}_{t+d})$. It is used for inference on x_t with information $\boldsymbol{y}_{t+d}$.

Figure 8.10 illustrates the delayed pilot sampling operation, with $\mathcal{A} = \{0, 1\}$ and $d = 2$, in which the pilot path for $(x_{t-1}^{(j)}, x_t = 0)$ is $(x_{t+1} = 1, x_{t+2} = 1)$ and the pilot path for $(x_{t-1}^{(j)}, x_t = 1)$ is $(x_{t+1} = 0, x_{t+2} = 1)$. Both are shown by the bold paths.

We can also use multiple pilots for each $x_t = i$ to increase the accuracy, but also possibly with additional computational cost.

This method has been successfully used in signal processing (Wang et al., 2002) and bioinformatics (Zhang and Liu, 2002). Lin et al. (2013) obtained certain theoretical properties of the delayed pilot method.

There are more advanced delayed methods, including the hybrid method, in which a combination of delayed sampling and delayed pilot methods are used, and the multi-level method, in which preliminary samples are first generated in a coarse state space using delayed SMC, then these samples are refined in the original state space to obtain the final samples. In addition, one can adaptively

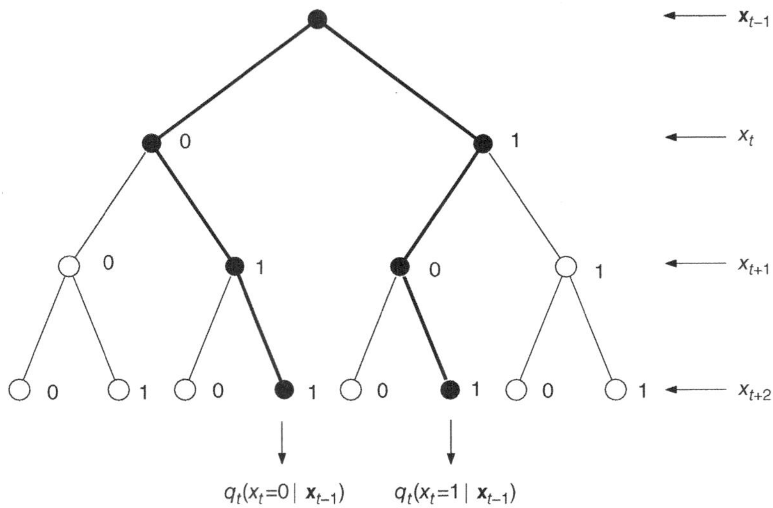

Figure 8.10 Illustration of a single-pilot delay method.

choose the length of delay, e.g. using long delay when information is weak and short/no delay when information is strong. See Lin et al. (2013) for more details.

8.4 DESIGN ISSUE II: RESAMPLING

The resampling step in Algorithm 8.5 is an indispensable component of SMC. As discussed before, the variance of w_t increases stochastically as t increases, resulting in a decrease in effective sample sizes. This is commonly known as the sample degeneracy problem. Hence, as the system evolves, it is preferable to insert a resampling/reallocation step between propagation recursions in order to stabilize the weight distribution and to prevent the samples from weight collapse. See Kong et al. (1994), Liu (2001), and Liu and Chen (1998). After a resampling step, the samples with a low priority score tend to be replaced by those with high priority scores. Priority scores reflect our preferences over different samples, according to certain objectives. See Chen et al. (2005), Fearnhead (2008), Chen et al. (2000), Wang et al. (2002), among others.

Some theoretical and heuristic arguments of resampling are given in Liu and Chen (1995). Note that if the weights $w_t^{(j)}$ are nearly constant, resampling only reduces the number of distinctive samples from the past states and incurs extra Monte Carlo variation when making inference about the past states. However, when the weights become very skewed, carrying many samples with small weights in propagation recursions is apparently wasteful. Resampling can provide chances for the good (i.e., important) samples to amplify themselves and hence rejuvenate the sampler to produce better samples for the *future* states.

8.4.1 The Priority Score

The selection of the priority score α_t is important. When it is chosen to be the weight, $\alpha_t^{(j)} = w_t^{(j)}$, the new weight of the selected samples becomes the constant one. These samples can be viewed as roughly distributed following the target distribution $p(x_t \mid \boldsymbol{y}_t)$, as discussed in Section 8.1.3. However, after resampling the samples become dependent. In many cases when w_t has a large variance, using $\alpha_t = w_t$ may result in a significant reduction in the number of distinct samples in the resampled set. Liu (2001) proposed using $\alpha_t = w_t^c$ $(0 < c < 1)$, since it is less greedy. Another possibility is the incremental-weight spreading priority score, where

$$\alpha_t = \left[\prod_{\ell=1}^{L} u_{t-\ell}\right]^{1/L}.$$

If resampling is carried out at every step, the running weight becomes

$$w_t = \prod_{\ell=2}^{L} u_{t-\ell}^{(\ell-1)/L},$$

an exponentially decaying geometric average of the incremental weights.

The delayed approach can also be used here. For example, we may use

$$\alpha_t = \frac{p(x_t \mid y_{t+d})}{p(x_{t-1} \mid y_{t+d-1}) g_t(x_t \mid x_{t-1})}$$

as the resampling priority score. This is not as good as the delayed sampling approach since the resampling reduces the diversity and increases the sample dependency. However, it may be easier to implement if $p(x_t \mid y_{t+d})$ is easy to evaluate, but difficult to generate samples from. It is also better than the delayed weighting approach, as resampling in the delayed weighting approach uses $\alpha_t = w_t$ as the priority score, which ignores the future information.

In many constraint problems, such as when we know *a priori* that the end state x_T is constrained to be at a fixed location $x_T = c$ (as in a diffusion bridge), it is crucial to utilize this information in generating samples of x_t $(t < T)$ in the early stages.

Zhang et al. (2003) proposed designing the priority scores to accommodate the functional shapes when estimating $\int h(\boldsymbol{x}_T)p(\boldsymbol{x}_T \mid \boldsymbol{y}_T)d\boldsymbol{x}_T$. They suggest using $\alpha_t = |\hat{\mu}_t(\boldsymbol{x}_t)w_t(\boldsymbol{x}_t)|$, where $\hat{\mu}_t(\boldsymbol{x}_t)$ is an estimate of $\int |h(\boldsymbol{x}_T)|p(\boldsymbol{x}_T \mid \boldsymbol{y}_T)dx_{t+1} \dots dx_T$.

In the prune-and-enrichment algorithm (Grassberger, 1997), α_t values are assigned as

$$\alpha_t^{(j)} = \begin{cases} 2 & \text{if } w_t^{(j)} \geq U, \\ 1 & \text{if } U > w_t^{(j)} \geq L, \\ 0.5 & \text{if } L > w_t^{(j)}, \end{cases}$$

where U and L are the upper and lower thresholds. In this case, a sample is four times more likely to be resampled if its weight is greater than U than those whose weight is less than L.

In the auxiliary particle filters (Pitt and Shephard, 1999) for SSMs, the priority score α_t is based on the future information y_{t+1}. Specifically, for each j, $j = 1, \ldots, m$, the predictive mean $\mu_{t+1}^{(j)} = E(x_{t+1} \mid x_t^{(j)}, y_t)$ is first calculated or estimated. Then the priority score is set as

$$\alpha_t^{(j)} = w_t^{(j)} p(y_{t+1} \mid x_{t+1} = \mu_{t+1}^{(j)}, x_t^{(j)}, y_t)$$

for resampling $x_t^{(j)}$ before sampling the new x_{t+1}. This score tries to evaluate the current sample $x_t^{(j)}$ based on the future information y_{t+1} without carrying out the sampling of x_{t+1}.

In discrete SSMs, we can use the priority score to combine propagation (sampling) and the resampling scheme. Specifically, if x_t takes values in $\{a_1, \ldots, a_k\}$, we can then evaluate $\alpha_t(x_{t-1}^{(j)}, a_i) \propto p(y_t \mid x_t = a_i, x_{t-1}^{(j)})$, $i = 1, \ldots, k$, $j = 1, \ldots, m$, then sample m distinct samples from $\{(x_{t-1}^{(j)}, a_i), i = 1, \ldots, k, j = 1, \ldots, m\}$ with probability proportional to $\alpha_t(x_{t-1}^{(j)}, a_i)$ and finally update the weights accordingly.

8.4.2 Choice of Sampling Methods in Resampling

There are various ways to perform resampling in SMC. The key is to ensure that the resampled samples remain properly weighted, with minimum computational cost. Suppose $S_t = \{(x_t^{(j)}, w^{(j)}), j = 1, \ldots, m\}$ is properly weighted with respect to $p(x_t \mid y_t)$. Suppose the priority scores $\alpha_t^{(j)}$ are normalized (sum up to one).

Algorithm 8.13: Simple Random Sampling

(A) Sample $I_j, j = 1, \ldots, m$ independently from the set $\{1, \ldots, m\}$ (with replacement) with probability $\alpha_t^{(j)}$.

(B) Let $\tilde{w}_t^{(j)} = w_t^{(j)} / \alpha_t^{(j)}$.

(C) Return $S_t' = \{(x_t^{(I_j)}, \tilde{w}^{(I_j)}), j = 1, \ldots, m\}$.

In Section 8.1.3 we showed that S'_t is properly weighted with respect to $p(x_t \mid \boldsymbol{y}_t)$ under mild conditions. This method is time consuming and has the largest Monte Carlo variation. However, because the resulting samples are iid given S_t, its theoretical properties are easy to investigate.

Algorithm 8.14: Residual Sampling (Liu and Chen, 1998)

(A) Obtain $k_j = [m\alpha^{(j)}]$ copies of j, where $[x]$ denotes the integer part of x, that is, $I_i = j$ if $\sum_{\ell=1}^{j-1} k_\ell < i \leq \sum_{\ell=1}^{j} k_\ell$, for $i = 1, \ldots, \sum_{j=1}^{m} k_j$.

(B) Let $m_r = m - \sum_{j=1}^{m} k_j$. Generate m_r iid samples $I_j, j = \sum_{j=1}^{m} k_j + 1, \ldots, m$ from $\{1, \ldots, m\}$ (with replacement) with probabilities proportional to $m\alpha^{(j)} - k_j = m\alpha^{(j)} - [m\alpha^{(j)}]$.

(C) Let $\tilde{w}_t^{(j)} = w_t^{(j)} / \alpha_t^{(j)}$.

(D) Return $S'_t = \{(x_t^{(I_j)}, \tilde{w}^{(I_j)}), j = 1, \ldots, m\}$.

This is the stratified sampling approach discussed in Section 8.1.2. It reduces the sampling variation as well as computation.

Algorithm 8.15: Stratified Sampling (Kitagawa, 1996)

(A) Set $c_0 = 0$, $c_j = c_{j-1} + \alpha_t^{(j)}, j = 1, \ldots, m$.

(B) Draw a starting point $u_1 \sim Unif[0, 1/m]$, and let $u_i = u_1 + (i - 1)/m$.

(C) If $c_{j-1} \leq u_i < c_j$, then $I_i = j$.

(D) Let $\tilde{w}_t^{(j)} = w_t^{(j)} / \alpha_t^{(j)}$.

(E) Return $S'_t = \{(x_t^{(I_j)}, \tilde{w}^{(I_j)}), j = 1, \ldots, m\}$.

In this approach only one random number u_1 needs to be generated. This method provides more systematic coverage of the space. For example, if all $\alpha^{(j)}$ are the same, then the resampled set is exactly the same as the original one, and no sample is lost. The computational complexity of this resampling method is of order $O(m)$.

Sometimes it is possible to combine sampling (propagation) with resampling.

Algorithm 8.16: Local Monte Carlo Method (Liu and Chen, 1998)

(A) Use a local Monte Carlo method to generate $(J_k, x_t^{(k)})$ from distribution

$$p(J, x_t) \propto \frac{p(x_{t-1}^{(J)}, x_t \mid \boldsymbol{y}_t)}{p(x_{t-1}^{(J)} \mid \boldsymbol{y}_{t-1})} \alpha_t^{(J)}.$$

(B) Let the new set of samples (for t) be $S_t = \{x_t^{(k)}, k = 1, \ldots, m\}$. The weight for each sample $x_t^{(k)}$ is $w_t^{(J_k)}/\alpha_t^{(J_k)}$.

5. Resampling with added diversification

Simple random sampling is based on the discrete approximation of $p(x_t \mid \boldsymbol{y}_t)$ using

$$\hat{F}(x) = \frac{1}{\sum_{j=1}^m w_t^{(j)}} \sum_{j=1}^m w_t^{(j)} I(x_t^{(j)} < x).$$

Liu and West (2001) suggested using a kernel smoother to approximate $p(x_t \mid \boldsymbol{y}_t)$. Specifically, let

$$\hat{p}(x \mid \boldsymbol{y}_t) = \frac{1}{\sum_{j=1}^m w_t^{(j)}} \sum_{j=1}^m w_t^{(j)} K_h(x_t^{(j)} - x).$$

If the kernel function $K_h(\cdot)$ is chosen as the normal density with mean zero and variance matrix $diag(h^2)$, the distribution is a mixture normal. Resampling using $\hat{p}(\cdot)$ is essentially sampling from the set $\{x_t^{(1)}, \ldots, x_t^{(m)}\}$ with probability proportional to $w_t^{(j)}$ (using one of the methods discussed above), then adding a Gaussian noise $e_t^{(I_j)} \sim N(0, \text{diag}(h^2))$ to each resampled sample. The resulting sample is more diversified, though choosing the bandwidth h becomes an issue.

8.4.3 Resampling Schedule

When to resample is also important. Frequent resampling can be *shortsighted* since resampling introduces strong dependency among the samples, as there are many duplicated samples in the resampled set. It also reduces the diversity among the samples in the sense that after frequent resampling, the survived samples tend to share common ancestors. In addition, frequent resampling makes the algorithm sensitive to outliers. For example, if y_t is an outlier, then many of the "good" samples tend to have small weights. Resampling using the weights as priority scores tends to eliminate those samples but retains the samples that happen to be closer to the outlying observation. One may completely lose track of the problem (e.g., no samples are good enough to recover in the future with non-outlying future observations). If resampling is done later, then the non-outlying future information $\{y_{t+1}, \ldots, y_{t+d}\}$ may correct the impact of y_t, hence one does not lose these "good" samples.

A deterministic resampling schedule performs resampling at time $t_0, 2t_0, 3t_0$, etc. A dynamic sampling schedule tries to maintain a minimum effective sample size. Specifically, given a sequence of thresholds $\{0 < c_t < 1\}$ in advance, do resampling when the effective sample size ESS in Equation (8.6) is less than mc_t. For further information, see Liu and Chen (1995). In Section 8.10.2 we demonstrate the importance of the resampling schedule.

In addition, the resampling step should be inserted before x_t is sampled if the full information proposal distribution of Algorithm 8.7 is used. This is because the weight $w_t^{(j)}$ at time t is a function of $x_{t-1}^{(j)}$ and does not depend on the sample $x_t^{(j)}$, as discussed in Section 8.3.1. Hence, by resampling $x_{t-1}^{(j)}$ first (with possible duplications) then sampling $x_t^{(j)}$ based on these resampled $x_{t-1}^{(j)}$, the samples $x_t^{(j)}$ are more diverse. Otherwise we may have duplicated $x_t^{(j)}$.

8.4.4 Benefits of Resampling

So why is resampling beneficial? Chopin (2004) provided a detailed analysis of the variance of

$$\frac{1}{m}\sum_{j=1}^{m} w_t^{(j)} h(x_t^{(j)}) \tag{8.19}$$

for estimating $\mu = E[h(x_t) \mid \boldsymbol{y}_t]$. To simplify the notations, denote $\pi_t(\boldsymbol{x}_s) = p(\boldsymbol{x}_s \mid \boldsymbol{y}_t)$ and $\pi_t(x_s \mid \boldsymbol{x}_{s-1}) = p(x_s \mid \boldsymbol{x}_{s-1}, \boldsymbol{y}_t)$. In addition, for $s \leq t$, define the conditional mean

$$\begin{aligned}
\mu_t(\boldsymbol{x}_s) &= E[h(x_t) \mid \boldsymbol{x}_s, \boldsymbol{y}_t] \\
&= \int h(x_t) p(\boldsymbol{x}_t \mid \boldsymbol{y}_t) dx_{s+1} \ldots dx_t \\
&= \int h(x_t) p(\boldsymbol{x}_{s+1:t} \mid \boldsymbol{x}_s, \boldsymbol{y}_t) dx_{s+1} \ldots dx_t.
\end{aligned}$$

Note that $E_{\pi_s}[\mu_t(\boldsymbol{x}_s)] = \mu$ for all $s = 1, \ldots, t$ and $\mathrm{Var}_{\pi_{s_1}}[\mu_t(\boldsymbol{x}_{s_1})] \geq \mathrm{Var}_{\pi_{s_2}}[\mu_t(\boldsymbol{x}_{s_2})]$ for $s_1 > s_2$. Roughly speaking, $\mu_t(\boldsymbol{x}_s)$ is the best estimate of μ given a known partial state path $\boldsymbol{x}_s$. It changes as $\boldsymbol{x}_s$ changes. Since $\boldsymbol{x}_{s_1}$ has more variates than $\boldsymbol{x}_{s_2}$ for $s_1 > s_2$, $\mu_t(\boldsymbol{x}_{s_1})$ has a larger total variance than $\mu_t(\boldsymbol{x}_{s_2})$ does. Another way to look at it is that $\mu_t(\boldsymbol{x}_{s_2})$ integrates out more terms of the overall integral $\int h(x_t)p(\boldsymbol{x}_t \mid \boldsymbol{y}_t)d\boldsymbol{x}_t$, hence it has a smaller variance.

Chopin (2004) showed that, if we use simple random resample at every time step, with $w_t^{(j)}$ as the priority score, then the variance of the estimator of Equation (8.19) is

$$\int \frac{\pi_t^2(x_1)}{g_1(x_1)}(\mu_1(x_1)-\mu)^2 dx_1 + \sum_{s=2}^{t} \int \frac{\pi_t^2(\boldsymbol{x}_s)}{\pi_s(\boldsymbol{x}_{s-1})g_s(x_s \mid \boldsymbol{x}_{s-1})}(\mu_s(\boldsymbol{x}_s)-\mu)^2 d\boldsymbol{x}_s. \tag{8.20}$$

This is to compare with the one without resampling

$$\int \frac{\pi_t^2(\boldsymbol{x}_t)}{\prod_{s=1}^{t} g_s(x_s \mid \boldsymbol{x}_{s-1})}(h(x_t)-\mu)^2 d\boldsymbol{x}_t. \tag{8.21}$$

To gain some insight into the variance in Equation (8.20), we note that the sth term,

$$\int \frac{\pi_t^2(\boldsymbol{x}_s)}{\pi_s(\boldsymbol{x}_{s-1})g_s(x_s \mid \boldsymbol{x}_{s-1})}(\mu_s(\boldsymbol{x}_s)-\mu)^2 d\boldsymbol{x}_s,$$

corresponds to the variance of the following estimator. Suppose we can generate iid samples $\boldsymbol{x}_{s-1}^{(j)}$ from $\pi_{s-1}(\boldsymbol{x}_{s-1}) = p(\boldsymbol{x}_{s-1} \mid \boldsymbol{y}_{s-1})$, the target distribution at time $s-1$, and we generate x_s from a trial distribution $x_s^{(j)} \sim g_s(x_s \mid \boldsymbol{x}_{s-1}^{(j)})$ with corresponding weight $w_s^{(j)} = \pi_t(\boldsymbol{x}_s)/[\pi_{s-1}(\boldsymbol{x}_{s-1})g_s(x_s \mid \boldsymbol{x}_{s-1})]$, and we are able to evaluate $\mu_s(\boldsymbol{x}_s^{(j)}) = E(h(x_t) \mid \boldsymbol{x}_s^{(j)}, y_t)$. We would estimate μ using

$$\frac{1}{m}\sum_{j=1}^{m} w_s^{(j)} \mu_s(\boldsymbol{x}_s^{(j)}).$$

The variance of this estimator is the sth term in Equation (8.20). Note this estimator differs from the SMC one in two places. First, the first $s-1$ terms $\boldsymbol{x}_{s-1}$ is sampled from the true target distribution $p(\boldsymbol{x}_{s-1} \mid \boldsymbol{y}_{s-1})$. It is in a way the best one can hope to achieve at time $s-1$. The second is that $\mu_s(\boldsymbol{x}_s^{(j)})$ is a much more accurate estimate of μ than $h(x_t)$, as it integrates out all the future terms from x_{s+1} to x_t. There is only one term $x_t^{(j)}$ involving IS. Hence the sth term is significantly better than that in Equation (8.21). Of course we have t terms in Equation (8.20). In addition, each term in the summation of Equation (8.20) has its advantages and disadvantages. For smaller s, $\mu_s(\boldsymbol{x}_s)$ has smaller variation, but typically the importance ratio

$$\frac{\pi_t(\boldsymbol{x}_s)}{\pi_{s-1}(\boldsymbol{x}_{s-1})g(x_s \mid \boldsymbol{x}_{s-1})} = \frac{p(\boldsymbol{x}_s \mid \boldsymbol{y}_t)}{p(\boldsymbol{x}_{s-1} \mid \boldsymbol{y}_{s-1})g(x_s \mid \boldsymbol{x}_{s-1})}$$

can be more divergent, since the trial distribution at time s typically depends only on $\boldsymbol{y}_s$, not $\boldsymbol{y}_t$.

8.5 DESIGN ISSUE III: INFERENCE

Standard inference with IS is in the form of Equation (8.5)

$$\hat{\mu}_t = \hat{E}[\varphi(\boldsymbol{x}_t) \mid \boldsymbol{y}_t] = \frac{\sum_{j=1}^m w_t^{(j)} \varphi(\boldsymbol{x}_t^{(j)})}{\sum_{j=1}^m w_t^{(j)}}.$$

There are several alternatives that may improve the efficiency of the estimator.

1. Estimation before resampling: A resampling step does not provide any extra information on the current and past states. It only adds additional Monte Carlo variations in the samples of the current and past states, as discussed in Section 8.1.3, even though it is critical in obtaining better samples of the future states. Hence, estimation should be done before carrying out a resampling step.

2. Rao–Blackwellization: If w_t does not depend on x_t, (as in the case when full information propagation step (8.14) is used as the propagation sampling distribution), then we can use

$$\hat{\mu}_t = \frac{\sum_{j=1}^m w_t^{(j)} E[\varphi(x_t) \mid x_{t-1}^{(j)}, y_t]}{\sum_{j=1}^m w_t^{(j)}}, \tag{8.22}$$

if $E[\varphi(x_t) \mid x_{t-1}^{(j)}, y_t]$ can be analytically carried out. This is because

$$\begin{aligned} E[\varphi(x_t) \mid \boldsymbol{y}_t] &= E[E\left\{\varphi(x_t) \mid x_{t-1}, \boldsymbol{y}_t\right\} \mid \boldsymbol{y}_t] \\ &\approx \frac{\sum_{j=1}^m w_t^{(j)} E[\varphi(x_t) \mid x_{t-1}^{(j)}, y_t]}{\sum_{j=1}^m w_t^{(j)}}. \end{aligned}$$

This estimator of $E[\varphi(x_t) \mid \boldsymbol{y}_t]$ is carried out without using the samples of x_t. It removes the sampling variation in x_t. In fact, this operation is equivalent to having an infinite number of iid copies of $x_t^{(j)}$ for each single $x_{t-1}^{(j)}$ and using their weighted averages.

3. Use delayed estimation if possible: This is because estimation of $E[\varphi(x_t) \mid \boldsymbol{y}_{t+d}]$ at time t is usually more accurate (closer to the true $\varphi(x_t)$) than $E[\varphi(x_t) \mid \boldsymbol{y}_t]$.

The former estimation uses more information, as shown in Lin et al. (2013), and it can be done with a simple delayed weighting method based on

$$\hat{E}[\varphi(x_t) \mid \boldsymbol{y}_{t+d}] = \frac{\sum_{j=1}^{m} w_{t+d}^{(j)} \varphi(x_t^{(j)})}{\sum_{j=1}^{m} w_{t+d}^{(j)}},$$

or with the delayed sampling method using $p(x_t \mid \boldsymbol{y}_{t+d})$ as the target distribution at time t.

8.6 DESIGN ISSUE IV: MARGINALIZATION AND THE MIXTURE KALMAN FILTER

When implementing Monte Carlo strategies, it is often good practice to carry out as much analytical computation as possible (Hammersley and Handscomb, 1964; Liu et al., 1994; MacEachern et al., 1998). In IS, it can easily be shown that the algorithm is more efficient after some components of the system are integrated out (marginalization).

Following the marginalization principle and the spirit of mixture Gaussian approximations, Chen and Liu (2000) developed a mixture Kalman filter (MKF) approach. It makes use of special properties of a conditional or partial conditional dynamic linear system and combines the SMC with the efficient Kalman filter. West (1992) pioneered the use of mixture Gaussian approximations in nonlinear SSMs.

8.6.1 Conditional Dynamic Linear Models

Conditional dynamic linear models (CDLMs) assume the form

$$\begin{aligned} x_t &= \boldsymbol{H}_{\Lambda_t} x_{t-1} + \boldsymbol{W}_{\Lambda_t} w_t \\ \Lambda_t &\sim \psi_t(\cdot \mid \Lambda_{t-1}) \\ y_t &= \boldsymbol{G}_{\Lambda_t} x_t + \boldsymbol{V}_{\Lambda_t} v_t, \end{aligned}$$

where $w_t \sim N(0, \boldsymbol{I})$ and $v_t \sim N(0, \boldsymbol{I})$ with proper dimensions and they are independent. The *indicator* Λ_t is an unobserved latent variable with transition probability density $\psi_t(\cdot)$. For any possible value λ in the support of Λ_t, the matrices $\boldsymbol{H}_\lambda, \boldsymbol{G}_\lambda, \boldsymbol{W}_\lambda$ and $\boldsymbol{V}_\lambda$ are known. The model appears in many applications, including those in Section 6.2.

8.6.2 Mixture Kalman Filters

Let $\boldsymbol{y}_t = (y_1, \dots, y_t)$ and $\boldsymbol{\Lambda}_t = (\Lambda_1, \dots, \Lambda_t)$. Note that

$$p(x_t \mid \boldsymbol{y}_t) = \int p(x_t \mid \boldsymbol{\Lambda}_t, \boldsymbol{y}_t) dF(\boldsymbol{\Lambda}_t \mid \boldsymbol{y}_t), \tag{8.23}$$

in which

$$p(x_t \mid \boldsymbol{\Lambda}_t, \boldsymbol{y}_t) \sim N(\mu_t(\boldsymbol{\Lambda}_t), \sigma_t^2(\boldsymbol{\Lambda}_t)). \tag{8.24}$$

Equation (8.24) is due to the fact that, given the trajectory of the indicator $\{\Lambda_1, \dots, \Lambda_t\}$, the system is linear and Gaussian, hence $p(x_t \mid \boldsymbol{\Lambda}_t, \boldsymbol{y}_t)$ is a Gaussian distribution. The corresponding $\mu_t(\boldsymbol{\Lambda}_t)$ and $\sigma_t^2(\boldsymbol{\Lambda}_t)$ in Equation (8.24) can readily be obtained through the Kalman filter discussed in Section 6.3.1. For simplicity, we denote

$$KF_t(\boldsymbol{\Lambda}_t) \equiv (\mu_t(\boldsymbol{\Lambda}_t), \sigma_t^2(\boldsymbol{\Lambda}_t)).$$

Because of Equation (8.24), $p(x_t \mid y_1, \dots, y_t)$ in Equation (8.23) is a mixture Gaussian distribution. If Λ_t takes values in a finite set of size K, then $p(x_t \mid \boldsymbol{y}_t)$ is a finite mixture of normal distributions, though the number of components is large (K^t). If Λ_t is a continuous random variable, then $p(x_t \mid y_1, \dots, y_t)$ is a continuous compound mixture of normal distributions. Sampling from such a distribution can be done with discrete samples directly from the distribution, but a more efficient method is to sample the Gaussian components. In our context, this is the same as sampling the unobserved latent indicator $\boldsymbol{\Lambda}_t$. It can be shown that, if

$$\{(\boldsymbol{\lambda}_t^{(1)}, w_t^{(1)}), \dots, (\boldsymbol{\lambda}_t^{(m)}, w_t^{(m)})\}$$

is a properly weighted sample with respect to $p(\boldsymbol{\Lambda}_t \mid \boldsymbol{y}_t)$, then we have

$$\begin{aligned} E[h(x_t) \mid \boldsymbol{y}_t] &= \int h(x_t) p(x_t \mid \boldsymbol{y}_t) dx_t \\ &= \int \left[\int h(x_t) p(x_t \mid \boldsymbol{\Lambda}_t, \boldsymbol{y}_t) dx_t \right] p(\boldsymbol{\Lambda}_t \mid \boldsymbol{y}_t) d\boldsymbol{\Lambda}_t \\ &= \int E[h(x_t) \mid \boldsymbol{\Lambda}_t, \boldsymbol{y}_t] p(\boldsymbol{\Lambda}_t \mid \boldsymbol{y}_t) d\boldsymbol{\Lambda}_t \\ &\approx \frac{\sum_{j=1}^m w_t^{(j)} E(h(x_t) \mid \boldsymbol{\lambda}_t^{(j)}, \boldsymbol{y}_t)}{\sum_{j=1}^m w_t^{(j)}}, \end{aligned}$$

where

$$E[h(x_t) \mid \boldsymbol{\lambda}_t^{(j)}, \boldsymbol{y}_t] = \int h(x) \phi(x; \mu_t(\boldsymbol{\lambda}_t^{(j)}), \sigma_t^2(\boldsymbol{\lambda}_t^{(j)})) dx,$$

due to Equation (8.24).

For example, to estimate the mean $E(x_t \mid \boldsymbol{y}_t)$ (the most encountered inference problem), we have $h(x) = x$ and

$$E(x_t \mid \boldsymbol{y}_t) \approx \frac{\sum_{j=1}^{m} w_t^{(j)} \mu_t(\lambda_t^{(j)})}{\sum_{j=1}^{m} w_t^{(j)}}.$$

The estimator of the variance is

$$Var(x_t \mid \boldsymbol{y}_t) \approx \frac{\sum_{j=1}^{m} w_t^{(j)} \sigma_t^2(\lambda_t^{(j)})}{\sum_{j=1}^{m} w_t^{(j)}}.$$

Note that this is equivalent to using a random mixture of normal distributions

$$\sum_{j=1}^{m} w_t^{(j)} N(\mu_t(\lambda_t^{(j)}), \sigma_t^2(\lambda_t^{(j)}))$$

to approximate the mixture Gaussian distribution $p(x_t \mid \boldsymbol{y}_t)$ in Equation (8.23). This approximation is more efficient than the discrete approximation using samples directly from the distribution. This can be seen from the fact that

$$\text{Var}[h(x_t) \mid \boldsymbol{y}_t] \geq \text{Var}[E(h(x_t) \mid \boldsymbol{\Lambda}_t, \boldsymbol{y}_t) \mid \boldsymbol{y}_t].$$

Formally, suppose that at time $t-1$ we have a properly weighted sample $(\boldsymbol{\lambda}_{t-1}^{(j)}, KF_{t-1}^{(j)}, w_{t-1}^{(j)})$ with respect to $\pi(\boldsymbol{\lambda}_{t-1} \mid \boldsymbol{y}_{t-1})$. At time t, we have the following algorithm.

Algorithm 8.17: Propagation Step in the MKF Algorithm

For $j = 1, \ldots, m$:

(A) Generate $\lambda_t^{(j)}$ from a trial distribution $g_t(\Lambda_t \mid \boldsymbol{\lambda}_{t-1}^{(j)}, KF_{t-1}^{(j)}, \boldsymbol{y}_t)$.

(B) Run one-step Kalman filter conditioning on $(\lambda_t^{(j)}, KF_{t-1}^{(j)}, y_t)$ and obtain $KF_t^{(j)}$.

(C) Calculate the incremental weight

$$u_t^{(j)} = \frac{p(\boldsymbol{\lambda}_{t-1}^{(j)}, \lambda_t^{(j)} \mid \boldsymbol{y}_t)}{p(\boldsymbol{\lambda}_{t-1}^{(j)} \mid \boldsymbol{y}_{t-1}) g_t(\lambda_t \mid \boldsymbol{\lambda}_{t-1}^{(j)}, KF_{t-1}^{(j)}, \boldsymbol{y}_t)}$$

and the new weight $w_t^{(j)} = w_{t-1}^{(j)} u_t^{(j)}$.

When Λ_t is a discrete random variable on a finite set, we can use the full information proposal distribution $p(\Lambda_t \mid \boldsymbol{\lambda}_{t-1}^{(j)}, KF_{t-1}^{(j)}, \boldsymbol{y}_t)$. Specifically, suppose Λ_t takes values in $\{1, \ldots, K\}$ and suppose that at time $t-1$ we have a properly weighted sample $(\lambda_{t-1}^{(j)}, KF_{t-1}^{(j)}, w_{t-1}^{(j)})$ with respect to $p(\lambda_{t-1} \mid y_{t-1})$. At time t, we have the following algorithm.

Algorithm 8.18: Full Information Propagation Step in the MKF Algorithm

For $j = 1, \ldots, m$:

(A) For each $k = 1, \ldots, K$, run the one-step Kalman filter to obtain $KF_t^{(j,k)}$ and

$$u_k^{(j)} = p(y_t \mid \Lambda_t = k, KF_{t-1}^{(j)})p(\Lambda_t = k \mid \lambda_{t-1}^{(j)}).$$

(B) Sample η from the set $\{1, \ldots, K\}$ with probability proportional to $\{u_1^{(j)}, \ldots, u_K^{(j)}\}$. Let $\lambda_t^{(j)} = \eta$. In this case $\lambda_t^{(j)} \sim p(\Lambda_t \mid \lambda_{t-1}^{(j)}, KF_{t-1}^{(j)}, y_t)$.

(C) Let $KF_t^{(j)} = KF_t^{(j,\eta)}$.

(D) The new weight is

$$w_t^{(j)} \propto w_{t-1}^{(j)} p(y_t \mid \lambda_{t-1}^{(j)}, KF_{t-1}^{(j)}) \propto w_{t-1}^{(j)} \sum_{k=1}^{K} u_k^{(j)}.$$

Note that $p(y_t \mid \Lambda_t = k, KF_{t-1}^{(j)})$ is a by-product of the Kalman filter in Equation (8.24).

When Λ_t is a continuous random variable, a simple (but not optimal) algorithm is as follows.

Algorithm 8.19: State Equation Propagation Step in the MKF Algorithm

For $j = 1, \ldots, m$:

(A) Sample a $\lambda_t^{(j)}$ from the state equation of Λ_t, $p(\Lambda_t \mid \Lambda_{t-1} = \lambda_{t-1}^{(j)})$.

(B) Run one-step Kalman filter conditioning on $(\lambda_t^{(j)}, KF_{t-1}^{(j)}, y_t)$ and obtain $KF_t^{(j)}$.

(C) The new weight is

$$w_t^{(j)} = w_{t-1}^{(j)} p(y_t \mid \lambda_t^{(j)}, KF_{t-1}^{(j)}).$$

8.7 SMOOTHING WITH SMC

Smoothing is one of the most important tasks of analyzing data with SSMs. In Sections 6.3.1 and 7.2 we discussed how a two-pass algorithm can be used to perform the smoothing task. The approach can also be adopted in association with SMC. Just as we adopt SMC and use properly weighted samples to present the underlying density in the filtering problem, we do the same for the smoothing densities $p(x_t \mid \boldsymbol{y}_T)$. Here we present several approaches.

8.7.1 Simple Weighting Approach

In Section 8.3 we showed that a pure propagation without resampling is a straightforward IS procedure, with the proposal distribution $g(\boldsymbol{x}_T) = g_1(x_1)g(x_2 \mid x_1) \ldots g_T(x_T \mid \boldsymbol{x}_{T-1})$ and its corresponding weight $w_T = u_1 u_2 \ldots u_T$ where u_t is the incremental weight in Equation (8.13). Since $(\boldsymbol{x}_T^{(j)}, w_T^{(j)})$ is properly weighted with respect to $p(\boldsymbol{x}_T \mid \boldsymbol{y}_T)$, the partially weighted samples $(x_t^{(j)}, w_T^{(j)})$ are properly weighted with respect to $p(x_t \mid \boldsymbol{y}_T)$ for all $1 \leq t \leq T$. That is, the weighted partial samples follow the smoothing distribution, and statistical inference such as estimating $E(h(x_t) \mid \boldsymbol{y}_T)$ can be done with the weighted averages of $h(x_t^{(j)})$.

However, this approach can be extremely inaccurate, especially for small t (the earlier states). Note that in SMC, the samples $x_t^{(j)}$ are generated based on information $\boldsymbol{y}_t$, not the full information $\boldsymbol{y}_T$. Even if $x_t^{(j)}$ is generated from $p(x_t \mid \boldsymbol{y}_t)$, it remains a poor trial distribution in view of the full information $p(x_t \mid \boldsymbol{y}_T)$. Its weight would be

$$w_t = \frac{p(x_t \mid \boldsymbol{y}_T)}{p(x_t \mid \boldsymbol{y}_t)} \propto p(y_{t+1}, \ldots, y_T \mid x_t).$$

This approach is not recommended for most applications.

Resampling steps are often inserted in SMC. Suppose at time t the resampling indices are $I_t^{(j)}$ $(j = 1, \ldots, m)$. When resampling is not carried out (at time t), let $I_t^{(j)} = j$. Starting with $\zeta_T^{(j)} = j$, recursively define

$$\zeta_t^{(j)} = \zeta_{t+1}^{I_t^{(j)}}$$

for $t = T-1, \ldots, 1$. The sequence $\{\zeta_t^{(j)}, t = 1, \ldots, T-1\}$ indicates the ancestors of the final sample $x_T^{(j)}$, that is, the final full sample is

$$\boldsymbol{x}_T^{(j)} = (x_1^{\zeta_1^{(j)}}, \ldots, x_T^{\zeta_T^{(j)}}).$$

Figure 8.11 illustrates the ancestor tree after the resampling steps. Resampling is carried out at $t = 2$ and $t = 4$. The arrows show the resampling path. The

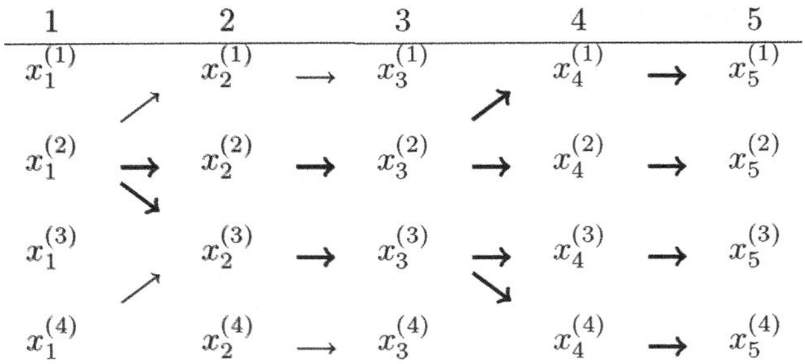

Figure 8.11 An illustration of sample ancestor tree after resampling.

ancestor lineage of $x_5^{(j)}, j = 1, \ldots, 4$ is shown by the bold arrows. For example, $\boldsymbol{x}_5^{(1)} = (x_1^{(2)}, x_2^{(2)}, x_3^{(2)}, x_4^{(1)}, x_5^{(1)})$ as one traces back following the bold arrows, starting at $x_5^{(1)}$. In this case all samples of $\{x_5^{(j)}, j = 1, \ldots, 4\}$ came from one ancestor $x_1^{(2)}$. Each resampling step reduces the number of unique ancestors. Suppose at time T each sample $\boldsymbol{x}_T^{(j)}$ is properly weighted with $w_T^{(j)}$ (taking into account of the resampling priority as shown in Section 8.4). As before, the estimation of $E(h(x_t) \mid \boldsymbol{y}_T)$ can be done with the weighted average

$$\hat{\mu}_t = \frac{\sum_{j=1}^{m} h(x_t^{\zeta_t^{(j)}}) w_T^{(j)}}{\sum_{j=1}^{m} w_T^{(j)}} = \frac{\sum_{v \in \Xi_t} h(x_t^{(v)}) [\sum_{\zeta_t^{(j)} = v} w_T^{(j)}]}{\sum_{j=1}^{m} w_T^{(j)}},$$

where Ξ_t is the set of unique values in the set $\{\zeta_t^{(j)}, j = 1, \ldots, m\}$. With frequent resampling and for $t \ll T$, the set Ξ_t can be very small. For instance, there is only one element in Ξ_1 in Figure 8.11. Hence the weighted average is done with a very small number of unique terms. Obviously the estimation error increases.

8.7.2 Weight Marginalization Approach

The Kalman smoother discussed in Section 6.3.3 uses the disentangled approach to link the filtering density $p(x_t \mid \boldsymbol{y}_t)$ and the future information $(y_{t+1}, \ldots, y_T)$ by inserting x_{t+1}. This approach can be used for SMC smoothing as well. The approach relies on the disentangle equation

$$\begin{aligned} p(x_t \mid \boldsymbol{y}_T) &= \int p(x_t, x_{t+1} \mid \boldsymbol{y}_T) dx_{t+1} \\ &= \int p(x_t \mid x_{t+1}, \boldsymbol{y}_T) p(x_{t+1} \mid \boldsymbol{y}_T) dx_{t+1} \end{aligned}$$

$$= \int p(x_t \mid x_{t+1}, \boldsymbol{y}_t) p(x_{t+1} \mid \boldsymbol{y}_T) dx_{t+1}$$

$$= \int \frac{p(x_{t+1} \mid x_t, \boldsymbol{y}_t) p(x_t \mid \boldsymbol{y}_t)}{p(x_{t+1} \mid \boldsymbol{y}_t)} p(x_{t+1} \mid \boldsymbol{y}_T) dx_{t+1}$$

$$= p(x_t \mid \boldsymbol{y}_t) \int \frac{p(x_{t+1} \mid x_t)}{\int p(x_{t+1} \mid x_t) p(x_t \mid \boldsymbol{y}_t) dx_t} p(x_{t+1} \mid \boldsymbol{y}_T) dx_{t+1}.$$

$$= p(x_t \mid \boldsymbol{y}_t) \int \frac{q_{t+1}(x_{t+1} \mid x_t)}{\int q(x_{t+1} \mid x_t) p(x_t \mid \boldsymbol{y}_t) dx_t} p(x_{t+1} \mid \boldsymbol{y}_T) dx_{t+1}. \quad (8.25)$$

Suppose we have completed the forward filtering process and obtained $(x_t^{(j)}, w_t^{(j)})$ properly weighted with respect to $p(x_t \mid \boldsymbol{y}_t)$. We can view $p(x_t \mid \boldsymbol{y}_t)$ as the trial distribution and $p(x_t \mid \boldsymbol{y}_T)$ as the target distribution. Equation (8.25) provides the weight calculation, if one can obtain an evaluation of

$$\tilde{w}_t = \int \frac{q_{t+1}(x_{t+1} \mid x_t)}{\int q(x_{t+1} \mid x_t) p(x_t \mid \boldsymbol{y}_t) dx_t} p(x_{t+1} \mid \boldsymbol{y}_T) dx_{t+1}. \quad (8.26)$$

Note that the two integrals here can both be estimated if we have a set of weighted samples from the smoothing density $p(x_{t+1} \mid \boldsymbol{y}_T)$ at time $t + 1$, and a set of weighted samples from the filtering density $p(x_t \mid \boldsymbol{y}_t)$ at time t. Since we do have a properly weighted samples with respect to $p(x_T \mid \boldsymbol{y}_T)$ from the filtering step, a backward iteration allows us to evaluate the integral, hence the smoothing weight Equation (8.26).

Specifically, suppose we have obtained the forward filtering (weighted) samples $\{(x_t^{(j)}, w_t^{(j)}), j = 1, \ldots, m, t = 1, \ldots, T\}$.

Algorithm 8.20: Weight Marginalization SMC Smoother

Let $\tilde{w}_T^{(j)} = w_T^{(j)}, j = 1, \ldots, m$. For $t = T - 1, T - 2, \ldots, 1$, for $j = 1, \ldots, m$:

(A) Calculate

$$\tilde{u}_t^{(j)} = \sum_{l=1}^{m} \frac{q_{t+1}(x_{t+1}^{(l)} \mid x_t^{(j)})}{\sum_{k=1}^{m} q_{t+1}(x_{t+1}^{(l)} \mid x_t^{(k)}) w_t^{(k)}} \tilde{w}_{t+1}^{(l)}$$

and the smoothing weight $\tilde{w}_t^{(j)} = w_t^{(j)} \tilde{u}_t^{(j)}$.

The resulting weighted samples $\{(x_t^{(j)}, \tilde{w}_t^{(j)}), j = 1, \ldots, m\}$ are properly weighted with respect to $p(x_t \mid \boldsymbol{y}_T)$.

This procedure was proposed by Tanizaki and Mariano (1994), Hrzeler and Knsch (1998), and Doucet et al. (2000). It does not regenerate samples during the smoothing step. It only adjusts the weights. Note that both $(x_t^{(j)}, w_T^{(j)})$ and $(x_t^{(j)}, \tilde{w}_t^{(j)})$ are properly weighted with respect to $p(x_t \mid \boldsymbol{y}_T)$. We have discussed this feature of IS in Section 8.1.3. The weights $\tilde{w}_t^{(j)}$ are marginal weights and, hence, are more efficient.

The procedure requires the recording of all forward filtering samples and weights $(x_t^{(j)}, w_t^{(j)})$ and the backward smoothing weight $\tilde{w}_t^{(j)}$. Resampling can be done in both the forward process and the backward process, though ancestors should not be resampled.

8.7.3 Two-filter Sampling

Briers et al. (2010) proposed a two-filter procedure for SMC smoothing. A standard SMC forward filter is carried out first to obtain $(x_t^{(j)}, w_t^{(j)})$ properly weighted with respect to $p(x_t \mid \boldsymbol{y}_t)$. Then, a second SMC backward filter is carried out, starting at $t = T$ to $t = 1$, using backward propagation distribution $\tilde{g}_t(x_t \mid x_{t+1}, y_t)$ and (backward) weight

$$\tilde{w}_t \propto \tilde{w}_{t+1} \frac{q_{t+1}(x_{t+1} \mid x_t) f_t(y_t \mid x_t)}{\tilde{g}_t(x_t \mid x_{t+1}, y_t)}$$

to get $(\tilde{x}_t^{(j)}, \tilde{w}_t^{(j)})$ properly weighted with respect to $p(x_t \mid \boldsymbol{y}_{t:T})$, where $\boldsymbol{y}_{t:T} = (y_t, \ldots, y_T)$. Because

$$\begin{aligned} p(x_t \mid \boldsymbol{y}_T) &\propto p(\boldsymbol{y}_{t:T} \mid x_t) p(x_t \mid \boldsymbol{y}_{t-1}) \\ &\propto p(\boldsymbol{y}_{t:T} \mid x_t) \int p(x_t \mid x_{t-1}) p(x_{t-1} \mid \boldsymbol{y}_{t-1}) dx_{t-1}, \end{aligned}$$

we have

$$\begin{aligned} \frac{p(x_t \mid \boldsymbol{y}_T)}{p(x_t \mid \boldsymbol{y}_t) p(\boldsymbol{y}_{t:T} \mid x_t)} &\propto \int p(x_t \mid x_{t-1}) p(x_{t-1} \mid \boldsymbol{y}_{t-1}) dx_{t-1} \\ &\approx \sum\nolimits_{k=1}^{m} q_t(\tilde{x}_t^{(j)} \mid x_{t-1}^{(k)}) w_{t-1}^{(k)}, \end{aligned}$$

where the last term is the result of integral approximation using the importance samples. Hence the sample $\tilde{x}_t^{(j)}$ should be weighted by

$$w_t^{*(j)} = \tilde{w}_t^{(j)} \sum_{k=1}^{m} q_t(\tilde{x}_t^{(j)} \mid x_{t-1}^{(k)}) w_{t-1}^{(k)},$$

and they are properly weighted with respect to $p(x_t \mid \boldsymbol{y}_T)$. This matching procedure is similar to that in IPF discussed in Section 8.3.

This procedure is similar to the weight marginalization algorithm. It adjusts the weights of the backward samples using the forward matching samples. If the backward sampling distribution $\tilde{g}_t$ is arbitrary, this procedure presents little advantage, as the backward samples $\tilde{x}_t$ only rely on the partial information $\boldsymbol{y}_{t:T}$. Hence the sample $\tilde{x}_T \sim p(x_T \mid y_T)$ is similar to $x_1 \sim p(x_1 \mid y_1)$. However, if one starts with a trial distribution $\tilde{g}_T(\cdot)$ that is close to $p(x_T \mid \boldsymbol{y}_T)$, the advantage can be significant. Unlike the weight marginalization algorithm, this is achievable, since we have the forward filtering samples $(x_T^{(j)}, w_T^{(j)})$ properly weighted with respect to $p(x_T \mid \boldsymbol{y}_T)$, which can be used to construct the sampling distribution $\tilde{x}_T \sim g_T(x_T \mid \boldsymbol{y}_T)$ that is close to $p(x_t \mid \boldsymbol{y}_T)$. Unfortunately, the samples themselves cannot be directly used, since one needs to be able to evaluate the trial distribution $g_T(\cdot)$ up to a normalizing constant. However, any reasonable approximation, such as a mixture normal distribution approximation, can be useful.

Fearnhead et al. (2010) proposed an extension of the two-filter algorithm. Again, a standard SMC forward filter is carried out to obtain $(x_t^{(j)}, w_t^{(j)})$ properly weighted with respect to $p(x_t \mid \boldsymbol{y}_t)$, and a second SMC backward filter is carried out to get $(\tilde{x}_t^{(j)}, \tilde{w}_t^{(j)})$ properly weighted with respect to $p(x_t \mid \boldsymbol{y}_{t:T})$. In addition, we also use a trial distribution $\bar{g}_t(x_t \mid x_{t-1}, \tilde{x}_{t+1}, y_t)$. At time t, we generate m^2 samples

$$\bar{x}_t^{(j,k)} \sim \bar{g}_t(x_t \mid x_{t-1}^{(j)}, \tilde{x}_{t+1}^{(k)}, y_t)$$

with weight

$$\bar{w}_t^{(j,k)} = \frac{f_t(y_t \mid \bar{x}_t^{(j,k)}) q_t(\bar{x}_t^{(j,k)} \mid x_{t-1}^{(j)}) q_{t+1}(\tilde{x}_{t+1}^{(k)} \mid \bar{x}_t^{(j,k)})}{\bar{g}_t(x_t \mid x_{t-1}^{(j)}, \tilde{x}_{t+1}^{(k)}, y_t)} w_{t-1}^{(j)} \tilde{w}_{t+1}^{(k)}.$$

Then $\{(\bar{x}_t^{(j,k)}, \bar{w}_t^{(j,k)}), j, k = 1, \ldots, m\}$ is properly weighted with respect to $p(x_t \mid \boldsymbol{y}_T)$. This is due to the fact that

$$\begin{aligned} p(x_t \mid \boldsymbol{y}_T) &\propto p(\boldsymbol{y}_{t:T} \mid x_t) p(x_t \mid \boldsymbol{y}_{t-1}) \\ &\propto p(y_t \mid x_t) \int p(x_t \mid x_{t-1}) p(x_{t-1} \mid \boldsymbol{y}_{t-1}) dx_{t-1} \times \\ &\quad \int p(x_{t+1} \mid x_t) p(x_{t+1} \mid \boldsymbol{y}_{t+1:T}) dx_{t+1}. \end{aligned}$$

Because this is an m^2-operation, computational cost can be high.

8.8 PARAMETER ESTIMATION WITH SMC

SSMs in practice often contain unknown parameters. Our previous discussion focuses on filtering and smoothing, with a completely specified SSM without unknown parameters. In this section we discuss how to use these techniques to estimate the parameters in a given model.

8.8.1 Maximum Likelihood Estimation

Maximum likelihood estimation requires the evaluation of the likelihood function

$$L(\theta \mid \boldsymbol{y}_T) = p_\theta(\boldsymbol{y}_T) = \int p_\theta(\boldsymbol{y}_T, \boldsymbol{x}_T) d\boldsymbol{x}_T. \tag{8.27}$$

This is the normalizing constant of

$$\pi_\theta^*(\boldsymbol{x}_T \mid \boldsymbol{y}_T) = \prod_{t=1}^{p} q_t(x_t \mid x_{t-1}, \theta) f_t(y_t \mid x_t, \theta).$$

Hence IS can be used, as discussed in Section 8.1.3, for estimating $Z = \int \pi^*(x)dx$. Specifically, if we generate samples $\boldsymbol{x}_T^{(j)}, j = 1, \ldots, m$ from a trial distribution $g(\boldsymbol{x}_T)$, then we can use

$$\hat{L}(\theta \mid \boldsymbol{y}_T) = \frac{1}{m} \sum_{j=1}^{m} \frac{\pi_\theta^*(\boldsymbol{x}_T^{(j)} \mid \boldsymbol{y}_T, \theta)}{g(\boldsymbol{x}_T^{(j)} \mid \boldsymbol{y}_T, \theta)}$$

as an estimator of $L(\theta \mid \boldsymbol{y}_T)$. Therefore, if one uses sequential construction of the trial distribution $g(\boldsymbol{x}_T \mid \boldsymbol{y}_T, \theta) = \prod_{t=1}^{T} g_t(x_t \mid \boldsymbol{y}_t, \theta)$ in (8.12), and if there is no resampling, the final weight $w_T^{(j)}$ obtained through SMC can then be used to estimate the likelihood

$$\hat{L}(\theta \mid \boldsymbol{y}_T) = \frac{1}{m} \sum_{j=1}^{m} w_T^{(j)}. \tag{8.28}$$

If we resample at time steps $t_1, \ldots, t_k$ (where $1 = t_0 < t_1 < \ldots < t_k < t_{k+1} = T$) using the weights $w_{t_i}^{(j)}$ as the priority scores, and use the fact that

$$p_\theta(\boldsymbol{y}_T) = p_\theta(y_1, \ldots y_{t_1}) \prod_{i=2}^{k} p_\theta(y_{t_i+1}, \ldots, y_{t_{i+1}} \mid \boldsymbol{y}_{t_i}),$$

we can obtain

$$\hat{L}(\theta \mid \boldsymbol{y}_T) = \prod_{i=1}^{k+1} \left[\frac{1}{m} \sum_{j=1}^{m} w_{t_i}^{(j)} \right]. \tag{8.29}$$

With the estimated likelihood function, one can obtain the MLE using optimization procedures. However, one critical problem with this approach is that the values of the estimated likelihood function are subject to Monte Carlo variation, as they are estimated using Monte Carlo samples. The resulting function may not be smooth, even though the true likelihood function is smooth. This creates significant difficulties for any optimization procedure to find stable estimates. Malik and Pitt (2011) proposed to use the same random numbers in evaluating the likelihood function for different θ values, hence obtaining a smooth estimator of the likelihood function. However, the procedure becomes difficult if the dimension of the state variable is high.

Since optimization algorithms such as the steepest ascent algorithm require the gradient (and curvature) of the function to be optimized, Poyiadjis et al. (2011) developed procedures to directly estimate the score and information matrix of the likelihood function of an SSM. Using Equation (8.27), we have

$$\nabla \log p_\theta(\boldsymbol{y}_T) = \int \nabla \log(p_\theta(\boldsymbol{x}_T, \boldsymbol{y}_T))p_\theta(\boldsymbol{x}_T \mid \boldsymbol{y}_T)d\boldsymbol{x}_T,$$

and

$$-\nabla^2 \log p_\theta(\boldsymbol{y}_T) = \nabla \log p_\theta(\boldsymbol{y}_T)\nabla \log p_\theta(\boldsymbol{y}_T)' - \frac{\nabla^2 p_\theta(\boldsymbol{y}_T)}{p_\theta(\boldsymbol{y}_T)},$$

where

$$\begin{aligned}\frac{\nabla^2 p_\theta(\boldsymbol{y}_T)}{p_\theta(\boldsymbol{y}_T)} &= \int \nabla \log p_\theta(\boldsymbol{x}_T, \boldsymbol{y}_T)\nabla \log p_\theta(\boldsymbol{x}_T, \boldsymbol{y}_T)' p_\theta(\boldsymbol{x}_T \mid \boldsymbol{y}_T)d\boldsymbol{x}_T \\ &\quad + \int \nabla^2 \log p_\theta(\boldsymbol{x}_T, \boldsymbol{y}_T)p_\theta(\boldsymbol{x}_T \mid \boldsymbol{y}_T)d\boldsymbol{x}_T,\end{aligned}$$

due to Louis' identity (Cappé et al., 2005). All aforementioned integrals can be approximated with a set of samples $\{(\boldsymbol{x}_T^{(j)}, w_T^{(j)}), j = 1, \ldots, m\}$ properly weighted with respect to $p(\boldsymbol{x}_T \mid \boldsymbol{y}_T)$. However, such an approximation suffers the same problem as using $w_T^{(j)}$ to estimate the likelihood.

Poyiadjis et al. (2011) refined the procedure using an alternative formulation based only on the integration with respect to the final marginal distribution of $p_\theta(x_T \mid \boldsymbol{y}_T)$. Specifically,

$$\nabla \log p_\theta(\boldsymbol{y}_T) = \int \nabla \log(p_\theta(x_T, \boldsymbol{y}_T))p_\theta(x_T \mid \boldsymbol{y}_T)dx_T$$

and

$$\frac{\nabla^2 p_\theta(y_T)}{p_\theta(y_T)} = \int \nabla \log p_\theta(x_T, y_T) \nabla \log p_\theta(x_T, y_T)' p_\theta(x_T \mid y_T) dx_T + \int \nabla^2 \log p_\theta(x_T, y_T) p_\theta(x_T \mid y_T) dx_T.$$

However, in this case $\nabla \log(p_\theta(x_T, y_T))$ and $\nabla^2 \log(p_\theta(x_T, y_T))$ are not immediately available. Poyiadjis et al. (2011) suggested a recursive formula:

$$\nabla \log(p_\theta(x_t, y_t)) = \frac{\nabla p_\theta(x_t, y_t)}{p_\theta(x_t, y_t)},$$

where

$$\nabla p_\theta(x_t, y_t) = p_\theta(y_{t-1}) f_t(y_t \mid x_t, \theta) \int q_t(x_t \mid x_{t-1}, \theta) \eta_t(x_{t-1}, x_t, y_t) p_\theta(x_{t-1} \mid y_{t-1}) dx_{t-1}$$

and

$$p_\theta(x_t, y_t) = p_\theta(y_{t-1}) f_t(y_t \mid x_t, \theta) \int q_t(x_t \mid x_{t-1}, \theta) \eta_t(x_{t-1}, x_t, y_t) dx_{t-1}.$$

The function $\eta_t(\cdot)$ in the prior equations is

$$\eta_t(x_{t-1}, x_t, y_t) = \nabla \log f_t(y_t \mid x_t, \theta) + \nabla \log q_t(x_t \mid x_{t-1}, \theta) + \nabla \log p_\theta(x_{t-1}, y_{t-1}).$$

To implement the procedure, suppose at time $t-1$ we have obtained a weighted sample $\{(x_{t-1}^{(j)}, w_{t-1}^{(j)}), j = 1, \ldots, m\}$ properly weighted with respect to $p(x_{t-1} \mid y_{t-1})$ and $a_{t-1}^{(j)} = \nabla p_\theta(x_{t-1}^{(j)}, y_t)$, the score function evaluated at $x_{t-1}^{(j)}, j = 1, \ldots, m$. With a standard propagation step we obtain a weighted sample $\{(x_t^{(j)}, w_t^{(j)}), j = 1, \ldots, m\}$ properly weighted with respect to $p(x_t \mid y_t)$. We can then approximate $a_t^{(j)} = \nabla p_\theta(x_t^{(j)}, y_t)$ with IS integration using the weighted sample. Specifically, let

$$a_t^{(j)} = \frac{\sum_{k=1}^m w_{t-1}^{(k)} q_t(x_t^{(j)} \mid x_{t-1}^{(k)}) \varphi(x_{t-1}^{(k)}, x_t^{(j)}, a_{t-1}^{(k)})}{\sum_{k=1}^m w_{t-1}^{(k)} q_t(x_t^{(j)} \mid x_{t-1}^{(k)})},$$

where

$$\varphi(x_{t-1}^{(k)}, x_t^{(j)}, a_{t-1}^{(k)}) = \nabla \log f_t(y_t \mid x_t^{(j)}, \theta) + \nabla \log q_t(x_t^{(j)} \mid x_{t-1}^{(k)}, \theta) + a_{t-1}^{(k)}.$$

Finally, at $t = T$, an estimate of the score $\nabla log p_\theta(y_T)$ can be obtained by

$$\hat{S}_T = \frac{\sum_{j=1}^m a_T^{(j)} w_T^{(j)}}{\sum_{j=1}^m w_T^{(j)}}.$$

An estimate of the information matrix can be obtained similarly.

8.8.2 Bayesian Parameter Estimation

1. Parameter as the initial state x_0: To make Bayesian inference, one needs to obtain the posterior distribution $p(\theta \mid \boldsymbol{y}_T)$ for a given prior distribution $p(\theta)$. The simplest way to accomplish this using SMC is to treat the unknown parameter θ as the initial state variable x_0 with the initial distribution $p(x_0) = p(\theta)$. Starting with a set of samples of θ from the prior distribution, $\theta^{(j)} \sim p(\theta)$, one proceeds with SMC to generate weighted samples $\{(\theta^{(j)}, \boldsymbol{x}_T^{(j)}, w_T^{(j)}), j = 1, \ldots, m\}$ properly weighted with respect to $p(\theta, \boldsymbol{x}_T \mid \boldsymbol{y}_T)$. Then the marginal sample $(\theta^{(j)}, w_T^{(j)})$ is properly weighted with respect to the posterior distribution $p(\theta \mid \boldsymbol{y}_T)$. As discussed in Section 8.7, such a procedure yields poor estimation of x_0 (or θ) due to weight degeneracy or less of diversity in the surviving samples of θ after resampling. Since this is essentially a smoothing problem, the approaches discussed in Section 8.7 can be used. However, since the SSM in this case is no longer Markovian (as all x_t depends on θ), the smoothing algorithms designed for Markovian SSMs do not produce accurate estimates.

2. MCMC approach: An alternative is to use an MCMC algorithm to generate samples from the joint posterior distribution $p(\theta, \boldsymbol{x}_T \mid \boldsymbol{y}_T)$. However, a vanilla implementation of MCMC algorithms is often ineffective in dealing with SSMs due to high dimensionality and poor mixing since $p(x_t \mid x_{t-1}, x_{t+1}, \boldsymbol{y}_T)$ is often sticky.

Andrier et al. (2010) proposed the *particle MCMC* algorithm that combines MCMC and SMC. To generate samples from the posterior distribution $p(\theta \mid \boldsymbol{y}_T)$ using the metropolis–Hastings algorithm discussed in Section 8.1.4, one would generate, at iteration i, a sample $\theta^* \sim g(\theta \mid \theta^{(i-1)})$ and accept $\theta^{(i)} = \theta^*$ with probability

$$p = \min\left\{1, \frac{p_{\theta^*}(\boldsymbol{y}_T)p(\theta^*)}{p_{\theta^{(i-1)}}(\boldsymbol{y}_T)p(\theta^{(i-1)})}\frac{g(\theta^{(i-1)} \mid \theta^*)}{g(\theta^* \mid \theta^{(i-1)})}\right\}.$$

Otherwise, set $\theta^{(i)} = \theta^{(i-1)}$. This requires the evaluation of the likelihood function $p_\theta(\boldsymbol{y}_T)$. Andrier et al. (2010) showed that using the SMC estimator in Equation (8.28) or (8.29) produces a valid Markov chain for $\theta^{(i)}$ with equilibrium distribution $p(\theta \mid \boldsymbol{y}_T)$. These authors also introduced a particle Gibbs sampler.

8.8.3 Varying Parameter Approach

The main difficulty in using SMC for parameter estimation is that the parameters are fixed while SMC is designed to deal with constantly changing latent variables. Treating the unknown parameters as state $x_0 = \theta$ at time $t = 0$ is not a good approach as we discussed before because we run into the sample degenerating problem. Once the samples of θ are generated, they will not be changed. The information from the observations are incorporated in the weight. As more and more information accumulates, the posterior distribution of θ is more and more different from the prior, making the weights more and more skewed. One possible alternative is to assume that the parameters are (slow) time-varying and to integrate them into the state variable, $\tilde{x}_t = (x_t, \theta_t)$. For example, let $\theta_t = \theta_{t-1} + \tau_t$. Then we can carry out the filtering and smoothing operations using SMC.

8.9 IMPLEMENTATION CONSIDERATIONS

SMC is a computational intensive method. Often, an efficient coding can significantly reduce the computation burden and, hence, can either obtain more accurate results by using the same computing time with more Monte Carlo samples or achieve the same estimation accuracy with less computing time. The following simple steps should always be kept in mind in programming SMC. They are simple to do, without engaging advanced programming methods, and can achieve better results than a straightforward implementation.

1. Avoid loops and use vector and matrix operations if possible: Most software, such as R or Matlab, is structured to take advantage of efficient linear vector and matrix operations, with built-in accelerators. Basic programming languages such as C++ and Python also have libraries that are optimized for these operations. On the other hand, using `for` loops can significantly slow down the computation. SMC has an inherit structure that is ideal for vector and matrix operations since the updating of Monte Carlo samples has the same structure and can often be vectorized. One can also use a graphical processing unit (GPU) to achieve parallelization (Lee at al., 2010).

2. Use log weight when possible: Notice that the weight updating formation $w_t = w_{t-1}u_t$ is in a product form. Repeated multiplications are not only computationally expensive, but also create numerical overflow and underflow. Using log weight in the updating step turns multiplications into summations, with much better numerical properties. In addition, in many situations the incremental weight is in an exponential form (e.g., normal or exponential family). Using log weight also saves the expensive `exp` operation.

3. Ignore common constants in weight calculation if possible: In most cases we use the weighted average for estimation and, hence, can ignore any multiplicative constant that is common to all Monte Carlo samples in weight calculation. If one retains the log weight $\ln(w_t)$, a common additive constant can be ignored.

4. Avoid overflow and underflow in calculating the normalized weight for resampling: To perform resampling, if the weight is used as the priority score, the log weights need to be transformed back to its original scale, then normalized. Taking the exp transformation often creates numerical instability, resulting in numerical overflow and underflow in computation. The problem can easily be solved by subtracting the maximum from each log weight before doing the exp transformation. This simple step ensures that there is at least one term equal to 1 in the summation in the denominator.

5. Efficient resampling procedure: Resampling can be one of the most expensive components in SMC implementation, especially when the Monte Carlo sample size is large. Simple random sampling using the priority scores is often used for theoretical investigation, but it can be slow in computation. In practice, a combination of residual sampling and stratified sampling is often used, with good properties. See Section 8.4 for more details.

Figures 8.12 and 8.13 show two implementations of performing a simple particle filter for the toy system

$$x_t = x_{t-1} + \varepsilon_t, \quad y_t = x_t + e_t$$

where $\varepsilon_t \sim N(0, \sigma^2)$ and $\varepsilon_t \sim N(0, \sigma_e^2)$. To fully use vector operations, the inner loop (lines B1 to B4) should be avoided in Implementation (II), shown in Figure 8.13, compared to its vectorized version (lines A1 and A2) in

```
        ## Implementation (I)
        ## observations in yy. state in xx. estimation in mu
        mm <- 10000
        tt <- 100
        xx <- 1:mm
        logw <- rep(0,mm)
        mu <- 1:tt
        for(t in 1:tt){
Line A1     xx <- rnorm(mm)*ss1
Line A2     logw <- logw+(xx*yy[t]-xx**2/2)/ss2
Line A3     logw <- logw-max(logw)
Line A4     w <- exp(logw)
            mu[t] <- xx*w/sum(w)
        }
```

Figure 8.12 Efficient implementation of a simple SMC.

```
          ## Implementation (II)
          ## observations in yy. states in xx. estimation in mu
          mm <- 10000
          tt <- 100
          xx <- 1:mm
          w <- rep(1,mm)
          mu <- 1:tt
          for(t in 1:tt){
Line B1      for(j in 1:mm){
Line B2         xx[j] <- rnorm(1)*ss1
Line B3         w[j] <- w[j]/sqrt{2*pi}*exp(-(xx-yy[t])**2/2/ss2)
Line B4      }
             mu[t] <- xx*w/sum(w)
          }
```

Figure 8.13 Poor implementation of a simple SMC.

Implementation (I), shown in Figure 8.12. The log weight operation in Line A2 is much simpler and faster than the multiplication operation in Line B3. Note that Line A2 ignores the constant $\sqrt{2\pi}$ and the common term y_t^2 in the calculation. The use of $-2x_t y_t + x_t^2$ is numerically more accurate than $(x_t - y_t)^2$. Lines A3 and A4 perform the stabilization (subtracting the maximum log weight) before the `exp` transformation.

8.10 EXAMPLES AND R IMPLEMENTATION

8.10.1 R Implementation of SMC: Generic SMC and Resampling Methods

8.10.1.1 Generic R Code for SMC Implementation In this section we introduce four generic R codes of SMC implementation. They are available in the `NTS` package. The first code is for a proposal distribution that is not the full information proposal distribution. The second code is for using the full information proposal distribution. The difference between the two is the resampling timing. For the full information proposal distribution, because the incremental weights $u_t^{(j)}$ do not depend on the sample of $x_t^{(j)}$, resampling should be done before the sampling of x_t. The third code uses R-B when full information proposal is used and R-B is possible. The fourth code is an MKF implementation using full information proposal and R-B estimation. We also include the generic R code for an

SMC smoother using the marginal weight method and MKF algorithm with full information proposal distribution when Λ_t takes a finite set of values.

```
##############
### function of generic SMC (not full information)
### with delay weighting method and estimating function funH()
### with Likelihood
##############
SMC <- function(Sstep,nobs,yy,mm,par, xx.init, xdim, ydim,
            resample.sch,delay=0,funH=identity){
#------------------------------------------------------------
   xxd <- array(dim=c(xdim,mm,delay+1))  # setting delay buffer
   delay.front = 1
   loglike=0
   xx <- xx.init
   if(xdim==1) xx=matrix(xx,nrow=1,ncol=mm)
   xxd[,,1] <- funH(xx)
   xhat <- array(dim=c(xdim,nobs,delay+1))
   logww <- rep(0,mm)
   for(i in 1:nobs){
      if(ydim==1){
         yyy <- yy[i]
      }else{
         yyy <- yy[,i]
      }
#------------ call single step propagation
      step <- Sstep(mm,xx,logww,yyy,par,xdim,ydim)
      xx <- step$xx
      logww <- step$logww-max(step$logww)
      loglike=loglike+max(step$logww)
      ww <- exp(logww)
      sumww=sum(ww)
      ww <- ww/sumww
#------------ managing delay buffer
      xxd[,,delay.front] <- funH(xx)
      delay.front <- delay.front%%(delay+1) + 1
      order <- circular2ordinal(delay.front, delay+1)
      xhat[,i,] <- tensor(xxd,ww,2,1)[,order]
#------------ resampling
      if(resample.sch[i]==1){
         r.index <- sample.int(mm,size=mm,replace=T,prob=ww)
         xx[,] <- xx[,r.index]
         xxd[,,] <- xxd[,r.index,]
         logww <- logww*0
         loglike=loglike+log(sumww)
      }
```

```
    }
    if(resample.sch[nobs]==0){
       ww <- exp(logww)
       sumww=sum(ww)
       loglike=loglike+log(sumww)
    }
#----------- rearranging delay output
    if(delay > 0){
       for(dd in 1:delay){
         xhat[,1:(nobs-dd),dd+1] <- xhat[,(dd+1):nobs,dd+1]
         xhat[,(nobs-dd+1):nobs,dd+1] <- NA
       }
    }
    return(list(xhat=xhat,loglike=loglike))
}
#------------------------end function ------------------------
```

Some explanations:

- *Sstep*: A function that performs one step propagation using a proposal distribution. Its input includes `(mm,xx,logww,yyy,par,xdim,ydim)`, where `xx` and `logww` are the last iteration samples and log weight. `yyy` is the observation at current time step. It should return `xx` (the samples x_t) and `logww` (their corresponding log weights).
- *nobs*: The number of observations T.
- *yy*: The observations, with T columns and *ydim* rows.
- *mm*: The Monte Carlo sample size m.
- *par*: A list of parameter values to pass to *Sstep*.
- *xx.init*: The initial samples of x_0.
- *xdim, ydim*: The dimension of the state variable x_t and the observation y_t.
- *resample.sch:* A binary vector of length *nobs*, reflecting the resampling schedule. resample.sch[i]= 1 indicating resample should be carried out at step i.
- *delay*: The maximum delay lag for delayed weighting estimation. Default is zero.
- *funH*: A user supplied function $h(\cdot)$ for estimating $E(h(x_t) \mid \boldsymbol{y}_{t+d})$. Default is identity($\cdot$) for estimating the mean. The function should be able to take vector or matrix as input and operates on each element of the input.

The function returns *xhat*, an array with dimensions $(xdim, nobs, delay + 1)$, and the scaled log-likelihood value *loglike*. If *loglike* is needed, the log weight calculation in the *Sstep* function should retain all constants that are related to the parameters involved. Otherwise, the *Sstep* function may remove all constants that are common to all the Monte Carlo samples. It needs a utility function *circular2ordinal*, also included in the `NTS` package, for efficient memory management. It requires the `tensor` package in R.

The following is the R function for SMC using the full information proposal. The inputs and outputs are the same as above, except the function *Sstep.Full* requires an additional binary indicator for resampling and it also returns an additional resampling indices. We skip the full R code here. It can be extracted from the `NTS` package.

```
##################
### function of generic SMC (full information) ### no R-B
### with delay weighting method and estimating function funH()
##################
SMC.Full <- function(Sstep.Full,nobs,yy,mm,par,xx.init,xdim,
      ydim, resample.sch,delay=0,funH=identity){...}
#-------------------------------------------------------------
```

The following is the R function for SMC using the full information proposal and R-B estimation of the mean only. No delay is allowed. The function *Sstep.Full.RB* should return the R-B estimate of the mean of the state at the current step.

```
################
### function of generic SMC (full information)
### with R-B estimation. No delay
################
SMC.Full.RB=function(Sstep.Full.RB,nobs,yy,mm,par,xx.init,xdim,
                    ydim,resample.sch){...}
#-------------------------------------------------------------
```

The following is a generic SMC smoother in Algorithm 8.20. It requires an additional user-specified function *Sstep.Smooth* for the backward smoothing step. The input of this additional function includes the samples $x_t^{(j)}$ and $x_{t+1}^{(j)}$ at times t and $t+1$, and their corresponding forward weight $w_t^{(j)}$ and backward weight $\tilde{u}_{t+1}^{(j)}$. Its output should be the new backward weight $\tilde{u}_t^{(j)}$ at time t. Note that resampling is only for future samples. Previous samples are not changed.

```
##################
### function of generic SMC Smoother
### with estimating function funH()
##################
SMC.Smooth=function(Sstep,Sstep.Smooth,nobs,yy,mm,par,
            xx.init,xdim,ydim,resample.sch,funH=identity){...}
#-----------------------------------------------------------------
```

The following is a generic function of an MKF implementation for discrete indicator Λ_t. It requires a function that performs one-step propagation under the MKF, with full information proposal distribution. Its input includes `(mm,II,mu,SS,logww,yyy,par,xdim,ydim)` where `II`, `mu`, `SS` are the samples of Λ_{t-1} and its corresponding mean and variance matrix of the Kalman filter components in the last iterations. `logww` is the log weight of the last iteration. It returns the R-B estimation of the mean and variance.

```
################
### function of generic MKF  (full information)
### with R-B estimation. No delay
################
MKF.Full.RB <- function(MKFstep.Full.RB,nobs,yy,mm,par,II.init,
                       mu.init,SS.init,xdim,ydim,resample.sch){...}
#-----------------------------------------------------------------
```

8.10.1.2 R Code for Resampling The following R codes implement the residual resampling and stratified resampling procedures.

```
##################
### Residual resample step
##################
resampling.res <- function(mm,ww){
   mj <- floor(mm*ww)
   index <- rep.int(1,mj[1])
   for(j in 2:mm) index <- c(index,rep.int(j,mj[j]))
   ww2 <- mm*ww-mj
   if(sum(ww2)>0){
      ww2 <- ww2/sum(ww2)
      mres <- mm-sum(mj)
      indexrex <- sample.int(mm,size=mres,replace=T,prob=ww2)
      index <- c(index,indexrex)
   }
   return(list(index=index))
}
#------------------------end function ------------------------
```

```
#################
### Stratified resample step
#################
resampling.str=function(mm,ww){
   u <- runif(1)
   cusumw <- cumsum(ww)
   mj <- floor(mm*cusumw-u)+1
   mj[mj<0] <- 0
   index <- rep(1,mj[1])
   for(j in 2:mm) index <- c(index,rep.int(j,mj[j]-mj[j-1]))
   return(list(index=index))
}
#-------------------------end function -------------------------
```

The R implementation of the function *sample.int* implements Walker's alias method (Walker, 1974; Ripley, 1987). It is optimized and much faster than the implements above, using rudimentary R codes. The following shows the CPU time of a simple trial. Clearly *sample.int* is much faster. Of course, it is possible to optimize the implementation of the residual sampling and stratified sampling methods. In all examples considered, we use *sample.int*.

```
> mm=8*10000
> ww=c(0,0,4,1,1,1,1,10)
> ww=rep(ww,10000)
> ww=ww/sum(ww)
> system.time(for(i in 1:10) resampling.res(mm,ww))
   user  system elapsed
  43.59    0.17   44.09
> system.time(for(i in 1:10) resampling.str(mm,ww))
   user  system elapsed
  70.93    0.55   71.73
> system.time(for(i in 1:10)
        sample.int(mm,size=mm,replace=T,prob=ww))
   user  system elapsed
   0.01    0.00    0.02
```

8.10.2 Tracking in a Clutter Environment

In this section we revisit the example of target tracking in a clutter environment discussed in Section 6.2.4. To simplify the demonstration, we set $\Delta T = 1$, the detection region $\Omega = [-A, A]$, and a fixed number of signals m_t. The probability that all the observed y_{ti} are noises is $1 - p_d$. It is the probability that the true signal is

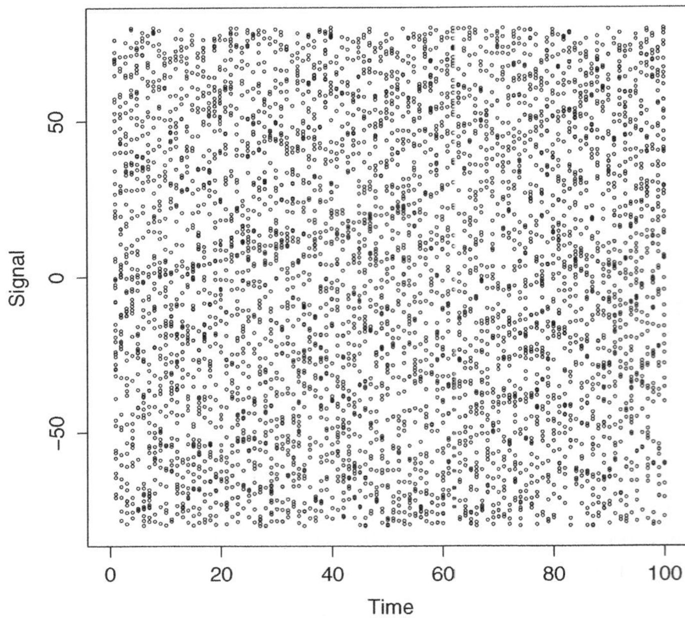

Figure 8.14 Plot of the signals observed in a clutter environment.

missing. Hence the model becomes

$$d_t = d_{t-1} + s_{t-1} + 0.5w_t, \tag{8.30}$$

$$s_t = s_{t-1} + w_t, \tag{8.31}$$

$$P(I_t = 0) = 1 - p_d \text{ and } P(I_t = i) = p_d/m_t, i = 1, \dots, m_t. \tag{8.32}$$

The observation equation is

$$y_{tI_t} = d_t + v_t, \quad \text{for } I_t \neq 0 \text{ and } y_{ti} \sim Unif(\Omega), \text{ for } i \neq I_t.$$

In simulation, we set the initial values to $d_1 = 0$ and $s_1 = 0.1$. The detection range is set to be $[-80, 80]$. $p_d = 0.95$ and $m_t = 50$. The noise variances are set as $\sigma_w^2 = 0.1^2$ and $\sigma_v^2 = 0.5^2$. Figure 8.14 shows one set of simulated data.

We consider three different proposal distributions.

1. Full state equation: Given $x_{t-1}^{(j)} = (d_{t-1}^{(j)}, s_{t-1}^{(j)}, I_{t-1}^{(j)})$, generate $x_t^{(j)} = (d_t^{(j)}, s_t^{(j)}, I_t^{(j)})$ using Equations (8.30) to (8.32). The incremental weight $u_t^{(j)} \propto p(y_t \mid x_t^{(j)})$

becomes

$$u_t^{(j)} \propto \frac{1}{\sqrt{2\pi}\sigma_v} \exp\left\{-\frac{(y_{ti}-d_t^{(j)})^2}{2\sigma_v^2}\right\} \text{ if } I_t^{(j)} = i \neq 0,$$

$$u_t^{(j)} \propto \frac{1}{|\Omega|} \text{ if } I_t^{(j)} = 0,$$

where $|\Omega|$ is the size of the monitoring region. In our simulation $|\Omega| = 80 - (-80) = 160$. Note that in both cases there is a common constant $|\Omega|^{-(m_t-1)}$, which is common to all $x_t^{(j)}$ and hence can be omitted.

2. Partial state equation: In this case we marginalize out the indicator I_t from the state equation to enhance efficiency. Given $x_{t-1}^{(j)} = (d_{t-1}^{(j)}, s_{t-1}^{(j)})$, we generate $x_t^{(j)} = (d_t^{(j)}, s_t^{(j)})$ using Equations (8.30) and (8.31). The incremental weight $u_t^{(j)} \propto p(y_t \mid x_t^{(j)}) = \sum_{i=0}^{m} p(y_t \mid x_t^{(j)}, I_t = i)p(I_t = i)$ becomes

$$u_t^{(j)} \propto \frac{1-p_d}{|\Omega|} + \frac{p_d}{m}\sum_{i=1}^{m} \frac{1}{\sqrt{2\pi}\sigma_v} \exp\left\{-\frac{(y_{ti}-d_t^{(j)})^2}{2\sigma_v^2}\right\}.$$

This weight evaluation is more expensive than that of the full state proposal, but clearly this is more efficient, since a randomly selected I_t has a small chance of being the correct one. The marginalization takes out this layer of variation.

3. Full information proposal: Given $x_{t-1}^{(j)} = (d_{t-1}^{(j)}, s_{t-1}^{(j)}, I_{t-1}^{(j)})$, generate $x_t^{(j)} = (d_t^{(j)}, s_t^{(j)}, I_t^{(j)})$ from the full information distribution that includes the observations:

$$\begin{aligned}
p(x_t \mid x_{t-1}^{(j)}, y_t) &\propto p(x_t \mid x_{t-1}^{(j)})p(y_t \mid x_t) \\
&\propto p(d_t, s_d \mid d_{t-1}^{(j)}, s_{t-1}^{(j)})p(I_t)p(y_t \mid I_t, d_t, s_d) \\
&\propto c_{tI_t}^{(j)}\phi(d_t \mid \mu_{tI_t}^{(j)}, \tau_{tI_t}^2)p(I_t)I(s_t = s_{t-1}^{(j)} + 2(d_t - d_{t-1}^{(j)} - s_{t-1}^{(j)})),
\end{aligned}$$

where $\phi(\cdot)$ is the normal density and

$$\tau_{ti}^2 = \frac{1}{\frac{4}{\sigma_w^2} + \frac{1}{\sigma_v^2}}, \quad \mu_{ti}^{(j)} = \frac{\frac{4(d_{t-1}^{(j)}+s_{t-1}^{(j)})}{\sigma_w^2} + \frac{y_{ti}}{\sigma_v^2}}{\frac{4}{\sigma_w^2} + \frac{1}{\sigma_v^2}} \quad \text{for } i \neq 0,$$

and $\tau_{t0}^2 = \sigma_w^2/4$, $\mu_{t0}^{(j)} = d_{t-1}^{(j)} + s_{t-1}^{(j)}$ for $i = 0$. The constants are

$$c_{ti}^{(j)} = \frac{\tau_{ti}}{\sqrt{\pi}\sigma_w\sigma_v} \exp\left\{ -\frac{2(d_{t-1}^{(j)} + s_{t-1}^{(j)})^2}{\sigma_w^2} - \frac{y_{ti}^2}{2\sigma_v^2} + \frac{\mu_{ti}^{(j)2}}{2\tau_{ti}^2} \right\} \text{ for } i \neq 0,$$

and $c_{t0}^{(j)} = 1$. In addition $P(I_t = i) = p_d/m$ for $i \neq 0$ and $P(I_t = 0) = (1 - p_d)/|\Omega|$.

To generate samples from $p(x_t \mid x_{t-1}^{(j)}, y_t)$, we first generate $I_t^{(j)}$ with probability

$$P(I_t = i \mid x_{t-1}^{(j)}, y_t) = \frac{p_d c_{ti}^{(j)}}{m C_t^{(j)}} \text{ for } i \neq 0 \text{ and}$$

$$P(I_t = 0 \mid x_{t-1}^{(j)}, y_t) = \frac{(1 - p_d)}{|\Omega| C_t^{(j)}},$$

where

$$C_t^{(j)} = \frac{1 - p_d}{|\Omega|} + \frac{p_d}{m} \sum_{i=1}^{m} c_{ti}^{(j)}.$$

Conditioning on $I_t^{(j)}$, we generate samples of $d_t^{(j)}$ and $s_t^{(j)}$. When $I_t^{(j)} \neq 0$, we have

$$d_t^{(j)} \sim N(\mu_{tI_t^{(j)}}, \tau_{tI_t^{(j)}}^2) \text{ and } s_t^{(j)} = s_{t-1}^{(j)} + 2(d_t^{(j)} - d_{t-1}^{(j)} - s_{t-1}^{(j)}).$$

When $I_t^{(j)} = 0$, there is no real signal and we use

$$d_t^{(j)} = d_{t-1}^{(j)} + s_{t-1}^{(j)} + 0.5\sigma_w\varepsilon_t^{(j)} \text{ and } s_t^{(j)} = s_{t-1}^{(j)} + \sigma_w\varepsilon_t^{(j)},$$

where $\varepsilon_t^{(j)} \sim N(0, 1)$. The incremental weight is

$$u_t^{(j)} \propto \int p(x_t \mid x_{t-1}^{(j)}) p(y_t \mid x_t) dx_t \propto C_t^{(j)},$$

which has already been obtained.

For full information proposal distribution, the incremental weight $u_t^{(j)} = C_t^{(j)}$ does not depend on the sampled $x_t^{(j)}$, hence resampling at time t should be done before x_t is sampled, and R-B in the form of Equation (8.22) can be used to estimate the location $\mu_t = E(x_t \mid y_t)$. Specifically, we have

$$\tilde{x}_t^{(j)} = E(x_t \mid x_{t-1}^{(j)}, y_t) = \frac{1}{C_t^{(j)}} \left(\frac{1 - p_d}{|\Omega|}(x_{t-1}^{(j)} + s_{t-1}^{(j)}) + \sum_{i=1}^{m} \mu_{ti}^{(j)} c_{ti}^{(j)} \right).$$

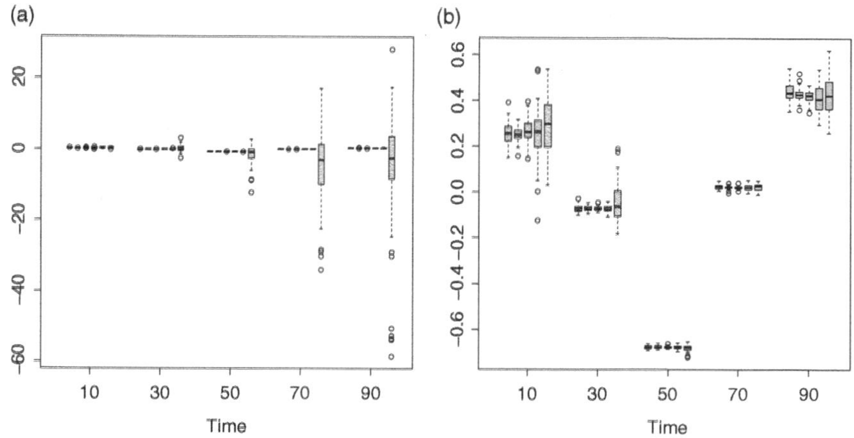

Figure 8.15 Boxplots for tracking error comparison between different resampling schedules.

Hence, the estimation of $\mu_t = E(x_t \mid \boldsymbol{y}_t)$ should be done before $x_t^{(j)}$ is sampled, but after the weight is calculated. We use

$$\tilde{\mu}_t = \frac{\sum_{j=1}^{m} \tilde{x}_t^{(j)} w_t^{(j)}}{\sum_{j=1}^{m} w_t^{(j)}}.$$

To demonstrate various properties of SMC, we first focus on the data set shown in Figure 8.14, with $\sigma_w^2 = 0.1^2$, $\sigma_v^2 = 0.5^2$, and $m_t = 50$. With Monte Carlo sample size $m = 10,000$, and using the marginal state equation as the proposal distribution, we run an SMC implementation 50 times (on the same data set) using each of the six resampling schedules sch=$(1, 2, 5, 10, 20, 100)$, where sch(1) means resample at every step and sch(100) means no resample (since the total time is 100). Using the properly weighted samples, the location d_t is estimated and compared with the true location. Figure 8.15(a) shows the boxplot of the estimation errors at times $t = (10, 30, 50, 70, 90)$. The six boxes at each time point show the errors of the six resampling schedules. The leftmost box corresponds to sch= 1. It is clear that resampling significantly improves the performance. Figure 8.15(b) is an enlarged version of Figure 8.15a and only shows the results of the first five resampling schedules. It seems that resampling once every two steps or resampling every step performs well for this particular data set. Note that the results are based on the same data set. Hence all variations shown are due to Monte Carlo variation.

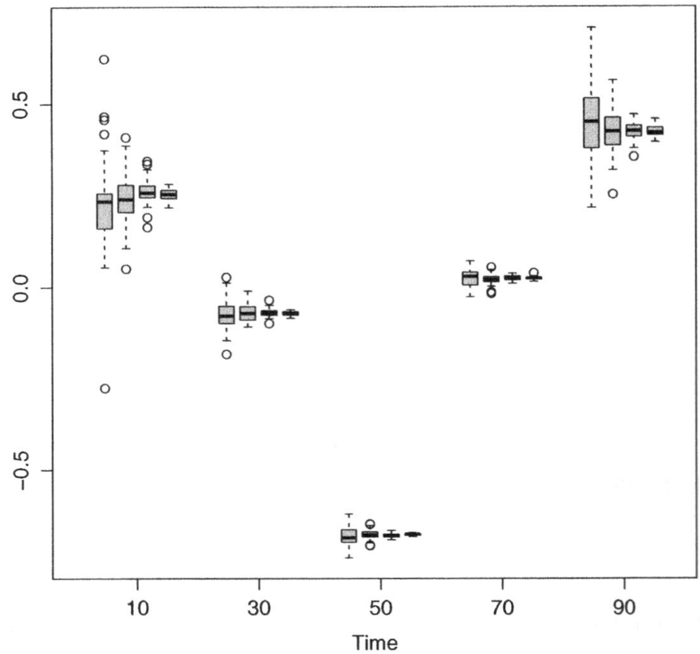

Figure 8.16 Boxplots for tracking error comparison between different Monte Carlo sample sizes.

To see the impact of the Monte Carlo sample size used in SMC implementation, using the same data and using the marginal state equation as the proposal distribution, we run an SMC implementation 50 times for each of the Monte Carlo simple sizes $m = (1000, 4000, 16000, 64000)$. Resampling is done at every step. Figure 8.16 shows the results also at times $t = (10, 30, 50, 70, 90)$. Roughly speaking, the variation reduces at the $\sqrt{m}$ rate as expected.

To see the impact of delay estimation, we use a smaller observation noise $\sigma_v^2 = 0.1^2$ and simulate the series 50 times and run an SMC implementation on each data set with the delayed weighting method using delay lags $d = 0$ to $d = 8$. Resampling is done at every step. Figure 8.17(a) shows the error distribution, where the errors are computed using the difference of the estimated location and the concurrent Kalman filter estimation with known signal indicator. Only the ones without losing tracks are shown. It is seen that for $d \geq 0$, the errors are larger. This is because the Kalman filter estimator is based on the current information $\boldsymbol{y}_t$, while the delayed estimation is based on $\boldsymbol{y}_{t+d}$. Figure 8.17(b) shows the difference between the estimated location and the true location. It can be seen that delay

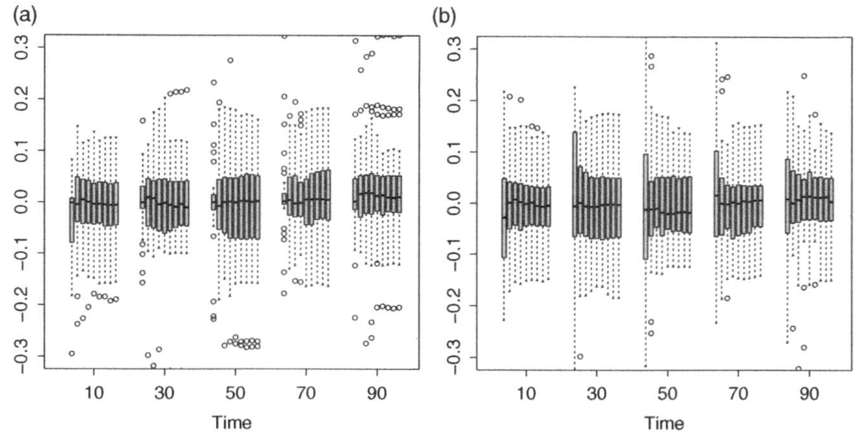

Figure 8.17 Boxplots for tracking the error comparison between the delayed estimation (using delay weighting method) using the partial state equation proposal with $m_t = 50$ and $\sigma_v^2 = 0.01$.

estimation improves the accuracy of estimating the true location, though longer delay does not help much.

To check the difference in general estimation accuracy between the partial state equation proposal and the full information proposal, we generated 50 data sets using the aforementioned parameter set. With $m = 10,000$ Monte Carlo sample size and resampling every step, the tracking errors are obtained. Figure 8.18 shows the boxplot of the differences in estimating the location. In each of the two graphs, the left-hand boxes show the errors using the partial state equation as the proposal distribution and the right-hand boxes shows that using the full information proposal distribution. From Figure 8.18(a) it is clear that the SMC filter fails to track the target in one of the simulated data sets, resulting in an outlying point. Figure 8.18(b) is an enlarged version of the boxplots by removing the extreme case. In this particular instance, the differences between the two proposal distributions are not significant.

We repeated the experiment with a different set of parameters. The detection range remains the same, but the number of total observations at each time m_t is set to 10 (instead of 50). The state equation noise level σ_v^2 is increased to 0.15^2 and the observational noise σ_v^2 is changed to 0.02^2. These changes make the identity of the true signals more important, since the signal now is sparse and with smaller measurement error. The Monte Carlo sample size is reduced to $m = 2000$. Figure 8.19 shows the difference between the location estimates of the SMC filter and that from a Kalman filter estimator with the true observation indicator I_t. Under

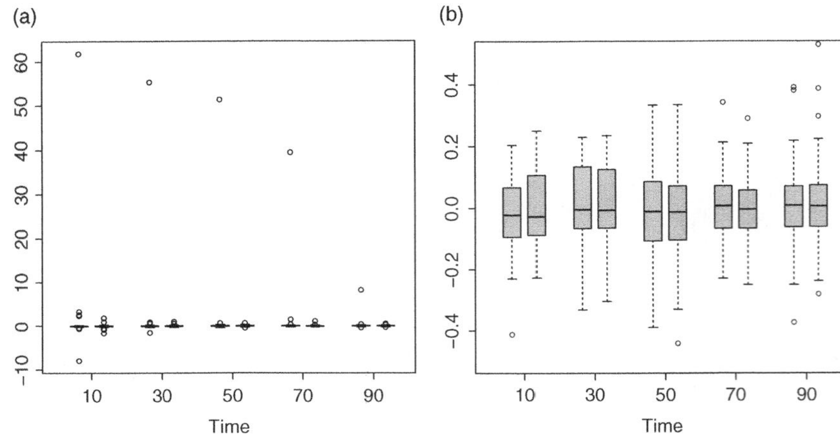

Figure 8.18 Boxplots for tracking the error comparison between the partial state equation proposal and the full information proposal, with $m_t = 50$ and $\sigma_v^2 = 0.01$.

the true observation indicators, the Kalman filter provides the optimal solution for a linear and Gaussian system. Figure 8.19(a) shows the results of the 50 simulated samples. The left, middle, and right boxes at each time point show the results using the partial state equation proposal, full information proposal, and full information proposal with R-B estimator, respectively. It is seen that in this case the SMC filter fails to track the target more frequently when the partial state equation proposal

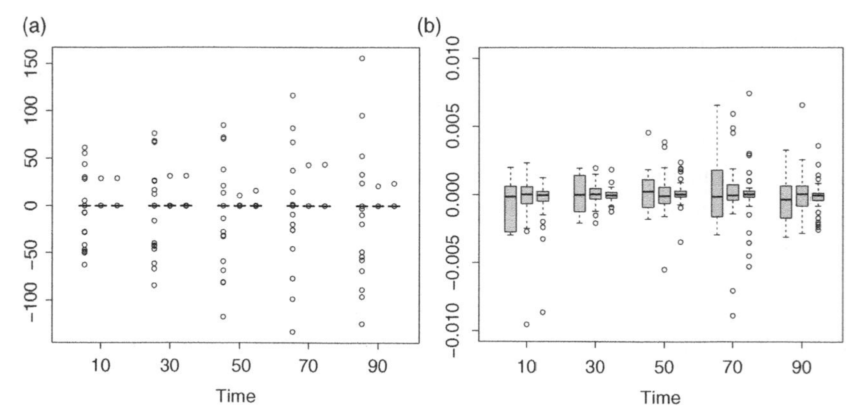

Figure 8.19 Boxplots for tracking the error comparison between the partial state equation proposal, the full information proposal, and the full information proposal with R-B. The boxplots show the differences between the SMC estimator and the Kalman filter estimator with true signal indicator I_i. The parameters used are $m_t = 10$, $\sigma_w^2 = 0.15^2$, and $\sigma_v^2 = 0.2^2$.

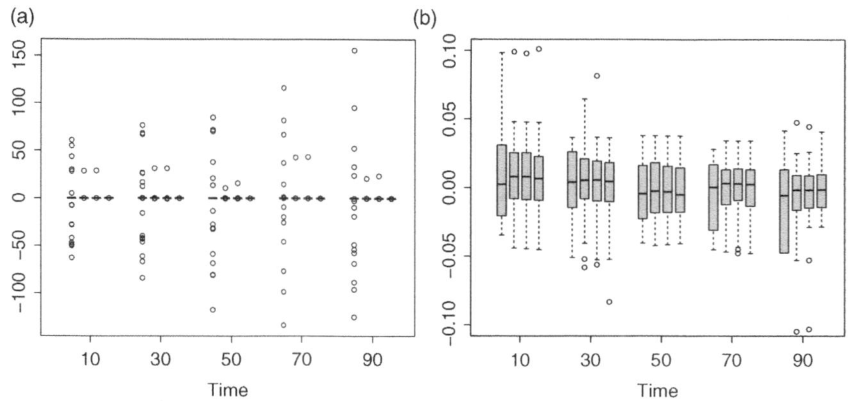

Figure 8.20 Boxplots for tracking the error comparison between the partial state equation proposal, the full information proposal, and the full information proposal with R-B. The boxplots show the difference between the SMC estimator and the Kalman filter estimator with true signal indicator I_i. The fourth boxplot shows the difference between the Kalman filter with true signal indicator I_i and the true location. The parameters used are $m_t = 10$, $\sigma_w^2 = 0.15^2$ and $\sigma_v^2 = 0.04$.

is used. The full information proposal fares much better. Figure 8.19(b) shows the tracking errors of the successful tracking data sets. It is clear now that the full information proposal improves the performance significantly. The R-B improves even further.

Figure 8.20 shows results of the same estimates as those of Figure 8.19. The errors are now calculated as the difference between the estimates and the true location. The additional fourth box shows the errors between the Kalman filter with true signal indicator and the true location. We can see that SMC with full information proposal works similarly to the Kalman filter with genie-aided true signals.

The following R functions for the propagation steps are used for the tracking in clutter problem, under partial state proposal, full information proposal, and full information with R-B estimation. We provide detailed codes for using the partial state proposal and the full information proposal, to illustrate how to construct the function as an input to the generic SMC main function. The full information propagation is more complicated not just to incorporate the y_t information, but also carry out resampling within the step before generating the new samples of $x_t^{(j)}$. We also implemented a Kalman filter estimation procedure, assuming that the true signal indicator I_t is known. The filter provides the optimal solution for the given series. Often we compare SMC estimators with this genie-aided solution.

R demonstration: Selected R code.

```
##################
### Function of propagation step: Clutter. partial state proposal.
Sstep.Clutter=function(mm,xx,logww,yyy,par,xdim,ydim){
#---------------------------------------------------------------
#------- state equation propagation
   ee <- rnorm(mm)*par$ssw
   xx[1,] <- xx[1,]+xx[2,]+0.5*ee
   xx[2,] <- xx[2,]+ee
   uu <- 1:mm
#------- for each MC sample. Vectorized on all yyy observations
  for(j in 1:mm){
   temp <- sum(exp(-(yyy-xx[1,j])**2/2/par$ssv**2))
   uu[j] <- temp/par$ssv/sqrt(2*pi)*par$pd/par$nyy+(1-par$pd)/par$yr
   }
#------- update weight
   logww <- logww+log(uu)-max(log(uu))
   return(list(xx=xx,logww=logww))
}
#-------------------------end function -------------------------
#################
### Function of Clutter Propagation step Full Information.
Sstep.Clutter.Full <- function(mm,xx,logww,yyy,par,xdim,ydim,
                                   resample.sch){
#---------------------------------------------------------------
#------- passing the parameters
   ssw2 <- par$ssw**2; ssv2 <- par$ssv**2
   nyy <- par$nyy; pd <- par$pd; yr <- par$yr
#------- setting needed temp variables
   mu.i <- matrix(nrow=nyy,ncol=mm)
   cc.i <- matrix(nrow=nyy,ncol=mm)
   CC <- 1:mm
   tau2 <- 1/(4/ssw2+1/ssv2); tau <- sqrt(tau2)
   aa <- 0.5*(-log(pi)+log(tau2)-log(ssw2)-log(ssv2))
   for(jj in 1:mm){
     mu.i[,jj] <- (4*(xx[1,jj]+xx[2,jj])/ssw2+yyy/ssv2)*tau2
     temp <- -2*(xx[1,jj]+xx[2,jj])**2/ssw2-yyy**2/2/ssv2
     cc.i[,jj] <- temp +mu.i[,jj]**2/2/tau2+aa
     cc.i[,jj] <- exp(cc.i[,jj])*pd/nyy
     CC[jj] <- sum(cc.i[,jj])+(1-pd)/yr
#------- updating weight
     logww[jj] <- logww[jj]+log(CC[jj])
   }
#-------- resampling
   if(resample.sch==1){
      logpp <- logww-max(logww)
```

```
      pp <- exp(logpp)
      pp <- pp/sum(pp)
      r.index <- sample.int(mm,size=mm,replace=T,prob=pp)
      xx[,] <- xx[,r.index]
      mu.i[,] <- mu.i[,r.index]
      cc.i[,] <- cc.i[,r.index]
      CC <- CC[r.index]
   }
#-------- generating new samples
 for(jj in 1:mm){
    ind <- sample.int(nyy+1,1,prob=c((1-pd)/yr,cc.i[,jj])/CC[jj])
    if(ind == 1){
       ee <- rnorm(1)
       xx[1,jj] <- xx[1,jj]+xx[2,jj]+0.5*ee*ssw
       xx[2,jj] <- xx[2,jj]+ee*ssw
     }
     if(ind >1){
        xx[2,jj] <- -2*xx[1,jj]-xx[2,jj]
        xx[1,jj] <- rnorm(1)*tau+mu.i[ind-1,jj]
        xx[2,jj] <- xx[2,jj]+2*xx[1,jj]
     }
   }
   return(list(xx=xx,logww=logww,r.index=r.index))
}
#----------------------end function ----------------------------
################
### Function of Clutter propagation step Full Information
### with R-B estimation of mean.
Sstep.Clutter.Full.RB=function(mm,xx,logww,yyy,par,xdim,ydim,
                                resample.sch){...}
################
 ### Function KF for tracking in clutter with known indicator I_t.
clutterKF <- function(nobs,ssw,ssv,yy,ii){...}
################
### Function of simulating target moving in clutter.
simuTargetClutter <- function(nobs,pd,ssw,ssv,xx0,ss0,nyy,
                                yrange){...}
#---------------------------------------------------------------
```

Some explanations are in order:

`Sstep.Clutter` performs one-step propagation using a partial state proposal. It requires a list in *par* that includes

```
par=list(ssw=ssw,ssv=ssv,nyy=nyy,pd=pd,yr=yr).
```

It returns the one-step propagation samples `xx` and `logww`.

`Sstep.Clutter.Full` performs one-step propagation using a full information proposal. In additional to `par`, it requires a binary indicator *resample.sch* to indicate whether a resampling is required at this step. It returns the one-step propagation samples `xx` and `logww`, and the resampling index, if a resampling is performed. Resampling is done before the current samples are drawn.

`Sstep.Clutter.Full.RB` performs one-step propagation using a full information proposal, and returns R-B estimate `xhatRB` for the mean for the current state. No delay is allowed. It also returns a non-Blackwellization estimate `xhat` for comparison.

`clutterKF` performs Kalman filter estimation with known indicator.

`simuTargetClutter` simulates the target signal under the clutter environment. A utility function `myboxplot` is also included in the NTS package.

The following R codes are used in the reported analysis.

R demonstration: Selected R code.

```
library("NTS")
#################
### simulation of one data set
#-----------------------------------------------------------------
set.seed(1)
#------ parameter
nobs <- 100; pd <- 0.95; ssw <- 0.1; ssv <- 0.5;
xx0 <- 0; ss0 <- 0.1; nyy <- 50;
yrange <- c(-80,80); xdim <- 2; ydim <- nyy;
#------ simulation
simu <- simuTargetClutter(nobs,pd,ssw,ssv,xx0,ss0,nyy,yrange)
plot(1,1,type='n',xlim=c(0,nobs+1),ylim=yrange,xlab='time',ylab='signal')
for(i in 1:nyy) points(1:nobs,simu$yy[i,],cex=0.4)
#fig_clutterSignal.pdf
##########
### KF with known indicators.
#-----------------------------------------------------------------
outKF <- clutterKF(nobs,ssw,ssv,simu$yy,simu$ii)
xhat.kf <- outKF$xhat

#########
### Check Resampling Schedule.  50 SMC runs. Same data
#-----------------------------------------------------------------
set.seed(1)
mm <- 10000
sch <- c(1,2,5,10,20,100)
nk <- 50
par <- list(ssw=ssw,ssv=ssv,nyy=nyy,pd=pd,yr=range)
xhat.r <- array(dim=c(nobs,6,nk))
for(m in 1:6){
```

```
    resample.sch <- rep.int(c(rep(0,sch[m]-1),1),nobs/sch[m])
    for(kk in 1:nk){
       xx.init <- matrix(nrow=2,ncol=mm)
       xx.init[1,] <- yrange[1]+runif(mm)*yr
       xx.init[2,] <- rep(0.1,mm)
       out <- SMC(Sstep.Clutter,nobs,simu$yy,mm,par,
              xx.init,xdim,ydim,resample.sch)
       xhat.r[,m,kk] <- out$xhat[1,,1]
    }
}
tstep <- c(10,30,50,70,90);
res.r <- array(dim=c(nobs,6,nk))
for(m in 1:6){
  for(ii in 1:nk) res.r[,m,ii] <- xhat.r[,m,ii]-simu$xx
}
myboxplot(res.r,tstep,xlab='time')          #fig_clutterR1.pdf
myboxplot(res.r[,1:5,],tstep,xlab='time')  #fig_clutterR2.pdf
##########
### Check MC sample size: resample every time step:
### 50 SMC runs. Same data
#-------------------------------------------------------------
set.seed(1)
par <- list(ssw=ssw,ssv=ssv,nyy=nyy,pd=pd,yr=yr)
resample.sch <- rep.int(1,nobs)
nk <- 50
xhat.m <- array(dim=c(nobs,4,nk))
mall <- c(1000,4000,16000,64000)
for(m in 1:4){
    for(kk in 1:nk){
       mm <- mall[m]
       xx.init <- matrix(nrow=2,ncol=mm)
       xx.init[1,] <- yrange[1]+runif(mm)*yr
       xx.init[2,] <- rep(0.1,mm)
       out <- SMC(Sstep.Clutter,nobs,simu$yy,mm,par,
                xx.init,xdim,ydim,resample.sch)
       xhat.m[,m,kk] <- out$xhat[1,,1]
    }
}
tstep <- c(10,30,50,70,90);
res.m <- array(dim=c(nobs,4,nk))
for(m in 1:4){
  for(ii in 1:nk) res.m[,m,ii] <- xhat.m[,m,ii]-simu$xx
}
myboxplot(res.m,tstep,xlab='time')     #fig_clutterM.pdf
##################
### Check Delay. resample every step. 50 SMC runs. Same data
#-------------------------------------------------------------
set.seed(1)
mm <- 10000; nk <- 50;
```

```
resample.sch <- rep.int(1,nobs)
delay <- 8
xhat.d <- array(dim=c(nobs,delay+1,nk))
for(kk in 1:nk){
   xx.init <- matrix(nrow=2,ncol=mm)
   xx.init[1,] <- yrange[1]+runif(mm)*yr
   xx.init[2,] <- rep(0.1,mm)
   out <- SMC(Sstep.Clutter,nobs,simu$yy,mm,par,
              xx.init,xdim,ydim,resample.sch,delay=delay)
   xhat.d[,,kk] <- out$xhat[1,,]
}
res1.d <- array(dim=c(nobs-8,8,nk))
res2.d <- array(dim=c(nobs-8,8,nk))
tt <- 1:(nobs-delay)
for(m in 1:delay){
  for(ii in 1:nk){
    res1.d[tt,m,ii] <- xhat.d[tt,m,ii]-xhat.kf[tt]
    res2.d[tt,m,ii] <- xhat.d[tt,m,ii]-simu$xx[tt]
  }
}
myboxplot(res1.d,tstep,xlab='time')      #fig_clutterD1.pdf
myboxplot(res2.d,tstep,xlab='time')      #fig_clutterD2.pdf
################
###---- check with different samples  nyy=50, ssv=0.1
################
nk <- 50
nobs <- 100; pd <- 0.95; ssw <- 0.1; ssv <- 0.1;
xx0 <- 0; ss0 <- 0.1; nyy <- 50; yrange <- c(-80,80);
xdim <- 2; ydim <- nyy; yr <- yrange[2]-yrange[1]
par <- list(ssw=ssw,ssv=ssv,nyy=nyy,pd=pd,yr=yr)
#--- simulation for repeated use
Yy <- array(dim=c(nyy,nobs,nk))
Xx <- matrix(ncol=nk,nrow=nobs)
Ii <- matrix(ncol=nk,nrow=nobs)
kk <- 0
seed.k <- 1
while(kk < nk){
  seed.k <- seed.k+1
  set.seed(seed.k)
  kk <- kk+1
  simu <- simuTargetClutter(nobs,pd,ssw,ssv,xx0,ss0,nyy,yrange)
  if(max(abs(simu$yy))>80){
    kk <- kk-1
  }
  else{
    Xx[,kk] <- simu$xx; Yy[,,kk] <- simu$yy; Ii[,kk] <- simu$ii
  }
}
#--- KF with known indicator
```

```
Xxhat.kf <- matrix(ncol=nk,nrow=nobs)
for(kk in 1:nk){
  outKF <- clutterKF(nobs,ssw,ssv,Yy[,,kk],Ii[,kk])
  Xxhat.kf[,kk] <- outKF$xhat
}
###---- ----- different delay
resample.sch <- rep.int(1,nobs)
delay <- 8
mm <- 10000;
Xxhat.d <- array(dim=c(nobs,delay+1,nk))
set.seed(1)
for(kk in 1:nk){
   xx.init <- matrix(nrow=2,ncol=mm)
   xx.init[1,] <- yrange[1]+runif(mm)*yr
   xx.init[2,] <- rep(0.1,mm)
   out <- SMC(Sstep.Clutter,nobs,Yy[,,kk],mm,par,
                xx.init,xdim,ydim,resample.sch,delay=delay)
   Xxhat.d[,,kk] <- out$xhat[1,,]
}
tstep <- c(10,30,50,70,90);
Xres1.d <- array(dim=c(nobs-8,9,nk))
Xres2.d <- array(dim=c(nobs-8,9,nk))
tt <- 1:(nobs-8)
for(m in 1:9){
  for(ii in 1:nk){
    Xres1.d[tt,m,ii] <- Xxhat.d[tt,m,ii]-Xxhat.kf[tt,ii]
    Xres2.d[tt,m,ii] <- Xxhat.d[tt,m,ii]-Xx[tt,ii]
  }
}
myboxplot(Xres1.d,tstep,xlab='time',ylim=c(-0.3,0.3))
#fig_clutterXD2.pdf
myboxplot(Xres2.d,tstep,xlab='time',ylim=c(-0.3,0.3))
#fig_clutterXD4.pdf

#---- compare Partial and Full ---
resample.sch <- rep.int(1,nobs)
delay <- 0; mm <- 10000;

Xxhat.full <- array(dim=c(nobs,1,nk))
Xxhat.p <- array(dim=c(nobs,1,nk))
Xres.com2 <- array(dim=c(nobs,2,nk))

set.seed(1)
for(kk in 1:nk){
   xx.init <- matrix(nrow=2,ncol=mm)
   xx.init[1,] <- yrange[1]+runif(mm)*yr
   xx.init[2,] <- rep(0.1,mm)
   out1 <- SMC(Sstep.Clutter,nobs,Yy[,,kk],mm,par,
                xx.init,xdim,ydim,resample.sch)
```

```
    Xxhat.p[,1,kk] <- out1$xhat[1,,1]
    out2 <- SMC.Full(Sstep.Clutter.Full,nobs,Yy[,,kk],mm,par,
                 xx.init,xdim,ydim,resample.sch)
    Xxhat.full[,1,kk] <- out2$xhat[1,,1]
}
tstep <- c(10,30,50,70,90);
for(ii in 1:nk){
  Xres.com2[,1,ii] <- Xxhat.p[,1,ii]-Xx[,ii]
  Xres.com2[,2,ii] <- Xxhat.full[,1,ii]-Xx[,ii]
}
myboxplot(Xres.com2,tstep,xlab='time')
#fig_clutterXCom3.pdf
myboxplot(Xres.com2,tstep,xlab='time',ylim=c(-0.5,0.5))
#fig_clutterXCom4.pdf

#################
#Check with different samples  nyy=10, ssv=0.02,ssw=0.15
#################
##--- generate samples for repeated use
nobs <- 100; pd <- 0.95; ssw <- 0.15; ssv <- 0.02;
xx0 <- 0; ss0 <- 0.1; nyy <- 10; yrange <- c(-80,80);
xdim <- 2; ydim <- nyy; nk <- 50
yr <- yrange[2]-yrange[1]
par <- list(ssw=ssw,ssv=ssv,nyy=nyy,pd=pd,yr=yr)

Yy10 <- array(dim=c(nyy,nobs,nk))
Xx10 <- matrix(ncol=nk,nrow=nobs)
Ii10 <- matrix(ncol=nk,nrow=nobs)

kk <- 0
seed.k <- 1
while(kk < nk){
  seed.k <- seed.k+1
  set.seed(seed.k)
  kk <- kk+1
  simu <- simuTargetClutter(nobs,pd,ssw,ssv,xx0,ss0,nyy,yrange)
  if(max(abs(simu$yy))>80){
    kk <- kk-1
  }
  else{
    Xx10[,kk] <- simu$xx; Yy10[,,kk] <- simu$yy; Ii10[,kk] <- simu$ii
  }
}
## KF with known indicators
Xxhat.kf10 <- matrix(ncol=nk,nrow=nobs)
for(kk in 1:nk){
  outKF <- clutterKF(nobs,ssw,ssv,Yy10[,,kk],Ii10[,kk])
  Xxhat.kf10[,kk] <- outKF$xhat
}
```

```
###---- compare Partial and Full --- mm=2000
resample.sch <- rep.int(1,nobs)
delay <- 0; mm <- 2000; nk <- 50
Xxhat.p10 <- array(dim=c(nobs,1,nk))
Xxhat.RB10 <- array(dim=c(nobs,2,nk))
Xres.com1.10 <- array(dim=c(nobs,3,nk))
Xres.com2.10 <- array(dim=c(nobs,4,nk))

set.seed(1)
for(kk in 1:nk){
   xx.init <- matrix(nrow=2,ncol=mm)
   xx.init[1,] <- yrange[1]+runif(mm)*yr
   xx.init[2,] <- rep(0.1,mm)
   out1 <- SMC(Sstep.Clutter,nobs,Yy10[,,kk],mm,par,
               xx.init,xdim,ydim,resample.sch)
   Xxhat.p10[,1,kk] <- out1$xhat[1,,1]
}

set.seed(1)
for(kk in 1:nk){
   cat('\n','iteration',kk);flush.console()
   xx.init <- matrix(nrow=2,ncol=mm)
   xx.init[1,] <- yrange[1]+runif(mm)*yr
   xx.init[2,] <- rep(0.1,mm)
   out3 <- SMC.Full.RB(Sstep.Clutter.Full.RB,nobs,Yy10[,,kk],mm,par,
               xx.init,xdim,ydim,resample.sch)
   Xxhat.RB10[,1,kk] <- out3$xhatRB[1,]
   Xxhat.RB10[,2,kk] <- out3$xhat[1,]
}

tstep <- c(10,30,50,70,90);
for(ii in 1:nk){
  Xres.com1.10[,1,ii] <- Xxhat.p10[,1,ii]-Xxhat.kf10[,ii]
  Xres.com1.10[,2,ii] <- Xxhat.RB10[,2,ii]-Xxhat.kf10[,ii]
  Xres.com1.10[,3,ii] <- Xxhat.RB10[,1,ii]-Xxhat.kf10[,ii]
  Xres.com2.10[,1,ii] <- Xxhat.p10[,1,ii]-Xx10[,ii]
  Xres.com2.10[,2,ii] <- Xxhat.RB10[,2,ii]-Xx10[,ii]
  Xres.com2.10[,3,ii] <- Xxhat.RB10[,1,ii]-Xx10[,ii]
  Xres.com2.10[,4,ii] <- Xxhat.kf10[,ii]-Xx10[,ii]
}
myboxplot(Xres.com1.10,tstep,xlab='time')
#fig_clutterXcom10_1.pdf
myboxplot(Xres.com1.10,tstep,xlab='time',ylim=c(-0.01,0.01))
#fig_clutterXcom10_2.pdf
myboxplot(Xres.com2.10,tstep,xlab='time')
#fig_clutterXcom10_3.pdf
myboxplot(Xres.com2.10,tstep,xlab='time',ylim=c(-0.1,0.1))
#fig_clutterXCom10_4.pdf
```

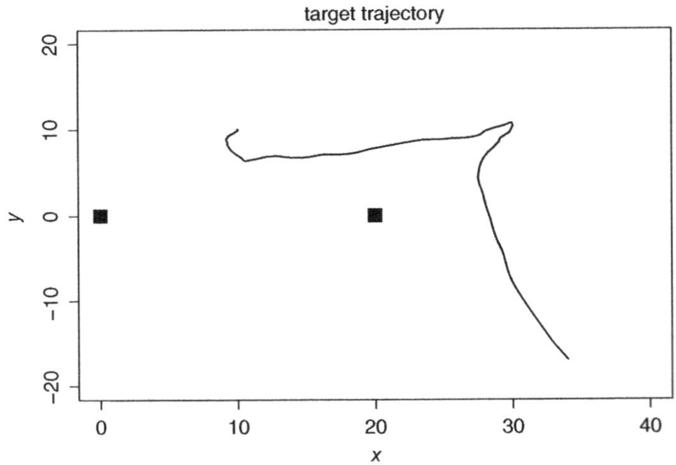

Figure 8.21 Simulated path of the bearing-only track example.

8.10.3 Bearing-only Tracking with Passive Sonar

For demonstration, we revisit the two-dimensional bearing-only tracking problem in Section 7.1.6. In simulation, we set the initial values to $x_0 = c(10, 10, 0.01, 0.01)$ The noise variances are set to $\sigma_w^2 = 0.05^2$, and $\sigma_v^2 = 0.5^2$. Figure 8.21 shows the same set of simulated data we used in Example 7.1.

The state equation can be written as

$$x_t = \boldsymbol{H} x_{t-1} + \boldsymbol{W} w_t \tag{8.33}$$

where

$$\boldsymbol{H} = \begin{bmatrix} 1 & 0 & 1 & 0 \\ 0 & 1 & 0 & 1 \\ 0 & 0 & 1 & 0 \\ 0 & 0 & 0 & 1 \end{bmatrix}, \quad \boldsymbol{W} = \begin{bmatrix} 0.5q_1 & 0 \\ 0 & 0.5q_2 \\ q_1 & 0 \\ 0 & q_2 \end{bmatrix}.$$

The observation equation is

$$\phi_{it} = \arctan\left(\frac{d_{2t} - \eta_{i2}}{d_{1t} - \eta_{i1}}\right) + e_{it} \quad \text{for} \quad i = 1, 2. \tag{8.34}$$

We consider using the state question as the proposal distribution. Specifically, given $x_{t-1}^{(j)}$, generate $x_t^{(j)}$ using (8.33). The incremental weight $u_t^{(j)} \propto p(y_t \mid x_t^{(j)}) = \sum_{i=0}^{m} p(y_t \mid x_t^{(j)})$ becomes

$$u_t^{(j)} \propto \exp\left\{-\frac{(\phi_{1t} - \hat{\phi}_{1t}^{(j)})^2 + (\phi_{2t} - \hat{\phi}_{2t}^{(j)})^2}{2\sigma_v^2}\right\}.$$

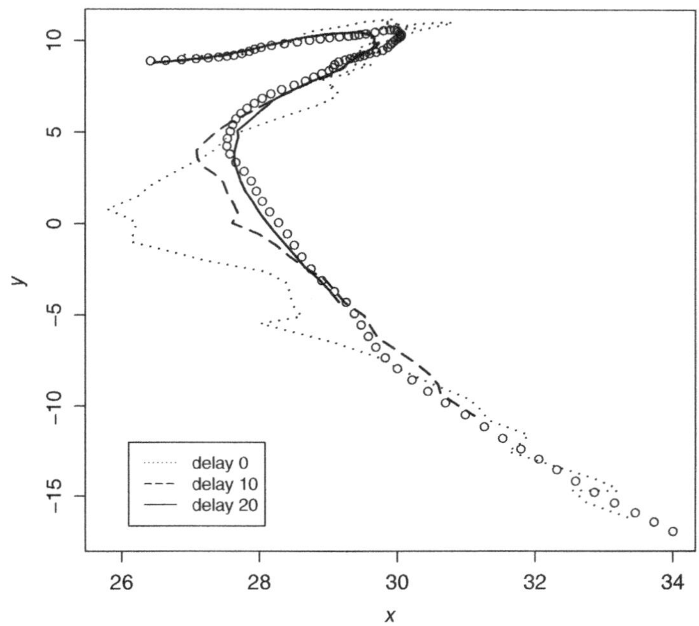

Figure 8.22 Delayed filtering results for the bearing-only track example.

Using the same data as in Example 7.1, Figure 8.22 shows the delayed estimation using the delay weighting method with delay $d = 0, 10$, and 20 and with Monte Carlo sample size $m = 10,000$. The benefit of using delayed estimation can be clearly seen. Figures 8.23 and 8.24 show the estimation error in the x and y directions. Again, in this particular case long delay is beneficial.

Figure 8.25 shows the smoothed estimation using the weight marginalization method with $T = 200$ as the end point. Due to intensive computation, $m = 5000$ Monte Carlo samples are used. We also include the EKF smoothing estimator shown in Example 7.1 as the dashed line for comparison. It is clear that the SMC method outperforms the EKF method.

The following R functions are used in the example.

```
################
#Function of single step propagation: passive sonar.
#                    proposal used: state equation
Sstep.Sonar=function(mm,xx,logww,yy,par,xdim=1,ydim=1){...}
####################
#Function of simple step smooth step: passive sonar
Sstep.Smooth.Sonar=function(mm,xxt,xxt1,ww,vv,par){...}
###################
```

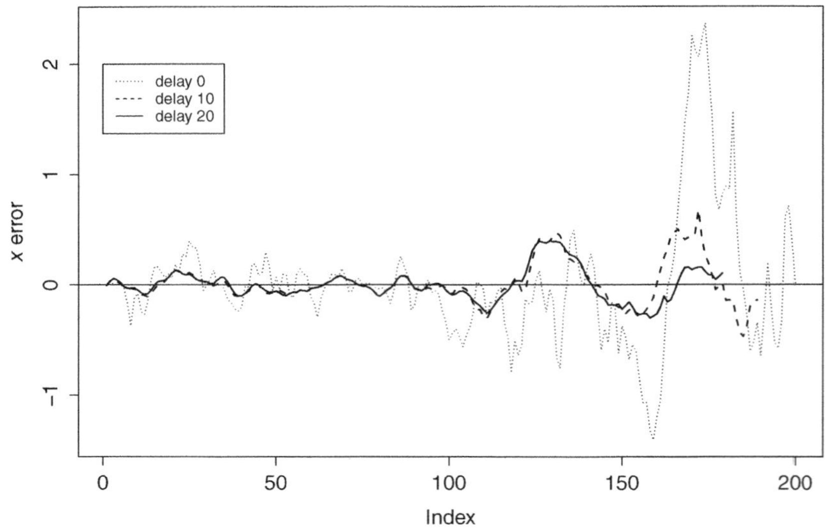

Figure 8.23 Tracking error in the x direction for the bearing-only track example.

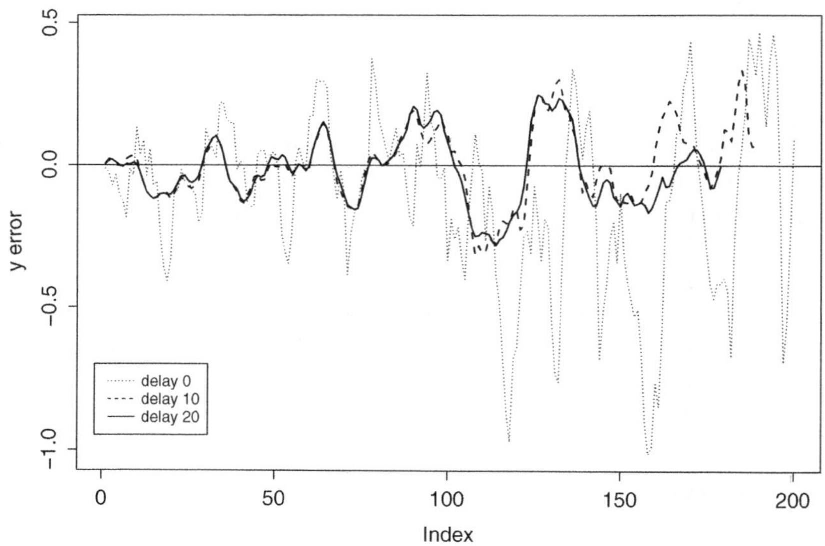

Figure 8.24 Tracking error in the y direction for the bearing-only track example.

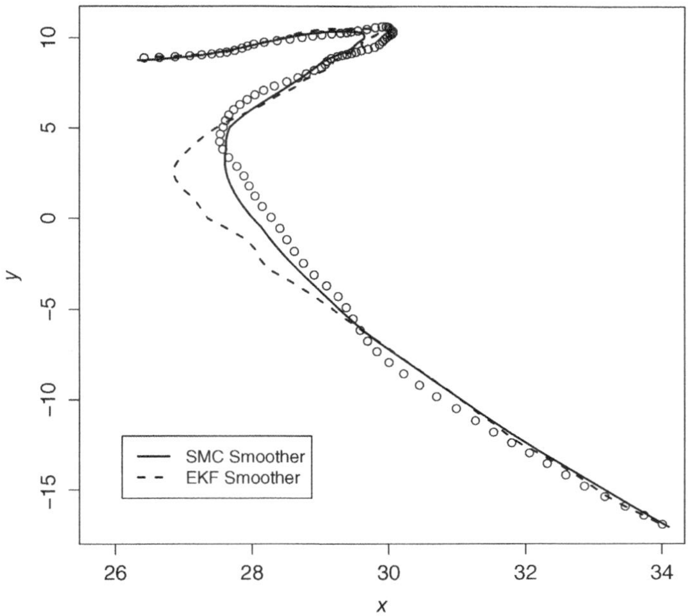

Figure 8.25 SMC smoothing and EKF smoothing for the bearing-only track example.

```
#Function to simulate the data
simPassiveSonar=function(nobs,q,r,start,seed=20){...}
#-------------------------------------------------------------
```

R demonstration: Two-dimensional passive sonar tracking.

```
#parameters
s2 <- 20  #second sonar location at (s2,0)
q <- c(0.03,0.03)
r <- c(0.02,0.02)
nobs <- 200
start <- c(10,10,0.01,0.01)
H <- c(1,0,1,0,
    0,1,0,1,
    0,0,1,0,
    0,0,0,1)
H <- matrix(H,ncol=4,nrow=4,byrow=T)
W <- c(0.5*q[1], 0,
    0, 0.5*q[2],
    q[1],0,
    0,q[2])
```

```
W <- matrix(W,ncol=2,nrow=4,byrow=T)
V <- diag(r)
mu0 <- start
SS0 <- diag(c(1,1,1,1))*0.01
#simulation
simu_out <- simPassiveSonar(nobs,q,r,start,seed=20)
yy <- simu_out$yy
tt <- 100:200
plot(simu_out$xx[1,tt],simu_out$xx[2,tt],xlab='x',ylab='y',xlim=c(25.8,34))
#--------------------------------------
#Run delayed filtering. Maximum delay 20
#--------------------------------------
mm <- 100000
set.seed(1)
resample.sch <- rep(1,nobs)
xdim <- 4;ydim <- 2;
xx.init <- mu0+SS0%*%matrix(rnorm(mm*4),nrow=4,ncol=mm)
par <- list(H=H,W=W,V=V,s2=s2)
delay <- 20
out <- SMC(Sstep.Sonar,nobs,yy,mm,par,xx.init,xdim,ydim,resample.sch,delay)

tt <- 100:200
plot(simu_out$xx[1,tt],simu_out$xx[2,tt],xlab='x',ylab='y',xlim=c(25.8,34))
for(dd in c(1,11,21)){
  tt <- 100:(200-dd)
  lines(out$xhat[1,tt,dd],out$xhat[2,tt,dd],lty=22-dd,lwd=2)
}
legend(26.1,-12,legend=c("delay 0","delay 10","delay 20"),lty=c(21,11,1))
## PassiveSonar.Fig1.pdf

plot(simu_out$xx[1,]-out$xhat[1,,1],ylab='x error',type='n')
lines(simu_out$xx[1,]-out$xhat[1,,1],lty=21,lwd=1)
lines(simu_out$xx[1,1:189]-out$xhat[1,1:189,11],lty=2,lwd=2)
lines(simu_out$xx[1,1:179]-out$xhat[1,1:179,21],lty=1,lwd=2)
abline(h=0)
legend(0,2,legend=c("delay 0","delay 10","delay 20"),lty=c(21,2,1))
## PassiveSonar.Fig2.pdf

plot(simu_out$xx[2,]-out$xhat[2,,1],ylab='y error',type='n')
lines(simu_out$xx[2,]-out$xhat[2,,1],lty=21,lwd=1)
lines(simu_out$xx[2,1:189]-out$xhat[2,1:189,11],lty=2,lwd=2)
lines(simu_out$xx[2,1:179]-out$xhat[2,1:179,21],lty=1,lwd=2)
abline(h=0)
legend(-2,-0.7,legend=c("delay 0","delay 10","delay 20"),lty=c(21,2,1))
## PassiveSonar.Fig3.pdf
#--------------------------------------
#Run smoothing
#--------------------------------------
set.seed(1)
mm <- 5000
par <- list(H=H,W=W,V=V,s2=s2)
```

```
xx.init <- mu0+SS0%*%matrix(rnorm(mm*4),nrow=4,ncol=mm)
out.s5K <- SMC.Smooth(Sstep.Sonar,Sstep.Smooth.Sonar,nobs,yy,mm,par,
xx.init,xdim,ydim,resample.sch)
tt <- 100:200
plot(simu_out$xx[1,tt],simu_out$xx[2,tt],xlab='x',ylab='y',xlim=c(25.8,34))
lines(out.s5K$xhat[1,tt],out.s5K$xhat[2,tt],lty=1,lwd=2)
##-- from Example 6.4.1  -- EKF smoothing results
lines(muTall[1,tt],muTall[2,tt],lty=2,lwd=2)
legend(26.1,-12,legend=c("SMC Smoother","EKF Smoother"),lty=c(1,2),lwd=2)
## PassiveSonar.Fig4.pdf
```

8.10.4 Stochastic Volatility Models

In Section 6.2.9 we discussed the stochastic volatility model of asset returns in the form

$$x_t = \phi_0 + \phi_1 x_{t-1} + \sigma\eta_t, \quad \eta_t \sim N(0,1)$$
$$y_t = \exp(x_t)v_t, \quad v_t \sim N(0,1),$$

where y_t is the observed asset return and x_t is the latent log volatility of y_t. This is an alternative approach to the volatility modeling discussed in Section 5.4.

In this section we demonstrate that SMC can be used in volatility modeling. We use the first 250 observations (in 2007) of daily log returns of the SPY index used in Example 5.3. Figure 8.26(a) shows the time plot of the series. Using the state equation as the proposal distribution $g_t(x_t) = p(x_t \mid x_{t-1})$ and Monte Carlo sample size $m = 20,000$, resampling once every five time steps, we estimated the log likelihood function of a set of parameters on a grid of $20 \times 20 \times 20$ in the range of $\mu \in (-5, -4.4)$, $\phi \in (0.9, 1)$, $\sigma \in (0.2, 0.5)$. The maximum is located at $\hat{\mu} = -4.779$, $\hat{\phi} = 0.9421$ and $\hat{\sigma} = 0.4053$. Figure 8.27 shows the estimated log likelihood value by changing one of the parameters while fixing the other two at their maximum estimates. We also perform an iterative grid search, starting on a coarse grid (10 grids on each dimension), finding the maximum, then setting a new set of $10 \times 10 \times 10$ grids around the maximum and reducing the grid size by a half. This procedure results in $\hat{\mu} = -4.7667$, $\hat{\phi} = 0.9500$, and $\hat{\sigma} = 0.3958$.

Using this set of parameters, we run SMC to obtain the filtered series and the smoothed series, shown in Figure 8.26(b) and (c), respectively.

We also checked local variability of the likelihood function. Without any resampling, we obtained the estimated log likelihood function using the same random seed for the SMC evaluation on all parameter grids, as well as using different random seeds. Figure 8.28 shows the results. Clearly, using the same seed produces a smoother log likelihood function estimator. Note that resampling would induce extra variability and, hence, the resulting estimator may not be smooth.

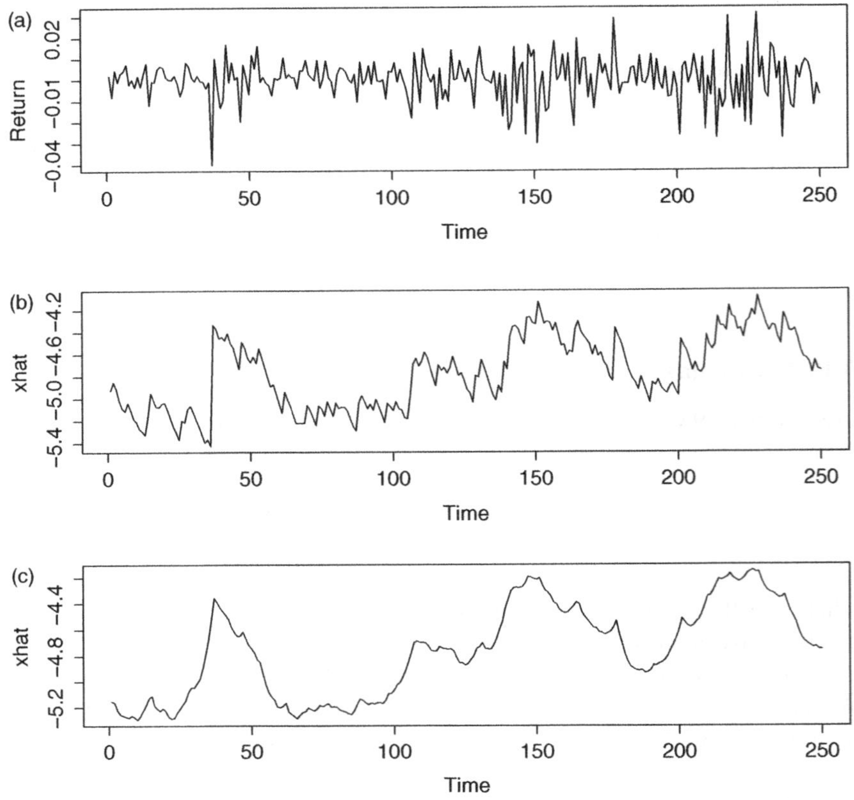

Figure 8.26 (a) Top panel: daily log returns of the SPDR S&P 500 ETF index (SPY) in 2007. (b) Middle panel: filtered estimates of stochastic volatility. (c) Bottom panel: smoothed estimates of stochastic volatility.

The following R functions are used for the stochastic volatility model. We used the `parallel` package for evaluating the likelihood function on a grid. It only works on linux/unix-based systems.

```
#######
#Function of single step propagation using state equation:  SV
Sstep.SV=function(mm,xx,logww,yyy,par,xdim,dim){...}
#######
#Function of SMC implementation for parallel processing
wrap.SMC=function(par.natural,setseed=T,resample=T){...}
#---------------------------------------------------------------
```

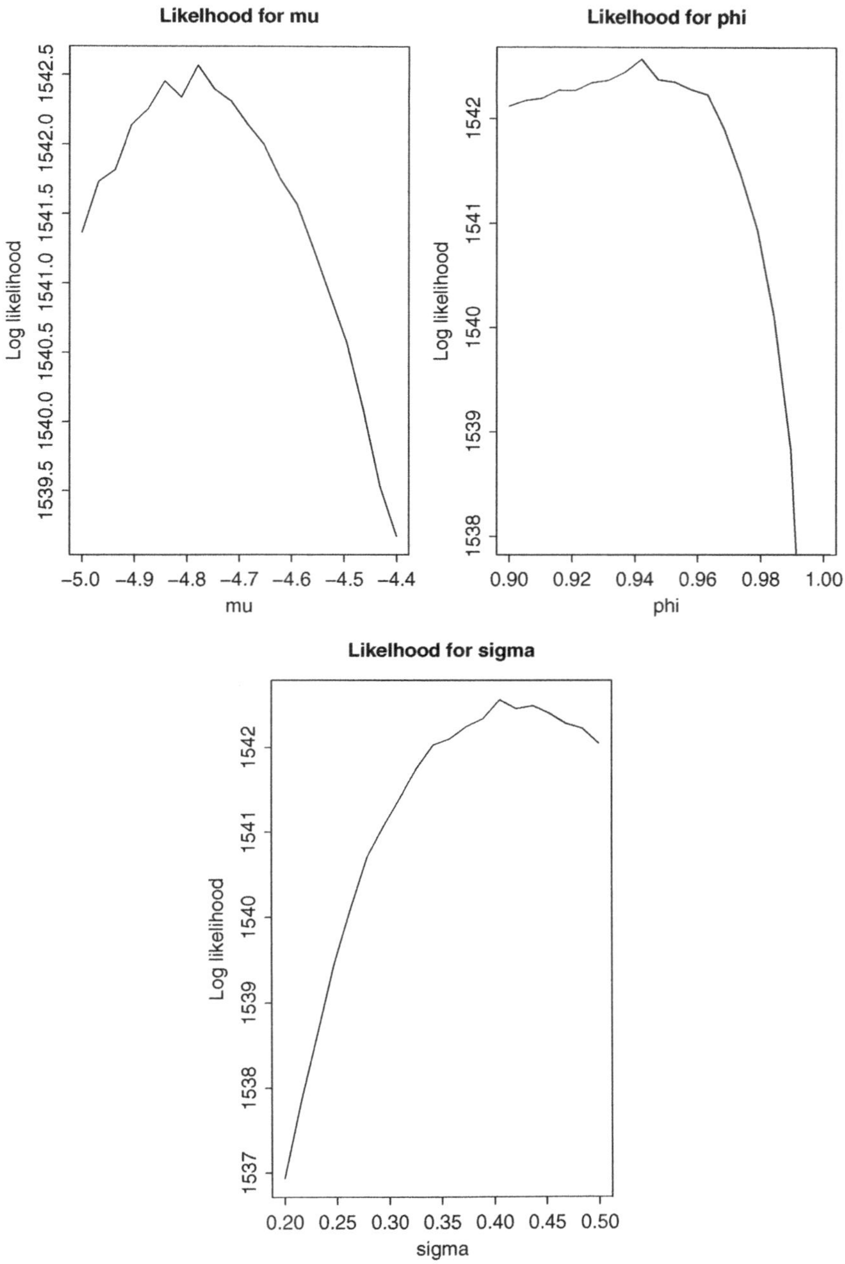

Figure 8.27 Estimated log likelihood functions on a grid, centered at the maximum of each parameter from a grid search.

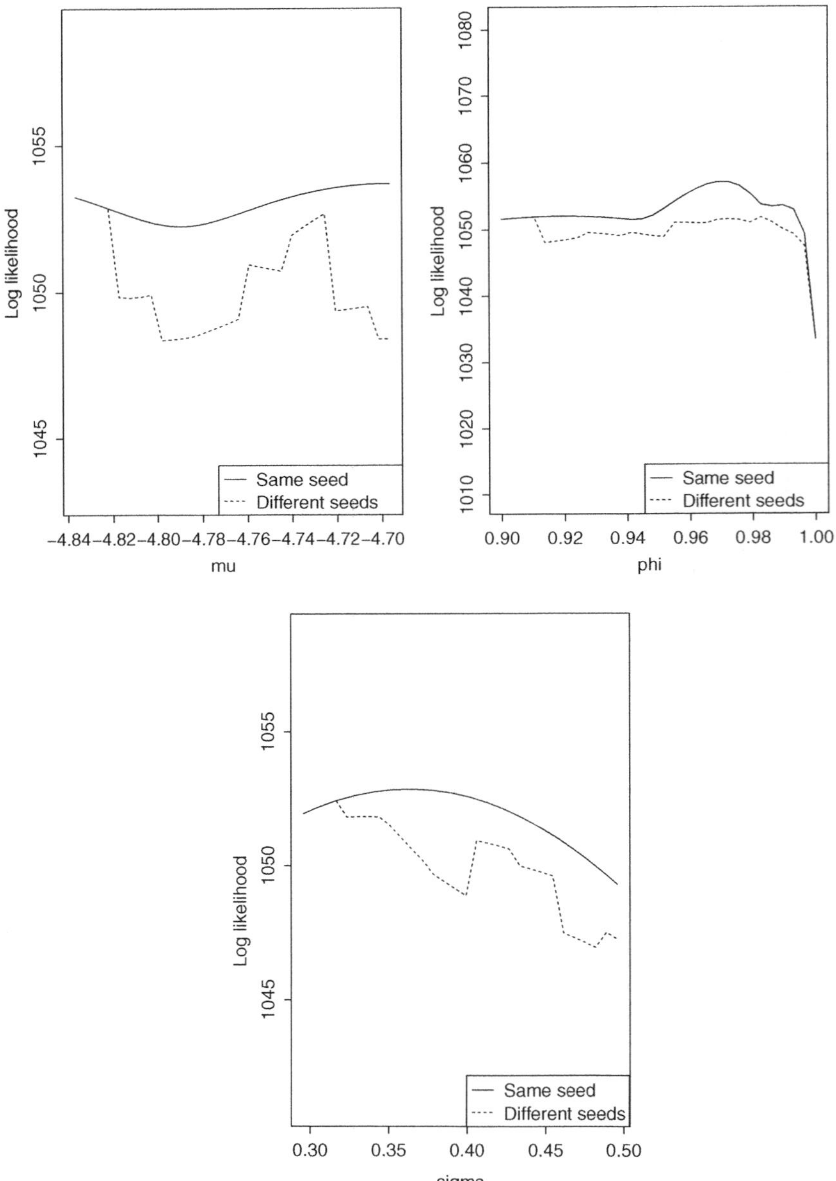

Figure 8.28 Comparison of the estimated log likelihood functions on a grid, using the same seed or different seeds. No resampling in the SMC implementation.

R demonstration: Stochastic volatility via SMC.

```
source("Rfunction.R")
library(tensor)
## Read in data
y <- matrix(scan("SPY0717.dat"),ncol=2,byrow=T)
yy <- y[,1]
yy <- yy-mean(yy)
yy <- yy[1:250]
## Define search grid and prepare for parallel run
##
range.mu <- seq(-5, -4.4, length=20)
range.phi <- seq(0.9, 1.0, length=20)
range.sigma <- seq(0.2, 0.5, length=20)
range <- list(range.mu, range.phi, range.sigma)
nobs <- length(yy)
mm <- 20000

grid <- expand.grid(range.mu, range.phi, range.sigma)
names(grid) <- c("mu", "phi", "sigma")
list.par <- as.list(data.frame(t(grid)))
## Run grid search parallel
###
library(parallel)
RNGkind("L'Ecuyer-CMRG")
### Use same seed for all computation
set.seed(1)
lll <- mcmapply(wrap.SMC, list.par,
              MoreArgs=list(setseed=T),
              mc.cores=detectCores(), mc.set.seed=T)
lll.array <- array(lll, dim=c(20,20,20))
lll.max <- max(lll.array)
max.index <- which(lll.array==lll.max, arr.ind=T)
### Plot
pdf('Loglike_grid.pdf', width=12, height=6)
par(mfrow=c(1,3))
plot(lll.array[,max.index[2],max.index[3]]~range.mu,
        xlab="mu", ylab="log likelihood",
        main="likelihood for mu", type='l',cex.lab=1.2)
plot(lll.array[max.index[1],,max.index[3]]~range.phi,
        xlab="phi", ylab="log likelihood",
        main="likelihood for phi", type='l',,cex.lab=1.2,
        ylim=c(1538,1542.5))
plot(lll.array[max.index[1],max.index[2],]~range.sigma,
        xlab="sigma", ylab="log likelihood",
        main="likelihood for sigma", type='l',,cex.lab=1.2)
```

```
dev.off()
## MLE: iterative grid search
##
### Define search box
LU.mu <- c(-5,-4.4)
LU.phi <- c(0.7,1.0)
LU.sigma <- c(0.2,0.5)
LU <- list(LU.mu, LU.phi, LU.sigma)
### Initial point
par.current <- c(-4.7, 0.9, 0.35)
grid.current <- matrix(rep(par.current, 10), ncol=3, byrow=T)
names(grid.current) <- c('mu','phi','sigma')
### Accuracy
epsilon <- c(0.01, 0.005, 0.005)
within.epsilon <- 0

while(within.epsilon<3){
   within.epsilon <- 0
   for(i in 1:3){
      sequence <- seq(LU[[i]][1], LU[[i]][2], length=10)
      grid.current[,i] <- sequence
      set.seed(1)
      lll.tmp <- mcmapply(wrap.SMC,
              as.list(data.frame(t(grid.current))),
               MoreArgs=list(setseed=T),
               mc.cores=detectCores(), mc.set.seed=T)
      par.current[i] <- sequence[which.max(lll.tmp)]
      grid.current[,i] <- par.current[i]
      if (LU[[i]][2]-LU[[i]][1] <= epsilon[i]){
          within.epsilon <- within.epsilon + 1
      }
      else {
           # Shrink the interval by half
           LU[[i]] <- 0.5*LU[[i]] + 0.5*par.current[i]
      }
   }
}
mle <- par.current
lll.mle <- vector('list', 3)
## Local variability around MLE without resampling
###
n.grid <- 30
range.local.mu <- seq(mle[1]-0.07, mle[1]+0.07, length=n.grid)
range.local.phi <- seq(mle[2]-0.05, mle[2]+0.05, length=n.grid)
range.local.sigma <- seq(mle[3]-0.1, mle[3]+0.1, length=n.grid)
range.local <- list(range.local.mu, range.local.phi,
                       range.local.sigma)
```

```
lll.local <- vector('list', 3)

for(i in 1:3){
   grid.tmp <- matrix(rep(mle, n.grid), ncol=3, byrow=T)
   grid.tmp[,i] <- range.local[[i]]
        ### Use same seed for all computation
   set.seed(1)
   lll.tmp <- mcmapply(wrap.SMC,
                   as.list(data.frame(t(grid.tmp))),
                   MoreArgs=list(setseed=T, resample=F),
                   mc.cores=detectCores(), mc.set.seed=T)
       ### Use different seeds for all computation
   set.seed(1)
   lll2.tmp <- mcmapply(wrap.SMC,
                  as.list(data.frame(t(grid.tmp))),
                  MoreArgs=list(setseed=F, resample=F),
                  mc.cores=detectCores(), mc.set.seed=F)
   lll.local[[i]] <- list(lll.tmp, lll2.tmp)
}
### Plot
pdf('Variability.pdf', width=12, height=6)
par(mfrow=c(1,3))
for(i in 1:3){
   lll.tmp <- lll.local[[i]][[1]]
   lll2.tmp <- lll.local[[i]][[2]]
   m <- min(lll.tmp,lll2.tmp)
   M <- max(lll.tmp,lll2.tmp)
   plot(range.local[[i]],lll.tmp,
        xlab=c("mu","phi","sigma")[i], ylab="log likelihood",
        type='l', ylim=c(2*m-M, 2*M-m), cex.lab=1.2)
   lines(range.local[[i]], lll2.tmp, lty=2)
        legend('bottomright', lty=c(1,2),
        legend=c('Same seed','Different Seeds'))
}
dev.off()
## Posterior Mean, Smoothing and filtering
###
ww <- exp(lll-max(lll))
ww <- ww/sum(ww)
###Posterior Mean
par.hat <- ww%*%as.matrix(grid)
###Filtering and Smoothing
par <- natural2ar(par.hat)
resample.sch <- rep(c(0,0,0,0,1), length=nobs)
set.seed(1)
xx.init <- par.hat[1] + par.hat[3]*rnorm(mm)
```

```
out <- SMC(Sstep.SV, nobs, yy, mm, par,
                xx.init, 1, 1, resample.sch, delay=nobs-1)
out.filtering <- out$xhat[1,,1]
out.smoothing <- diag(out$xhat[1,, nobs:1])
### Plot
pdf('Filtering.pdf', width=10, height=8)
par(mfrow=c(3,1))
par(mar=c(0,6,4,1))
plot(yy, xaxt='n', ylab='Y', type='l', main='Filtering')
par(mar=c(6,6,0,1))
plot(out.filtering,xlab='Time',ylab='xhat',type='l')
plot(out.smoothing, xlab='Time', ylab='xhat', type='l')
dev.off()
```

8.10.5 Fading Channels as Conditional Dynamic Linear Models

In Section 6.2.5 we discussed using SSM in modeling and solving digital communication problems. In particular, we discussed recovering signals from a system with Rayleigh flat-fading channels. Here we demonstrate the use of conditional dynamic linear models and the MKF of Section 8.6 for the recovery of signals via the Butterworth filter of order $r = 3$. Specifically, we consider the model used in Chen et al. (2000), namely

$$\begin{aligned}&\alpha_t - 2.37409\alpha_{t-1} + 1.92936\alpha_{t-2} - 0.53208\alpha_{t-3}\\&\quad = 0.01(0.894909u_t + 2.58227u_{t-1} + 2.58227u_{t-2} + 0.894909u_{t-3}),\end{aligned}$$

and rewrite the model in the form of

$$\begin{aligned}\text{State equations:}&\quad \begin{cases} x_t = \boldsymbol{H}x_{t-1} + \boldsymbol{W}w_t,\\ d_t \sim \mathit{Bernoulli}(\{-1,1\},0.5),\end{cases}\\ \text{Observation equation:}&\quad y_t = \boldsymbol{G}_{d_t}x_t + \boldsymbol{V}v_t,\end{aligned}$$

where

$$\boldsymbol{H} = \begin{bmatrix} 2.37409 & -1.92936 & 0.53208 & 0\\ 1 & 0 & 0 & 0\\ 0 & 1 & 0 & 0\\ 0 & 0 & 1 & 0\end{bmatrix},\quad \boldsymbol{W} = \begin{bmatrix}1\\0\\0\\0\end{bmatrix},$$

and

$$\begin{aligned}\boldsymbol{G}_1 &= 10^{-2} \times [0.894909, 2.58227, 2.58227, 0.894909],\\ \boldsymbol{G}_{-1} &= -10^{-2}[0.894909, 2.58227, 2.58227, 0.894909].\end{aligned}$$

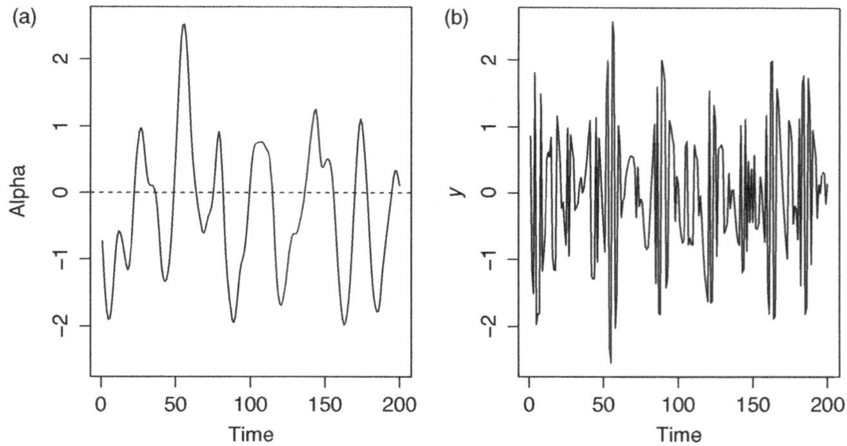

Figure 8.29 (a) Underlying changing channel coefficient α_t and (b) received signal y_t of a simulated sample of a fading channel.

In communication y_t is often a complex variable. For simplicity, here we consider the case that y_t only takes real values. The resulting error rate we encounter is higher than that in Chen et al. (2000), since they essentially observed two independent series of y_t (the real part and the imaginary part of the variable). In addition, we only consider the case of independent signals and simple Gaussian state and observational noises.

In our study x_t is a four-dimensional state variable, $d_t \in \{-1, 1\}$ is the differential encoded signals, and we are interested in estimating the information bit $s_t = d_t d_{t-1}$ (see Section 6.2.5 for details). In addition, $w_t \sim N(0, 1)$ and $v_t \sim N(0, \sigma_v^2)$. We will vary σ_v to control the signal to noise ratio.

Figure 8.29(a) shows the underlying channel coefficient $\alpha_t = \boldsymbol{G}x_t$. The absolute values of the roots of the AR polynomial $\phi(z) = 1 - \phi_1 z - \phi_2 z^2 - \phi_3 z^3$ are 1.149, 1.149, and 1.42. They are close to the unit circle and, hence, the series is slow moving. The long-run frequency for the α_t to change its sign is 0.0732 in the stationary case. This is the lower error bound for differential decoder estimator $\hat{s}_t = \widehat{d_t d_{t-1}} = sign(y_t y_{t-1})$ as it assumes that α_t does not change sign. Figure 8.29(b) shows the observed y_t, with the signal-to-noise ratio around 25.8 dB, defined as $10 \ln_{10}(var(\alpha_t)/\sigma_v^2)$.

We used an MKF with R-B with full information proposal to estimate the differential code $s_t = d_t d_{t-1}$. In all cases, we used $m = 100$ Monte Carlo samples. The performance improvement of using a larger m is small. Only the results of concurrent estimation (i.e., estimation s_t based on $\boldsymbol{y}_t$ with no delay) are obtained, as an illustration. Resampling is done at every step.

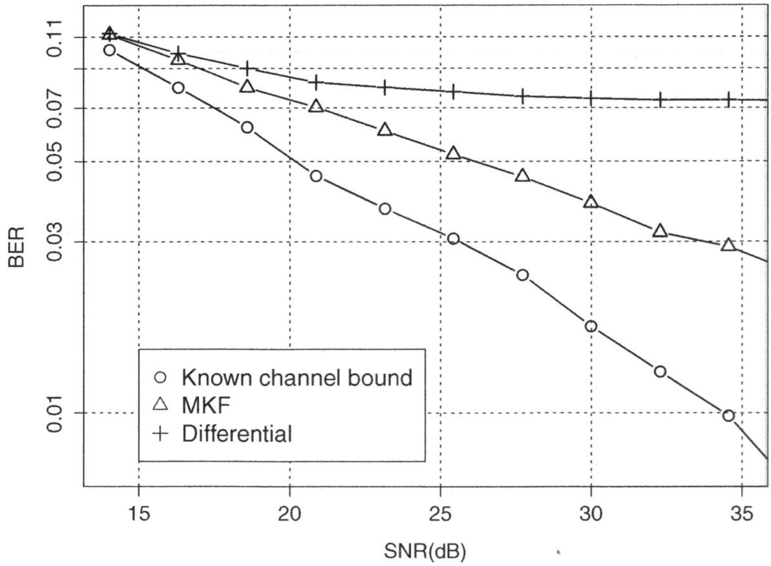

Figure 8.30 Bit error rates of the signal recovery of a fading channel.

Figure 8.30 shows the bit error rates of the MKF estimator under different signal-to-noise ratios. The performance of the simple differential decoder and the known channel error bound are also shown. The known channel error bound is obtained by using the true channel coefficient α_t and the estimator $\hat{s}_t = sign(y_t\alpha_t)sign(y_{t-1}\alpha_{t-1})$. It can be seen that the MKF estimator is able to break the error floor of the differential decoder.

To see the benefit of treating the system as a conditional dynamic linear model and the use of MKF, we compare it with the standard SMC approach. Ignoring the conditional dynamic linear model structure, the state variable is now (x_t, d_t). To use the generic SMC implementation, we also include the information bit s_t as part of the state variable (for estimation purposes only). We use the partial state equation of x_t as the proposal distribution, but use the observation to determine the signal d_t, as we did in the tracking in the clutter environment of Section 8.10.2. Specifically, at time t, the samples of $x_t^{(j)}$ are generated using

$$x_t^{(j)} = \boldsymbol{H}x_{t-1}^{(j)} + \boldsymbol{W}w_t, \quad w_t \sim N(0, 1),$$

and

$$P(d_t^{(j)} = 1) = \frac{\exp\{-0.5(y_t - \boldsymbol{G}x_t^{(j)})^2/\sigma_v^2\}}{\exp\{-0.5(y_t - \boldsymbol{G}x_t^{(j)})^2/\sigma_v^2\} + \exp\{-0.5(y_t + \boldsymbol{G}x_t^{(j)})^2/\sigma_v^2\}}.$$

MKF						
m	5	10	20	50	100	300
BER	0.0499	0.0486	0.0475	0.0475	0.0473	0.0475
CPU time	11.75	22.78	44.80	110.78	220.92	653.84
SMC						
m	100	200	500	1000	4000	8000
BER	0.0746	0.0578	0.0493	0.0515	0.0479	0.0480
CPU time	4.43	6.85	13.95	26.22	111.22	225.33

Table 8.2 Bit error rate (BER) and processing time comparison between MKF and standard SMC for fading channels.

The incremental weight is

$$u_t^{(j)} = \exp\{-0.5(y_t - Gx_t^{(j)})^2/\sigma_v^2\} + \exp\{-0.5(y_t + Gx_t^{(j)})^2/\sigma_v^2\}.$$

Care needs to be taken to avoid underflow in the weight calculation in this case, especially when σ_v^2 is small.

We estimate the differential signal s_t by

$$\hat{s}_t = \frac{\sum_{j=1}^m d_t^{(j)} d_{t-1}^{(j)} w_t^{(j)}}{\sum_{j=1}^m w_t^{(j)}}.$$

For the case of SNR= 25.4dB, using the same data set of length 10,050 (and detecting the last 10,000 signals), Table 8.2 shows the bit error rates of using the two algorithms with different Monte Carlo sample sizes.

We treat the results of using MKF with $m = 300$ as the benchmark. It is seen that MKF is able to obtain good results using only $m = 20$ samples (mixture of 20 Kalman filters). To achieve the same result, the standard SMC would need to use $m = 4000$ samples, doubling the processing time.

The following R functions are needed for the demonstration.

```
##############
### function of Mixture Kalman Filter step for fading channels
##############
MKFstep.fading=function(mm,II,mu,SS,logww,yyy,par,xdim,ydim,resample){
    HH <- par$HH; WW <- par$WW;
    GG1 <- par$GG1; GG2 <- par$GG2; VV <- par$VV;
    prob1 <- 1:mm; prob2 <- 1:mm; CC <- 1:mm
    mu.i <- array(dim=c(xdim,2,mm)); SS.i <- array(dim=c(xdim,xdim,2,mm));
    xxRB <- matrix(nrow=xdim,ncol=mm)
    IpRB <- 1:mm
    for(jj in 1:mm){
```

```
      out1 <- KFoneLike(mu[,jj],SS[,,jj],yyy,HH,GG1,WW,VV)
      out2 <- KFoneLike(mu[,jj],SS[,,jj],yyy,HH,GG2,WW,VV)
      prob1[jj] <- exp(-out1$like)
      prob2[jj] <- exp(-out2$like)
      CC[jj] <- prob1[jj]+prob2[jj]
      mu.i[,1,jj] <- out1$mu
      mu.i[,2,jj] <- out2$mu
      SS.i[,,1,jj] <- out1$SS
      SS.i[,,2,jj] <- out2$SS
      xxRB[,jj] <- (prob1[jj]*mu.i[,1,jj]+prob2[jj]*mu.i[,2,jj])/CC[jj]
      IpRB[jj] <- prob1[jj]/CC[jj]*(2-II[jj])+(1-prob1[jj]/CC[jj])*(II[jj]-1)
      logww[jj] <- logww[jj]+log(CC[jj])
    }
    ww <- exp(logww-max(logww))
    ww <- ww/sum(ww)
    xhatRB <- sum(xxRB*ww)
    IphatRB <- sum(IpRB*ww)
    if(resample==1){
      r.index <- sample.int(mm,size=mm,replace=T,prob=ww)
      mu.i[,,] <- mu.i[,,r.index]
      SS.i[,,,] <- SS.i[,,,r.index]
      prob1 <- prob1[r.index]
      CC <- CC[r.index]
      II <- II[r.index]
      logww <- logww*0
    }
    II.new <- (runif(mm) > prob1/CC)+1
    for(jj in 1:mm){
      mu[,jj] <- mu.i[,II.new[jj],jj]
      SS[,,jj] <- SS.i[,,II.new[jj],jj]
    }
    xhat <- sum(mu*ww)
    Iphat <- sum(((2-II)*(2-II.new)+(II-1)*(II.new-1))*ww)
    return(list(mu=mu,SS=SS,II=II.new,logww=logww,xhat=xhat,
                xhatRB=xhatRB,Iphat=Iphat,IphatRB=IphatRB))
}

#################
### function of SIS step: Fading
#################
SISstep.fading=function(mm,xx,logww,yyy,par,xdim2,ydim){
    HH <- par$HH; WW <- par$WW; VV <- par$VV; GG <- par$GG
    xxx <- xx[1:4,]
    xxx <- HH%*%xxx+WW%*%matrix(rnorm(mm),nrow=1,ncol=mm)
    alpha <- GG%*%xxx
    pp1 <- 1/(1+exp(-2*yyy*alpha/VV**2))
    temp1 <- -(yyy-alpha)**2/2/VV**2
    temp2 <- -(yyy+alpha)**2/2/VV**2
    temp.max <- apply(cbind(temp1,temp2),1,max)
    temp1 <- temp1-temp.max
    temp2 <- temp2-temp.max
    loguu <- temp.max+log(exp(temp1)+exp(temp2))
```

```
    logww <- logww+loguu
    logww <- as.vector(logww)-max(logww)
    II.new <- (runif(mm) > pp1)+1
    xx[6,] <- (2-xx[5,])*(2-II.new)+(xx[5,]-1)*(II.new-1)
    xx[5,] <- II.new
    xx[1:4,] <- xxx
    return(list(xx=xx,logww=logww))
}
```

R demonstration: Fading channel example.

```
### Fading channel analysis
HH <- matrix(c(2.37409, -1.92936, 0.53028,0,
            1,0,0,0,
            0,1,0,0,
            0,0,1,0),ncol=4,byrow=T)
WW <- matrix(c(1,0,0,0),nrow=4)
GG <- matrix(0.01*c(0.89409,2.68227,2.68227,0.89409),nrow=1)
VV <- 1.3**15*0.001
par <- list(HH=HH,WW=WW,GG=GG,VV=VV)
set.seed(1)
par(mfrow=c(1,2))
simu <- simu_fading(200,par)
plot(simu$alpha,type='l',ylim=range(simu$yy),ylab='alpha',xlab='time')
abline(h=0,lty=2)
plot(simu$yy,type='l',ylab='y',xlab='time')
#fig_fading0

#------- get SNR and root of AR polynomial
SNR0=10*log(var(simu$alpha)/VV**2)/log(10)
print(SNR0)
abs(polyroot(c(1,-2.37409, +1.92936, -0.53028)))

#------- MKF setup
GG1 <- GG; GG2 <- -GG
xdim <- 4; ydim <- 1
nobs <- 10050
mm <- 100; resample.sch <- rep(1,nobs)

kk <- 20
SNR <- 1:kk; BER1 <- 1:kk; BER0 <- 1:kk; BER.known <- 1:kk
tt <- 50:(nobs-1)
for(i in 1:kk){
   VV <- 1.3**i*0.001
   par <- list(HH=HH,WW=WW,GG=GG,VV=VV)
   par2 <- list(HH=HH,WW=WW,GG1=GG1,GG2=GG2,VV=VV)
```

```
    set.seed(1)
    simu <- simu_fading(nobs,par)
    mu.init <- matrix(0,nrow=4,ncol=mm)
    SS0 <- diag(c(1,1,1,1))
    for(n0 in 1:20){
      SS0 <- HH%*%SS0%*%t(HH)+WW%*%t(WW)
    }
    SS.init <- aperm(array(rep(SS0,mm),c(4,4,mm)),c(1,2,3))
    II.init <- floor(runif(mm)+0.5)+1
    out <- MKF.Full.RB(MKFstep.fading,nobs,simu$yy,mm,par2,II.init,
                  mu.init,SS.init,xdim,ydim,resample.sch)
    SNR[i] <- 10*log(var(simu$yy)/VV**2-1)/log(10)
    Strue <- simu$ss[2:nobs]*simu$ss[1:(nobs-1)]
    Shat <- rep(-1,(nobs-1))
    Shat[out$IphatRB[2:nobs]>0.5] <- 1
    BER1[i] <- sum(abs(Strue[tt]-Shat[tt]))/(nobs-50)/2
    Shat0 <- sign(simu$yy[1:(nobs-1)]*simu$yy[2:nobs])
    BER0[i] <- sum(abs(Strue[tt]-Shat0[tt]))/(nobs-50)/2
    S.known <- sign(simu$yy*simu$alpha)
    Shat.known <- S.known[1:(nobs-1)]*S.known[2:nobs]
    BER.known[i] <- sum(abs(Strue[tt]-Shat.known[tt]))/(nobs-50)/2
}

plot(SNR,BER.known,panel.first=grid(abline(v=c(15,20,25,30,35),
    h=0.01*c(1,3,5,7,9,11),lty=2)),log='y',xlim=c(14,35),yaxt="n",
    ylim=c(0.007,0.12),type='b',ylab='BER',xlab='SNR(dB)',pch=1,
    lwd=2,cex=1.5)
axis(2,0.01*c(1,3,5,7,9,11))
lines(SNR,BER1,type='b',pch=2,lwd=2,cex=1.5)
lines(SNR,BER0,type='b',pch=3,lwd=2,cex=1.5)
legend(15,0.015,pch=c(1,2,3),cex=1.5,
    legend=c('known channel bound','MKF','differential'))

## comparison of MKF and SMC with partial state equation proposal
## get processing time

VV <- 1.3**15*0.001
par <- list(HH=HH,WW=WW,GG=GG,VV=VV)
par2 <- list(HH=HH,WW=WW,GG1=GG1,GG2=GG2,VV=VV)
nobs <- 10050
set.seed(2)
simu <- simu_fading(nobs,par)

time.M=NULL
mm.all <- c(5,10,20,50,100,300)
kk <- 6
BER.M <- 1:kk
```

```
GG1 <- GG
GG2 <- -GG
xdim <- 4; ydim <- 1
for (i in 1:kk){
   ptm <- proc.time()
   mm <- mm.all[i]
   mu.init <- matrix(0,nrow=4,ncol=mm)
   SS0 <- diag(c(1,1,1,1))
   for(n0 in 1:20){
     SS0 <- HH%*%SS0%*%t(HH)+WW%*%t(WW)
   }
   SS.init <- aperm(array(rep(SS0,mm),c(4,4,mm)),c(1,2,3))
   II.init <- floor(runif(mm)+0.5)+1
   print(c(i));flush.console()
   out <- MKF.Full.RB(MKFstep.fading,nobs,simu$yy,mm,par2,II.init,
                  mu.init,SS.init,xdim,ydim,resample.sch)
   Strue <- simu$ss[2:nobs]*simu$ss[1:(nobs-1)]
   Shat <- rep(-1,(nobs-1))
   Shat[out$IphatRB[2:nobs]>0.5] <- 1
   BER.M[i] <- sum(abs(Strue[tt]-Shat[tt]))/(nobs-50)/2
   time.M=cbind(time.M,proc.time()-ptm)
}
BER.M
time.M

time.S=NULL
mm.all <- c(100,200,500,1000,4000,8000)
delay <- 0
kk <- 6
xdim2 <- 6
BER.S <- 1:kk
for (i in 1:kk){
   ptm <- proc.time()
   mm <- mm.all[i]
   SS0 <- diag(c(1,1,1,1))
   for(n0 in 1:20){
     SS0 <- HH%*%SS0%*%t(HH)+WW%*%t(WW)
   }
   xx.init <- matrix(nrow=xdim2,ncol=mm)
   xx.init[1:4,] <- sqrt(SS0)%*%matrix(rnorm(4*mm),nrow=4,ncol=mm)
   xx.init[5,] <- sample.int(2,mm,0.5)
   xx.init[6,] <- rep(1,mm)
   print(c(i));flush.console()
   out <- SMC(SISstep.fading,nobs,simu$yy,mm,par,xx.init,
                  xdim2,ydim,resample.sch)
   Strue <- simu$ss[2:nobs]*simu$ss[1:(nobs-1)]
   Shat <- rep(-1,(nobs-1))
```

```
    Shat[out$xhat[6,2:nobs,1]>0.5] <- 1
    BER.S[i] <- sum(abs(Strue[tt]-Shat[tt]))/(nobs-50)/2
    time.S=cbind(time.S,proc.time()-ptm)
}
BER.S; time.S
```

8.11 EXERCISES

8.1 Box–Muller algorithm. Show that, if X_1 and X_2 are independent, following *Unif*[0, 1], then

$$Y_1 = \sqrt{-2\ln(X_1)}\cos(2\pi X_2), \quad Y_2 = \sqrt{-2\ln(X_1)}\sin(2\pi X_2)$$

are independent, following $N(0, 1)$.

8.2 Compare the execution time of the two implementations shown in Figures 8.12 and 8.13 in Section 8.9 by generating a series of length `tt` = 100 from the simple model $x_t = x_{t-1} + w_t$ and $y_t = x_t + v_t$, where $w_t \sim N(0, 1)$ and $v_t \sim N(0, 1)$ independent of w_t, and record the execution time. Change the Monte Carlo sample size `mm` and series length `tt` to see the differences.

8.3 For the simple model $x_t = x_{t-1} + \sigma_w w_t$ and $y_t = x_t + \sigma_v v_t$, where $w_t \sim N(0, 1)$, $v_t \sim N(0, 1)$, and they are independent. Implement the following algorithms:

- (A.1) SMC using the full information proposal distribution.
- (A.2) Bootstrap particle filter (state equation as proposal distribution).
- (A.3) Independent particle filter (observation equation as proposal distribution) with single matching $L = 1$.
- (A.4) Kalman filter.

Use the following variances when simulating data:

- (B.1) $\sigma_w = 1$ and $\sigma_v = 1$.
- (B.2) $\sigma_w = 4$ and $\sigma_v = 1$.
- (B.3) $\sigma_w = 1$ and $\sigma_v = 4$.

Use the following resampling schedules:

- (C.1) Resample every step.
- (C.2) Resample every five steps.
- (C.3) No resampling.

Simulate 500 series, each of 200 observations, starting at $x_0 = 0$ under the three variance settings, and run the four algorithms under the three resampling schedules (for SMC algorithms). Use initial distribution $x_0 \sim N(0, 1)$. Make comparisons.

8.4 Use SMC to re-analyze Example 1.4.

(a) Use the estimated model in Example 1.4 to obtain the filtered and smoothed estimate of λ_t and compare.

(b) Treat the estimated γ_0, γ_1 and δ_{13} obtained in Example 4 as known, and estimate the dispersion parameter α by evaluating the log-likelihood function using SMC on a grid between 0.05 to 0.15. Obtain estimates and compare them with those obtained in Example 1.4.

REFERENCES

Andrier, C., Doucet, A., and Holensterin, R. (2010). Particle Markov chain Monte Carlo methods. *Journal of the Royal Statistical Society, Series B* **72**: 269–341.

Berzuini, C., Best, N., Gilks, W., and Larizza, C. (1997). Dynamic conditional independence models and Markov chain Monte Carlo methods. *Journal of the American Statistical Association* **92**: 1403–1412.

Briers, M., Doucet, A., and Maskell, S. (2010). Smoothing algorithms for state-space models. *Annals of the Institute Statistical Mathematics* **62**: 61–89.

Brunnermeier, M. K. (2009). Deciphering the liquidity and credit crunch 2007–2008. *Journal of Economic Perspectives* **23**: 77–100.

Campbell, J. Y. and Shiller, R. J. (1988). The dividend-price ratio and expectations of future dividends and discount factors. *Review of Financial Studies* **1**: 195–228.

Cappé, O., Moulines, E., and Ryden, T. (2005). *Inference in Hidden Markov Models*. Springer-Verlag, New York.

Chen, R. and Liu, J. S. (2000). Mixture Kalman filter. *Journal of the Royal Statistical Society, Series B* **62**: 493–508.

Chen, R., Wang, X., and Liu, J. S. (2000). Adaptive joint detection and decoding in flat-fading channels via mixture Kalman filtering. *IEEE Transactions on Information Theory* **46**: 2079–2094.

Chen, Y., Xie, J., and Liu, J. S. (2005). Stopping-time resampling for sequential Monte Carlo methods. *Journal of the Royal Statistical Society, Series B* **67**: 199–217.

Chopin, N. (2004). Central limit theorem for sequential Monte Carlo methods and its applications to Bayesian inference. *Annals of Statistics* **32**: 2385–2411.

Douc, R., Moulines, E., and Stoffer, D. (2014). *Nonlinear Time Series: Theory, Methods and Applications with R Examples*. Taylor & Francis, CRC Press, Boca Raton, FL.

Doucet, A., Godsill, S., and Andrieu, C. (2000). On sequential Monte Carlo sampling methods for Bayesian filtering. *Statistics and Computing* **10**: 197–208.

Emond, M., Raftery, A., and Steele, R. (2001). Easy computation of Bayes factors and normalizing constants for mixture models via mixture importance sampling. Technical report, Department of Statistics, University of Washington, Seattle.

Evans, G. W. (1991). Pitfalls in testing for explosive bubbles in asset prices. *American Economic Review* **81**: 922–930.

Fan, S., Chenney, S., Hu, B., Tsui, K., and Lai, Y. (2006). Optimizing control variate estimators for rendering. *Computer Graphics Forum* **25**: 351–357.

Fearnhead, P. (2008). Computational methods for complex stochastic systems: a review of some alternatives to MCMC. *Statistics and Computing* **18**: 151–171.

Fearnhead, P., Wyncoll, D., and Tawn, J. (2010). A sequential smoothing algorithm with linear computational cost. *Biometrika* **97**: 447–464.

Fischer, V. and Drutarovský, M. (2003). True random number generator embedded in reconfigurable hardware. In *Cryptographic Hardware and Embedded Systems – CHES 2002*, B. S. Kaliski, K. Ko, and C. Paar (eds), Lecture Notes in Computer Science, vol. 2523, pp. 415–430. Springer, Berlin.

Ford, E. and Gregory, P. (2007). Bayesian model selection and extrasolar planet detection. In *Statistical Challenges in Modern Astronomy IV*, G. J. Babu and E. D. Feigelson (eds), vol. 371, pp. 189–205.

Fox, D., Thrun, S., Burgard, W., and Dellaert, F. (2001). Particle filters for mobile robot localization. In *Sequential Monte Carlo in Practice*, A. Doucet, J. F. G. de Freitas, and N. J. Gordon (eds), Springer-Verlag, New York.

Gelman, A. and Meng, X. (1998). Simulating normalizing constants: From importance sampling to bridge sampling to path sampling. *Statistical Science* **13**: 163–185.

Giordani, P. and Kohn, R. (2010). Adaptive independent Metropolis–Hastings by fast estimation of mixtures of normals. *Journal of Computational and Graphical Statistics* **19**: 243–259.

Givens, G. and Raftery, A. (1996). Local adaptive importance sampling for multivariate densities with strong nonlinear relationships. *Journal of the American Statistical Association* **91**: 132–141.

Gordon, N., Salmon, D., and Smith, A. (1993). A novel approach to nonlinear/non-Gaussian Bayesian state estimation. *IEE Proceedings on Radar and Signal Processing* **140**: 107–113.

Grassberger, P. (1997). Pruned-enriched Rosenbluth method: Simulation of polymers of chain length up to 1,000,000. *Physical Review E* **56**: 3682–3693.

Hammersley, J. M. and Handscomb, D. C. (1964). *Monte Carlo Methods*. Methuen's Monographs on Applied Probability and Statistics. Wiley, London.

Hammersley, J. M. and Morton, K. W. (1954). Poor man's Monte Carlo. *Journal of the Royal Statistical Society, Series B* **16**: 23–38.

Hesterberg, T. (1988). *Advances in Importance Sampling*, PhD thesis, Stanford University.

Hesterberg, T. (1995). Weighted average importance sampling and defensive mixture distributions. *Technometrics* **37**: 185–194.

Hrzeler, M. and Knsch, H. R. (1998). Monte Carlo approximations for gen eral state-space models. *Journal of Computational and Graphical Statistics* **7**: 175–193.

Isard, M. and Blake, A. (2004). ICONDENSATION: Unifying low-level and high level tracking in a framework. Technical Report, Oxford University, http://robots.ox.ac.uk/ab.

Kitagawa, G. (1996). Monte Carlo filter and smoother for non-Gaussian nonlinear State space models. *Journal of Computational and Graphical Statistics* **91**: 1–25.

Kong, A., Liu, J., and Wong, W. (1994). Sequential imputations and Bayesian missing data problems. *Journal of the American Statistical Association* **89**: 278–288.

Lee, A., Yau, C., Giles, M. B., Doucet, A., and Holmes, C. C. (2010). On the utility of graphics cards to perform massively parallel simulation of advanced Monte Carlo methods. *Journal of Computational and Graphical Statistics* **19**: 769–789.

Li, W., Tan, Z., and Chen, R. (2013). Two-stage importance sampling with mixture proposals. *Journal of the American Statistical Association* **108**: 1350–1365.

Li, W., Chen, R., and Tan, Z. (2016). Efficient sequential Monte Carlo with multiple proposals and control variates. *Journal of the American Statistical Association* **111**: 298–313.

Liang, F., Liu, C., and Carroll, R. J. (2007). Stochastic approximation in Monte Carlo computation. *Journal of the American Statistical Association* **102**: 305–320.

Lin, M., Zhang, J., Cheng, Q., and Chen, R. (2005). Independent particle flters. *Journal of the American Statistical Association* **100**: 1412–1421.

Lin, M., Chen, R., and Liu, J. S. (2013). Lookahead strategies for sequential Monte Carlo. *Statistical Science* **28**: 69–94.

Liu, J. S. (2001). *Monte Carlo Strategies in Scientific Computing*, Springer, New York.

Liu, J. S. and Chen, R. (1995). Blind deconvolution via sequential imputations. *Journal of the American Statistical Association* **90**: 567–576.

Liu, J. S. and Chen, R. (1998). Sequential Monte Carlo methods for dynamic systems. *Journal of the American Statistical Association* **93**: 1032–1044.

Liu, J. and West, M. (2001). Combined parameter and state estimation in simulation-based filtering. In *Sequential Monte Carlo in Practice*, A. Doucet, J. F. G. de Freitas, and N. J. Gordon (eds). Springer-Verlag, New York.

Liu, J., Wong, W., and Kong, A. (1994). Covariance structure of the Gibbs sampler with applications to the comparisons of estimators and augmentation schemes. *Biometrika* **81**: 27–40.

MacEachern, S., Clyde, M., and Liu, J. (1998). Sequential importance sampling for nonparametric Bayes models: The next generation. *Canadian Journal of Statistics* **27**: 251–267.

Majzoobi, M., Koushanfar, F., and Devadas, S. (2011). FPGA-based true random number generation using circuit metastability with adaptive feedback control. In *CHES*, pp. 17–32. Springer, Berlin.

Malik, S. and Pitt, M. K. (2011). Particle filters for continuous likelihood evaluation and maximization. *Journal of Econometrics* **165**: 190–209.

Marshall, A. (1956). The use of multi-stage sampling schemes in Monte Carlo computations. In *Symposium on Monte Carlo Methods*, M. Meyer (ed.), pp. 123–140. Wiley, New York.

Oh, M. and Berger, J. (1993). Integration of multimodal functions by Monte Carlo importance sampling. *Journal of the American Statistical Association* **88**: 450–456.

Owen, A. B. and Zhou, Y. (1999). Adaptive importance sampling by mixtures of products of beta distributions. Technical Report, Department of Statistics, Stanford University.

Owen, A. and Zhou, Y. (2000). Safe and effective importance sampling. *Journal of the American Statistical Association* **95**: 135–143.

Pitt, M. and Shephard, N. (1999). Filtering via simulation: auxiliary particle filters. *Journal of the American Statistical Association* **94**: 590–599.

Poyiadjis, G., Doucet, A., and Singh, S. (2011). Particle approximations of the score and observed information matrix in state space models with application to parameter estimation. *Biometrika* **98**: 65–80.

Raghavan, N. and Cox, D. (1998). Adaptive mixture importance sampling. *Journal of Statistical Computation and Simulation* **60**: 237–260.

Ripley, B. (1987). *Stochastic Simulation.* John Wiley and Sons, New York.

Robert, C. and Casella, G. (2004). *Monte Carlo Statistical Methods.* Springer, New York.

Rosenbluth, M. N. and Rosenbluth, A. W. (1955). Monte Carlo calculation of the average extension of molecular chains. *Journal of Chemical Physics* **23**: 356-359.

Rubinstein, R. and Kroese, D. (2008). *Simulation and the Monte Carlo Method.* Wiley, Hoboken, NJ.

Smith, P., Sha, M., and Gao, H. (1997). Quick simulation: A review of importance sampling techniques in communications systems. *Selected Areas in Communications, IEEE Journal* **15**: 597–613.

Tanizaki, H. and Mariano, R. S. (1994). Prediction, filtering and smoothing in non-linear and non-normal cases using Monte Carlo integration. *Journal of Applied Econometrics* **9**: 163–179.

van der Merwe, R., Doucet, A., de Freitas, N., and Wan, E. (2002). The unscented particle filter. In *Advances in Neural Information Processing Systems* (NIPS13), T. K. Leen, T. G. Dietterich, and V. Tresp (eds). MIT Press, Cambridge, MA.

Veach, E. and Guibas, L. (1995). Optimally combining sampling techniques for Monte Carlo rendering. In *Proceedings of the 22nd Annual Conference on Computer Graphics and Interactive Techniques*, pp. 419–428. ACM.

Walker, A. J. (1974). New fast method for generating discrete random numbers with arbitrary frequency distributions. *Electronics Letters* **10**: 127–128.

Wang, X., Chen, R., and Guo, D. (2002). Delayed pilot sampling for mixture Kalman filter with application in fading channels. *IEEE Transaction on Signal Processing* **50**: 241–264.

West, M. (1992). Mixture models, Monte Carlo, Bayesian updating and dynamic models. *Computer Science and Statistics* **24**: 325–333.

West, M. (1993). Approximating posterior distributions by mixture. *Journal of the Royal Statistical Society. Series B* **55**: 409–422.

Zhang, J. L. and Liu, J. S. (2002). A new sequential importance sampling method and its application to the two-dimensional hydrophobic Chydrophilic model. *Journal of Chemical Physics* **117**: 3492–3498.

Zhang, J., Chen, R., Tang, C., and Liang, J. (2003). Origin of scaling behavior of protein packing density: A sequential Monte Carlo study of compact long chain polymer. *Journal of Chemical Physics* **118**: 6102–6109.

INDEX

Nonlinear Time Series Analysis, First Edition. Ruey S. Tsay and Rong Chen.

Companion website: www.wiley.com/go/tsay/nonlineartimeseries

Printed and bound by CPI Group (UK) Ltd, Croydon, CR0 4YY

17/09/2026

14964312-0007